160 873262 5

HISTORY OF MATHEMATICS

HISTORIES OF PROBLEMS

The Inter-IREM Commission

Epistemology and History of Mathematics

Translated by Chris WEEKS

Preface by
John FAUVEL
The Open University

ISBN 2-7298-4730-8

© ellipses / édition marketing S.A., 1997
   32 rue Bargue, Paris (15e).

Preface

In recent years mathematics education at school and college has changed dramatically. New technologies have entered the class room, fresh guidance on the curriculum has been provided at all levels, and the very reasons for teaching and learning mathematics have been reassessed and revisited in an institutional context where teachers are more publically accountable, subjected to greater pressures from parents and politicians, than ever before.

One of the most significant, though perhaps quietest, changes in educational practice has been the growing enthusiasm of mathematics teachers for a historical dimension to their teaching and to their pupils' learning. Many teachers have found that understanding the historical development of the subjects they teach adds to their repertoire of classroom skills, and that students themselves can benefit from coming to see mathematics as a developing subject created to solve particular problems in specific cultures over the centuries.

In seeking to develop their own historical understanding and that of their pupils, teachers come up against the problem of how to put this general insight into practice in their own lives and classrooms. How have other teachers made use of history of mathematics? What particular pedagogical insights has the use of history led to? What difficulties of learning mathematics can be resolved through careful exploration of the historical development of the subject? What different understanding of the nature of mathematics can be reached through sharing with young people the creative, humanistic aspect which has been so prominent a feature of mathematics in the past? Can history be used to make mathematics learning more efficient, at a time when there are ever-closer curricular constraints on a teacher's freedom of action?

Here English-speaking teachers have a great deal to learn from the example of France, where for several years now the IREM has coordinated a growing understanding of how the history of mathematics may be and has been used in the mathematics classroom. As the papers in this volume make clear, there is a great range of ways in which history can be invoked to help in the growing understanding of pupils through their school career, and a range of different purposes it serves. Teachers working within an English-speaking educational culture will warmly welcome these examples of what can be achieved, in such a variety of ways, by mathematics teachers in France sensitive to the historical dimension.

In the coming years there will be various projects to develop and spread knowledge of the use of history in the teaching of mathematics. A major international study is underway to promote a careful examination of this many-faceted topic and lead to a sharing of good practice between teachers in many lands. This book is a seminal contribution towards the growing international understanding of these issues.

Recently the British people have reaffirmed their wish that Britain should be amongst its friends in the heart of Europe, not standing isolated on the sidelines. This volume is a valuable harbinger of things to come, when teachers across Europe, and indeed across the world, will share their experiences of history and teaching in the service of providing a better mathematics education for children everywhere.

John Fauvel

Foreword

The inter-IREM Commission "Epistemology and History of Mathematics" has come together over the past fifteen years consisting of members of IREM who are engaged in research in the history of mathematics. The IREMs (Institutes of Research into Mathematics Education) are University Institutes whose objectives are: research into the teaching of mathematics, in-service training for teachers, the writing and publishing of documents, and involvement in initial training of teachers. There are twenty-five IREMs in France.

The "Epistemology and History of Mathematics" inter-IREM Commission consists of teachers from collèges (11 – 15 yr.), lycées (15 + yr.), teacher training institutions and universities teaching mathematics and also teachers of philosophy, history and the physical sciences, as well as researchers into the history of science. It is, therefore, a place for the exchange of ideas and reflection among people whose background and experience is very varied. It organises, at both national and international level, multi-disciplinary colloquia and university summer schools on the history of mathematics, and has published many works on the history and teaching of mathematics.

The idea for the starting point for this new project, proposed by Gilles Bonnefoy of the Lyons IREM was to introduce the history of mathematics through the consideration "great problems" as themes occurring in the development of mathematics. The idea was to present a history of mathematics that was not divided up into different historical periods, nor into the different fields of mathematical knowledge. The different chapters of this work deal with the birth and development of problems showing how this led to the creation and transformation of mathematical tools needed for their solution.

The "Epistemology and History of Mathematics" inter-IREM Commision had already shown its interest in the role of problems in the history and activity of mathematics through the organisation of a Colloquium on this theme at Montpelier in 1985.This interest was the concern of teachers who were stimulated by a wish to oppose a dogmatic view of their subject and who wished to give a sense of understanding ot those they taught. Now, the history of mathematics shows that it is by the solution of problems that meaning can be given to mathematical concepts and theories. As Gaston Bachelard wrote in his *La formation de l'esprit scientifique*, (The development of the scientific spirit), "*It is precisely this sense of the problem which is the true mark of the scientific spirit. For a scientific mind all understanding comes from an answer to a problem. If there is no questioning, there can not be any scientific understanding*".

The fifteen histories that compose this work are, then, the stories of fifteen "great problems" in the history of mathematics. Let it be clearly understood that the present work does not claim to deal with all the great problems in the history of mathematics.

These histories set out to be an introduction to mathematics that is both cultural and historical. They have been written with the intention of "giving a taste for intellectual adventure", to use an expression of Jean-Pierre Le Goff of the Basse Normandie IREM. They are aimed at teachers of mathematics and other disciplines for use as a means of introducing a historical perspective into the teaching of mathematics. They are also aimed at final year pupils as well as students, particularly those in the I.U.F.M., (University Institutes of Teacher Training), who intend to become teachers.

The histories contain historical material and referenced quotations that will allow the reader to learn in what way the problems were posed and solved at different periods of time. They also contain exercices to be solved according to ancient and modern methods: working through these will let the reader himself enter into an adventure. Each chapter has a bibliography which contains, in addition to the historical sources that have been used, a certain number of other books recommended for further study of the subject. Each chapter also contain hints for the solution of the exercises.

The different histories are sometimes linked together and, while each one is a history in itself, the whole taken together contains an interwoven thread. The whole work is the collective effort of around thirty IREM members who were the authors and collaborators of the different chapters. In alphabetical order, giving their IREM, they were: Evelyne Barbin (Le Mans), Monique Belet and André Belet (Toulouse), Didier Bessot (Basse-Normandie), Rudolf Bkouche (Lille), Gilles Bonnefoy (Lyon), Anne Boyé (Nantes), Martine Bühler (Paris VII), Jean-Luc Chabert, Michel Crubellier (Lille), Denis Daumas (Toulouse), Jöelle Delattre (Lille), Jean-Pierre Friedelmeyer (Strasbourg), Michèle Grégoire (Paris VII), Marianne Guillemot, Michel Guillemot (Toulouse), Yvette Horain (Lille), Gilles Itard (Le Mans), François Jaboeuf (Montpellier), Claudine Kahn (Strasbourg), Michèle Lacombe (Paris VII), Françoise Lalande (Montpellier), Jean-Luc Le Chevalier (Lille), Jean-Pierre Le Goff (Basse Normandie), Xavier Lefort (Nantes), Frédéric Métin (Reims), Anne Michel-Pajus (Montpellier), Henry Plane (Dijon), Jacky Sip (Lille).

Evelyne Barbin

For the inter-IREM Commission

Contents

1

En Route for Infinity

Michel GUILLEMOT
Denis DAUMAS
IREM Toulouse

All this shows that the human mind asks itself questions that are so strange when infinity enters in, that one must not be surprised if there is difficulty in coming to an answer.

Leibniz, *Nouveaux Essais,* IV, 3.

The road to infinity has no end: and to follow it we have to make some choices. We could start in the 19th century, the time when the mathematical idea of infinity could be said to have the right to exist. We could, alternatively, follow a route through various disciplines of mathematics to show how the idea of infinity has gradually infiltrated them. We have instead chosen a different path, leaving to one side certain exceptional monuments on the way, such as the infinitesimal calculus. We have decided to look at infinity in number, and to try to show how, throughout the course of time, men and women answered these two fundamental questions about cardinality and ordinality:

– how do we count?

– how do we order?

Unfortunately, we shall have to leave by the way side the contributions of some very prestigious civilizations, such as those of the Arabs or the Chinese. But think of the journey as a holiday journey: we could take certain motorways that lead to places of great reputation, but we have chosen an alternative route in order to visit a place of special interest. From small beginnings we shall make our journey towards the infinite, enumerable or continuous, and reach right up to the foothills of the transfinite.

Setting off for the islands: just a little will do

Passing by the magnificent beaches of Tahiti or the Mururoa Atoll, we sail on to the Murray Islands, situated between New Guinea and the Cap York peninsula of Australia. There

> *if Hunt is to be believed, certain of the natives knew only the following numbers: netat for "one", neis for "two", neis-netat for "three" (2 + 1) and neis-neis for "four" (2 + 2); beyond that they used a word meaning "multitude".* [1]

Do you think we've made a good start on our long journey? Haven't we got there quickly?

> *the few words used to express numbers in the language of aborigines or islanders and the fact that they rarely go beyond six, does not in any way mean that they are unable to count. Even less that their civilization is at a prehistoric stage.* [2]

Studies have shown that these people are the descendants of peoples who, at another period of time, had a more developed language or a better understanding of number. It is very difficult to make a return to a past when such little evidence has come down to us here in the 20th century. Let us simply be content to marvel at the fact that, so many thousands of years ago, women and men, so often rather abusively referred to as "primitive", had grasped the first abstraction in the history of mathematics, namely the number 2. It is of little importance that they stopped at 2, others stopped at 4 and others at 6. Did they need to go further? Were not they satisfied with this little? Do we not today often try to change the unit of measure precisely because we do not want to deal with larger numbers? For example, a declared income of $550 000 000 represents an enormous wealth but the fact that the director of Disneyland earns more than the 4000 gardeners that work for him carries more meaning. [3] But before we come to such large numbers we have yet to get to three. [4]

The early Egyptian hieroglyphics and Chinese writing contain the first indications of three to mean many or plural. (See Fig. 1.)

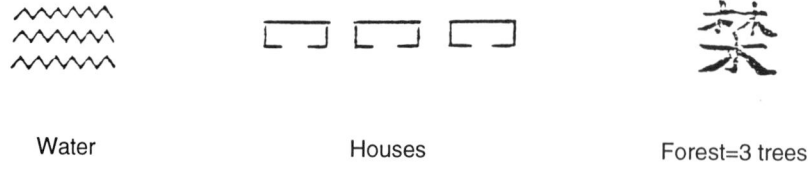

Water   Houses   Forest=3 trees

Figure 1

> *The Step of Three is the decisive one, which introduces the infinite progression into the number sequence.* [5]

1.   Ifrah, *Histoire universelle des chiffres*, p. 13.
2.   Roux, *L'homme et son nombre*, p. 16.
3.   Albert, *Capitalisme contre capitalisme*, p. 96.
4.   NDT. In French, play on words "guerre de Troie" (the Troyan War) and "guerre de trois" (war of three).
5.   Menninger, *Number Words and Number Symbols*, p. 17.

Even though it can not be established for certain, it appears that language contains evidence of the struggle to pass beyond the old barrier of two. "Three" is often associated with many as, for example, *three* and *through,* or the Latin *tres* and *trans,* or the French *trois* and *très.* After one and two, the ordinals were formed "third, fourth, fifth,..." Formerly, first kept its meaning, which is before all the others, while we find second used for "the other" or "the one that comes after", compare the Latin "secundus" [from *sequi, secutus* to follow]. It is difficult to draw a distinction between second and twice. Let us go on further, leaving the few for the many.

By the pyramids the scribes counted many

The origin of the hieroglyphic writing of numbers is unknown. They appear on the earliest documents that have come down to us and they remain unchanged throughout the three thousand years of Egyptian civilisation. Notice, in passing, that they are not the same in cursive, hieratic or demotic, Egyptian script. As for the number system that was used, it is similar to that used by most peoples who want to be able to deal with large numbers, that is, it is a base ten additive system. But guided by their care for purity and their artistic taste, the Ancient Egyptians used different figures to refer to different powers of ten.

1	10	100	1 000	10 000	100 000	1 000 000

bar	handle ?	rope	lotus flower	finger	tadpole	sitting god

Figure 2

The last sign is a god sitting with arms uplifted to a star filled sky: a million represents an astronomical number. The notation contains evidence of the "three barrier" perhaps, in the way in which the symbols are laid out in group of three. For example, 5 and 9 were often shown as

            III   III

            II   III

                                       III

Nor was grouping by four excluded and we find the hieroglyphic forms of 4 and 8 as follows:

            IIII   IIII

                                       IIII

No other sign is used, nor did they use a subtraction principle (compare the Roman IV and IX), so if an Egyptian scribe should have had to write 1992, it would appear like this

Figure 3

But artistic flair predominated and the lotus flowers were often displayed as a magnificent bouquet.

Exercise 1   How many goats are there?

Figure 4

Hieroglyphic symbolism was consistent but this consistency was not found in the language nor in the construction of the ordinals:

> *[as to gender], the number two is naturally dual. The integers, from 3 onwards, must originally have had a masculine sense which most of them lost; from the start, plurals, they were afterwards treated as singular.*[1]

We find evidence of various ways of representing ordinals:

– two written forms for twice or second

– one way of writing numbers from three to ten followed by a different way after ten.

Numbers after ten were indicated by the addition of a participle "which completes" represented by the sign of a whip. We may speculate on this meaning but it could be that the scribe is remembering the harshness of his apprenticeship:

> *You beat upon my back and your teaching went into my ear.*[2]

Besides integers, the Egyptians also used fractions. They had special signs for 2/3 and 1/2 but otherwise fractions like 1/n were written with the sign for a mouth above the number (perhaps a reference to sharing food.)

$$\frac{2}{3} \qquad\qquad \frac{1}{2} \qquad\qquad \frac{1}{3} \qquad\qquad \frac{1}{12}$$

Figure 5

1.   Lefèvre, *Grammaire de l'Égyptien classique*, p. 107.
2.   Erman, *La civilisation égyptienne*, p. 421.

But there was no other signs for representing other fractions and so the Egyptian scribe would write 2/7 as:

Figure 6

which corresponds to

$$\frac{2}{7} = \frac{1}{4} + \frac{1}{28}.$$

Exercise 2	The Rhind Papyrus contains lists of 2/n "fractions" written out as sums of different 1/n fractions with n always less than 1000. Express 2/71 as a sum of different 1/n fractions.

In order to carry out multiplications, (see chapter 15) which was done by doubling, fraction representation had to be chosen with care; firstly it was essential to avoid large numbers, and secondly there was always the possibility of going round in circles.

For example if we write

$$(1)\ \frac{2}{n} = \frac{1}{n} + \frac{1}{2n} + \frac{1}{3n} + \frac{1}{6n}$$

then by doubling we get:

$$\frac{4}{n} \ =\ \frac{2}{n} + \frac{1}{n} + \frac{2}{3n} + \frac{1}{3n}$$

$$=\ \left(\frac{1}{n} + \frac{1}{2n} + \frac{1}{3n} + \frac{1}{6n}\right) + \frac{1}{n} + \left(\frac{1}{2n} + \frac{1}{6n}\right) + \frac{1}{3n},\ \text{using (1) and the "}\frac{2}{3n}\text{ rule "}$$

$$=\ \frac{2}{n} + \frac{1}{n} + \frac{2}{3n} + \frac{1}{3n},\ \text{on regrouping.}$$

So we have got nowhere! The Ancient Egyptians were obliged to use many ingenious methods for carrying out their calculations. These finite decompositions must have reinforced their taste for exactness and finiteness. Is it possible that a different form of number notation might have allowed to escape finite bounds?

Between two rivers: the Babylonians adopt a definitive position[1]

History starts at Sumer is the title of Samuel Kramer's excellent book. Can not we claim that the history of the writing of numbers began at Sumer, between the Tigris and Euphrates rivers? We are gradually beginning to understand the long chain of events which led the proto-Sumerian scribes of the 4th millenium B.C. to adopt a hybrid system of numbering deriving from a variety of units of measure and based upon 2, 3, 6 or 10. There is no concern here for arithmetic purity, the system being at the same time both additive and multiplicative, nor for aesthetics, the signs being simply the imprints of a reed or a piece of ivory in clay. When small and pressed into the clay at an angle, the

---

1.   NDT. In French play on words "prendre position" (adopt a definitive position) and "système numérique de position" (numeric system of position).

mark represented the number one: when small but held perpendiculary to the tablet the mark was a small circle and represented ten. A larger size reed produced sixty and 600 resulted from combining marks according to a multiplicative principle.

1	10	60	600	3 600	36 000
∪	○	⌴	⌴	◖	◉
small reed "oblique"	small reed "perpendicular"	large reed "oblique"	large reed "oblique" "small reed perpendicular"	large reed, "perpendicular"	large reed, "perpendicular" small reed "perpendicular"

Figure 7

Exercise 3   The front of this tablet shows numbers of sacks and the back gives their total. Check that the total is correct.

Front   Back

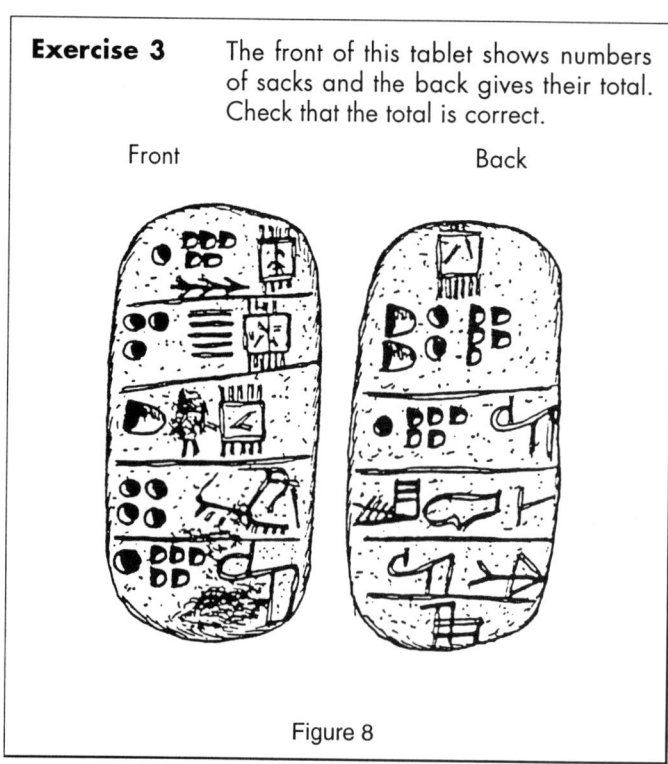

Figure 8

Even though the Sumerian scribes did not go very far in their writing of numbers, they could, nonetheless, represent numbers of any size whatsoever by using reeds of different size. The additive and multiplicative number system could be extended indefinitely. Don't we make use of a similar process when we use different signs to indicate "magnitudes" of different statistical populations?

Exercise 4   Show how Sumerian scribes would have written very large numbers.

The development of writing gradually led to cuneiform writing using a stylus. The multiplicative principle for writing numbers was abandoned and instead they simply used different sizes of stylus to represent different magnitudes.

1	10	60	600	3 600	36 000
Y	$\mathsf{\prec}$	 Y	 $\mathsf{\prec}$	 Y	 $\mathsf{\prec}$
small stylus "vertical"	small stylus "oblique"	large stylus "vertical"	large stylus "oblique"	very large stylus, "vertical"	very large stylus, "oblique"

Figure 9

Exercise 5   What are the numbers shown on this clay tablet?

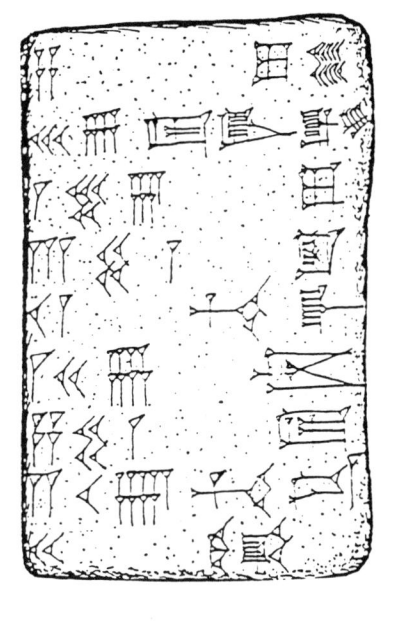

Figure 10

How much further could we go? Maybe be the answer to this is what caused the Babylonians to introduce the greatest invention of mathematics, namely a positional number system. The vertical stylus mark would now represent, not only one, but any power of 60, positive or negative. Thus 1992 in the additive system would be written:

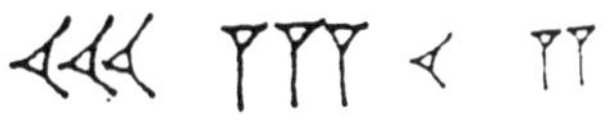

Figure 11

and in the positional system:

would represent
$$3 \times 60 \times 10 + 3 \times 60 + 1 \times 10 + 2 \times 1.$$

In fact, for the Babylonians, this last expression for 1992 could just as well stand for

$$30 \times 60^3 + 3 \times 60^2 + 10 \times 60 + 2 \times 1$$

or, not having a zero,

$$30 \times 60^4 + 3 \times 60^3 + 10 \times 60 + 2 \times 60^{-1}$$

or, in general,

$$30 \times 60^p + 3 \times 60^q + 10 \times 60^r + 2 \times 60^s$$

where p,q,r,s are positive or negative integers written in decreasing order.

The Greeks and the Arabs were certainly aware of the value of such a system of numeration. Alongside their own additive systems they also used the Babylonian base sixty system. For example, Theon of Alexandria (c. 370 A.D.) gives an approximation to

$\sqrt{4500}$ as 67° 4' 55''. And today, our own use of 60 in time and angle measure derives from this extraordinary invention. We can also note a hint of the presence of the infinite since all integers could be written down, no matter how large.

But what about everyday fractions? If 2/7 had been expanded finitely by the Egyptians, this was not possible for the Babylonians. The number 7 is prime and is not a divisor of 60, so 2/7 can only be written as an approximation in sexagesimals just as it can be in decimals.

Exercise 6	a) Find an approximation to 2/7 in sexagesimals to three terms. b) Find the period of the sexagesimal form of 2/7. c) Find the period of the decimal form of 2/7.

When we read the result of the division of 2 by 7 on a calculator as 0.285 714 286, we are not made aware of the infinite nature of the decimal expansion nor of its periodicity: in this we are not so very far away from our distant Babylonian ancestors. They certainly sometimes improved approximations, but there are no surviving documents that suggest that they had any algorithm that would always lead to better approximations. Both the Egyptian methods and the Babylonian sexagesimal methods had this in common: they used finite processes that fitted the needs of those who calculated. The time for "theorising" had yet to come. We shall leave this strictly numerical domain if we are to reflect upon the nature of its elements. The idea of number itself may help us better to understand what concerns us.

Crossing the Mediterranean: Pythagoreans think[1] everything comes down to numbers

Sailing across the waters of the Mediterranean, we may well follow the very route, as legend has it, that Pythagoras took. Going to Egyptian and Babylonian sources, he learned from them first, a certain mystique of numbers, and secondly their astronomical knowledge.

> *Order and disorder, this bipolarity can be found alike in the serious scientific study of the heavens as well as in popular works. If Aristotle, whose philosophy ruled western scientific throughout two thousand years, did not deal with comets in his* De Caelo *but in his* Meteorologica, *it was because, by the disorder that they introduced, they could only belong to the lower levels of the atmosphere, hardly higher than where storms were formed and winds were born, certainly not to the higher spheres where the stars moved according to immutable laws.[2]*

Well before Aristotle, the Pythagoreans had laid down the immutable laws of number. Their view that "all is number" led them to put representation, harmony and laws on the same footing. As order and disorder seemed to be mutually opposed in the heavens, might not number, whole and necessarily finte, be the means of better understanding the infinite?

---

1.   NDT. In French, play on words "se figurer" (to imagine something) and "nombre figuré " (figurate number).
2.   Verdet, *Le ciel, ordre et désordre*, p. 15-17.

Each whole, non-zero, number was associated with a visible figure which was not a symbol: 2 was heteromorphic, of the form n (n + 1), 3 was triangular, 4 a square and n was n-gonal. To be valid, the shape of the figure had to be of a similar form that could be generated indefinitely by the addition of shapes of the same form called gnomons, a term that referred to an instrument used for astronomical observation. So, for example, the square with its shadow produces another square:

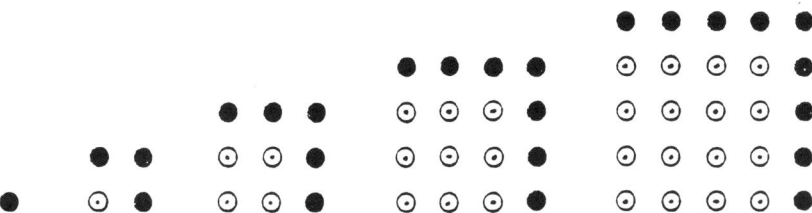

Figure 12

and similarly for the triangular numbers:

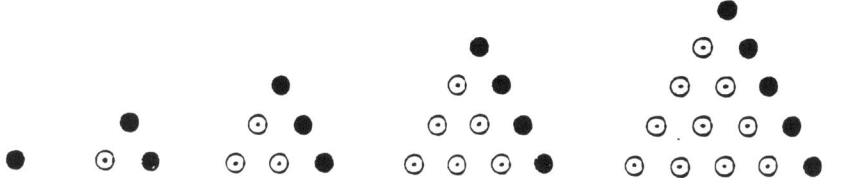

Figure 13

Exercise 7   What is the general expression for a triangular number?

Exercise 8   Find the first seven pentagonal numbers and a general expression.

This geometric definition served as a way of visualising properties which today we would prove by recurrence relations. This is what the second century A.D. neo-Pythagorean Theon of Smyrna has to say:

> *The odd numbers added one to the other produce square numbers. The successive odd numbers are 1, 3, 5, 7, 9, 11. Adding these successively in order gives square numbers. For example, One is the first square number because one One [is] One. Next the odd number 3, if one adds this gnomon to the unit, one gets a square equal to a number equal to its number of times for it will be 2 according to its length and 2 for its width. Next in succession comes the odd number 5. If one adds this to the square, it will produce a new square 9 which has 3 for its length and 3 for its width. Next, the odd number 7, which if added to 9 produces 16, which is 4 long and 4 wide. And the same reasoning to infinity*[1]

In other words, the permanent shape of the figure, with its associated gnomon, is the justification for this "reasoning to infinity". The fact is that this is just "showing" the result. What are the principles that we would need to use in order to make it into a proof?

---

1.   Theon of Smyrna, *Expositio*, p. 15 [Fr. tr. Joelle Delattre].

In the Rose City: it is not a descent into hell

Might not infinity be a sort of hell into which it is best not to descend? Fermat, who was Councillor to the Toulouse Parliament of the Rose City, had sufficient time to devote himself to his many activities. In arithmetic he showed a taste for seeking out and proving difficult problems that touched upon the taboo of the infinite. In August 1659 he wrote to Carcavi, curator of the Royal Library:

> *And in the cases where ordinary methods given in Books prove insufficient for handling such difficult propositioins, I have at last found an entirely singular way of dealing with them. I call this method of proving infinite descent or indefinite descent.*[1]

What was this? Suppose we wish to prove that a proposition $P(n)$ is true for all whole numbers n. Fermat argued by *reductio ad absurdum*: he supposed there to be an integer n_0 for which the proposition was not true. He then established a *descent*: there exists an integer n_1, strictly less than n_0, for which $P(n_1)$ is also not true. He then obtains a strictly decreasing infinite sequence of integers (n_p) for which $P(n_p)$ is not true:

> *Being given some number, on deseending there is no infinis less than that.*[2]

and hence a contradiction.

To put it in other words, Fermat has identified the fundamental property of the set of natural numbers: there is no infinite strictly decreasing sequence of natural numbers. Prior to Fermat, other mathematicians had stated this property in a similar way and had put it to use. For example, the 13th century Campanus of Novare, one of the first translators of Euclid into Latin, had used an argument similar to Fermat's in proving the irrationality of the Golden Ratio.

The Golden Ratio is defined by

$$\phi = 1 + \frac{1}{\phi} \text{ giving } \phi = \frac{1 + \sqrt{5}}{2} = 1,618...$$

Reasoning by *reductio ad absurdum*, Campanus assumed that there existed two integers x_1, x_2 such that

$$(1) \frac{x_1}{x_2} = 1 + \frac{x_2}{x_1} = \frac{x_1 + x_2}{x_1} \text{ and } x_2 < x_1 < 2 x_2$$

Subtracting, he obtained

$$\frac{x_1}{x_2} = \frac{x_1 + x_2 - x_1}{x_1 - x_2} = \frac{x_2}{x_1 - x_2}$$

Putting x_3 for $(x_1 - x_2)$ we get an expression of the same type as (1) but whose terms are strictly smaller:

$$\frac{x_1}{x_2} = \frac{x_2}{x_3} = \frac{x_2 + x_3}{x_2} \text{ with } x_3 < x_2 < 2 x_3 < x_1$$

Continuing in this way produces a strictly decreasing sequence of natural numbers, which is absurd.

The great theoretician of numbers Euler was, later, to take up and prove some of Fermat's claims which he had left unproved. He also used the method of infinite

---

1.   Fermat, *Œuvres*, vol. III, p. 431.
2.   *Ibid.*

descent. Here, as an example, is the first theorem of his *Theorematum quorandam arithmeticorum demonstrationes* of 1738:

> *The sum of two fourth powers such as $a^4 + b^4$ cannot be a square unless one of them is zero.* [1]

Assuming that $(a^4 + b^4)$ is a square with a and b are prime to one another, there will be numbers p and q, also prime to one another, one odd and the other even, such that

$$a^2 = p^2 - q^2 \text{ and } b^2 = 2 pq.$$

> *But, since $a^2 = p^2 - q^2$, it must be that p is odd since otherwise $p^2 - q^2$ cannot be square. So p must be odd and q even. However, since 2 pq has to be a square, it follows that both p and 2q are squares since p and 2q are self prime. Hence, in order that $p^2 - q^2$ should be a square, we must have $p = m^2 + n^2$ and $q = 2$ mn, and a second time there needs to exist two numbers m, n self prime with one odd and the other even. But since 2q is a square, then 4 mn and so mn will be a square; whence both m and n must be squares. I therefore put $m = x^2$ and $n = y^2$, and get $p = m^2 + n^2 = x^4 + y^4$ which must also be a square. Thus if $a^4 + b^4$ is a square then it follows that $x^4 + y^4$ must also be a square. It is clear that the numbers x and y are less than a and b. Hence, starting again with the fourth powers $x^4 + y^4$, ...* [2]

Exercise 9   Prove by the method of infinite descent that, for all positive integer n
$$2(n + 2) < 2^{n + 2}$$

The method of infinite descent only requires a finite number of steps. In fact we only need two, using the idea of well ordering, the centre piece of Cantor's view of sets. The set of natural numbers is well ordered since every non-empty subset of the naturals has a least member. From this, the contradiction argument of the method of infinite descent, comes to asserting that the set E of natural numbers, for which the property $P(n)$ is not valid, is non empty. There must therefore be a least member n_0. The descent process allows one to find an integer n_1, strictly less than n_0, for which $P(n_1)$ is not valid, which contradicts the fact that n_0 is the least member of E.

Exercise 10   Prove that these propositions are equivalent:

I) there cannot exist an infinite sequence of strictly decreasing natural numbers

II) the set of natural numbers is well ordered, that is that every non-empty subset has a least member.

Although they involve a finite process, proofs that use the well ordering principle or infinite descent, are far removed from a naive sense of the natural numbers, namely that it is a sequence in which each number has a successor. Instead of this sometimes difficult, though finite, method of descent, perhaps we would be better to proceed by a more natural method of induction, taken one step at a time.

---

1.   Euler, *Theorematum*, p. 42.
2.   *Ibid.*, p. 44.

Off to the Capital: the challenge[1] of going forward

In Paris we meet another of Fermat's correspondents: Pascal. They often disagreed, as much in their approach to the theory of probability as in their treatment of the infinite. Fermat found the method of descent, Pascal took up the challenge of going forward. In his *Traité du triangle arithmétique* he set out clearly what has come to be known as the principle of mathematical induction. (In Germany this is known as complete induction, *"vollständige Induktion"*, and in France, since Poincaré in 1902, recurrence argument, *"raisonnement par récurrence"*.) This is what Pascal has to say:

> *Even though this proposition may have an infinite number of cases, I shall give a very short proof of it, assuming two lemmas.*
> *The first, which is self evident, is that the proposition is valid in the second base. [...]*
> *The second is that if the proposition be found to be valid in any base then it must necessarily be valid for the following base.*
> *From this it can be seen that it is necessarily valid for all the bases: for it is valid for the second base by the first lemma: then by the second lemma it must be true for the third base, and hence for the fourth, and so on to infinity.*
> *Hence, it is only necessary to prove the second lemma.[2]*

By base, Pascal meant the bases of the different triangles within his arithmetic triangle. Referring back to Theon of Smyrna, the Pythagorian gnomon has left behind its visual form and set it itself up as a principle. But is anything to be gained by this? In his well known letter to Carcavi on *The History of the Roulette*, Pascal wrote:

> *Now, every triangular number taken two times and reduced by the size of its exponent is the same as the square of its exponent; as for example, the third triangular number 6, being doubled is 12, which being reduced by its exponent 3, the result is 9, which is the square of 3.*
> *This can easily be seen from Maurolic and the truth of my proposition follows.[3]*

Exercise 11   Prove Pascal's claim.

This concerns Maurolycus who had also used the recurrence principle. His *Arithmetic Book* of 1557, which was not published until just after his death in 1575, contained a number of propositions. After proving proposition 13 that *"every square number with the next odd number added to it gives the next square,"* he states proposition 15, that the sum of consecutive odd numbers will be a square.

> *We can see that 1 together with the next odd number makes 4. Similarly, the second square number together with the third odd number 5 makes the third square number 9. Again, the third square number 9, together with the fourth odd number 7, makes the fourth square number 16 and so on to infinity, the proof being by repeated use of proposition 13.[4]*

Exercise 12   a) Show that every square together with the succeeding odd number is a square.
                    b) Prove by induction that the sum of consecutive odd numbers is a square.

---

1.   NTD. In French play on words "Paris", the capital and the "pari" (challenge or bet) of Pascal.
2.   Pascal, *Œuvres complètes*, p. 53.
3.   *Ibid.*, p. 137.
4.   Cassinet, *Le premier livre de l'arithmétique de Francisco Maurolycus*, p. 17.

Studying Pythagorian figurate numbers has proved to have been the seed bed from which has grown the principle of mathematical induction. But is that sufficient for us to "believe" in the validity of such an argument? Pascal himself only asked us to accept that it can be "seen" that it must be valid for all integers.

When Peano published his *Arithmetices Principia* at the end of the last century, he showed that he was quite clear about this. He set up Pascal's principle as an axiom which, in his symbolic language became:

Figure 14

We can transcribe this as

> If k is a class of integers such that 1 belongs to k, and such that for each integer x belonging to k, (x + 1) belongs to k, then N is contained in k.

A symbolic expression, but one heavy with meaning, as Poincaré observed in his *La Science et l'Hypothèse*

> The essential characteristic of induction argument, what it consists of, condenses here an infinity of syllogisms into a unique formula. [1]

Exercise 13   Prove that the method of infinite descent and mathematical induction are logically equivalent.

But some refused this infinity of syllogisms as whole. Using this process as a way of moving from the finite to the infinte was to grant to the infinite a certain status. Let us go back somewhat.

The Pharos at Alexandria: does it throw light upon the infinite?

Alexandria: Aristotle had been teacher to the founder of this city and many famous mathematicians had connections with it, such as Euclid, Hero, Diophantus, Ptolemy, Pappus, Menelaus and others. How did Greek mathematicians and philosophers understand the notion of infinity?

The discovery of irrationality, the opposition between the finite and the infinite, had proved a fatal blow for Pythagorian philosophy. Zeno, philosopher of the second half of the 5th century B.C., had set forward certain paradoxes to show the impossibility of deciding between a finite or atomistic view and a non finite or continuistic view. The paradox of Achilles and the tortoise is the best known.

> [It] amounts to this, that in a race the quickest runner can never overtake the slowest, since the pursuer must first reach the point where the pursued started, so that the slower must always hold the lead [...], but it proceeds along the same lines as the bisection-argument, for in both a division of the space in a certain way leads to the result that the goal is not reached, [...]. And the axiom that which holds a lead is never overtaken is

---

1.   Poincaré, *La Science et l'Hypothèse*, p. 38-39.

false: it is not overtaken, it is true, while it holds a lead: but it is overtaken nevertheless if it is granted that it traverses the finite distance prescribed.[1]

Aristotle, critical of these paradoxes, set up a theory of the infinite in distinguishing between the actual infinite and the potential infinite (see chapter 15):

> *Our account does not rob the mathematicians of their science, by disproving the actual existence of the infinite in the direction of increase, in the sense of the untraversable. In point of fact they do not need the infinite and do not use it. They postulate only that the finite straight line may be produced as far as they wish.*[2]

We have the potential of being able to write down all the natural numbers. But we can also see that the collection of all the natural numbers cannot be envisaged as whole, it is impossible to write them all down. In other words, to use modern terminology, they do not form a set. This was to be the fixed position for almost two thousand years. As for Euclid, his use of the infinite was at a number of levels. Not being able to use the idea of an actual infinite set, he had to set out his statements in the form of the potential infinite. So, *Elements* IX, 20 reads:

> *Prime numbers are more than any assigned multitude of prime numbers.*[3]

Today we would prefer to say *"the set of primes is infinite."* This realisation or actualisation of the infinite makes no fundamental change to the Euclidean proof. If one takes the product of a finite number of primes and adds 1, the result is either another prime, or has prime factors which are different from the original primes.

Exercise 14   Prove that the set of prime numbers is infinite.

In fact language here is illusory: there is no calculating procedure or theory that will allow us to understand this property. What, for example, is the (10^{1992}) th prime number? Shall we be obliged to wait until the arrival of even more powerful computers before we can know?

Leaving the strictly arithmetical domain, we come to the main field of concern of the infinite, namely magnitudes and what we today call analysis. To put it briefly, the swift footed Achilles was able to catch up with the Tortoise because "summing to infinity" was valid. In mathematical terms, this corresponds in its essence, to the Archimedean sense of magnitude. Euclid states this as a definition in Book V of the *Elements*:

> *Magnitudes are said to have a ratio to one another which are capable, when multiplied, of exceeding one another.*[4]

Leave ratios aside (see chapter 2): what this means is that given any two magnitudes, then adding a number a sufficient number of times to the smaller, we shall obtain a magnitude that is greater than the larger.

Exercise 15   Express in formal language the fact that the set of real numbers is
                       Archimedean.

---

1.   Ross, W.D. (ed.), *The Works of Aristotle*, vol. II, Physica, V 239b.
2.   *Ibid.*, III, 207b.
3.   Heath, *The Thirteen Books of Euclid's Elements*, vol. 2, p. 412.
4.   *Ibid.*, p. 114.

The Archimedean character of Euclidean magnitudes has many advantages. It guarantees the existence of a certain number, even if it should turn out to be extremely large.

Further, it can be used as the justification for the "potential limit" contained in *Elements* X, 1:

> Two unequal magnitudes being set out, if from the greater there be subtracted a magnitude greater than its half, and from that which is left a magnitude greater than its half, and if this process be repeated continually, there will be left some magnitude which will be less than the lesser magnitude set out.[1]

Without loss of generality, we can assume subtractions of parts equal to a half. Let AB and CD be two magnitudes such that AB is greater than CD. There exists an integer n such that n CD is greater than AB. Extend CD n times, and each time find half of AB or half of the previous half *"until the divisions in AB are equal in multitude with the [n – 1]divisions in [CD extended]"* In other words, in our figure we have:

$$4CD > AB$$

$$CD = DD_1 = D_1D_2 = D_2D_3$$

$$\text{and } AB_1 = \frac{AB}{2}, AB_2 = \frac{AB_1}{2} = \frac{AB}{2^2}, AB_3 = \frac{AB}{2^3}$$

And so, AB_3 is less than CD. Note that Euclid, as was his custom, uses a particular n, here n = 3.

Exercise 16   Prove Euclid's proposition X 1

a) using Euclid's method

b) directly

The potential of the infinite is no longer merely in the language, it has become operational. It involves a justification for the "exhaustions" of the method of exhaustion (see chapter 3). It is sufficient to be able to find a process that leads to "subtracting a magnitude greater than its half." The Aristotelian principle of the potential infinite is going to have a hard life. To stricter mathematical objections there were added religious ones. To quote Descartes, for example,

> and because we do not know how to imagine how many more stars God may create, we assume that their number is indefinite. And we call these things indefinite rather than infinite in order that the word infinite should be kept only for God.[2]

Is it, indeed, possible to challenge these philosophical-religious taboos in order finally to deal with the actual infinite?

---

1.   *Heath, The Thirteen Book of Euclid's Elements*, vol. 3, p. 14.
2.   Descartes, *Œuvres philosophiques*, p. 108.

From Prague to Braunschweig: the gods have gone mad

Prague was once one of the great cities of cultural and scientific tradition. But by the end of the 18th century it had become rather a backwater, far removed from centres of political and intellectual decision making. This to some extent explains the lack of influence and the little that was known of one of its sons, Bolzano. He had been ordained a priest in 1804 and was much occupied with religious, ethical and social matters, the latter casting him as a radical and leading to his dismissal from the chair of philosophy. He was attracted by philosophy, methodology of science and particularly, by mathematics and logic. In this context it was natural that the ideas concerning the infinite should find a place. Indeed it was the subject of one of his most celebrated writings, the posthumously published *Paradoxien des Unendlichen* (*Paradoxes of the Infinite*). Bolzano was the first to "actualise" the infinite, considering collections as a whole:

> *I call a set a collection where the order of its parts is irrelevant and where nothing essential is changed if only the order is changed.* [1]

But it is as if he drew back somewhat when he wished to give an example of an infinite set: what he offers is an example of the infinite that arises from continued increasing:

> *It can easily be seen that the set of all absolute and true propositions is an infinite set. [...] For any number, no matter how large, there exist the same number of distinct propositions, and beyond these we are able to construct new propositions or, it is better to say, there are such propositions whether we construct them or not.* [2]

Could it be otherwise? The actual infinite presupposes a good acquaintance with the infinite and therefore with, on the one hand, the finite and on the other, with the set of natural numbers. In this way numbers become a free creation of the human mind. In 1872 in Braunschweig, Dedekind applied himself along these lines. He had just published his construction of the reals from the rationals. In his second publication, *Was sind und was sollen die Zahlen?* (*The Nature and Meaning of Numbers*), he stated:

> *numbers are free creations of the human mind, they serve as a means of apprehending more easily and more sharply the difference of things.* [3]

But Kronecker took a different view:

> *The integer numbers were made by God, everything else is the work of man.* [4]

Here, in the difference between the view that numbers are a creation of the human mind and the view that numbers are God given, lies the essence of the two mathematical views of the potential and the actual infinite. Dedekind, for the first time, was bold enough to offer a definition of an infinite set: it is a set for which there is a bijection (one-one correspondence) between it and a proper subset:

> *A System S is said to be infinite when it is similar to a proper part of itself; in the contrary case S is said to be a finite system.* [5]

---

1.   Bolzano, *Paradoxien des Unendlichen*, § 4, p. 77.
2.   *Ibid.*, § 13, p. 14.
3.   Dedekind, *Essays on the Theory of Numbers*, Preface, p. 31.
4.   Weber, *Leopold Kronecker*, p. 19.
5.   Dedekind, *Essay on the Theory of Numbers*, § 64, p. 63.

Exercise 17   Prove that the set of natural numbers is an infinite set.

In order to complete his construction of the set of natural numbers, he was led to use Bolzano's example of the set of thoughts. In 1908 in his *Untersuchungen über die Grundlagen der Mengenlehre* (Researches on the Foundations of Set Theory), Zermelo set out a system of seven axioms for set theory: here is his axiom for the infinite:

> *Axiom of infinity: There exists in the domain at least one set Z that contains the null set as an element and is so constituted that to each of its elements a there corresponds a further element of the form* {a}, *in other words, that with each of its elements a it also contains the corresponding set* {a} *as an element.*[1]

In other words the domain of objects considered to be sets (given that they satisfy Zermelo's axioms) contains a set Z which contains the empty set and also contains the set {a} whenever it contains the element a. The smallest set Z_0 having the properties of Z allows us, purely symbolically, to define

> *the simplest example of a countable infinite set.*[2]

This is the set whose first members are ϕ, {ϕ},{{ϕ}} and, in general, each {a} is the successor of a. In other words we have:

$$0 = \phi, 1 = \{\phi\}, 2 = \{\{\phi\}\}, ..., n + 1 = \{n\} = \{\{...\{\phi\}...\}\}$$

Other definitions were proposed for infinite sets or sets isomorphic to Z_0. A defintion that is more ordinal is, for example:

$$0 = \phi, 1 = \{\phi\} = \{0\}, 2 = \{\phi,\{\phi\}\} = \{0,1\}, ..., n + 1 = n \cup \{n\} = \{0, 1, 2, ..., n\}$$

where the integer n contains n distincts elements.

Now that the potential infinite has become actual, there remain other questions to face:

– how can infinite sets be counted?

– how can infinite sets be ordered?

A correspondence between Braunschwieg and Halle: the infinite is transgressed

Hardly hundred miles lies between Braunschwieg and Halle where Cantor spent his whole life as a teacher, and correspondence between Cantor and Dedekind was rapid. In 1872 both published their constructions for the real numbers and consequently came to introduce the actual infinite in their work. Cantor wanted to be able to count these new infinite sets, both the rationals and the reals. To do this he used the idea of a one-one correspondence. This was a natural extension of countability from finite sets to infinite ones. But the task was not without difficulties and came up against many firm convictions, not least the Euclidean adage:

> *the whole is greater than the part.*[3]

---

1.   Zermelo, *Untersuchungen über die Grundlagen der Mengenlehre*, p. 266-267, trad. p. 204.
2.   *Ibid.*, p. 267, trad. p. 204.
3.   Heath, *The Thirteen Books of Euclid's Elements*, vol. 1, p. 155.

On the 29th November 1873, Cantor wrote to Dedekind:

> *Allow me to present to you a question which has a certain theoretical interest for me, but to which I can find no answer. [...] Suppose we take the set of all the individual positive integers n and represent it by (n): then take the set of all the real positive numbers x and represent it by (x): the question is simply to know whether (n) can be put into a correspondence with (x) in such a way that to each element of one of the sets there corresponds an element and one only of the other.*[1]

Cantor was in fact perplexed, since at first sight it appears impossible since (n) is discrete and (x) is continuous. On the other hand, it is possible to set up a one-one corespondence between (n) and the set of rationals and even with the set of integer n-tuples.

Exercise 18   Show that a one-one correspondence exists between N and N^2.

On the 2nd December, Dedekind added to Cantor's confusion when he showed that the set of algebraic numbers, real or complex, were also countable. Having established this result, it seems rather less surprising when one considers that all algebraic numbers are the roots of polynomials with integer coefficients and for any given degree they are enumerable. Consequently, we obtain a countable union of countable sets. It remains to show that this union is also a countable set. Now this is not at all evident unless we introduce a further axiom for the infinite, the axiom of choice, which was not formulated until 1904 by Zermelo!

We are not there yet, and Dedekind had to proceed another way by introducing a new concept of height. To the polynomial

$$a_0 x^n + a_1 x^{n-1} + \dots + a_{n-1} x + a_n$$

he assigned a height

$$N = n - 1 + |a_0| + |a_1| + \dots + |a_n|$$

and he ordered algebraic equations using this device. To each height there corresponds a finite number of polynomials and consequently a finite number of algebraic numbers. For example:

for $N = 1$, the polynomial x and the algebraic number 0.

for $N = 2$, the polynomials $x + 1$ and $x - 1$ and the new algebraic numbers 1 and -1

for $N = 3$, the polynomials $2x + 1$, $2x - 1$, $x + 2$, $x - 2$ and $x^2 + 1$, giving new algebraic numbers $-\frac{1}{2}, \frac{1}{2}, -2, 2, i$, and $-i$. This allows us to classify the first algebraic numbers in the following order: $0, -1, +1, -2, 2, -\frac{1}{2}, \frac{1}{2}, -i$ and i.

Exercise 19   What new algebraic numbers are produced from $N = 4$?

Despite offering this masterful proof, Dedekind did not give much encouragement to Cantor to continue to pursue the question of the countability of the reals. This question :

> *Does not justify spending too much time on it, since it is of no practical importance*[2].

---

1.   Cavailles, *Philosophie mathématique*, p. 187-188.
2.   *Ibid.*, p. 194.

In no way discouraged, Cantor sent Dedekind his first proof that the reals were uncountable, in a letter of the 7th December. This was a scornful rejection of practical importance! There are many more transcendental numbers than algebraic numbers yet only a few are known to us. In other words, Cantor had just discovered that there are two sorts of infinities: countable and continuous. Stung by curiosity, Dedekind sent a simplified proof by return of post, and this was the one that Cantor later published.

Suppose that an interval [a,b] is countable: then it is possible to construct a sequence of enclosed intervals ([a_n, b_n]): a_0 and b_0 are respectively equal to a and b and for each integer n, a_{n+1} and b_{n+1} are equal to the first two numbers of [a,b] which belong to the open interval]a_n, b_n[. The convergence of the sequences (a_n) and (b_n) leads to a contradiction.

Exercise 20   Prove that [a, b] is not countable.

Having established two sorts of infinities, countable infinities and the infinity of the continuum, Cantor went on to develop the idea of cardinality for infinite sets using the idea of *power*.

> *If it is possible to set up a one way correspondence, element by element, between two well defined sets M and N [...] it will be convenient to express this by saying that these two sets have the same power.*[1]

He already knew of the powers of countable sets and of the continuum. Were there others? Cantor was able to give a positive reply. In that the continuum has the same power as the set of subsets of N and a power strictly greater than that of N. More generally, the set P(X) of subsets of X has a power strictly greater than that of X. In other words, X and P(X) do not have the same power and there exists a mapping of X to P(X) which is into and not onto. But hardly had this property been discovered when there arose one of the first paradoxes of set theory which Cantor identified. If the collection of all sets is some set A, then the power set of A, P(A) must also be an element of A yet it has a power strictly greater than that of A. In 1908 Zermelo found a way round this paradox by introducing the axiom of selection. This said that a new set could only be made up from a selection of elements of an existing given set, in other words the collection of sets could not be a set.

This paradox being resolved, other questions arose. Did another power exist between the countable infinite and the continuum? More generally, was there a power between that of a set and that of its power set? Can any two arbitrary sets be compared in terms of their powers?

Since Cohen's work in 1963 we now know that we cannot give an answer to all these questions. Cantor had asserted in 1878 that

> *If two sets M and N are not of the same power then M has the same power as an integral part of N, or N has the same power as an integral part of M.*[2]

In other words, for him, two arbitrary sets could always be compared. Now this assertion requires the axiom of choice, as Hartogs showed in 1915. Cohen showed precisely that the axiom of choice is unprovable. It is just the same with the hypothesis of the continuum and with the generalised hypothesis of the continuum which state

---

1.   Cantor, *Une contribution à la théorie des ensembles*, p. 311.
2.   *Ibid.*

respectively that there does not exist any power between that of the countable and the continuous and between that of a set and its set of subsets. In other words we can claim that all sets have powers that can be compared. But we could just as easily claim the opposite. Mathematics is not one and indivisible!

Well before any serious study of the powers of sets, that is the cardinal aspect of sets, Cantor had become acquainted with the ordinal aspect of infinite sets. For our stay in Halle, the hotel Hilbert is appropriate. It is always full and yet any newcomer can always be found accommodation. How is that possible? The answer is very simple: it has an infinite number of rooms. Every time a new guest arrives, all that has to be done is to give him the first room and move all the other occupants to the next room along. We are using the fact that the set of non zero natural numbers has the same power as the set of naturals including zero. But we can also consider the problem from an ordinal point of view. Assume that the manager has already assigned a number to each guest and that he has used up all the natural numbers. When a new guest arrives the manager wishes to assign him a number and so creates a new number ω which comes after all the natural numbers. This is the first transfinite number.

This transfinite labelling or indexation can be extended. Knowing that there are as many integers as there are even integers or odd integers, we can assign the rank p to the even integer 2p. We can dispose of the integers by classifying the odd numbers, following Cantor, by creating new transfinite numbers ($\omega + p$) which can be used to assign the rank $\omega + p$ to the odd number 2p + 1. After this a new arrival could be given the number $\omega + \omega$, labelled 2ω. Cantor went on, step by step, to define a countable infinity of transfinite numbers:

$$\omega, \omega + 1, \omega + 2, ..., \qquad \omega + p, ...$$
$$\text{then} \quad 2\omega, 2\omega + 1, 2\omega + 2, ..., \qquad 2\omega + p, ...$$
$$\text{then} \quad \omega, n\omega + 1, n\omega + 2, ..., \qquad n\omega + p, ...$$
$$\text{then} \quad \omega^2, \omega^2 + 1, \omega^2 + 2, ..., \qquad \omega^2 + p, ...$$
$$\text{then} \quad \omega^n, \omega^n + 1, \omega^n + 2, ..., \qquad \omega^n + p, ...$$
$$\text{then} \quad \omega^\omega, \omega^\omega + 1, \omega^\omega + 2, ..., \qquad \omega^\omega + p, ...$$
$$\text{then}...$$

Exercise 21   Use transfinite numbers to classify

a) the set of natural number pairs,

b) the set of natural number triples.

The set of these transfinite numbers is countable: writing out all the real numbers is not enough. Cantor did not stop here, he extended the schema by which he had defined ω. Following the class of all transfinite numbers as here defined, there exists a new transfinite number, as there did after all the integers. All that is needed is a transfinite extension using the previous procedure. Empowered by this transfinite indexation procedure, we are now in a position to ask ourselves the following question, as did Cantor: for any given arbitrary set, are transfinite numbers sufficient to provide an indexation of all its elements? In other words, since the collection of transfinite numbers is well ordered, does it follow that the set is also well ordered?

From Heidelberg we return in good order

At the second Mathematical Congress held in Paris in 1900, Hilbert delivered his famous lecture *On future problems in mathematics.* The first problem related to the power of the continuum.

> *The question now arises whether the totality of all numbers may not be arranged in another manner so that every partial assemblage may have a first ellement, i.e., whether the continuum cannot be considered as a well ordered assemblage—a question which Cantor thinks must be answered in the affirmative. It appears to me most desirable to obtain a direct proof of this remarkable statement of Cantor's, perhaps by actually giving an arrangement of numbers such that in every partial system a first number can be pointed out.*[1]

Exercise 22   Show that the lexicographic ordering on N x N defined by (x, y) < (x', y') if and only if [x < x' or (x = x' and y < y')] is a well ordering.

At the Third Congress held at Heidelberg in 1904, König claimed to be able to give a negative reply. This bolt from the blue shook the Cantorian community and on 24th September 1904, Zermelo showed that every set could be well ordered. But he used what came later to be called the axiom of choice. We are today aware of more than two hundred statements that are equivalent to the axiom of choice, among which are:

Every set can be well ordered

For any two sets their powers can be compared

Every infinite set is of equal power to its square

Every infinite product of non empty sets is itself non empty

Every vector space has at least one basis

In every set of non empty subsets it is possible to choose an element in such a way as to constitute a new set, that is to define a function of choice.

Exercise 23   f is called a function of choice over E, if for all non empty subsets of E in E, f(X) belongs to X.

a) Prove that if E is well ordered, a function of choice can be defined on E.

b) Prove that if a function of choice exists over E the infinite product of a family of non empty subsets of E is non empty.

This possibility of choice is apparent when finite sets are being considered but it is not at all the same thing when infinite sets are considered. The quarrels were lively and

---

1.   Hilbert, *On future problems in mathematics,* p. 71, english tr. p. 446-447.

Zermelo thought himself to have dealt with the problem in 1908 when he included the axiom of choice in his system of set theory axioms. But should this axiom be accepted? It has been used in many branches of mathematics. Both Gödel in 1938, and then Cohen in 1963, showed that it was compatible with the other set theory axioms and independent of them. Should we not therefore be content?

The infinite quest is eternal. If Zorn's lemma is the most frequently used form of the axiom of choice and if Kuratowski has been able to offer his *"method of elimination of the transfinite numbers of mathematical arguments,"* we know also that the worm is in the bud. The Banach-Tarski paradox is there to remind us of the need for a certain vigilance. In fact, in using the axiom of choice, we can show

> *that it is possible, for example, in cutting up a pea into a finite number of bits, then rearranging them without deformation, to reconstitute, not the initial pea, that would be too easy, but two peas of the same size as the first one.*[1]

Astounding! And yet it is not just the miracle of the multiplication of peas, but also that of our questions about the infinite. We have certainly not come to an end of questions, nor of possible answers to them.

---

1.   Guinot, *Le paradoxe de Banach-Tarski*, p. 5.

Bibliography

Sources texts

ALBERT, *Capitalisme contre capitalisme*, Seuil, Paris, 1991.

ARISTOTLE, *Works of...*: Ross, W.D. (ed.), Oxford 1930.

BOLZANO, *Paradoxien des Unendlichen*, Prihonsky, Leipzig, 1851. English tr. Steele, *Paradoxes of the Infinite*, Routledge and Kegan, London, 1950.

CANTOR, "Ein Beitrag zur Mannigflatigkeitslehre", *Journal für die reine und angewandte Mathematik*, **84**, (1878), 242-258: = "Une contribution à la théorie des ensembles", *Acta mathematica*, **2**, (1883), 311-328.

CASSINET, "Le premier livre de l'arithmétique de Fransciso Maurolycus (1557)", *Cahiers du Séminaire d'Histoire des Mathématiques de Toulouse*, **3**, (1981), 1-27.

CASSINET & GUILLEMOT, *L'axiome du choix dans les mathématiques de Cauchy (1821) à Gödel (1940)*, Thèse d'Etat, Toulouse, 1983.

CAVAILLES, *Philosophie Mathématique*, Hermann, Paris, 1962.

DEDEKIND, *Was sind und was sollen die Zahlen*, Vieweg, Braunschweig, 1888. English tr. Beman, Dover, New York, 1962.

DESCARTES, *Œuvres philosophiques*, Ed. Alquier Garnier, Paris, 1973.

ERMAN & RANK, *La civilisation égyptienne*, Payot, Paris, 1986.

EULER, "Theorematum quorundam arithmeticocum demonstrationes", *Œuvres Commentationes Arithmeticae*, t. 1, Ed. Rudio Teubner, Leipzig, 1940, 38-58. French tr. Mathe, *Méthodes démonstratives heuristiques dans les Commentaires arithmétiques d'Euler*, Thèse 3e cycle, Toulouse, 1984.

FERMAT, *Œuvres*, t. III, Gauthier Villars, Paris, 1894.

GUINOT, *Le paradoxe de Banach Tarski*, Aleas, Lyon, 1991.

HEAT, *The Thirteen Books of Euclid's Elements*, 2nd edition, 3 vols., Cambridge University Press, Cambridge 1926, repr. Dover, New York, 1956.

HILBERT, "Sur les problèmes futurs des mathématiques", *Compte Rendu du Deuxième Congrès International des Mathématiciens tenu à Paris du 6 au 12 Août 1900*, Gauthier Villars, Paris, 1902. English tr. in *Bulletin of the mathematical society* **8** (1902) 437-479.

IFRAH, *Histoire universelle des chiffres*, Seghers, Paris, 1981.

KRAMER, *L'histoire commence à Sumer*, Arthaud, Paris, 1986.

LEFEVRE, *Grammaire de l'Egyptien classique*, Institut français d'archéologie orientale, Cairo, 1955.

MENNINGER, *Number Words and Number Symbols*, MIT Press, Cambridge, Mass., 1969.

PASCAL, *Œuvres complètes*, Seuil, Paris, 1963.

PEANO, *Arithmetices Principia*, Bocca, Turin, 1889.

POINCARÉ, *La Science et l'hypothèse*, Rpr. Flammarion, Paris, 1988.

ROUX, *L'homme et son nombre*, IREM CRDP Besançon, 1988.

THEON OF SMYRNA, *Exposition des connaissances mathématiques utiles pour la lecture de Platon*, Ed. Dupuis, Paris, 1898. Rpr. Culture et civilisation, Brussels, 1966.

VERDET, *Le ciel, ordre et désordre*, Gallimard, Paris, 1987.

WEBER, "Léopold Kronecker", *Jahresbericht der Deutschen Mathematiker-Vereinigung*, **2**, (1893), 5-31.

ZERMELO, "Untersuchungen über die Grundlagen der Mengenlehre", *Mathematische Annalen*, **65**, (1908), 261-281. English tr. in Van Heijenhort (ed.), *From Frege to Gödel: A Source Book in Mathematical Logic, 1879-1931*, Harvard University Press, Cambridge, Mass., 1967, p. 198-215.

General works for further reading

DAVIS & HERSCH, *The Mathematical Experience*, Birkhaüser, Boston, 1980: Penguin Books, 1983.

DIEUDONNÉ, *Abrégé d'histoire des mathématiques 1700-1900*, Hermann, Paris, 1978.

DOUGLAS & HOFSTADTER, *Gödel, Escher, Bach*, Basic Books, New York, 1979.

GUILLEN, *Bridges to infinity*, Tarcher, 1983.

MAOR, *To infinity and beyond. A cultural history of the infinite*, Princeton University Press, 1991.

MOORE, *The Infinite*, Rootledge, London, 1990.

SALANKIS, *L'herméneutique formelle: L'Infini- Le Continu- L'Espace*, CNRS, Paris, 1991.

SINACEUR, *Corps et modèles*, Vrin, Paris, 1991.

How did you get on?

Ex. 1   1 422 000

Ex. 2 $\frac{2}{71} = \frac{1}{40} + \frac{1}{568} + \frac{1}{710}$ (from the Rhind Papyrus)

this solution is not unique. Another solution, for example, is

$\frac{1}{42} + \frac{1}{426} + \frac{1}{497}$

Ex. 3   15 + 30 + 60 + 40 = 145. There are also 15 fowl.

Ex. 4

36 000 × 6	216 000 × 10	2 160 000 × 6
216 000	2160 000	12 960 000

very large stylus oblique	very large stylus oblique small stylus vertical	very large stylus vertical

Ex. 5   4, 38, 117, 221, 11, 88, 281, 139, 20.

Ex. 6 $\frac{2}{7} \sim \frac{17}{60} + \frac{8}{60^2} + \frac{34}{60^3}$

$\frac{2}{7} = \left(\frac{17}{60} + \frac{8}{60^2} = \frac{34}{60^3}\right) \left(1 + \frac{1}{60^3} + \frac{1}{60^6} + \dots + \frac{1}{60^{3n}} + \dots\right)$

$\frac{2}{7} = 0{,}285714 \left(1 + \frac{1}{10^6} + \frac{1}{10^{12}} + \dots + \frac{1}{10^{6n}} + \dots\right)$

Ex. 7   The nth triangular number is the sum of the first n consecutive integers which comes to
$\frac{n(n + 1)}{2}$

Ex. 8

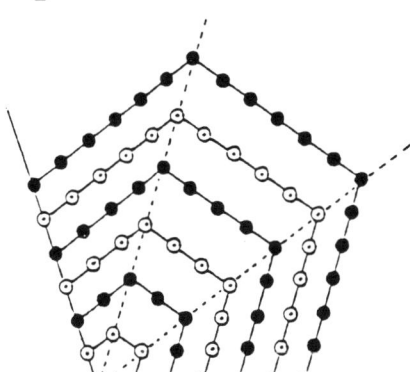

$P_n^5 = \frac{n(3n - 1)}{2}$

1, 5, 12, 22, 35, 51, 70

Figure 8

Ex. 9   If $2(n + 2) > 2^{n + 2}$ then $2(n + 1) > 2^{n + 2} - 2 = 2^{n + 1} + 2^{n + 1} - 2 \geq 2^{n + 1}$

Ex. 10   All this required is to show that (not i) and (not ii) are equivalent.

Ex. 11 $2 T_n - n = 2 \frac{n(n + 1)}{2} - n = n^2$

Ex. 12 $1 + 3 + 5 + 7 + \dots + (2n + 1) = (n + 1)^2$ and $(n + 1)^2 + (2n + 3) = (n + 2)^2$

Ex. 13   It is sufficient to transpose:

Infinite descent: $(\forall\, n \in \mathbb{N})\, [\text{not } P(n) \Rightarrow (\exists\, n' \in \mathbb{N})\, (n' < n \text{ and not } P(n'))]$

Induction: $(\forall\, n \in \mathbb{N})\, \{[(\forall\, q \in \mathbb{N})\, (q < n \Rightarrow P(q))] \Rightarrow P(n)\,\}$

Ex. 14 Argue by reductio ad absurdum considering the number $p_1 p_2 \ldots p_n + 1$

Ex. 15 $(\forall a \in \mathbb{R}), (\forall b \in \mathbb{R}), [(0 < a < b) \Rightarrow (\exists n \in \mathbb{N})(n\,a > b)]$

Ex. 16 $n\,CD > AB$ and $AB_{n-1} = \dfrac{AB}{2^{n-1}} < CD$

Ex. 17 The mapping $n \to n + 1$ is a one-one mapping from $\mathbb{N}$ to $\mathbb{N}^*$ where $\mathbb{N}^*$ is a proper subset of $\mathbb{N}$: $\mathbb{N}$ being the natural numbers including zero, $\mathbb{N}^*$ the natural numbers excluding zero.

Ex. 18 Consider the mapping $f: \mathbb{N}^2 \to \mathbb{N}$ where $f(x, y) = y + (x + y)(x + y + 1)/2$

Ex. 19 New equations are

$3x \pm 1 = 0,\ x \pm 3 = 0,\ 2x^2 \pm 1 = 0,\ x^2 \pm x \pm 1 = 0,\ x^2 \pm 2 = 0$

giving new algebraic numbers

$\pm\dfrac{1}{3},\ \pm 3,\ \pm\dfrac{\sqrt{2}}{2} \pm i\,\dfrac{\sqrt{2}}{2},\ \dfrac{+1 \pm \sqrt{5}}{2},\ \dfrac{-1 \pm i\sqrt{3}}{2},\ \dfrac{+1 \pm i\sqrt{3}}{2},\ \pm\sqrt{2},\ \pm i\sqrt{2}.$

Ex. 20 Simply follow Dedekind's argument.

Ex. 21 Let $(n.\ p)$ correspond to $(n\omega + p)$

Let $(0, n, p)$ correspond to $(p\omega + q)$ and let (n, p, q) correspond to $(\omega^{n+1} + p^\omega + q)$

Ex. 22 Assume E is a non empty subset of $\mathbb{N} \times \mathbb{N}$ then

$\min E = (\min \{x: (x, y) \in E\},\ \min \{y: (\min \{x: (x, y) \in E\}, y) \in E\}$

Ex. 23 a) $f(X) = \min X$

b) $(f(X_i))_{i \in I} \in \displaystyle\prod_{i \in I} x_i$

2

Must we always be rational? From incommensurable magnitudes to real numbers

Denis DAUMAS
Michel GUILLEMOT
IREM Toulouse

One day Socrates had a slave brought before him, and drew a square of side two feet upon the ground. He then asked him to find the side of a square of twice the area.

The slave's first attempt was to double the length of the side. Under Socrates' guidance he was led to see that doubling the side resulted in quadrupling the area instead of doubling it. So a square of side four feet would be too great, but it would need to be greater than two feet, so he could try a square of side three feet. This did not work either and the slave became slightly unnerved: "It's no use, Socrates, I just don't know."

We note that Socrates himself suggested that numbers might have to be abandoned to find a solution: "Try to tell us exactly. If you don't want to count it up, just show us on the diagram." Socrates drew three squares side by side, an initial square, one four times as large and a copy of it for solving what had now become a geometrical problem: to draw a square half the area of the enlarged square.

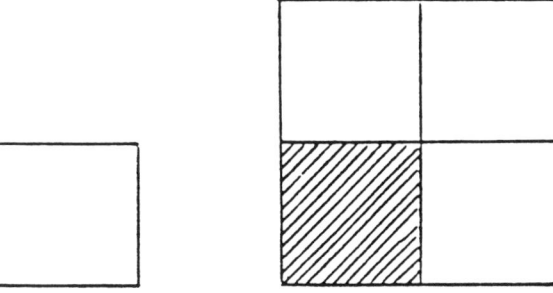 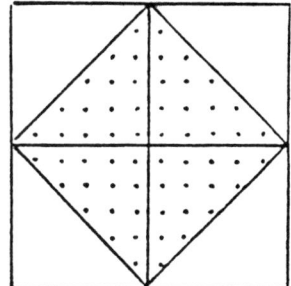

Figure 1

Since a diagonal cuts a square into two equal parts, the solution comes from cutting in half each of the four small squares by a suitable diagonal. The slave was then able to see the solution to the problem.

This scene comes from a passage of Plato's *Meno*[1], a dialogue that was written at the beginning of the 4th century BC. From a modern perspective, the mathematical content of this episode can be summarised thus: the square constructed on the diagonal d of a square of side a is certainly double the area of the square constructed on the side a (that is $d^2 = 2a^2$); but the ratio between the diagonal and the side is not rational, it is $\sqrt{2}$. And therein lies the arithmetic stumbling block for the slave, in his attempt to solve the problem. However, in the scene between Socrates and the slave, the problem is not posed in those terms: it concerns the impossibility of measuring the diagonal of a square using a portion of one of its sides (in this case, a half, but one could equally well take a third, a quarter,...). We are no longer in a situation that is purely numerical: we have here the idea of a measure assuming a role as an intermediary between numbers and geometric magnitudes. While magnitudes remain commensurable, that is they can be measured with a common unit, the result of the measure will be a whole number, and there is a guaranteed correspondence between geometry and arithmetic. But the matter becomes much more difficult when we are dealing with incommensurable magnitudes, and we need to be able to answer two related questions:

– how can measures be used when lengths are not directly measurable? Is there a way of resolving the differences, between the infinite process of approximation, and the problem of incommensurability?

– what sort of new "rationality" would allow us to treat the irrational ratios of incommensurable magnitudes as numbers?

We shall see how, during the course of history, these questions sometimes made demands on our three protagonists: numbers, magnitudes and measures.

An untroubled infancy?

Measuring is an elementary activity which has been practised ever since the first civilisations of the 4th millenary BC. Given an economy that was largely dependent upon agriculture, the Mesopotamians were not only aware of measures for length and area. They also measured volumes, in locally defined units, or using an arbitrary unit, and we know that weighing was practised by gold and silver smiths.

The use of measurement was noted by Herodotus. Following his visit to Egypt in the 5th century BC, he wrote:

> *This king [the semi-mythical Sesostris] moreover (so they said) divided the country among all the Egyptians by giving each an equal square parcel of land, and made this his source of revenue, appointing the payment of a yearly tax. And many man who was robbed by the river of a part of his land would come to Sesostris and declare what had befallen him; then the king would send men to look into it and measure the space by which the land was diminished, so that thereafter it should pay in proportion to (kata logon) the tax originally imposed. From this, to my thinking, the Greeks learnt the art of geometria.*[2]

We should not overlook the etymology of the Greek word "geometry" which means to measure the Earth.

1.   Plato, *Protagoras and Meno*, 82a to 85b.
2.   Herodotus, *Œuvres Complètes* II 109, p. 183, english tr. in Fowler, *The mathematics...*, p. 283.

In this context, how should we interpret the significance of the Babylonian tablet YBC 7289 (Figure 2)?

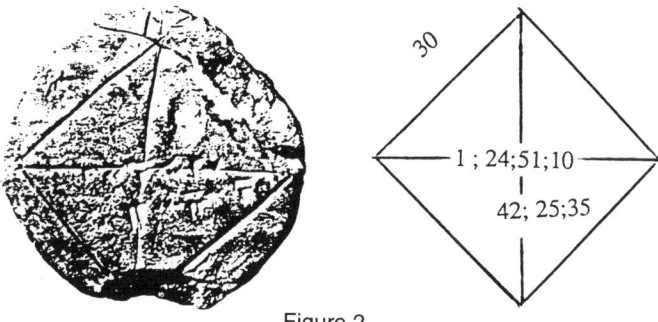

Figure 2

A Babylonian scribe drew the square and its diagonals on a clay tablet in about 1800 BC. In sexagesimal notation, there appears on one side of the square the number 30, and across a diagonal are written the numbers: 1; 24; 51; 10 and 42; 25; 35. Now,

$$30 \left(1 + 24/60 + 51/60^2 + 10/60^3\right) = 42 + 25/60 + 35/60^2$$

In other words, we can conclude that the scribe knew that the ratio of the diagonal of a square to its side was equal to $1 + 24/60 + 51/60^2 + 10/60^3$.

Exercise 1	Find the order of this approximation for $\sqrt{2}$:
	– in base ten
	– in base sixty

The approximation for "$\sqrt{2}$" normally found in Babylonian documents is $1 + 25/60$. The approximation in the tablet YBC 7789 is a better one, but that does not mean that the Babylonians had asked themselves questions about the commensurability or incommensurability of magnitudes. The same can be said of the Egyptians when they used the value $(D - 1/9\, D)^2$ for the area of a disc of diameter D. Economic needs, and perhaps more theoretic studies of which we are ignorant, led to better approximations for these ratios, without their authors being aware of the character of the processes they were using, whether finite or not.

Exercise 2	Find the order of this Egyptian approximation:
	– in base ten
	– using the inverses of natural numbers

Plato is said to have inscribed over the entrance to his Academy the well-known maxim: "Let no one enter here who is not a geometer": are we to understand that earlier than that, through the use of the intermediary of measurement, geometric magnitudes could have had a numerical aspect? After all, the Babylonians (using sexagesimals) and the Egyptians (using fractions of the type 1/n) certainly had tools which were well adapted for finding precise, or even exact, numerical expressions for the ratios of geometrical magnitudes. Today, there is general agreement that the discovery of incommensurability was due to the Greeks, and we identify the Pythagoreans, in the 6th century BC, with the key to that discovery. What view did the Pythagoreans have of our three characters: numbers, magnitude and measure? How was the discovery of incommensurability made, and what were its consequences?

Number assumes the lead of our trio ... but it trips up over the square!

Legend has it that Pythagoras knew the secrets of the Egyptians well before Herodotus. Little is known about Pythagoras, who appears as a largely mythical person. But we know rather more about the philosophy of Pythagoreanism which is founded upon the idea of "all is number": the universe, music, harmony and opposition, all is regulated by whole numbers and the ratios between them. The intrusion of the incommensurability of the diagonal and side of a square, (or the diagonal and side of a regular pentagon), into this world was felt as a veritable crisis. A number of stories grew up around this subject, concerning the secrecy in which the discovery had to be kept and the divine anger that would be brought to bear on the first who divulged the secret.

We also know the way in which the Pythagoreans associated number and spatial position through the use of figurate numbers (see Chap. 1). The arithmetic unit, represented by a geometric point, plays the role of a material atom, and sequences of numbers are associated with geometric shapes, as with the triangular numbers:

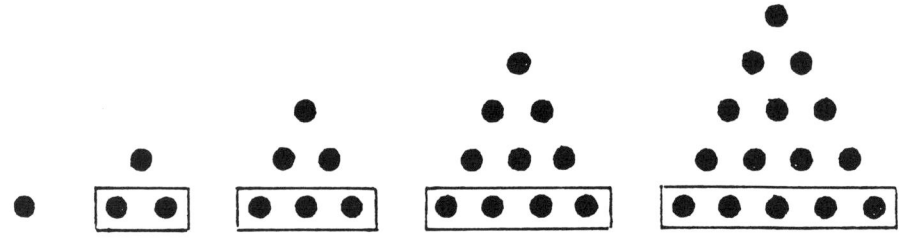

Figure 3

The idea that the geometrical line was made up of a finite number of point-atoms certainly existed at the time of Pythagoras, giving rise, at the beginning of the 5th century BC, to what we know as Zeno's paradoxes. Now, if p is the number of points on the side of a square and q is the number of points on its diagonal then, by Pythagoras, q^2 must be equal to $2p^2$. Perhaps it was the arithmetical impossibility of the relation $q^2 = 2p^2$, that gave the Pythagoreans their first proof of the existence of incommensurability, in this case the incommensurability of the side and diagonal of a square. It is impossible to say. However, at the end of the fourth century BC, Aristotle gave this proof as the archetype of proof by reductio ad absurdum:

> For all who effect an argument per impossibile infer syllogistically what is false, and prove the original conclusion hypothetically when something impossible results from the assumption of its contradictory; e.g. that the diagonal is incommensurate, because odd numbers are equal to evens if it is supposed to be commensurate. One infers syllogistically that odd numbers come out equal to evens, and one proves hypothetically

the incommensurability of the diagonal, since a falsehood results through contradicting this.[1]

Aristotle assumes that the proof is well known to his readers, and does not give the details.

Presented as the 117th and last proposition of Book X of Euclid's *Elements*, Peyrard published a detailed version of such a proof. The proof is not by Euclid and has been interpolated: it is also rather heavy and clumsy.

We prefer to offer the reader the opportunity of constructing his own proof using the exercises that follow:

Exercise 3　　a) Prove that the square of an odd number is odd.

b) Suppose that p and q are two non zero natural numbers such that q^2 is equal to $2p^2$ and that p and q have no common factors. Prove that it follows that p is at the same time both odd and even. Draw a conclusion.

Exercise 4　　Is it possible to construct an analogous proof to show that $\sqrt{3}$ is irrational?

The discovery of incommensurability struck a blow at the idea of an indivisible unit point, which was the irreducible principle upon which all geometrical magnitudes were built. However, does everything have to be rejected? The unit is the foundation of number: might it be possible to envisage a unit of geometrical figures, through the construction of unlimited sequences of triangular numbers, square numbers, pentagonal numbers, ..., and might it not also be possible to express the relation between a side and a diagonal by constructing other unlimited sequences of numbers? This is a possible interpretation of a text by Theon of Smyrna, a Neo-Pythagorean from the second century AD.

Theon attempts a reconciliation, but it was incomplete!

Just as numbers potentially contain triangular, square, and pentagonal ratios, and ones corresponding to the remaining figures, so also we can find side and diagonal ratios appearing in numbers in accordance with the generative principles; for it is from these that the figures acquire balance. Therefore since the unit, according to the supreme generative principle, is the starting-point of all the figures, so also in the unit will be found the ratio of the diagonal to the side.[2]

The task of constructing these numbers is less easy for the square and diagonal, than it is for figurate numbers, since the side and diagonal numbers need to be constructed at the same time. Nevertheless, the unit being the foundation of all things, the side and diagonal numbers must not evade this principle, and so Theon took the unit to be the

---

1.　　Aristotle, *Premiers analytiques* I. 23, 23-40, 41a, p. 121-122, english tr. Fowler, *The mathematics...*, p. 297.
2.　　Theon of Smyrna, *Exposition. . .,* XXXI, p. 71. This translation taken from D.H. Fowler, *The Mathematics of Plato's Academy*, p. 58.

first number for both side and diagonal. There follows the "generative principle", that is a recurrence process:

> *Now there are added to the side a diagonal and to the diagonal two sides...[1]*

This allowed him to obtain the first three pairs of side and diagonal numbers: 2 and 3, then 5 and 7 and finally 12 and 17.

In general, we can say that this algorithm generates two sequences of whole numbers (λ_n) and (δ_n) which are defined by the recurrence relations:

$$\lambda_1 = \delta_1 = 1$$

and for all non zero natural numbers,

$$\lambda_{n+1} = \lambda_n + \delta_n \qquad \text{and} \qquad \delta_{n+1} = 2\lambda_n + \delta_n.$$

Exercise 5   Calculate the first thirteen terms of the sequences (λ_n) and (δ_n) and also the value of $(\delta_n/\lambda_n)^2$. What can you conclude? How far would you have to go before your calculator gives 0 for the value of $(\delta_n/\lambda_n)^2 - 2$?

The unit is also used to integrate the idea of incommensurability into the Pythagorean philosophy. In fact, Theon of Smyrna notes:

> *The square on the diagonal will be now greater by a unit, now less by a unit, than twice the square on the side...[2]*

That is,

$$(1) \quad \delta_n{}^2 = 2\lambda_n{}^2 + (-1)^n.$$

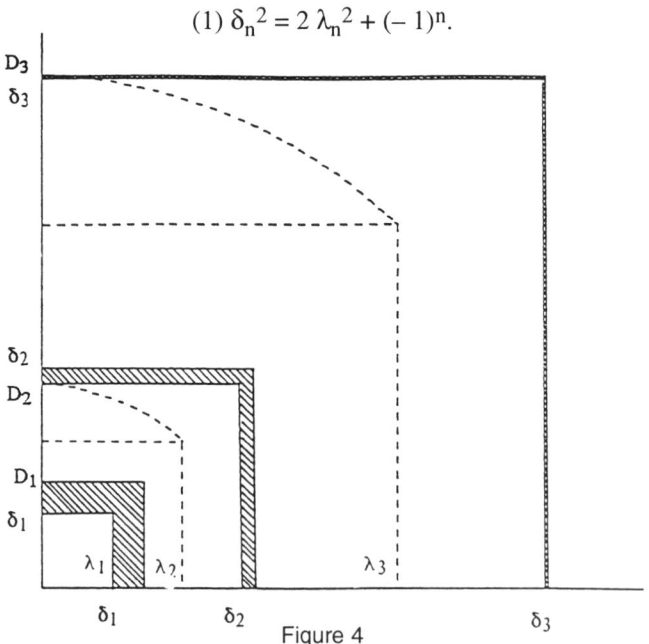

Figure 4

In figure 4 we have constructed squares on the diagonals D_1, D_2 and D_3 of the squares of side λ_1, λ_2 and λ_3. We have also drawn squares of side δ_1, δ_2 and δ_3, the

---

1.   *Ibid.*
2.   *Ibid.*

diagonal numbers associated with the lateral numbers λ_1, λ_2 and λ_3. The differences, showing the excess or deficit of one over the other, have been shaded on the diagram and represent the values of $\delta_n^2 - 2\lambda_n^2$ for n equal to 1, 2 and 3. Each has the value of one unit of area: with the chosen starting unit, and taking account of the thickness of the drawn lines, we have to stop here; it is impossible to distinguish between δ_4 and D_4!

Exercise 6   Prove the result (1).

It appears then, from figure 4, that this difference of one unit remains constant, (apart from its sign), as the dimensions of the squares, that is λ_n and δ_n, become increasingly large. Theon was unable to achieve the infinite value of the ratios δ_n/λ_n (the actual infinite) but he was able to content himself with the fact that he could make the ratio δ_n/λ_n as close as he wished to the ratio of the diagonal to the side of a square (the potential infinite) in order to consider that "numbers potentially contain" this latter ratio, although irrational, and thus managed to save the essentials of Pythagoreanism. And more than that, it becomes possible to generalise Theon's algorithm for other irrationals of the type $\sqrt{k}$ by constructing the sequences (u_n) and (v_n) with the defining relations:

$$u_1 = v_1 = 1$$

and for all non zero natural numbers,

$$u_{n+1} = u_n + v_n \text{ and } v_{n+1} = k\,u_n + v_n.$$

Exercise 7   Given (u_n) and (v_n) as the sequences just defined, prove that:

a) for all non zero natural numbers n, $v_n^2 - k\,u_n^2 = (1-k)^n$

b) $(v_n/u_n)^2$ converges to k.

c) Let $k = 4$ and prove that in this case $v_n = 2u_n$ is equivalent to $v_{n-1} = 2u_{n-1}$.

Hence show that Theon's algorithm does not provide us with a criterion for irrationality.

Following the proof of the impossibility of being able to find two numbers p and q such that q^2 is equal to $2p^2$, we find ourselves now, with these lateral and diagonal numbers, being able to find closer and closer approximations to the ratio of two incommensurable magnitudes, yet with a method that leaves the ratio itself, at infinity, inaccessible. Rather than letting numbers and magnitudes engage in this endless pursuit, might it not be better to let them bury the hatchet and allow each to go his own way? But what role does measure play in all this, whose vocation it is to act as intermediary?

Divorce à la Euclidienne

Euclid's *Elements*, the first Bible of mathematics, was produced at the beginning of the third century BC. Books VII to IX concern numbers (arithmos) and the others, magnitudes (megethos). Even where there are similar propositions for numbers and magnitudes, the propositions are proved independently, each in its own way!

Measure is present in both number and magnitude, but we shall see that in passing from one to the other it undergoes a slight change in its features. The first three definitions of Book VII specify the concept of number and the place that measure occupies:

> *1. An **unit** is that by virtue of which each of the things that exist is called one.*
> *2. A **number** is a multitude composed of units.*
> *3. A number is **a part** of a number, the less of the greater, when it measures the greater.*[1]

Numbers can be used for counting and so, for measuring. But before we can count we must say what is the unit for counting (in a book, it could the chapter, the page, the word, ...), and before beginning to measure we must say what the unit of measure is. The "unit", in the sense in which it is defined by Euclid, is there to enable us to count, and is not in itself a number. Aristotle makes this very clear:

> *For one is not a real object in its own right. And this is only reasonable: one means a measure of some plurality, and number means a measured plurality and a plurality of measures. (Thus there is good reason for one not to be a number: a measure is not itself measures...)*[2]

At the same time, the unit sometimes takes on the role of number in the *Elements*: when subtracting two numbers, Euclid allows that the result could be the unit, and he also proposes adding the unit to a number. Finally, in place of our verb to divide, Euclid uses the term "to measure". It is, therefore, inconceivable that "the unit" could itself be divisible. Above all, the technique of measuring is at the heart of numerous Euclidean number processes. We have in mind, for example, the technique for finding the greatest common divisor (we should call it the greatest common measure, but we abbreviate to GCD) and this process is referred to as the Euclidean algorithm. The process is found as the first proposition of Book VII of the *Elements*:

> *Two unequal numbers being set out, and the less being continually subtracted in turn from the greater, if the number which is left never measures the one before it until an unit is left, the original numbers will be prime to one another.*[3]

Consequently, if the two numbers are not relatively prime, the same process will yield a remainder different from 1 which measures the preceding number. This remainder is the greatest common divisor of the two original numbers, and this is proved in proposition 2 of Book VII.

To show how the algorithm works, consider this example: to find the GCD of 119 and 85, first take 85 from 119, which leaves 34. Now, 34 does not divide exactly into 85, so subtract 34 from 85. This can be done twice, to leave a remander of 17. Now, 17 divides exactly into 34, therefore it must divide into 85 and 119, and 17 is the GCD.

Exercise 8	a) Use the Euclidean algorithm to find the GCD of

$$a = 71\ 755\ 875 \text{ and } b = 61\ 735\ 500$$

b) The values of a and b given above, arise in a calculation by the astronomer Aristarchus of Samos (c. 300 BC). Aristarchus gives 43/37 as an approximation to a/b. Show that this approximation is obtained if the Euclidean algorithm is terminated when the remainder is negligible. Is it an over-approximation or an under-approximation?

---

1.   Heath, *The Thirteen Books of Euclid's Elements*, VII Def. 1 to 3, vol. 2. p. 277.
2.   Aristotle, *Metaphysics* N 1088a.
3.   Heath, *The Thirteen Books of Euclid's Elements*, VII, 1, vol. 2. p. 296.

Let us now consider magnitudes. This term was never defined by Euclid, but the context allows us to be clear about its meaning. Magnitudes are geometric: lines (segments, circles, arcs of circles, ...), surfaces (polygons, discs, ...) or solids (polyhedra, spheres, ...). One of the characteristics of magnitudes is that they can be subdivided into parts of the same nature (a line into lines; a surface into surfaces; ...), which in their turn can themselves be further subdivided. However, the point is not a magnitude:

> A **point** is that which has no part.[1]

Numbers are not magnitudes: they are assemblages of units, and the unit itself cannot be subdivided. Number is discrete whereas magnitude is a continuous quantity, and between the discrete and the continuous there is a difference which had not escaped Aristotle:

> Plurality is that which is potentially divisible into non-continuous parts. Magnitude is that which is divisible into continuous parts.[2]

The Euclidean algorithm, referred to by some authors as "anthyphairesis", a term derived from the Greek word used by Euclid to describe subtracting alternately, can apply to magnitudes. It only requires the geometrical construction of the differences and successive remainders. In fig. 5 we show B subtracted twice from A, the remainder R_1 subtracted from B to leave a remainder R_2, and R_2 subtracted three times from R_1 to leave a remainder R_3 ...

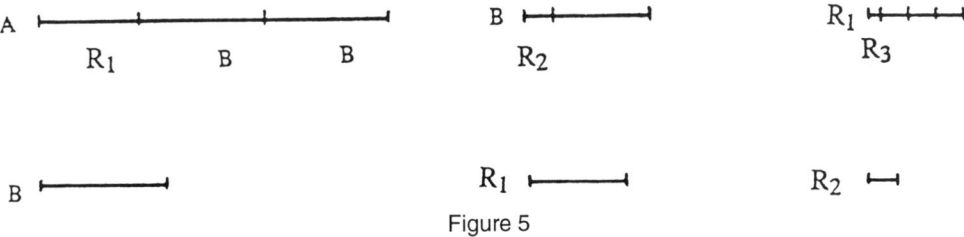

Figure 5

Whenever a remainder exactly measures the preceding one, the algorithm terminates: just as for numbers, we obtain the greatest common measure of the two original magnitudes. But the difference between numbers and magnitudes, identified by Aristotle, is seen here in full measure: where some smaller magnitude cannot be found that will divide all the others, the process will continue indefinitely. Euclid proves, in proposition 2 of Book x, that in this case, the two magnitudes are incommensurable:

> If, when the less of two unequal magnitudes is continually subtracted in turn from the greater, that which is left never measures the one before it, the magnitudes will be incommensurable.[3]

Anthyphairesis derives from the property of measure and allows us to determine whether or not two magnitudes are commensurable. However, nowhere in the *Elements* does Euclid offer us an example of the application of this criterion. Euclid shows little interest in its practical use: rather he was following a train of thought whose theoretical aspect lay beyond the practice of calculations. However, we certainly have repeated

---

1.   Heath, *The Thirteen Books of Euclid's Elements*, I Def. 1, vol. 1, p. 153.
2.   Aristotle, *Metaphysics* Δ 13 1020a.
3.   Heath, *The Thirteen Books sof Euclid's Elements*, X 2, vol 3. p. 17.

subtractions here which never finish. How might we recognise, after a finite number of operations, that we are engaged in a process that has no end to it?

The path followed by Proclus

In his *Commentary on Plato's Republic*, Proclus (5th century AD) suggests that the justification for the construction of diagonal and lateral numbers is to be found in proposition 10 of Book II of Euclid's *Elements*. This proposition asserts, in effect, that if a segment BD is added to a segment AB whose mid-point is C then we have the relation:

$$AD^2 + DB^2 = 2\,(AC^2 + CD^2)$$

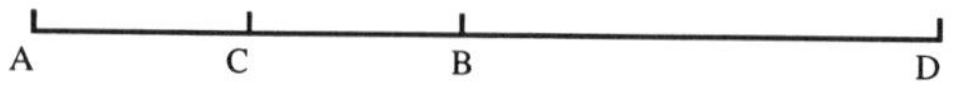

If we let AC = x and CD = y, proposition II. 10 becomes the identity

$$(y + x)^2 + (y - x)^2 = 2\,(y^2 + x^2)$$

Figure 6

When AC and BD represent, respectively, the side and diagonal of a square, BD^2 is equal to $2\,AC^2$, and proposition II. 10 then allows us to write:

$$AD^2 = 2\,CD^2.$$

AD and CD are then the side and diagonal, respectively, of a larger square. If we put AC = λ and BD = δ, we obtain the relations:

$$CD = \lambda + \delta \quad \text{and} \quad AD = 2\lambda + \delta$$

which we recognise as "Theon's formulae".

Conversely, if the side CD is taken away from the diagonal AD, there remains AC. If AC is now taken away from CD, the remainder is BD; AC and BD are, respectively, the side and diagonal of a smaller square, and we can now take AC from BD...

Given Euclid's algorithm, the property of squares that is derived from proposition II.10 allows us to illustrate geometrically the successive remainders. We are now in a position to answer the question: how can we prove the incommensurability of the diagonal and side of a square by anthyphairesis?

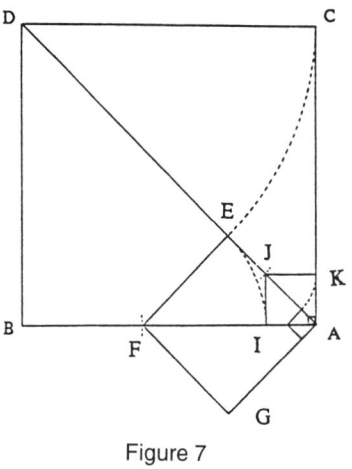

Figure 7

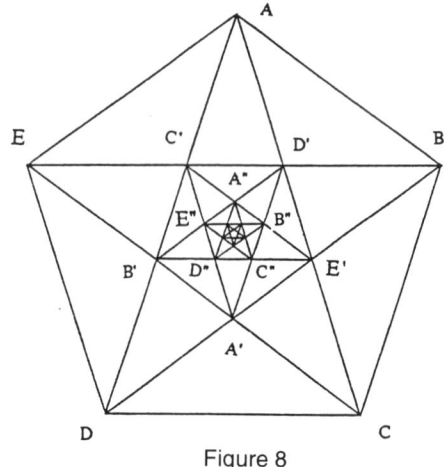

Figure 8

Given a square ABCD (Figure 7), take the side CD from the diagonal AD using a construction to find the point E on AD where DE is equal to DC: the remainder is AE which is smaller than CD.

We must now take the remainder AE from CD, or from AB. To do this, construct on AB the point F where BF is equal to AE. The remainder AF is, as we have already seen, the diagonal of a square of side AE, that is the square AEFG.

We can subtract AE once more from AF obtaining AI, which we can take away from AE to obtain AJ, the diagonal of the square AIJK ... and we observe successively reducing squares arranged around a common vertex A, up to the point where they become indistinguishable from A itself. And beyond that? What we have is a never ending algorithmic process: the remainders are the sides and diagonals of squares and yet, no matter how small it becomes, the side can never measure the diagonal.

What we have done is to use the mathematics of Ancient Greece to show how proposition 2 of Book X can be used to prove the incommensurability of the side and diagonal of the square. The infinite appears here in the form of a periodic phenomenon.

There is an anthyphairesis that is even more spectacular, namely that of the side and diagonal of the regular pentagon. The figure of unfathomable depth, which is obtained when the diagonals of a regular pentagon ABCDE are drawn (Figure 8), and then the diagonals of A'B'C'D'E' ... provides a stunning representation of another process towards the infinitely small, in which the pentagon periodically re-appears. The star pentagon (pentagram) was the emblem of the Pythagoreans. The ratio between the diagonal and the side of a regular pentagon, which we call the "Golden Number", plays an important part in Greek mathematics (the section in extreme and mean ratio). It is possible that the pentagon provided one of the first cases of incommensurability.

Exercise 9	Show that in a regular pentagon ABCDE (figure 8), AD – AE = AD – AB' = B'D = B'E' and then, that AE – B'D = B'C'.
	Show how this leads to the incommensurability of the diagonal and side of a regular pentagon.

With the break away from numbers, and the attempt by magnitudes to gain their independence, these latter have made their mark: anthyphairesis has become accepted, and it allows us to establish whether or not magnitudes are commensurable. But a further, yet more decisive battle remains, that of proportionality.

It would be best to present this problem by starting with an example: 2 is contained three times in 6 and five times in 10, 3 is contained three times in 9 and five times in 15. The numbers 6, 10, 9 and 15 are, therefore, proportional or, to put it in Euclid's terms: 6 is to 10 as 9 is to 15.

In the same way, one could define the proportionality of four commensurable magnitudes, two by two: if there exists a common measure contained three times in a segment BC and five times in a segment CD, that is 5.BC is equal to 3.CD, and if we know that triangles on equal bases and having the same height are equal, it is easy to show that in figure 9, three times the area of triangle ACD is equal to five times the area of triangle ABC and so ABC is to ACD as BC is to CD.

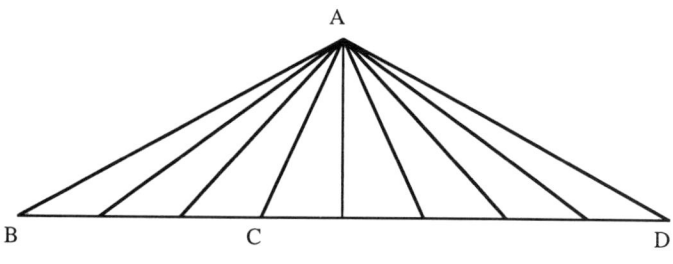

Figure 9

When Euclid stated as the first proposition of Book VI, that

> *triangles and parallelograms which are under the same height are to one another as their bases.*[1]

he was considering the proportionality of four magnitudes, which includes the case where they are not commensurable two by two. In this case, the ratio of two magnitudes cannot be considered as the ratio of two numbers. Is it hardly reasonable to think of a ratio that is incalculable?

The solution offered by Eudoxus

The ratio of two magnitudes is defined in Book V of the *Elements* and, although there is no direct proof, a number of scholars have attributed it to Eudoxus of Cnide (c.350 BC).

The definition of ratio seems rather round-about to the modern reader:

> A ***ratio*** *is a sort of relation in respect of size between two magnitudes of the same kind. Magnitudes are said to **have a ratio** to one another which are capable, when multiplied, of exceeding one another*[2].

It would appear that the first definition is there in order to forewarn the reader that "ratio" (logos), which he has so far been used to use in connection with numbers, is now going to be used with magnitudes. These magnitudes must be of the same type; ratios can only be used to compare lines with lines, surfaces with surfaces or solids with solids. We are concerned here with quantities, and it is here that we are at the heart of the problem: ratio resides in the quantitative domain yet it is not always calculable. We cannot know any more about this "ratio". The second definition describes a test: for if two magnitudes a and b have a ratio between them, then there exists at least one pair of integers (m, n) such that

$$ma > b \text{ and } nb > a$$

This test is attributed to Archimedes, who himself attributed it to Eudoxus. Today we describe magnitudes that satisfy this criterion as Archimedean.

Since Euclide is unable to describe this ratio, and it can not be calculated, he must find a way of showing how these ratios could be equal, and how to compare unequal ratios.

---

1.   Heath, *The Thirteen Books of Euclid's Elements,* VI, 1, vol.2. p. 191.
2.   *Ibid.* V, Def., 3, 4, vol. 2. p. 114.

We note, further, that a statement such as:

$$\text{"a/b = c/d if and only if ad = bc"}$$

is meaningless, since we do not have, in general, a way of multiplying magnitudes. Certainly we can have:

> *If four straight lines be proportional, the rectangle contained by the extremes is equal to the rectangle contained by the means*[1]

but it is impossible to think of a geometrical meaning for the product of surfaces. Faced with this difficulty, Eudoxus overcame the problem by proposing these definitions:

> *Magnitudes are said to **be in the same ratio**, the first to the second and the third to the fourth, when, if any equimultiples whatever be taken of the first and third, and any equimultiples whatever of the second and fourth, the former equimultiples alike exceed, are alike equal to, or alike fall short of, the latter equimultiples respectively taken in corresponding order.*

> *Let magnitudes which have the same ratio be called **proportional.***

> *When, of the equimultiples, the multiple of the first magnitude exceeds the multiple of the second, but the multiple of the third does not exceed the multiple of the fourth, then the first is said to **have a greater ratio** to the second than the third has to the fourth.*[2]

To put it in modern terms, equality is defined by:

$$a/b = c/d$$

if and only if, for any given pair (m, n) of non-zero natural numbers, we have one of the following three cases:

(1)   ma < nb and mc < nd,

(2)   ma = nb and mc = nd,

(3)   ma > nb and mc > nd.

The second case can only occur when both a and b are commensurable and c and d are commensurable. In fact, if ma is equal to nb, the nth part of a is equal to the mth part of b. In this case, the existence of at least one pair of non-zero natural numbers m, n such that

$$ma = nb \text{ and } mc = nd$$

is sufficient to prove proportionality, but for incommensurable magnitudes only the cases (1) and (3) are possible.

Exercise 10   Prove, using the definitions of Eudoxus, that if there exist two natural numbers m, n such that ma = nb and mc = nd, then the magnitudes a, b, c, d are proportional.

Based on these definitions, Book V established a large number of general properties of proportions, such as:

> *If four magnitudes be proportional, they will also be proportional alternately.*[3]

1.   Heath, *The Thirteen Books of Euclid's Elements*, VI 16, vol. 2., p. 221.
2.   *Ibid.*,V Def. 5, 6, 7. vol. 2. p. 114.
3.   *Ibid.*,V 16, vol. 2., p. 164.

The proof is as follows: Let a, b, c, d be four proportional magnitudes. Then for all non-zero natural numbers m, n we have

$$ma/mb = nc/nd$$

since

$$ma/mb = a/b \text{ and } nc/nd = c/d.$$

It follows that, if ma < nc then mb < nd (if not, a/b < c/d), and similarly, if ma = nc then mb = nd, and if ma > nc then mb > nd. Euclid concludes, following the definition of the equality of ratios of magnitudes, that a/c = b/d. Euclid therefore considers as equivalent, the formulation

(ma < nc and mb < nd) or (ma = nc and mb = nd) or (ma > nc and mb > nd)

which appears in the definition, and what appears in the proof, namely:

(if ma < nc then mb < nd) and (if ma = nc then mb = nd) and (if ma > nc then mb > nd)

Exercise 11   1) Prove the equivalence of the two forms used by Euclid when the set of magnitudes is ordered.

2) Prove, in the manner of Euclid, proposition 12 of Book V which is: if a/b = c/d then a/b = (a + c)/(b + d)

3) Prove that "triangles which are under the same height are to one another as their bases." (Book VI 1).

Being unable to envisage the ratio of two incommensurable magnitudes as the ratio of their measures, Eudoxus proposed a new set, that of the ratios of magnitudes. For Euclid, this set contains the ratios of numbers:

Commensurable magnitudes have to one another the ratio which a number has to a number.[1]

We have also here, in embryo, the multiplication of "ratios": proposition 22 of Book V states that if a/b = c/d and b/e = d/f, then a/e = c/f; Euclid uses the term "ratio *ex aequali*" for a/e, which is the result of the composition of the two ratios a/b and b/e and this is like our own multiplication. At the same time, in order to "multiply" a/b and c/d, we need to find, that is to construct geometrically, a magnitude e such that c/d = b/e: hence a/e = a/b × c/d. We will illustrate this using proposition 23 of Book VI:

Equiangular parallelograms have to one another the ratio compounded of the ratios of their sides.[2]

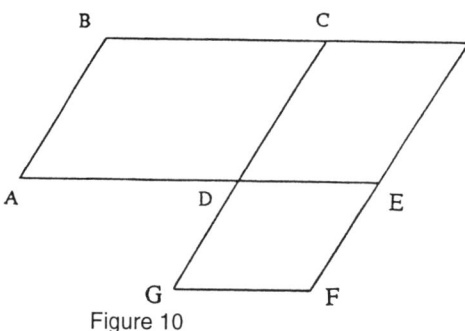

Figure 10

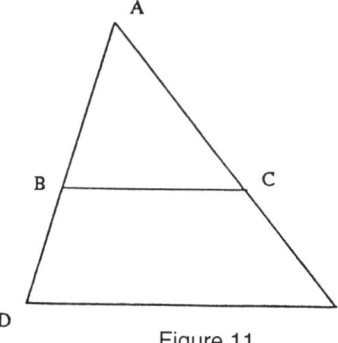

Figure 11

---

1.   Heath, *The Thirteen Books of Euclid's Elements,* X 5, vol. 3, p. 24.
2.   *Ibid.,* VI 23, vol. 2, p. 247.

Euclid is able to state that the ratio of the parallelograms (AC) to (DF) (figure 10), is composed of the ratios of their sides, that is:

$$AD/DE \times CD/DG$$

for, he had already shown, at proposition 12, the construction

> To three given straight lines to find a fourth proportional.[1]

If the three given straight lines a, b, c are represented by AB, BD and AC respectively (Figure 11), then the line through D, parallel to BC, will cut (AC) at E such that

$$AB/BD = AC/CE.$$

There are other cases of the "multiplication" of ratios to be found in the *Elements* but there is no explicit postulate of the existence of a fourth proportional to three magnitudes by which it becomes possible to "multiply" any two ratios.

With their separation from arithmetic, magnitudes have established an existence in their own right, as geometric objects, and with it incommensurability becomes respectable. But their ratios are derived from the geometric nature of magnitudes and need geometric constructions, and they do not form a set which is closed under the operation of multiplication. Furthermore, the need for calculations which arose from the development of astronomy, and problems such as quadrature and cubature, did not fit well with this rupture between numbers, magnitudes, ratios of numbers and ratios of magnitudes, each of which had its own rules of operation. Little by little, because it was easier and became accepted as legitimate, magnitudes and their ratios came to be treated as if they were numbers. But we are still left with the fundamental question: how can we give a numerical value to the ratio of two incommensurable magnitudes?

Omar Khayyam: an attempt to make numbers rational

In 1077 Omar Khayyam wrote his *Commentary on the difficulties to be found in the introductions to Euclid's book*. In it, he expresses the ratio of a magnitude A to a magnitude B, by considering a magnitude G which has the same ratio to some chosen unit as A has to B, and then says that G should be conceived as

> not like a line, a surface, a body or a time interval, but as a magnitude which is non material and which belongs to the domain of numbers, but not to absolute and veritable numbers, for the ratio of A to B may often not be numerically measurable, that is that it may not be possible to find two numbers which have a ratio equal to this ratio.[2]

This number G is not "veritable", just like the measure with which it is associated, but it is to measure that Omar Khayyam appeals in order to define the ratio of two magnitudes. He defines the equality of two ratios between four magnitudes as follows:

> We transfer all multiples of the first to the second such that the remainder becomes less than the first; we transfer all multiples of the third to the fourth such that the remainder becomes less than the third; let the multiple of the first in the second be equal to the multiple of the third in the fourth.[3]

1. Heath, *The Thirteen Books of Euclid's Elements*, VI 12, vol. 2, p. 215.
2. Youschkevitch, *Les mathématiques arabes*, p. 88.
3. *Ibid.*, p. 85.

He then starts again with these two remainders, and says

> *the number of the remainders of the first and of the second is equal to the number of the remainders of the third and of the fourth, and this being so indefinitely. In this case the ratio of the first to the second is necessarily equal to the ratio of the third to the fourth*[1].

We recognise here the process of alternate subtractions which typifies anthyphairesis, and in the passage we have just quoted, the subtraction process continues indefinitely, which we recall as characterising the incommensurability of two magnitudes. The equality of the ratios is linked to the equality of the partial quotients at each stage of the anthyphairesis. To put it in modern notation: If we have four magnitudes A, B, C, D such that A < B and C < D, we subtract A from B as many times as possible, and if q_1 is this number of times, we have

$$B = q_1 A + R_1 \text{ and } R_1 < A$$

We now start again with A and R_1 to obtain a second integer q_2 where $A = q_2 R_1 + R_2$, and then start again with R_1 and R_2 ... The ratio A/B becomes associated with an infinite sequence q_1, q_2, q_3, ..., and in the same way another infinite sequence q'_1, q'_2, q'_3, ... is associated with the ratio C/D. The equality of the ratios A/B, C/D is assured whenever we have q_n and q'_n equal for all n. In the case where the sequence of partial quotients is finite (for commensurable magnitudes or for the ratio of integers), the two ratios will only be equal if the two sequences have the same number of terms, and if at each stage the partial quotients are equal.

When A/B is not equal to C/D there is a stage m where the sequence of partial quotients becomes different for the first time, either q_m differs from q'_m or one of the sequences terminates at stage m − 1. The inequality between q_m and q'_m and the parity of m allows us to determine the nature of the inequality between A/B and C/D.

Exercise 12   Prove that if m is even and $q_m > q'_m$ then A/B > C/D.

Prove that if A/B = $(q_1 , q_2 , q_3 , ..., q_m)$ and C/D = $(q'_1, q'_2, q'_3, ..., q'_m, q'_{m+1}, ...)$, and for all n, $1 \le n \le m$, $q_n = q'_n$, and m is even, then A/B < C/D.

Using his definition for the equality of ratios, Omar Khayyam is able to prove many of the properties of proportions, often quite easily. For example, that the equality of ratios is transitive:

$$\text{if } A/B = C/D \text{ and } C/D = E/F, \text{ then } A/B = E/F$$

which is derived from

$$\text{if } q_n = q'_n \text{ and } q'_n = q''_n \text{ then } q_n = q''_n,$$

the transitive property of equality of numbers.

Exercise 13   Using Omar Khayyam's definition prove the result: if A/B = C/D then A/B = (A + C)/(B + D)

However, proposition 16 of Book V (that if A/B = C/D then A/C = B/D) cannot be proved with the anthyphairetic definition of the equality of ratios. Omar Khayyam got

---

1.   *Ibid.*, p. 85.

out of the difficulty by establishing that his definition and that of Eudoxus were equivalent. The proof depends on the existence of a fourth proportional to three given magnitudes A, B, C, that is a magnitude D such that A/B is equal to C/D. In fact, he explains, it is possible to find by duplication (multiplication by 2) a magnitude N sufficiently large so that C/N < A/B, and by dichotomy (division by 2), a magnitude M sufficiently small so that C/M > A/B. He then appeals to a "principle of continuity" of magnitudes, to conclude that between M and N there must be a magnitude D such that C/D is equal to A/B. The continuity to which Omar Khayyam refers is the same as is found in Aristotle's *Metaphysics* quoted earlier, and consists of the idea of being able to divide magnitudes indefinitely. In fact, our set Q of rationals is continuous in Omar Khayyam's sense. But his proof, according to which, in going from C/N to C/M we pass through all intermediary values, and in particular A/B, is not however rigorous. Nonetheless, the idea of basing the new numbers he was trying to create, on the principle of continuity is not without merit. The existence of a fourth proportional allowed Omar Khayyam to go further than Euclid in generalising the composition (in the sense of multiplication) of ratios.

If Eudoxus had found a solution to the problem set by the discovery of incommensurability by constructing a theory of magnitudes and their "ratios" distinct from numbers, the balance has swung back the other way with Omar Khayyam, whose ratios of magnitudes depend on numbers, different from integers, which are used as a measure of the magnitudes.

However, Omar Khayyam's efforts in trying to establish a different basis for the ratios of magnitudes, than that found in Book V of the *Elements*, was not followed up by other mathematicians to any extent. In fact, most of them did not see the need, and gradually they became used to thinking of Euclidean ratios as if they were, in effect, truly numbers. Without it being noticed, integers, magnitudes and their ratios took on an existence within the "mansion of real numbers", together with a brand new measure which could be used to associate a number with every point of a straight line. And yet, a question remains: how is such a union conceivable?

Is it a problem of the foundations of mathematics? ...

It was only at the beginning of the 19th century that certain criticisms were aired, principally against the inclusion within analysis of results that had been derived from other disciplines, such as geometry and kinematics. These criticisms were clearly stated by Bolzano in his 1817 *Rein analytischer Beweis, etc.* [*A purely analytical proof of the theorem: between any two values of an equation which have opposite signs there exists at least one real root*]:

> For in fact, if one considers that the proofs of the science should not merely be confirmations [...] but rather justifications, i.e., presentations of the objective reason for the truth concerned, then it is self-evident that the strictly scientific proof, or the objective reason, of a truth which holds equally for all quantities, whether in space or not, cannot possibly lie in a truth which holds merely for quantities which are in space.[1]

We cannot fail to approach the theorem which preoccupied Bolzano, called today the intermediate value theorem, except from the "principle of continuity" which had

---

1.   Bolzano, *Démonstration...*, p. 137, english tr. p. 161.

been appealed to by Omar Khayyam. Bolzano, who was dissatisfied with the geometrical image of continuity, was led to make a clear formulation of the concept of a continuous function. Later, he attempted the construction of the reals, but this task was to remain incomplete for more than a half of century.

When he started his studies at the Dijon Faculty of Sciences, Charles Meray also became aware that certain propositions of analysis were *"regarded as axioms"*. In 1869 he wrote his *Remarks on the nature of quantities defined by being the limits of given variables* in which he gives the first construction of real numbers using Cauchy sequences. Unfortunately, the rather confidential nature of the *Revue des Sociétés Savantes* where it appeared, together with the peculiar language style of the author, caused the theory to remain unknown for a considerable time.

Georg Cantor and Eduard Heine, disciples of Weierstrass, had rather more success. Like their master, they developed a theory, which soon became widely known. It formed part of the "arithmetisation of analysis", a subject dear to Weierstrasss, and just as with Meray, the theory was based on Cauchy sequences. We quote here from Heine's *Die Elemente der Functionenlehre* [*Elementes of the Theory of Functions*]:

> We use the term number-sequence for a sequence of numbers a_1, a_2, etc., a_n, where for each non zero number η, being as small as one wishes, there exists a value n, such that $a_n - a_{n+\nu}$ is less than η for all positive integer ν [...]
> We generally use the term number or number-symbol for the symbol attached to a number-sequence.[1]

Like Meray, *"for the first time obliged to lecture upon the elements of the differential calculus"*[2], Richard Dedekind *"felt more keenly than ever before the lack of a really scientific foundation of mathematics."*[3] This was in 1858, and he recapitulates an idea that had been developed almost half a century earlier by Bolzano, of which Dedekind was at that time unaware:

> The most rigorous expositions of the differential calculus do not base their proofs upon continuity but, with more or less consciousness of the fact, they either appeal to geometric notions.[4]

Bolzano had put the emphasis on the intermediate value theorem, while Dedekind, like Meray, preferred

> the theorem that every magnitudes which grows continually, but not beyond all limits, must certainly approach a limiting value.[5]

An assortment of preoccupations, pedagogic and logico-philosophic, have thus led mathematicians, sometimes in isolation, to give to real numbers a theoretic foundation. We shall not examine a wide range of such offerings, but choose Dedekind's solution. His approach is innovative since, more than the other theories, it illustrates the fundamental essence of real numbers: that they are continuous.

> The statement is so frequently made that the differential calculus deals with continuous magnitude, and yet an explanation of this continuity is nowhere given.[6]

---

1.   Heine, *Die Elemente der Functionenlehre*, p. 174.
2.   Dedekind, *Essays on the theory of numbers*, p. 1.
3.   *Ibid.*
4.   *Ibid.*, p. 2.
5.   *Ibid.*, p. 1.
6.   *Ibid.*, p. 2.

... *A solution: the cuts*[1]

The idea that underlies Dedekind's creation is a cut across the set of rationals. 'Creation' does not overstate the case, and the word was used by Dedekind himself.

> *If now any separation of the system R (of rational numbers) into two classes A_1, A_2, is given which possesses only this characteristic property that every number a_1 in A_1 is less than every number a_2 in A_2 then for brevity, we shall call such a separation a cut and designate it by (A_1,A_2).*[2]

For any given cut (A_1,A_2), there are two possibilities. In the first case, either A_1 has a greatest element or A_2 has a least element. Certainly, in this case, this rational number is "produced by the cut (A_1,A_2)" but, on the other hand, the cut (A_1,A_2) does not produce any new number: the set of rationals being considered, hypothetically, as known. In the second case, this does not happen and A_1 does not have a greatest element and A_2 does not have a least element: only these types of cuts will produce new numbers:

> *Whenever, then, we have to do with a cut (A_1,A_2) produced by no rational number, we create a new irrational number α, which we regard as completely defined by the cut (A_1,A_2)*[3]

Divine inspiration! This creation allows us to define *"the domain $\mathfrak{R}$ of real numbers"* containing both the rationals and the new irrational numbers that have been created.

Exercise 14   Show that "the class of rational numbers" whose square is less than 3, does not have a greatest element.

To complete his creation, Dedekind went on to define, on the one hand an order relation on $\mathfrak{R}$ and, on the other hand, operations that were a natural extension of the known operations for the rationals. The rationals, in Dedekind's construction, become a subset of the reals. Nowadays, adopting a more set-theoretic approach, our starting point is the set of all rational cuts which form a subset. But for Dedekind, this addition to the set of rationals was not available. It guarantees continuity of the set of real numbers created in this way.

> *The domain $\mathfrak{R}$ possesses also continuity, i.e., the following theorem is true: if the system $\mathfrak{R}$ of all real numbers breaks up into two classes A_1 and A_2 such a that every number a_1 in the class A_1 is less than every number a_2 in the class A_2 then there exists one and only one number a by which this separation is produced.*[4]

In other words, no cut defined on $\mathfrak{R}$ will produce a new number: the set $\mathfrak{R}$ is closed under the operation of Dedekind cuts. The proof is simple. To each cut (A_1,A_2) of $\mathfrak{R}$ we can associate a cut (A_1,A_2) of the set of rationals defined as follows:

$A_1 = \{q \in Q : q \in A_1\}$

$A_2 = \{q \in Q : q \in A_2\} = Q - A_1$

---

1.   In French, play on words: "coupure" means "cut" but also "expedient".
2.   *Ibid.*, p. 12-13.
3.   *Ibid.*, p. 15.
4.   *Ibid.*, p. 20.

The real number a defined by (A_1,A_2) satisfies these conditions. We have shown in fact, that:

> *a itself is either the greatest number in A_1 or the least number in A_2.*[1]

In other words, we come back to the first case with cuts on the rationals: the cut (A_1,A_2) does not produce a "new number" since the product of the cut is the real number a. To create a structure for the set of real numbers, Dedekind does not, in a sense, add on the new numbers to the set of rationals. In fact, it is simpler to identify a rational number with the cut it produces, and to use the cuts to determine whether or not they produce rational numbers. Thus,

> *to obtain a basis for the orderly arrangement of all real, i.e., all rational and irrational numbers, we must investigate the relation between any two cuts (A_1,A_2) and (B_1,B_2), produced by any two numbers α and β.*[2]

There are three cases to consider:

1) $A_1 = B_1$ and so $A_2 = B_2$.

> *The two cuts are perfectly identical, which denote in symbols by*

$$\alpha = \beta \ or \ \beta = \alpha\text{"}[3]$$

2) The difference between A_1 and B_1 results in a singleton. For example,

$A_1 = B_1 \cup \{a'_1\}$, and so $B_2 = A_2 \cup \{b'_2\}$, with $a'_1 = b'_2$.

and the cuts (A_1,A_2) and (B_1,B_2)

> *are produced by the same rational number $\beta = b'_2 = a'_1 = \alpha$ "*[4]

3) The difference between A_1 and B_1 contains two distinct elements and, consequently, an infinite number. This happens where B_1 is contained in A_1, we say that α is greater than β, that β is less than α, which we express in symbols by

$$\alpha > \beta \ as \ well \ as \ \beta < \alpha[5]$$

And so Dedekind was able to state:

> *It is to be noticed that this definition coincides completelys with the one given earlier, when α, β are rational.*[6]

Exercise 15   Show that a complete ordering relation can be defined on $\Re$ in this way.

As for calculations, they appear as an extension of calculations with the rationals:

> *To reduce any operation with two real numbers α, β to operations with rational numbers, it is only necessary from the cuts (A_1, A_2), (B_1, B_2) produced by the numbers α and β, in the system R to define the cut (C_1, C_2) which is to correspond to the result of the operation, γ.*[7]

1.   *Ibid.*, p. 21.
2.   *Ibid.*, p. 15.
3.   *Ibid.*, p. 16.
5.   *Ibid.*, p. 17.
5.   *Ibid.*
6.   *Ibid.*
7.   *Ibid.*, p. 21.

So, for example, we can define the sum of (A_1, A_2) and (B_1, B_2) by the cut (C_1, C_2) which corresponds to it, as follows:

$C_1 = \{c \in Q: a_1 \in A_1 \text{ and } b_1 \in B_1 \text{ and } a_1 + b_1 \geq c\}$

$C_2 = Q - C_1$

Exercise 16   Prove that the set of reals also possesses a commutative group structure.

Taking care, and being a touch provocative, Dedekind could conclude:

> *Just as addition is defined, so can the other operations of the so-called elementary arithmetic be defined, viz., the formation of differences, products, quotients, powers, roots, logarithms, and in this way we arrive at real proofs of theorems, (as, e.g., $\sqrt{2}\sqrt{3} = \sqrt{6}$), which to the best of my knowledge have never been established before.*[1]

Some writers have drawn a parallel between Eudoxus and Dedekind in their approach to the equality of ratios: all that is needed is to transform the definitions of Eudoxus to:

$a/b = c/d$ if and only if, for all rational m/n,

($a/b < m/n$ and $c/d < m/n$) or ($a/b = m/n$ and $c/d = m/n$) or ($a/b > m/n$ and $c/d > m/n$)

and to regard the ratio a/b as a "cut" of the set of rationals m/n.

If such a parallel between the works of Eudoxus and Dedekind can be made, it is only fair to point out that their two approaches to the foundation of what we now call the set of real numbers, were different, and that each was original. Eudoxus freed geometry from the inadequacies of number by constructing ratios of magnitudes that were independent of the ratios of numbers. Dedekind laid the foundations for the reals, starting with the integers, and in a way managed to achieve the dream of the Pythagoreans.

By the use of Dedekind cuts, (or by using Cauchy sequences), the foundations of elementary mathematics is assured. Meray and Dedekind were able to carry out their studies with the rigour that they desired. Today, the construction of the reals has no place in the secondary mathematics curriculum, and is often ignored in more advanced courses. Instead, the reals are defined by the means of a set of axioms.

But, other writers offer a different way of looking at the reals. They argue for a new type of analysis, non-standard analysis, which gives a new status to infinitesimals. The integers are no longer alone: as Reeb has it, there are the naive numbers, and there are others. But that is another story…

---

1.   *Ibid.*, p. 22.

Bibliography

Sources texts

ARISTOTLE, *La métaphysique,* tr. Tricot, Vrin, Paris, 1981.

ARISTOTLE, *Les premiers analytiques*, tr. Tricot, Vrin, Paris, 1966.

ARISTOTLE, *Aristotle's Metaphysics: Books M and N*, tr. Julia Annas, Clarendon Press, Oxford, 1976.

BOLZANO, "Démonstration purement analytique du théorème: entre deux valeurs quelconques qui donnent deux résultats de signes opposés se trouve au moins une racine réelle de l'équation", French tr. Sebestik in *Revue d'Histoire des Sciences,* 17 (1964), 136-164.

There is an English translation from the German of Bolzano's 1817 paper *Rein analytischer Beweis*: S B Russ, "A translation of Bolzano's Papar on the Intermediate Value Theorem" in *Historica Mathemtica,* **7** (2) (1980) pp. 156-185. Extracts from it appear in J. Fauvel and J. Gray, *The History of Mathematics, A Reader,* Macmillan (1987) but not the extract quoted in this chapter.

DEDEKIND, *Stetigkeit und irrationale Zahlen*, Vieweg, Braunschweig, 1872. tr. Milner & Sinaceur in *Les Nombres. Que sont-ils et à quoi servent-ils?* Bibliothèque de l'Ornicar, Paris, 1978, English tr. Beman, *Essays on the theory of numbers*, Dover, New York, 1968.

EUCLID, Books I to IV: Vitrac, *Euclide, Les Eléments*, PUF, Paris, 1990; Books VII to IX: Itard, *Les livres arithmétiques*, Hermann, Paris, 1961; other books: Peyrard, *Les œuvres d'Euclide*, Blanchard, Paris, 1966.

All quotations from Euclid's *Elements* here, have been taken from the Dover reprint of the edition by Heath, and referenced accordingly.

HEAT, *The Thirteen Books of Euclid's Elements*, 2nd edition, 3 vols., Cambridge University Press, Cambridge 1926, repr. Dover, New York, 1956.

HEINE, "Die Elemente der Functionenlehre" in *Journal für die reine und angewandte Mathematik,* **74,** (1872) 172-188. French tr. Friedelmeyer & Guillemot, IREM de Toulouse, 1988 (unpub.).

HERODOTE, Thucydide, *Œuvres complètes*, tr. Barguet, La Pléiade, Paris, 1964.

NEUGEBAUER, O., *The Exact Sciences in Antiquity*, 2nd ed. Brown University Press, 1957. Repr. Dover, New York, 1969, tr. Souffrin, *Les sciences exactes dans l'anquité*, Actes Sud, Arles, 1990.

PLATON, *Ménon*, tr. Canto-Sperber, Garnier-Flammarion, Paris, 1991.

The text used in this English translation was: Plato, *Protagoras and Meno*, tr. W K C Guthrie, Penguin, Harmondsworth, 1956.

PROCLUS DE LYCIE, *Les commentaires sur le premier livre des Eléments d'Euclide*, tr. Ver Eecke, Desclee de Brouwer, Bruges, 1948.

THÉON DE SMYRNE, *Exposition des connaissances mathématiques utiles pour la lecture de Platon*, tr. Dupuy, reprint: Culture et Civilisation, Brussels, 1966.

YOUSCHKEVITCH, *Les mathématiques arabes (VIII – XVe s.)*, tr. Cazenave & Jaouiche, Vrin, Paris, 1976.

Reference texts

CAVEING, *La constitution du type mathématique de l'idéalité dans la pensée grecque*, Presses Université Lille III, 1992.

DESANTI, "Une crise de développement exemplaire: la 'découverte' des nombres irrationels" in *Logique et connaissance scientifique,* La Pléiade, Paris, 1967, 439-464.

DHOMBRES, *Nombre, mesure et continu*, Cedic, Nathan, Paris, 1980.

General works for further reading

BARREAU & HARTHONG, *La mathématique non standard*, Ed. du CNRS, Paris, 1989.

CLAPIE, DOPFER, GUILLEMOT & SPIESSER, *Quelques aspects de l'arithmétique pythagoricienne*, IREM Toulouse, Toulouse, 1987.

DAHAN & PEIFFER, *Une histoire des mathématiques*, Seuil, Paris, 1986.

FOWLER, *The Mathematics of Plato's Academy: A New Reconstruction*, Clarendon Press, Oxford, 1987, pbk reprint with corrections OUP, Oxford, 1990.

GARDIES, *L'héritage épistémologique d'Eudoxe de Cnide*, Vrin, Paris, 1988.

GUILBAUD, *Leçons d'à peu près*, Bourgois, Paris, 1985.

LOBRY, *Et pourtant ils ne remplissent pas N*, Aleas Editeurs, Lyon, 1989.

MICHEL, *De Pythagore à Euclide,* Les Belles Lettres, Paris, 1950.

SERRES, "Gnomon: les débuts de la géométrie en Grèce" in *Eléments d'histoire des sciences*, Bordas, Paris, 1989.

How did you get on?

Ex. 1 $5.10^{-7} < 2 - (1 + 24/60 + 51/60^2 + 10/60^3) < 6.10^{-7}$ and the same difference lies between $7/60^4$ and $8/60^4$.

Ex. 2 $(8/9)^2$ approximates to $\pi/4$ from above, and the difference lies between 4.10^{-3} and 5.10^{-3} or between $1/212$ and $1/211$.

Ex. 3 a) $(2n + 1)^2 = 2(2n^2 + 2n) + 1$
b) $q^2 = 2p^2$ so q^2 is even: q must be even and p odd.
Putting $q = 2q'$, then $q^2 = 2p^2$ becomes $4q'^2 = 2p^2$ and so $p^2 = 2q'^2$. This makes p^2 even and so p is even. Since p cannot be, at the same time, both even and odd, there cannot be any non zero natural numbers p, q such that $q^2 = 2p^2$.

Ex. 4 First prove that if a number is not a multiple of 3 (it must be $3n + 1$ or $3n + 2$) then its square is not a multiple of 3. Then, if $q^2 = 3p^2$, it can be proved that p must be, at the same time, both a multiple of 3 and not a multiple of 3...

Ex. 5 $\lambda_{13} = 33\ 461$ and $\delta_{13} = 47\ 321$. The value of $(\delta_{13} / \lambda_{13})^2$ may show 2 on your calculator but if you evaluate $(\delta_{13} / \lambda_{13})^2 - 2$ it will not be zero. For example, a TI 80 gives $(\delta_{13} / \lambda_{13})^2 = 1.999\ 999\ 999$, and $(\delta_{13} / \lambda_{13})^2 - 2 = - 8.94.\ 10^{-10}$; it does not give 0 for the calculation until $(\delta_{17} / \lambda_{17})^2 - 2$.

Ex. 6 $\delta_n^2 - 2\lambda_n^2 = \delta_{n-1}^2 + 4\delta_{n-1}\ \lambda_{n-1} + 4\lambda_{n-1}^2 - (2\delta_{n-1}^2 + 4\delta_{n-1}\ \lambda_{n-1} + 2\lambda_{n-1}^2) = -\delta_{n-1}^2 + 2\lambda_{n-1}^2 = - (\delta_{n-1}^2 - 2\lambda_{n-1}^2)$.
Since $\delta_1^2 - 2\lambda_1^2 = -1$, we have $\delta_n^2 - 2\lambda_n^2 = (-1)^n$.

Ex. 7 $v_{n + 1}^2 - k\ u_{n + 1}^2 = (1-k)(v_n^2 - k\ u_n^2)$ and $v_1^2 - k\ u_1^2 = 1 - k$, so $v_n^2 - k\ u_n^2 = (1-k)^n$.

By using the recurrence relations, we can show that
$u_{2p + 1} = k^p + ...$ and $v_{2p + 1} = (2p + 1)\ k^p + ...$ and
$u_{2p + 2} = (2p + 2)\ k^p + ...$ and $v_{2p + 2} = k^p + 1 + ...$

Hence $\left(\dfrac{(k - 1)^{2p + 1}}{u_{2p + 1}^2}\right) < \dfrac{(k - 1)^{2p + 1}}{k^{2p}} = \left(\dfrac{(k - 1)}{k}\right)^{2p}(k - 1)$

and $\dfrac{(k - 1)^{2p + 2}}{u_{2p + 2}^2} < \dfrac{(k - 1)^{2p + 2}}{(2p + 2)^2 k^{2p}} = \left(\dfrac{k - 1}{k}\right)^{2p} \left(\dfrac{k - 1}{2p + 2}\right)^2$.

Since k is strictly greater than 1, then
$0 < \dfrac{k - 1}{k} < 1$, and so $\lim \left(\dfrac{k - 1}{k}\right)^{2p} = 0$.

From the first part, $\left(\dfrac{v_n}{u_n}\right)^2 = k + \dfrac{(1 - k)^n}{u_n^2}$, and so $\left(\dfrac{v_n}{u_n}\right)^2$ converges to k.

Ex. 8 $r_1 = a - b = 10\ 020\ 375$, $r_8 = 20\ 250$, $r_9 = 3375$, and $20\ 250 = 6 \times 3375$,
$a/b = 1 + r_1 /b = 1 + 1/(b/r_1) = 1 + 1/(6 + r_2 / r_1) = 1 + 1/(6 + 1/(6 + r_3 / r_2))$.
If we ignore r_3 , we obtain $1 + 1/(6 + 1/6) = 43/37$ and $a/b > 43/37$.

Ex. 9 The triangles EAB' and DB'E' are isosceles (the sizes of their angles being $\pi/5$, $2\pi/5$, $2\pi/5$ and $\pi/5$, $\pi/5$, $3\pi/5$ respectively) and so AE = AB' and DB' = B'E'. The remainder of the first subtraction is the diagonal of the pentagon A'B'C'D', the remainder of the second, is its side B'C'. So after two subtractions we have again the side and diagonal of a pentagon, and the remainders have ratios which are periodic.

Ex. 10 Let a, b, c, d be four magnitudes such that ma = nb and mc = nd where m, n are two non zero natural numbers. For all non zero natural numbers k, l, kn < lm or kn = lm or kn > lm. Take the case kn < lm and multiply both sides by a and then by c to obtain kna < lma and knc < lmc. Now replace ma by nb, and mc by nd, to get kna < lnb and knc < lnd. Dividing by n yields: ka < lb and kc < ld. The other two cases are found by replacing < with = , and then with > throughout. So we have, for all non zero naturals k, l, [(ka < lb and kc < ld) or (ka = kb and kc = ld) or (ka > lb and kc > ld)] as required.

Ex. 11 1) For a completely ordered set of magnitudes, we have
not (ma < nb) $\Leftrightarrow$ (ma $\geq$ nb) and
not (ma > nb) $\Leftrightarrow$ (ma $\leq$ nb).
[(ma < nb $\Rightarrow$ mc < nd) and (ma = nb $\Rightarrow$ mc = nd) and (ma > nb $\Rightarrow$ mc > nd)]
can be written:
[(ma $\geq$ nb or mc < nd) and (ma $\neq$ nb or mc = nd) and (ma $\leq$ nb or mc > nd)]
By distributing "and" with respect to "or", and rejecting impossible combinations such as (ma $\geq$ nb and ma $\neq$ nb and ma $\leq$ nb), we obtain:
[(ma $\geq$ nb and ma $\neq$ nb and mc > nd) or (ma $\geq$ nb and mc = nd and ma $\leq$ nb) or (mc < nd and ma = nd and ma $\leq$ nb)], that is [(ma > nb and mc > nd) or (ma = nb and mc = nd) or (ma < nb and mc < nd)].
2) For all non zero natural numbers m, n, if ma < nb then mc < nd and so ma + mc < nb + nd; the same argument applies to = and >.
3) If EC = mBC, then $\triangle$ AEC is m times $\triangle$ ABC; if FC = nCD, then $\triangle$ AFC is n times $\triangle$ ABC.
If EC = FC then $\triangle$ AEC = $\triangle$ AFC
If EC > FC then $\triangle$ AEC > $\triangle$ AFC
If EC < FC then $\triangle$ AEC < $\triangle$ AFC

Ex. 12 $q_m > q'_m + 1/(q'_m + 1 + 1/...)$, next: $q_{m-1} + 1/q_m < q'_m 1 + 1/(q'_m + 1/...)$ so that at each successive stage the sense of the inequality changes. In retracing the steps from stage m to come to $1/(q_1 + 1/(q_2 + 1/(... + 1/q)))$ or A/B the sense changes an even number of times and so A/B > C/D.

Ex. 13 If $A = q_1 B + R_1$ and $C = q_1 D + R'_1$ then $A + C = q_1 (B + D) + (R_1 + R'_1)$
next: $B = q_2 R_1 + R_2$ and $D = q_2 R'_1 + R'_2$ and so
$B + D = q_2 (R_1 + R'_1) + (R_2 + R'_2) + ...$
we get $(A + C)/(B + D) = (q_1, q_2, ...)$ and so A/B = (A + C)/(B + D).

Ex. 14 Suppose that the class of rational numbers whose square is less than 3 has a greatest member P/Q. We have P > 1, Q $\geq$ 1 and $(P/Q)^2 < 3$. Let P' = 2P + 3Q and Q' = P + 2Q then we have $(P'/Q')^2 - 3 = ((2P + 3Q)^2 - 3(P + 2Q)^2) / (P + 2Q)^2 = (P^2 - 3Q^2) / (P + 2Q)^2 = (P^2/Q^2 - 3) / Q^2 (P + 2Q)^2$ and so if $(P/Q)^2 - 3$ is negative then so is $(P'/Q')^2 - 3$ and P'/Q' is in the same class as P/Q. Also, $P^2/Q^2 - 3 < P'^2/Q'^2 - 3$ so P'/Q' > P/Q contrary to hypothesis and so the class cannot have a greatest member.

Ex. 15 For all real numbers a and b the cuts which define a and b must satisfy one of the three cases envisaged by Dedekind, therefore $a = b$ or $a < b$ or $a > b$. This means that we have a complete ordering relation.

Ex. 16 Be bold!

3

Measuring the Pyramid

Michèle GRÉGOIRE
Group M:A.T.H.
IREM Paris VII

At the beginning of the twentieth century the second international mathematics congress was held in Paris at which, on the 8th August 1900, David Hilbert presented his famous contribution *"On future problems in mathematics"*. For all the principle disciplines of mathematics he posed problems that would concern research during the coming century. Just one of them, the third, dealt with elementary mathematics and its title was *"On the equality of two tetrahedra having equal bases and altitudes."* In order to calculate the volume of a pyramid and in order to prove that two pyramids having the same base and height were equal, mathematicians have always had to have recourse to complicated methods. The analogous problem for the plane, to prove that two triangles with the same base and height are equal in area, can be solved in an elementary manner: for example, one triangle may be cut up into a number of pieces and reassembled as the other triangle. (cf. exercise 1.) Might it not be possible to do the same for two tetrahedra, that is to cut up one into a finite number of pieces which may be reassembled to form the other? This suggests the idea of an elementary theory of volumes. As can be seen in the following extract from his communication, Hilbert's intuition was that this is not possible; he proposed to his colleagues that they find two tetrahedra having equal bases and heights which could not both be composed of the same pieces :

> In two letters to Gerling, Gauss expresses his regret that certain theorems of solid geometry depend upon the method of exhaustion [...] (or upon the axiom of Archimedes); Gauss mentions in particular the theorem of Euclid, that triangular pyramids of equal altitudes are to each other as their bases. Now the analogous problem in the plane has been solved. Gerling also succeeded in proving the equality of volume of symmetrical polyhedra by dividing them into congruent parts. Nevertheless, it seems to me probable that a general proof of this kind for the theorem of Euclid just mentioned is impossible, and it should be our task to give a rigourous proof of its impossibility. This would be obtained, as soon as we succeeded in specifying two tetrahedra of equal bases and equal altitudes which can in no way be split up into congruent tetrahedra. [1]

(We can replace the use of "congruent" by "superposition", or "the image of each is a positive isometry of the other".)

---

1.   Hilbert, *Sur les problèmes futurs…* , trad. Laugel, p. 74-75 [English tr. in *Bulletin of the American Mathematical Society* (2), **8** (1902), 437-479, p. 449.]

The problem is a difficult one, but his pupil Max Dehn produced an example of two such tetrahedra in the same year 1900.

And so, it is not possible to calculate the volume of the pyramid in an elementary way. In this chapter we shall consider some of the ways that mathematicians since the Ancient Greeks have used for calculating the volumes of pyramids. Contrary to other problems, it is not the search for a solution that has concerned mathematicians: it is thought by some historians of mathematics that a formula for finding the volume of the truncated pyramid was already known to the Egyptians as early as 1850 B.C. and in the third century B.C. Greek geometers proved that the pyramid is a third of the prism with the same base and the same height. It is rather the search for a proof more in accord with the spirit of the times, more satisfactory, or more rigourous, which is the motor for the changes that we shall meet. We shall see how the pyramid resists subdivision and requires us to resort to ideas of the infinite and the infinitely small.

But first of all we shall describe elementary methods of calculating areas and volumes and identify those shapes that can be measured.

Exercise 1	Let ABC and BCD be two triangles having the same height and drawn on the same base BC. Let (EF) be the straight line equidistant from (BC) and (AD) and let (BE), (AH), (DH') and (CF) be perpendiculars to (BC). Make the triangle BCD from the pieces that make up the triangle ABC.

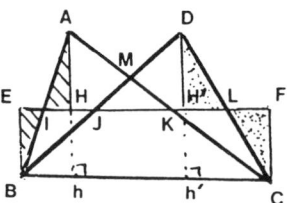

Figure 1

Elementary calculations of areas and volumes

Euclid's 'method of areas'.

We must bear in mind what the Greeks meant by 'measuring' an area or a volume. It did not mean to associate a number with any length or surface or volume, but only to compare plane figures with other plane figures or solids with solids, that is to compare magnitudes of the same nature and to show their equality or find their ratio. To find the quadrature of a plane figure meant to construct a square having the same area as the given figure: to find the cubature of a solid meant to construct a cube having the same volume.

The 'method of areas' allows us to construct a parallelogram of equal area to any rectilineal figure. It relies on the following results proved in Book I of the *Elements*[1] :

1) In the parallelogram ABCD the triangles ABD and CDB can be superposed and so have the same area (Figure 2); we shall now say that two figures having the same area are "equivalent."

2) Two parallelograms having the same base and the same height are equivalent (Figure 3): the triangles ABE and CDF being superposable, it is sufficient to cut off DGE and then add BCG to the two triangles ABE and DCF to obtain the two parallelograms.

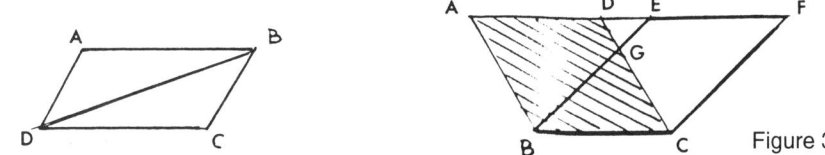

Figure 2   Figure 3

3) Two triangles on the same base and with the same height are equivalent: ABC is half of the parallelogram AEBC, BCD is half of the parallelogram BCFD which is equivalent to AEBC (Figure 4).

4) Euclid gives a method for the construction of a parallelogram equivalent to any given triangle; he can use a specified side and angle for the equivalent parallelogram and hence he is able to build up parallelograms equivalent to any given polygon (Figure 5).

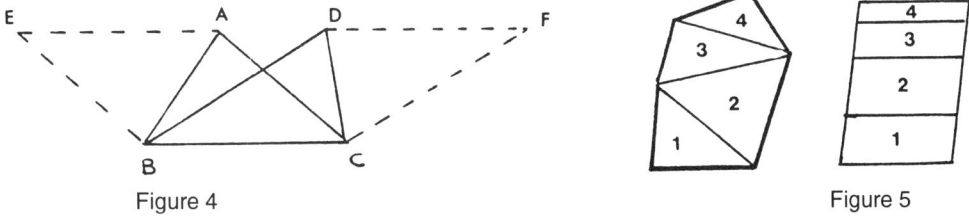

Figure 4   Figure 5

It is certainly the case that two figures that can be subdivided into a (finite) number of the same pieces or into a number of equivalent pieces are equivalent figures. We note that Euclid also uses the following principle: if two figures F and F' can be filled in by using equivalent pieces which have been obtained from equivalent pieces then F and F' are equivalent.

Euclid proceeded in the same way, as far as he was able, in considering solids: thus a triangular prism is half a parallelepiped[2] and two parallelepipeds on the same base and having equal heights can be dissected using pieces with the same dimensions and so are equivalent:

"Parallelepipedal solids which are on the same base and of the same height, and in which the extremities of the sides which stand up are on the same straight lines, are equal to one another." (Figure 6) and *"Parallelepipedal solids which are on the same base and of the same height, and in which the extremities of the sides which stand up are not on the same straight lines, are equal to one another."* (Figure 7)[3]

1.   Heath, *The Thirteen Books of Euclid's Elements,* I, 34-38, 42, 44 and 45, p. 323-346.
2.   *Ibid.*, XI, 28, p. 330.
3.   *Ibid.*, XI, 29 and 30, p. 333.

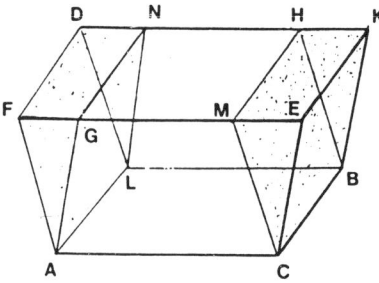

Figure 6

Exercise 2   Prove these results using Euclid's method.

1) The parallelepipeds ACBLMHDF and ACBLNKEG of figure 6 are equivalent.

2) In figure 7 the shaded parallelepiped ACBLNGEK is equivalent to the parallelepiped ACBLPOQR and this parallelepiped is equivalent to the hatched parallelepiped ACBLMFDH.

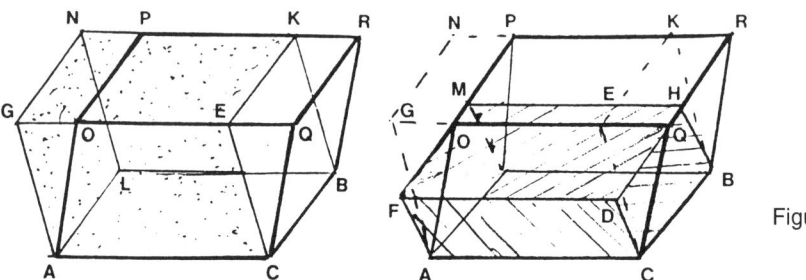

Figure 7

Finally, by using the theory of proportions of Book V, Euclid was able to go on to compare plane figures and to the compare solid figures. He showed that triangles (or parallelograms) having the same height were in the ratio of their bases and that two parallelograms whose bases and heights were inversely proportional were equivalent.[1] For solids he showed that parallelepipeds (or prisms) with the same height are in the same ratio as their bases and that two parallelepipeds whose bases and heights are inversely proportional are equivalent.[2] He also proved, both for plane figures and solids, the classical results for the dimensions of similar figures. For example, *"Similar parallelepipedal solids are to one another in the triplicate ratio of their corresponding sides."*[3] (the ratio of volumes is the cube of the ratio of the sides.)

The method of areas, with its extension into solids, together with the theory of proportions allowed Euclid to compare rectilineal figures and to compare certain solids, namely parallelepipeds and prisms.

The "manipulative methods" of ancient China

Contrasting with the treatment by the Greeks the Chinese treatment from the beginning of our era, and perhaps even before, generated algorithms by which areas and volumes might be calculated. The proofs, which were not always given, involved the process of building up (smaller) pieces. Chinese geometry places importance on the

---

1.   *Ibid.,* VI, 1 and 14, p. 191, p. 216.
2.   *Ibid.*, XI, 25, 31, 32 and 34, p. 325-349.
3.   *Ibid.,* XI, 33, p. 342.

idea of manipulation of material objects analogous to pieces of a puzzle or a construction kit, and on the principle that area and volume do not change under rearrangement of the pieces. *"The figures may take strange shapes but the numbers (which measure their areas or volumes) remain equal."* wrote Liu Hui in the third century A.D. in his commentary on the first century B.C. work *Arithmetic in Nine Chapters on Mathematical Art {Chiu-Chang Suan-Shu}*[1]. This work is one of the most ancient Chinese mathematical texts and the classical work par excellence which is referred to, not only by numerous authors of the Chinese tradition, but also by Japanese, Korean and Vietnamese authors; it is the required manual of mathematical instruction, much in the same way as Euclid's *Elements* has been in the West. It contains calculations of plane rectilineal figures by means of their subdivision and reassembly as rectangles. Volumes are found by using blocks of wood of standard shape and size which both author and reader have to hand. Often the manipulations concern pieces with specified dimensions but, where it is appropriate, the author indicates that the derived formulas will be valid for arbitrary dimensions.

The following exercise involves following the methods of *Arithmetic in Nine Chapters on mathematical art* in evaluating the volume of a truncated pyramid.

Exercise 3	The regular truncated pyramid on a square base is called *fang ting* (square pavilion). Its dimensions are a, b, and h. (The height h is here taken to be equal to a.)
	1) Show that it is possible to subdivide the fang ting into a cube of side a, four right prisms standing on right-angled triangle bases and four right pyramids with square bases; the prisms are isometric as are the pyramids.
	2) Show how to build a cube of side a with three of the pyramids.
	3) Show, either by building the solids, or by showing carefully on the figures, that by using the pieces obtained from a subdivision of three fang tings of the same dimensions it is possible to build three solids of volume a^2h, abh and b^2h respectively. Hence deduce a formula for the volume of the fang ting.

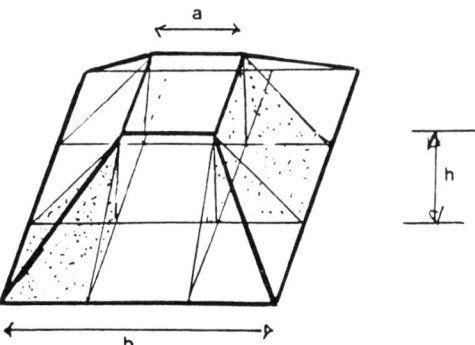

Figure 8

1.   Marztloff, *Histoire des mathématiques chinoises*, p. 260.

Equi-subdivision of figures having the same area or the same volume

Two plane figures which can be subdivided using the same pieces are, of course, equivalent. In the 19th century mathematicians posed the converse problem: if two rectilineal figures have the same area, can they always be subdivided using the same (finite) number of polygonal pieces? The proof that it was possible was given in 1832 by F.Bolyai, the father of J.Bolyai, and also, independently, in 1833 by the Prussian officer and mathematician P. Gerwien.[1]

Gerwien and Bolyai considered subdivisions to be equivalent if they were symmetrical to each other with respect to a straight line (irrespective of orientation). A century later, around 1950, the Swiss mathematician Hadwiger asked himself the following question: is it possible to subdivide two figures of equal area using pieces that are strictly superposable by displacement, without having to use pieces of different orientation? He proved that it was indeed possible.[2]

Exercise 4   Show how two triangles which are each the reflection of the other may be subdivided using the same six pieces which may be superposed using only positive isometries. (Use the centres of the inscribed circles.)

Another formulation of the same question as used in the USSR Mathematics Olympiad: A baker made a cake in the shape of a triangle (of any shape). A box was made for it which had the correct dimensions but was the wrong way up. Suggest a dissection of the cake so that the pieces could be placed in the box so that the icing remains on the top.

It was the analogous problem for solids that Hilbert had proposed in 1900. If two polyhedra have the same volume is it possible to subdivide them using the same finite number of polyhedral pieces? (We shall refer to such cases as equi-subdivisible.) In particular, are two pyramids having the same height and with equivalent bases equi-subdivisible? Since Euclid no mathematician has succeeded in exhibiting such a subdivision. Is that solely due to a lack of imagination or skill, or is it in fact impossible? In 1900 Max Dehn replied to the question by showing that a regular tetrahedron is not equi-subdivisible with a cube nor with a right-angle isosceles tetrahedron of the same volume, and that there exist an infinite number of pairs of non equi-subdivisible tetrahedra.[3] In contrast to the result for plane figures, equality of volume is not a sufficient condition for equi-subdivisibility. Dehn found that a necessary condition for equi-subdivisibility of two polyhedra P and P' was the existence of a linear relation, modulo π, between their dihedral angles. The integer coefficients of this relation depend upon the lengths of the edges of the two polyhedra.

This condition cannot be verified in general for the pyramids obtained from the subdivision of a prism into three pyramids. Dehn's condition is also a sufficient condition for the equi-subdivisibility of two solids with the same volume but this was not established until 1965 by Sydler.[4]

---

1.   Gerwien, *Journal für die reine und angewandte Mathematik* (Journal de Crelle), 1833, p. 228-234.
2.   Hadwiger and Glur, Element. Math. 6, p. 97-106.
3.   Dehn, *Ueber raumgleiche Polyeder*, p. 345-354.
4.   Sydler, *Commentarii mathematici helvetici* 40, 1965, p. 43-80.

In order to find the volume of a pyramid we shall have to start all over again with dissections which are infinitely many. We shall see how mathematicians developed sophisticated methods in order to overcome this problem, by trying to cheat the infinite. Greek mathematicians developed processes that avoided the infinitely many and the infinitely small. At other times and in other civilisations mathematicians were content to dissect the pyramid into an infinite number of infinitely small pieces...

How to avoid using the idea of the infinite: Euclid's method

The oldest proof is that contained in Book XII of *The Elements*. According to Archimedes, it was discovered by Eudoxus, a mathematician of the 4th century B.C. whose works, long since lost, formed the starting point for certain books of *The Elements*. This proof consists in noticing that a triangular based prism can be subdivided into three pyramids. By taking them in pairs Euclid states that they are of the same height on equal bases. The pyramids CABD and CEDB of figure 9 have the same height (the distance from C to the face ABED) and their bases are each half the parallelogram ABED. The same can be said for DCBE and DECZ, whose bases are half the parallelogram BEZC.

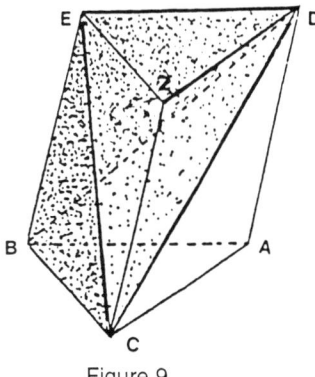

Figure 9

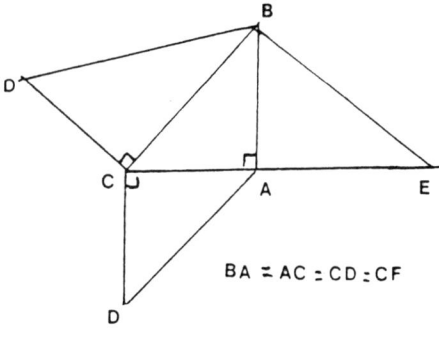

BA = AC = CD = CF

Figure 10

Exercise 5	Cut out three nets as shown in figure 10: ABC is a right-angled isosceles triangle with a right-angle at A. CF = CD = AC, AE = AD and the triangles CAD, ABE and CFB have right-angles at C, A and C respectively. Fold two of the nets in the same way and the third in the opposite way. Show they make three tetrahedra which together make up a right prism with base ABC. (These three tetrahedra are called Hill's tetrahedra after the English mathematician who studied them in 1895.[1])

The essential point of the proof, as we have already shown, lies in proving that two pyramids of the same height are in the ratio of their bases, for the dissection once started seems to go on infinitely. In the 5th century B.C. atomistic theories, which tended to consider magnitudes as being a collection of an infinite number of indivisibles, were vigorously attacked by the Eleates, numbering among them

---

1.   Hill, *Proceedings of the London Mathematical Society*, 27, 1896, p. 39-53.

Parmenides and his disciple Zeno. The paradox of dichotomy, the first of Zeno's famous paradoxes, questioned whether movement was possible since *"that which is in locomotion must arrive at the half-way stage before it arrives at the goal"*[1] then to half the remaining distance and so *ad infinitum*. He raised in particular the problem of thinking about a magnitude composed of an infinite sum of parts. Aristotle's response, which dominated Greek thought and even Western thinking for almost two millenia, was dictated by 'common sense'. Aristotle rejected what could not be physically thought of, he rejected the ideas of infinity and the infinitely small in mathematics and relegated these concepts to metaphysics. Greek mathematics, then, developed a way of approaching geometric magnitudes through using successive divisions which remained finite in number. There was no end to subdividing the pyramid but the reasoning used by Eudoxus, and followed by Euclid, avoided the problem by not wanting to continue as far as an end to divisions. A magnitude will only be thought of in terms of a finite number of magnitudes of the same type (surfaces for a surface, solids for a solid...)

Euclid did not show the equality of volumes directly; he only worked with and proved inequalities, $V > V'$ or $V < V'$. He calculated and reduced the size of the difference between the magnitudes under consideration. He had recourse to that famous proof, called 'proof by exhaustion' since the 17th century, which is attributed to Eudoxus and which was used to such effect by Archimedes a century later to find the area of a disk and the area of a segment of the parabola.

The Principle of Proof by Exhaustion

This method of proof was used most often in solving problems of comparison of areas. But we see it here being used to compare solids. Euclid showed, not the 'equality' (in the sense of volume) of two pyramids, but he showed that the ratio of two pyramids of the same height was as the ratio of their bases. In order to give a general idea of the method we shall show, by a similar method, that two pyramids with equal bases and heights are equivalent. First, we shall step aside from Euclid's argument and to return to it again in the next part.

In order to show that a solid P has the same volume as a solid P', a solid S is constructed inside P in such a way that the difference between P and S can be made as small as we please. We also construct a solid S' inside P', S' being in this example equivalent to S, and we go on to prove that the two hypotheses $P < P'$ and $P > P'$ lead to contradictions. Note that we are able to make the difference between the two volumes as small as we please and this fact allows us to avoid repeated infinite dissections and the use of the infinitely small. Euclid's argument derived from X 1 of *The Elements* which can be considered as Zeno's paradox 'turned round' and which says, in substance: "Take away from a magnitude more than half of it and then take away from the remainder more than its half, and so on; it is possible to carry on like this until what remains is smaller than any given magnitude." The proof of this proposition rests on what we call Archimedes axiom which, in modern terms, asserts that for any two positive real numbers a, b there exists an integer n such that $nb > a$. The solid S inscribed within P is constructed in stages such that, from each stage to the next, the difference between P and S is reduced by more than its half.

1.   Aristotle, *Physica*, Book VI, 9, 239b, line 13.

How can we show that the hypotheses P < P' and P > P' should both be rejected? Suppose first of all that P < P' and define a solid S' inscribed within P'. S' can have a known volume since it can be made from a number of prisms in such a way that the difference P' – S' should be less than the difference P' – P. From this we get the inequality (i) S' > P. In like manner we can construct another solid S inside the pyramid P where S is made up in the same way as S' and consequently S = S' < P. This inequality contradicts (i) and so we must reject the hypothesis P < P'. Interchanging P and P' we can similarly find that we need to reject P > P'.

Euclid's process for comparing two pyramids having the same height

Euclid uses a *reductio ad absurdum* argument to show that two pyramids of the same height are in the ratio of their bases (what we would write as $\frac{P}{P'} = \frac{B}{B'}$). He assumed firstly that $\frac{P}{P'} < \frac{B}{B'}$ and then that there was a solid X smaller than P' that satisfied $\frac{P}{X} = \frac{B}{B'}$. Inside the pyramid P' he constructs a solid S' made up from a number of prisms. Euclid points out that the centres of the six edges of a triangular pyramid can be used to define two pyramids (shaded in figure 11), which are similar to the initial pyramid, together with two equivalent prisms TKZBEH and TKLCZH. The two prisms have the same volume because they are each half of the same parallelepiped and each prism contains within it a pyramid equal to the smaller pyramids; P_1, for example, contains KEBZ equal to TEAH. The sum of the two prisms is therefore greater than half of the initial pyramid.

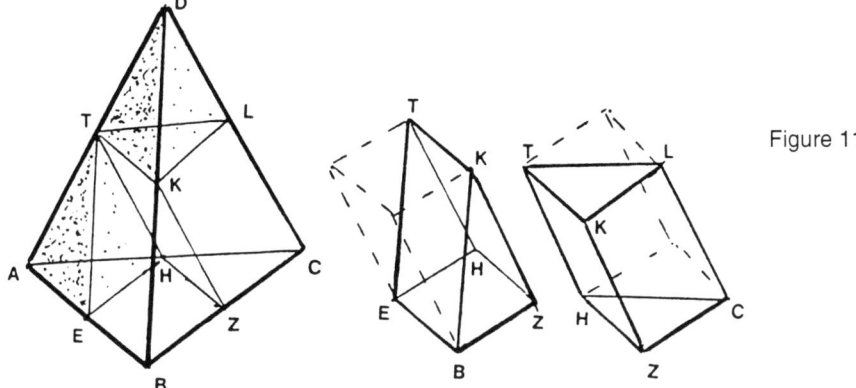

Figure 11

Using the centres of the edges of the smaller pyramids to begin another dissection, Euclid defines new prisms and new smaller pyramids within which he again carries out dissections (figure 12). He uses as many of the prisms as are needed, which step by step fill up the pyramid, to the point where the total volume of the sum of the smaller pyramids left behind becomes less than the assumed difference P' – X. (This volume left behind being equal to the difference P' – S' between the volume of the initial pyramid and the solid S'.) In effect, he has used proposition X 1 since he has shown that at each stage of the dissection he subtracts more than half of the remaining pyramid. What is left behind, that is all the small pyramids, can be made smaller than P' - X. After a sufficient number of steps we have P' – S' < P' – X and so establish the inequality (i) S' > X.

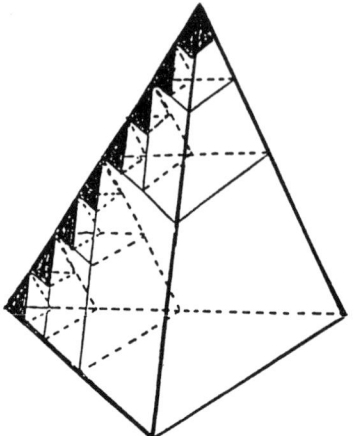

Figure 12

The same construction can be carried out inside the pyramid P. The prisms constructed within P and P' are the same in number and taken in pairs are equal in height. Their volumes are in the same ratio as the bases of the two pyramids[1]; and the same is true for the two collections of prisms S and S'. We are able to write $\frac{S}{S'} = \frac{B}{B'} = \frac{P}{X}$. And so $\frac{S}{P} = \frac{S'}{X}$. As the solid S is within P then S < P and this last equality leads to S' < X which contradicts the inequality (i). The hypothesis $\frac{P}{P'} < \frac{B}{B'}$ is therefore rejected. Euclid repeats the same reasoning to show that the hypothesis $\frac{P}{P'} > \frac{B}{B'}$ must also be rejected. The pyramids must therefore be in the same ratio as their bases.

Now that we have looked at the underlying structure of Euclid's argument, here is the sequence of propositions in the order in which they are stated:

> *Proposition 3: Any pyramid which has a triangular base is divided into two pyramids equal and similar to one another, similar to the whole and having triangular bases, and into two equal prisms; and the two prisms are greater than the half of the whole pyramid.*

> *Proposition 4: If there be two pyramids of the same height which have triangular bases, and each of them be divided into two pyramids equal to one another and similar to the whole, and into two equal prisms, then, as the base of the one pyramid is to the base of the other pyramid, so will all the prisms in the one pyramid be to all the prisms, being equal in multitude, in the other pyramid.*

> *Proposition 5: Pyramids which are of the same height and have triangular bases are to one another as the bases.*

> *Proposition 7: Any prism which has a triangular base is divided into three pyramids equal to one another which have triangular bases.*

> *Porism: From this it is manifest that any pyramid is a third part of the prism which has the same base with it and equal height.[2]*

Euclid's method, then, consists in considering a finite number of pieces built up by a finite number of steps but sufficiently many to almost completely fill the pyramid. What

---

1.   Heath, *The Thirteen Books of Euclid's Elements*, XII 4, vol. 3, p. 382.
2.   *Ibid.*, XII 3,4,5,7 vol. 3 p. 378 et seq.

remains is manipulated in such a way as to produce a contradiction. This reasoning by *reductio ad absurdum* argument avoids any reference to the infinite.

The Chinese method or The infinitely small need not concern us

Little is known about the relations that existed at the beginning of our era between the Chinese and Greek civilisations. However, a century after Zeno the nominalist Chinese school also enunciated the paradoxes concerning continued division of a finite quantity and constructing the finite from an infinite number of small pieces. *"If you take a stick one foot in length and cut it in two every day something will still be left to cut after ten thousand generations"* ... *"That which has no thickness cannot be put together to make thickness but may be spread out to cover a thousand lilies."* [1] In contrast to the Ancient Greeks, Chinese mathematicians principally concerned themselves with concrete problems and developed pragmatic methods to solve them. In his commentary on *Nine Chapters on Mathematical Art,* Liu Hui explained a method for calculating the volume of a pyramid in which he makes no attempt to avoid using 'extremely small' quantities, yet there is also evidence of a Euclidean approach. The pyramid considered, called Yangma, has a rectangular base and an edge perpendicular to the base.[2]

The calculations were as follows.

To find the volume Y of the pyramid Yangma add to it a tetrahedron (Bienao) of volume B to make up a triangular prism whose base AEP is half a rectangle. The volume C of the prism is already known ($C = \frac{1}{2}$ abh). Liu Hui shows that Y = 2B and, since C = Y + B, he obtains

$$Y = 2B = \frac{2C}{3} = \frac{2}{3}\left(\frac{abh}{2}\right) = \frac{abh}{3} \text{ (see figure 13)}$$

In order to prove the result Y = 2B, Liu Hui (just like Euclid) dissects both the Yangma and Bienao using the centres of their edges. The Yangma is a rectangular pyramid which yields a rectangular parallelepiped, two prisms and two rectangular pyramids. The Bienao, which is a triangular pyramid, is subdivided into two prisms and two triangular pyramids. The total volume of the two prisms and the parallelepiped of the Yangma is twice the volume of the two Bienao prisms.

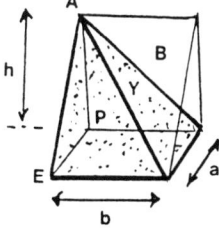

Figure 13

1.   Hui Shi, quoted by Dhombres in *Nombre, mesure et continu,* p. 291.
2.   An English translation and commentary on the corresponding extract of the *Amerithmetic in Nine Chapters on Mathematical Art* can be found in D.B. Wagner: "A early derivation of the volume fo a pyramid: Liu Hui, third century A.D.", *Historia mathematica* **6**, 1979, p. 164-188.

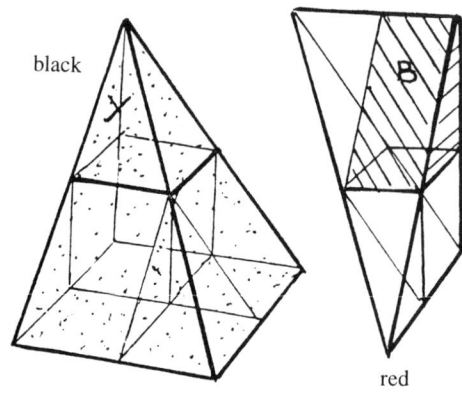

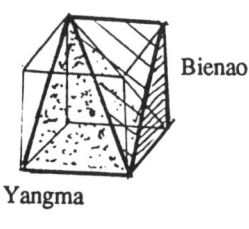

Figure 14   Figure 15

It is easy for Liu Hui to establish this relation for the Yangma pyramid which has the three equal dimensions a = b = h which is the case shown in figure 15 where Yangma and Bienao make up half a cube. Here two blocks B can be put together to make the block Y. If the dimensions are not equal then the black prism shaped pieces no longer make a parallelepiped when put together. But Liu Hui is able to show in a different way that the volume of each pair of prisms taken together is half the volume of a parallelepiped of dimensions $\frac{a}{2}$, $\frac{b}{2}$, and $\frac{h}{2}$. He assumes, without explicitly stating it, that two symmetrical pieces will have the same volume. Liu Hui, who apparently had black blocks to make up the Yangma and red pieces for the Bienao, expresses this property by stating: The black volumes are double the red volumes.

It remained for him to show that for the black and red pieces left over the ratio would be the same. Now, there remained two small black pyramids in the Yangma and two small red pyramids in the Bienao, each of which had the same shape as the original pyramids from which they had been cut, although they were smaller. Liu Hui then recuts each of the pyramids with the same type of dissection obtaining black prisms (from the rectangular pyramid) and red prisms (from the triangular pyramid) the ratio of whose volumes is 2:1. Liu Hui continues with these dissections until the volumes left over are "so small that they have lost all form". He says, in substance; "The smaller they are halved, the finer are the remaining (dimensions). The extreme of fineness is called "subtle". That which is subtle is without form. When it is explained in this way, why concern oneself with the remainder? (Chiu-Chang, 168).

It is worth noting here the difference between the pragmatic Chinese approach, which neglects the small excess of the volume of the pyramid over the sum of the prisms, compared with the Greek approach which attempts to deal with the excess through 'enclosing' them. We may also notice the important difference given to discourse and dialectic. Taoist philosophy which developed in the 5th century B.C. claimed that it was impossible to communicate through discourse: it was better to "show by example." In the Greek city states of the same period, by contrast, discourse held prime position and equally so in scientific practice. The way in which later developments in Western thought were influenced by this characteristic of Greek philosophy is well known.

Are indivisibles infinitely small?

Throughout the Middle Ages and right up to the 16th century, both in the West and in the Arab world, methods for finding areas and volumes followed Greek principles. By contrast, from the beginning of the 16th century mathematicians were bold enough to use new tools to attack the problems of quadrature and cubature and also to find tangents to curves and to determine centres of gravity ... The new procedures presupposed the use of the infinitely small and infinite aggregates, and indivisibles, procedures which, as we have seen, were excluded by classical Greek geometry. The mathematician most often associated with these new ideas was Bonaventura Cavalieri whose *Geometria indivisibilibus continuorum* appeared in 1635. Others, like Kepler for example, had also used comparable procedures. Cavalieri's work presents a systematic construction and the appearance of a rigourously based theory: the model provided by Euclid's *Elements* enjoys an honoured place. But in order to follow his method we have to imagine that a line is composed of an infinite number of points (like a string of pearls), that a surface is made up of an infinite number of lines (like wire gauze) and that a solid is made up from an infinite number of surfaces (like pages of a book.)

The Principle of Cavalieri: comparison of indivisibles

Notwithstanding the metaphor that he proposed, Cavalieri did not go so far as to suggest that the continuum was composed of indivisibles. He did not claim, like Galileo and the atomists, the existence of indivisible atoms; for him, indivisibles did not 'make up' the figure, rather were they the result of dissections of the figure according to a strict "rule".

In the plane, two parallel straight lines enclose a figure and the "rule" is to allow one of the lines to slide towards the other, remaining always parallel to it. Cavalieri's method for comparing two figures uses the principle that if the segments of each of the figures cut by the moving line are always in a certain ratio k:1 then the areas of the figures will be in the ratio k:1. This became known as the Principle of Cavalieri. A similar procedure was used to compare two solids by considering parallel plane sections.

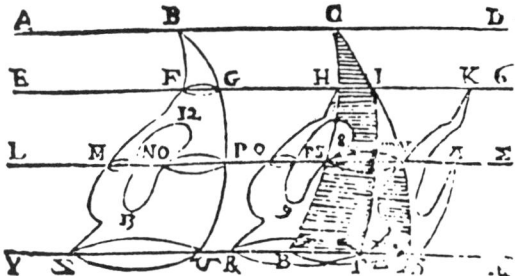

Figure 16

Calculating the volume of a pyramid using Cavalieri's Principle

Cavalieri, like Euclid before him, used the idea of cutting up a prism into three pyramids each pair having the same height with bases of equal area. The originality in Cavalieri's approach lies in his method of showing that pyramids of the same height on bases of equal area are equal in volume. Cavalieri states and proves in proposition 7 of *Geometria Indivisibilibus* that *"conical solids of the same height are to each other as their bases"* and uses this in proposition 8 to deduce that a cylindrical solid is three times the conical solid constructed on the same base and with the same height. By conical solid or cylindrical solid he meant cones or cylinders on any base, circular or polygonal; the figures used for the proofs were, however, triangular pyramids. To prove proposition 7, Cavalieri considered some plane cutting the heights AE and BF of the solids at C and D and producing plane sections GIO and XNVP.

Hence $\dfrac{LM}{IO} = \dfrac{MA}{AO} = \dfrac{AE}{AC} = \dfrac{BF}{BD} = \dfrac{QR}{NP}$ and, since similar figures are in the ratio of the squares of their corresponding sides: $\dfrac{KLM}{GIO} = \left(\dfrac{LM}{IO}\right)^2 = \left(\dfrac{QR}{NP}\right)^2 = \dfrac{SQTR}{XNVP}$. Hence by transposition $\dfrac{KLM}{SQTR} = \dfrac{GIO}{XNVP}$ and so the ratio of the figures GIO and XNVP, cut by any plane parallel to the base is constant. The figures are, according to Cavalieri, proportional "analogues" and so the whole figures, solid conics of the same height, are to each other as are their bases.

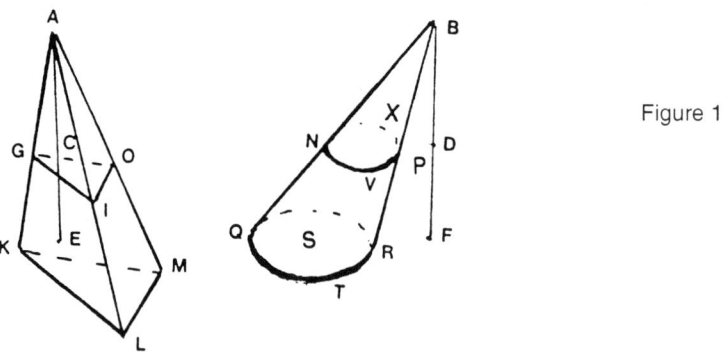

Figure 17

There is no French translation of *Geometria Indivisibilibus*[1] but Cavalieri's Principle is used by Father Lamy in his *Eléments de Géométrie*. Lamy was a member of the Oratory and an ardent defender of Descartes and the new scientific spirit. His book was a pedagogical work in which he popularised Cavalieri's methods. Lamy posits that solids are made up of *"an infinite number of planes having a certain thickness, which cannot be felt, and which lie parallel and are placed one upon the other."* (First Demand) and that *"two solids of the same height and cut by planes of equal thickness have the same number of planes"* (Second Demand). In order to anticipate objections that might be raised against the use of "an infinite number of planes", the author draws back a bit and remarks: *"by an infinite number of planes, I simply mean a large number."* Lamy recalls that sections of a pyramid produced by planes parallel to the base will be similar to two bases B and B' in the same ratio:

> *Fifth Lemma. If two pyramids of the same height are cut, wheresoever between the parallel planes, by planes parallel to their base, the sections of one will be to the other of the same height as the base of the one is to the other.*

1.   Extracts, in English, can be found in Struik, p. 209 and in Smith, p. 605.

Then he states the theorem:

> *Pyramids of the same height are in the ratio of their bases.*
> *1° By the first Demand two pyramids can be considered as being composed of parallel planes placed one upon the other. 2° By the second Demand two pyramids of the same height have an equal number of parallel planes. It only remains to prove that all the planes of which these pyramids are composed and which are corresponding, being at the same height, are in the same ratio as their bases. A fact that was proved by the last Lemma: Hence, if the bases are equal then these two pyramids will be equal, since all the planes taken at the same height in both pyramids are equal; if the base of one is a third of the base of the other, as each plane of the one will be a third of the other taken at the same height, the one pyramid will be a third of the other. Thus pyramids of the same height are in the ratio of their bases.*[1]

It is evident that, once the Principle of Cavalieri has been adopted, this new and powerful method proves economic and proofs are established in an elegant and simple fashion.

Roberval's method: summation of indivisibles

In the 1630s, and independently of Cavalieri, Roberval developed a method of handling indivisibles. His approach, differing from Cavalieri's, was set out in his *Traité des indivisibles,* which was published posthumously in 1693. Roberval considered that *"all lines, straight or curved, can be divided up into an infinite number of parts or short lines all of which are equal to each other or can be put into some progression at will, such as squares to squares, cubes to cubes, squared-squares to squared-squares or according to some other power."*[2] In the same way *"areas can also be divided up into an infinite number of small areas which are equal or which have equal differences [...]; solids can be divided up into an infinite number of small solids..."*[3] He represents the small lines which make up a line segment by points, which he numbered; each line is then associated with a whole number. He represents the infinity of small areas which make up a surface by lines, and the surface, being the sum of all the small areas is considered as the sum of the small lines which represent these small areas, or the sum of the numbers associated with each small line. The small solids which make up the total solid are represented by an infinity of small areas, to which numbers may also be associated, whose sum can thus be associated with the total solid. A pyramid with a square base can be represented by the sum of the first square whole numbers, the different small solids making up the pyramid being represented by small square surfaces of area 1^2, 2^2, 3^2, 4^2, ... If the side of a cube is n, its volume will be n^3, and likewise the volume of the pyramid will be $1^2 + 2^2 + 3^2 + ... + n^2$.

This sum, known since Archimedes[4], has the value $\dfrac{n^3}{3} + \dfrac{n^2}{2} + \dfrac{n}{6}$ (Roberval evaluates this below for n = 4.)

Exercise 6	Prove that $1^2 + 2^2 + 3^2 + ... + n^2 = \dfrac{n^3}{3} + \dfrac{n^2}{2} + \dfrac{n}{6}$

1. Lamy, *Elémens de Géométrie,* 1685, p. 217-219.
2. Roberval, *Traité des Indivisibles,* p. 247.
3. *Ibid.,* p. 249.
4. Archimedes, *On Spirals,* X 1 in Dijksterhuis, p. 122.

Following this, he considers the comparison of magnitudes of different order which leads him to reject the terms $\frac{n^2}{2}$ and $\frac{n}{6}$ which he justifies as follows: "*So the line or the side are not in the same ratio as the cube, [...] for lines taken to infinity can only make up a square and there are an infinity of squares in a cube, if a single square is added or taken away it will have no effect.*"[1]

> "*In the same way, if lines or the points which represent them are of the order of squares, then taken a sufficient number of times, they will be as squares are to cubes, or as the pyramid is to the column, namely as 1 to 3; for taking whatever finite sum of squares their sum will be greater than a third of the greatest corresponding cube by at most the greatest square, nevertheless in an infinite division there will only be that third; for the said sum will never exceed the $\frac{1}{3}$ of the cube by more than half of the largest square $+ \frac{1}{6}$ of the side. Now, in a cube there are an infinite number of sides, and adding on half of one of them is not a great matter, and it is even less so for $\frac{1}{6}$ of a line or a side of the same cube.*
>
> *So for a cube being 64, to get the sum of the squares whose largest is the [square of the] side of the said cube, you take its third, namely $21\frac{1}{3}$, to which add half of the greatest square, namely 8, and you get $29\frac{1}{3}$, to which must also be added of the 4 which is the side, namely $\frac{2}{3}$, and you will have 30 for the sum of the first four squares.*"[2]

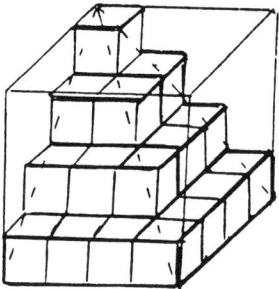

Figure 18

Exercise 7   Prove the equivalence of the two problems: quadrature of the parabola and cubature of the pyramid, by showing that they are each solved through use of the sum of $1^2 + 2^2 + 3^2 + ... + n^2$.

It can be argued that Roberval's approach involved a methodological leap forward which was truly revolutionary. Even if this mathematician did not claim to describe nature, or the essence of the mathematical objects, he carried out calculations of figures as if they were composed of an infinite number of elementary indivisibles. His method could be used for more than the simple comparison of two figures, which had been the case with Cavalieri's method. He offered the means by which quadrature or cubature of a figure could be found directly by the summation of the infinite number of small parts which compose it. He introduced, further, ways of comparing infinitely small quantities between themselves, treating them in the same way as magnitudes. Finally, he allowed

1.   Roberval, *Traité des Indivisibles*, p. 249.
2.   *Ibid.*, p. 248.

himself to neglect terms which were small in comparison with other terms. The Ancients always avoided this procedure through the use of inequalities and 'enclosing'.

The Jesuits against indivisibles

The method of indivisibles, used without caution, can lead to paradoxes. Paradoxical results were brought to light notably by Father Guldin, a Jesuit who taught in Rome and in Graz in Austria. He criticised the *Geometria indivisibilibus* in his work *De centro gravitatis* (1635-1641). In contrast to the priests of the Oratory, like Lamy, who were very open to new ideas, the Jesuits were trenchant opponents of the use of indivisibles for reasons that were more theological than mathematical. They were afraid of opening the door to a return to an atomistic theory of nature which would, as Sforza Pallavicino wrote *"upset the Church's teaching on the Mysteries of the Eucharist."*[1] Infinite division of a geometric figure offended against Roman Catholic conceptions of the idea of continuity. The Roman College rejected the method of indivisibles in 1649 as "contrary to common opinion" and banned its use in Jesuit colleges.

The Jesuit teacher Father Tacquet of Anvers also used the study of paradoxes to support his rejection of indivisibles. For example, he compared half of a cone abcdec (a being the vertex) and the triangle abc (figure 19). The curved surface is able to be made up of the aggregate of the half-circles (like imq) obtained by cuting the cone by different planes parallele to the base, and the triangle abc is able to be made up of aggregate of segments (like iq). The ratio of half the circle imq to the segment iq is constant and equal to $\frac{\pi}{2}$. Tacquet argues that by use of the "Method of indivisibles" the ratio of the half cone lateral area to the area of the triangle abc should also be $\frac{\pi}{2}$; which is evidently false. Cavalieri would have objected, had he been around, that the two surfaces to be compared were not situated in the same plane according to the requirements of his rule.

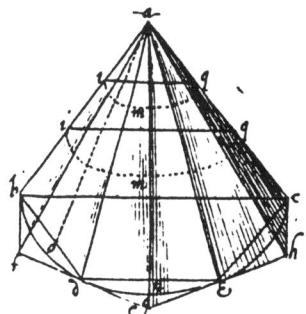 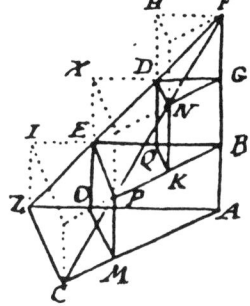

Figure 19   Figure 20

Father Tacquet considered that the sole means of proof acceptable was proof by exhaustion which, however, he modified in an important way. What he proposed witnesses somewhat of a return to Euclidean methods but is, however, influenced by the use of the 'slices' of the method of indivisibles. In his *Elementa geometriæ planæ ac solidæ* of 1654, in order to show that two triangular pyramids of the same height are to each other as their bases, Tacquet inscribed and circumscribed a triangular pyramid

1.   Festa, *La Recherche*, Sept. 1990, p. 1046.
2.   Tacquet, *Cylindria et Annularia*, p. 38-39.

with prisms of the same height. He notes that the difference between the sum of the circumscribed prisms and the sum of the inscribed prisms (which in the figure are the sum of the volumes FHDN, DNPEX, EPCZ, ...) is equal to the lowest prism ABOCIZ (which he identifies as CIBA) (Figure 20).

In taking the number of prisms to infinity, the common height of the prisms, and so the height AB of the lowest prism, will become smaller than any given quantity. The volume of this prism, which is the difference between the inscribed and circumscribed prisms, is smaller than any given quantity; therefore the difference between the pyramid itself and the sum of the interior prisms is smaller than any given quantity. The inscribed prisms "are finally" the pyramid. In order to compare two pyramids of the same height, he constructs prisms within both of them, the volumes of which he shows to be in the ratio of their bases. Proceeding in a way that was analogous to that of Descartes or Grégoire de Saint-Vincent when they studied curved areas, he asserted that "in the end" the ratio between the volumes of the two pyramids will be the same as that between the interior prisms and that it is therefore equal to the ratio of the bases, since the difference between the volume of each of the two pyramids and the sum of the corresponding interior prisms can be made as small as is wished. Tacquet, without precisely stating his assertions, considers that the inscribed prisms "tend towards" the pyramid as their number increases to infinity, even though he refuses to consider that each of them becomes infinitely small!

Infinitesimals, tools of the differential and integral calculus

In the 17th century methods for quadrature and cubature blossomed, but minds were divided over the acceptability of the use of infinitesimal quantities for finding results and their role in proofs. These reflections were carried forward on the innovatory current by the founders of differential and integral calculus. A new result, giving rich benefits, was the recognition that the problems of quadrature and that of finding tangents were each the inverse of the other, what is today called the fundamental theorem of the calculus. This discovery allowed one to calculate integrals (and therefore areas and volumes) by looking for primitives, that is to say by inverting the operation of differentiation. Newton and Leibniz proposed new techniques for studying complex curves, finding their tangents, studying their maxima or minima, determining points of inflexion, cusps and evolutes, solving problems of quadrature and cubature ..., and of course solving numerous problems that arise in mechanics and physics. They were able to unify methods of solution of problems that had hitherto been considered as different. Thanks to these new techniques, Newton was able to lay the foundations of modern physics and astronomy. Infinitesimals formed part of the techniques forged by Leibniz; Newton also made use of them, before he went beyond them.

Isaac Newton and his Fluxions

We shall say a few words about Newton's fundamental work in so far as it concerns his treatment of area. In his early writing he carried out operations with infinitely small quantities which he called moments. In order to determine the area under a curve, he no longer added up infinitesimal areas, like his predecessors, but considered the ratio of the moment (or the infinitely small increment) of the area to the moment of the abscissa at a particular point of the curve. Here is the germ of the idea of the indefinite integral. From 1671, with his *Method of Fluxions* (not published until 1736), and then with his *Principia,* published in 1687, he abandoned the use of infinitesimals, by considering all variable quantities to depend on time as a universal referent. He went on to define the idea of fluxion as a measure of instantaneous variation of a quantity.

This aspect was developed by Colin Maclaurin, one of Newton's most fervent disciples, who strove to make explicit the foundations of Newton's method. In his *Treatise of fluxions* 1742, Maclaurin stated: *"The fluxion of a solid, that can be conceived to be generated by any plane surface moving parallel to itself and perpendicular to a given axis, is measured by a prism that has the generating surface for its base and its altitude equal to the right line which measures the fluxion of the axis ..."*[1] In particular he considered the fluxion of a pyramid with vertex A generated by the uniform movement of a variable triangular surface PMm. (Figure 21) A small increment in the volume of the pyramid ADEe, as the point P moves from D to G, is exactly the truncated pyramid EDehGH. The prism *P* with base DEe and height BD represents the fluxion of the pyramid ADEe; this being only a part of that truncated pyramid. Maclaurin finds the other parts of the truncated pyramid by using the ideas of second and third fluxions. He established a Taylor expansion in geometric form. By using the fluxion of a fluent quantity, the fluent can be found, but Maclaurin did not do this for the pyramid.

Exercise 8 Verify, by finding the volumes of the solids being considered as a function of AD and DG, that the following statement by Maclaurin is equivalent to a Taylor expansion of the volume of ADEe in terms of AD:

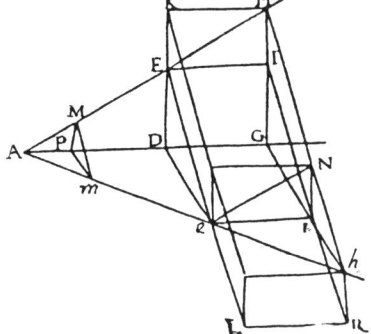

Figure 21

Thus the solids EDeIGi, NHEe, NhLe respectively measure the first, second and third Fluxions of the pyramid ADEe when AD flows uniformly, and its fluxion is represented by DG. The three parts which constitute the frustrum EDeHGh (or the increment of the pyramid that is generated while AD acquires the augment DG) are the prism EDeIGi, the prism EHIeNi and the pyramid eNih; of which the first measures the first Fluxion, the second measures one half of the second Fluxion, the last measures one sixth part of the third Fluxion of the pyramid ADEe; and this part eNih being invariable, the pyramid has no fourth Fluxion.[2]

1.   Maclaurin, *A treatise of Fluxions*, Book, I, ch. IV, prop. 7, p. 142.
2.   *Ibid,* p. 145-146.

Leibniz and differences

Leibniz worked with infinitesimals in a more confident way: for him the fundamental notion was that of difference (or differential). The difference of a variable quantity corresponded with an infinitely small increment of that quantity. In figure 22, taken from l'*Analyse des infiniment petits pour l'intelligence des lignes courbes* by the Marquis de l'Hospital, AP is equal to x, Pp (or dx) is the difference of x, y is equal to PM, dy or Rm is the difference of y; if S is the area of the mixed-line region APM then dS is the area of PMmp ... Leibniz considers as equal those quantities which differ by only an infinitely small amount, and so for an infinitely small increment dx, the portion of the curve Mm and the corresponding portion of the tangent at M, the lengths AP and Ap, and the mixed-line area MPpm and the rectangle MPpR, respectively merge together (The difference between two infinitely small areas being a higher order infinitesimal). Leibniz, and above all his disciples, thought of areas and volumes as sums of infinitesimal elements (sums of rectangles MPpm ...) and evaluated them by inverting the operations of differentiation; the notation "integral" was introduced. It is the definite integral that emerges.

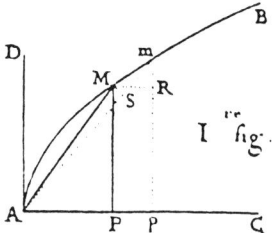

Figure 22

Leibniz' writings were abundant but also very dispersed so that they were never organised into treatises. The success of his methods on the continent were principally secured through the publication of a treatise on the differential calculus, l'*Analyse des infiniment petits,* by the Marquis de l'Hospital in 1696. No treatise on the integral calculus existed, a calculus which *"consists in building up infinitely small quantities to magnitudes or to all those things of which they are differences."*[1] The philosopher Malebranche, close to Lamy and the Oratorians, teachers who were receptive to new theories, encouraged mathematicians of his entourage to write elementary works on the integral calculus. He first of all approached his secretary Louis Carré, who later became a member of the Royal Academy of Sciences. The latter made a compilation of existing material and in 1700 published a *Method for measuring surfaces, the dimension of solids, their centres of gravity, percussion and oscillation by the application of the integral calculus.* This was the first treatise on the integral calculus and the only one that concerned itself with evaluating the volume of the prism and the pyramid. The work was republished in 1750 but was not widely distributed (and contained some errors). Important works on the integral calculus appeared later, during the course of the 18th century, and were more concerned with finding methods for evaluating more and more complex integrals than with identifying and laying the foundations of elementary calculus. A characteristic of the spirit of mathematicians of this time was to push further forward, to seek out solutions to new problems, rather than to lay a firm foundation for

1.   Marquis de l'Hospital, *Analyse des infiniment petits,* Preface.

their methods. Very few writers can be found who would wish to explore the use of the integral calculus for the evaluation of simple volumes. In geometric works of the 18th century (Clairaut, Wolf, Lacaille, in d'Alembert's *Encyclopédie...*) the volume of the pyramid was presented by methods inspired by the use of indivisibles. These authors compared "slices" cut by parallel planes without concerning themselves with the fact that the truncated pyramids had differing forms (the inclinations of the sides of the pyramids differed one to another, even if their bases were equivalent).

Using Integral Calculus to find the volume of a pyramid, according to L. Carré

Louis Carré bears witness to the enthusiasm generated by Leibniz' discoveries: "*one discovers thereby, with a marvellous facility, things which one could not find by ordinary geometry without great work and difficulty.*"[1] His work starts by setting out the postulates of integral calculus. Concerning volumes he wrote: "*It has been the custom for geometers to think of most Solids as composed of an infinite number of parallel Surfaces [...] being straight or curved, all of which they took to have a height which was an infinitely small magnitude, and which were the elements of a proposed figure. Hence, conceiving the height of any figure to be divided into an infinity of equal small parts, they were regarded as so many differences which were expressed by the characteristic d, being the height of the infinitely small solids which made up the figure.*"[2] Carré follows this with rules for evaluating the integrals of various differentials.

In the second section are considered dimensions of solids and, at proposition II, the volume of the pyramid. He sets out the calculations for a regular pyramid; his calculations remain valid for a pyramid with a regular polygon as base but whose vertex does not lie on the axis of the base, but Carré does not mention this.

The base of the pyramid has perimeter c and area $\frac{bc}{2}$ (b being the distance from the centre B to one of the sides of the base) (figure 23). A plane, parallel to the base, passes through a point P on the height AB where AP = x and AB = a. The section is a polygon similar to the base, the ratio between them being equal to the square of the ratio between the lengths x and a; the area of the section will be $\frac{bcx^2}{2a^2}$. The differential of the pyramid is thus $\frac{bcx^2dx}{2a^2}$ whose integral is $\frac{bcx^3}{6a^2}$, which is the value for the pyramid of height AP, and so gives a value of $\frac{abc}{6}$ for the whole pyramid.

Problem: TO EVALUATE a pyramid.
Let there be a pyramid of height AB = a, its part AP = x, & Pp = dx, and let there be a radius BD = r from B & a radius PM = y from P; it is evident that if the pyramid is cut by a plane parallel to the base and passing through P, the section will be a polygon similar to the base: And letting c stand for the sum of the sides of the base, & b stand for the height of the triangles, the base polygon $= \frac{bc}{2}$; *and to find the area of the section the reasoning is as follows. Similar polygons are to each other as double ratio of their radii, so the base is to the section as rr is to yy: but rr . yy :: aa . xx; hence a a. xx::* $\frac{bc\ xx}{2\ a\ a}$ = *to the polygon with radius P.M, and multiplying by dx we get* $\frac{bc\ xx\ dx}{2\ a\ a}$ = *to the small*

1.   Carré, *Méthode pour la mesure des surfaces. . .*, Preface.
2.   *Ibid.*, Section première, p. 1-2.

fragment which has height Pp, & is the differential of the pyramid, whose integral = \F(bc

x^3;6 a a) is the value for the pyramid of height AP, which gives $\dfrac{abc}{6}$ for that of the whole

pyramid, because at B, x becomes = a. [...]

If the base of the pyramid is not a regular polygon, the reasoning is almost the same; for similar polygons are to each other in double ratio of their corresponding sides, or rather the differentials or elements of the pyramid decrease in the double ratio of their heights.[1]

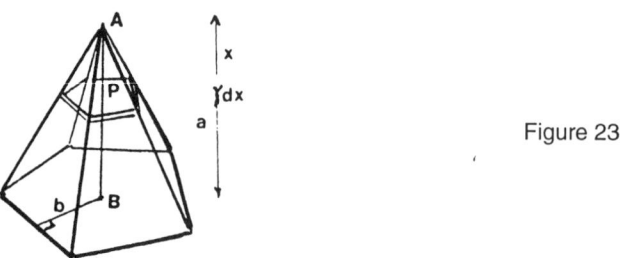

Figure 23

The integral calculus is a much more powerful method for cubature than its forerunners but it has recourse to the use of infinitesimals. We note that Leibniz neither affirmed them nor sought to prove their existence: *"I no more admit infinitely small magnitudes than infinitely large magnitudes – I hold both to be as a shorthand way of speaking, to the benefit of fictions of the mind which serve calculus in much the same way as imaginary roots serve Algebra."*[2] He advised mathematicians not to concern themselves with their actual existence but to press on to new discoveries. That was what most mathematicians of the 18th century did. In the 19th century, by contrast, there developed an opposing movement.

Avoiding the Infinite through Arithmetisation

The 19th century opened to a period of research into foundations and axiomatics. The need for scientists to teach in the numerous schools that had been opened since the French revolution was perhaps one of the reasons that impelled French mathematicians, most influential at this time, to place theories on a rigourous foundation. The *Eléments de Géométrie* by Adrien-Marie Legendre, published in 1794, already bore witness to this effort. Legendre, reacting against 18th century treatises returned to a Euclidean, axiomatic, approach in order, as he wrote in his preface, *"to satisfy the mind by composition using very rigourous elements."* His treatise dominated the teaching of geometry in the whole of the Western world for a century. Twenty-one successive editions appeared between 1794 and 1878 and translations appeared in many languages. To calculate the volume of a pyramid, he proposed a proof that did not rely upon the method of indivisibles, nor on the differential and integral calculus, which, like Euclid's, avoided all use of infinitesimals or recourse to the infinite. He used the same, previously explained, dissection that Euclid used (into prisms and smaller pyramids), but short circuited the classical approach: he did not make use of the decomposition of the pyramid into a prism and three pyramids, nor did he compare the volumes of pyramids with equal bases and equal heights. He showed that the small pyramids

---

1.   *Ibid.*, Section seconde, proposition II, p. 41-42.
2.   Leibniz, Dutens II, 267 in Cléro and Le Rest, *La naissance du calcul infinitésimal . . .*, p. 178.

produced by the dissection (see figure 12 and figure 24) have their volumes in a geometric progression with common ratio 1/4 and which, therefore, can be made as small as is wished. By use of reasoning analogous to reasoning by exhaustion, he showed that the volume of a pyramid can be neither greater than, nor less than, a third of Bh.

Let SABC be any triangular pyramid with base ABC and height SO; I say that the solidity of the pyramid SABC will be a third of the product of the surface ABC and the height SO such that one has

$$SABC = \frac{1}{3} ABC \times SO \text{ or } = SO \times \frac{1}{3} ABC.$$

For if this equality is denied, it must be that the solidity SABC will be the product of SO with a quantity greater or less than $\frac{1}{3} ABC$.

Let 1° this quantity be greater, such that according to the preceding proposition, the pyramid SABC will be subdivided into two equivalent prisms AGHFDE, EGICFH and two equal pyramids SDEF, EGBI. Now, the solidity of the prism AGHFDE is DEF x PO, and that of two prisms will therefore be DFE x 2 PO or DFE x SO. Subtracting the two prisms from the whole pyramid, what remains will be equal to twice the pyramid SDEF, whence

$$2 \, SDEF = SO \times (\frac{1}{3} ABC + M - DEF).$$

But since SA is twice SD, the surface ABC is quadruple DFE and therefore

$$\frac{1}{3} ABC - DFE = \frac{4}{3} DFE - DFE = \frac{1}{3} DFE; \text{ and so}$$

$$2 \, SDEF = SO \times (\frac{1}{3} DEF + M).$$

And so, taking half of both sides, $SDEF = SP \times (\frac{1}{3} DEF + M)$.

From this it can be seen that, to find the solidity of the pyramid SDEF, add to a third of the base the same surface M which had been added to the base of the large pyramid and multiply by the height of the small pyramid SP.

If SD is bisected at K and through K a plane KLM parallel to DEF is constructed, meeting the perpendicular SP at Q, the same argument will show that the solidity of the pyramid SKLM is equal to $SQ \times (\frac{1}{3} KLM + M)$.

Continuing in the same way to produce a sequence of pyramids whose sides are decreasing in double ratio, and whose bases are decreasing in quadruple ratio, one arrives finally at a pyramid Sabc whose base abc will be less than 6M: let Sp be the height of this last pyramid, then its solidity, deduced from that of the preceding pyramids, will be; then since, and so it follows that the solidity of the pyramid Sabc will be $Sp \times (\frac{1}{3} abc + M)$, hence, because $M > \frac{1}{6} abc$, and therefore $\frac{1}{3} abc + M > \frac{1}{2} abc$ it would follow that the solidity of the pyramid Sabc would be $> Sp \times \frac{1}{3} abc$. Which is absurd since it was proved in corollary II of the preceding proposition that the solidity of a triangular pyramid is always less than half the product of its base and height; hence 1° it is impossible for the solidity of the pyramid to be greater than. "SO $\times \frac{1}{3} ABC$.[1]

---

1.   Legendre, *Eléments de Géométrie*, Livre VI, proposition 17, first editions, p. 205-207.

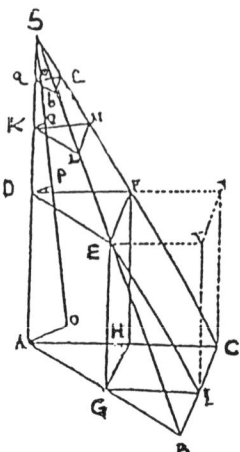

Figure 24

Exercise 9   Prove, with the aid of a dissection of the pyramid SABC by the centres of the edges, corollary II of proposition XVI of Book VI of Legendre's *Eléments de Géométrie,* namely: *"The solidity of a pyramid is less than half of the product of its base and height."*

Legendre completes his proof in showing that the hypothesis *"the solidity of SABC is less than $\frac{1}{3}$ ABC x SO "* is in contradiction with the first corollary of proposition XVI, which states that *"the solidity of a pyramid is greater than a quarter of the product of its base and height.".* He is therefore able to conclude that *"the solidity of a triangular pyramid is equal to a third of the product of its base and height."*[1]

Exercise 10   Prove, by an argument analogous to the preceding text, that the following is impossible: SABC = SO x ($\frac{1}{3}$ ABC – M) where M stands for a positive quantity.

Through out the course of numerous editions, Legendre never ceased to revise his text and in 1823, for the 12th edition, he proposed a new proof which appeared to him to be the simplest and clearest he could produce. It had been suggested to him by Querret, a headmaster at St. Malo, as he explains in his preface. But the chief idea of the proof had already been present in Lacroix's *Eléments de Géométrie,* published in 1811 and also in the proof given in Father Tacquet's *Eléments de Géométrie.*

Legendre pushed the need for rigour further than Tacquet and Lacroix, not allowing any "taking it to the limit." He makes explicit all the details and particularly all the inequalities necessitated by his argument. Like Tacquet and Lacroix he built up prisms of thickness k, which could be appropriately chosen, in the interior of two pyramids of the same base and height that he wished to compare. But instead of considering that the accumulation of prisms could almost become merged with the pyramid, he used a *reductio ad absurdum* argument to prove that the difference between the two pyramids being compared could not be other than zero. Legendre chose the height k so that the volume of the first exterior prism (which we shall denote by v_p) would be less than the supposed difference V – v between the two pyramids. This difference V – v is less than

---

1.   *Ibid.*

the difference between the exterior prisms of the first pyramid and the interior prisms of the second pyramid (which we shall denote by $V_e - v_i$). Now, $V_e - v_i$ is exactly equal to the volume v_p of the first exterior prism. Legendre thereby obtains the two contradictory inequalities $V - v < V_e - v_i$ and $V - v > V_e - v_i$.

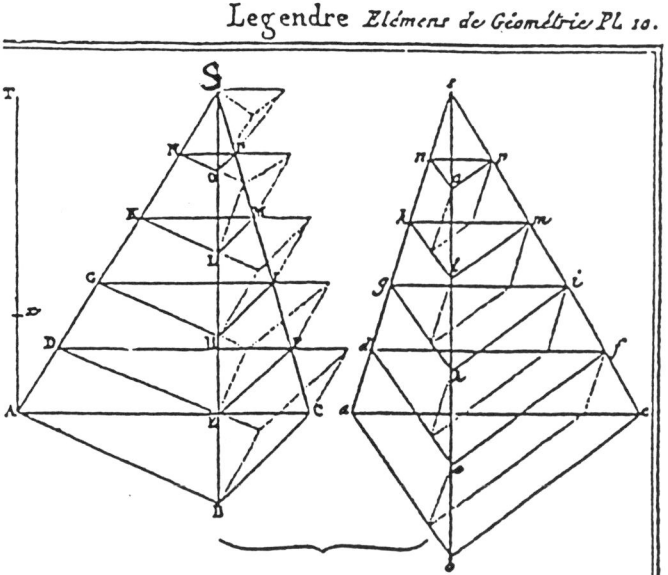

Legendre *Élémens de Géométrie Pl. 10.*

Figure 25

These calculations through building up prisms in "staircases" anticipates approximating areas under curves by rectangles, "Riemann summation", which would not be defined until several years later. They announced the idea of the formalisation of the concept of limit which Bolzano, and then Weierstrass, would set up using their procedures for the arithmetisation of the infinite. Legendre's proof inspired later proofs in many different geometrical treatises up to the middle of the 20th century. Witness Hadamard's proof, exemplary for its conciseness and simplicity, that appeared in his *Leçons de Géométrie* published in 1898 and reissued right up to 1949. The staircase, certainly diabolical for some, was present in all the manuals. Hadamard, moreover, uses the language of limits.

> *If [...] we have taken the number of divisions to be equal to n, we should construct, in each pyramid, n − 1 interior prisms [and] n exterior prisms, and the preceding argument would show us that the difference between V and V' (the volumes of the two pyramids) is less than surface ABC x H / n, that is to say less than the nth part of the prism of base ABC and height H. But this quantity can be made as small as is wished by taking n sufficiently large. The conclusion to which we come cannot be if V and V' are not equal.*
>
> *Corollary: The preceding argument shows that **the volume of the pyramid SABC is the common limit of the volumes S_n, s_n as n increases indefinitely**, since that volume V differs from each of the quantities S_n, s_n by less than they differ from each other, and the difference $S_n - s_n$ tends to zero.* [1]

---

1.   Hadamard, *Leçons de Géométrie*, Chapter III, § 406, p. 106.

What is the ultimate secret of a pyramid?

Theories based on the method of exhaustion, indivisibles, fluxions, differentials, infinitesimals – these theories blossomed, were in opposition and sometimes came back into favour. The method of infinitesimals is incontestably the most rapid and direct, but certain mathematicians, towards the end of the 18th century and in the 19th century, were dissatisfied with the lack of rigourous foundations. They searched for what they called "the true metaphysic" of the calculus. Lazare Carnot, in his *Réflexions sur la métaphysique du calcul infinitésimal* of 1797, considered all methods to be equivalent and that a principle of compensation of errors validated the correctness of the results of the calculus. He picked up on the ideas of Euler and Lagrange. On the other hand, d'Alembert had been the first to affirm that the notion of limit was *"the true metaphysic of the calculus"*. He did not, however, put the idea into practice when he calculated the volume of the pyramid in his *Encyclopédie*. His idea was defended by Laplace and put into practice this time in his *Cours de l'Ecole Normale de l'An III*. But the idea of limit, the germ of which was in the proof by Tacquet, then in the language of Newton, and made explicit by d'Alembert, would need to be made precise by Cauchy in his *Cours d'Analyse de l'Ecole polytechnique,* and then arithmetised by Weierstrass. It was through the language of limits that Cauchy defined the idea of a continuous function, then the derivative, then the integral of a continuous function on an interval, and this language can explain in a unifying way all the methods for evaluating the volume of a pyramid that had been used since Euclid. Each of two pyramids being compared has for its volume the limit of the sequence of volumes of prisms defined, either by Euclid's iterative dissections, or by the prisms defined by Tacquet or Lagrange. The integral of the function, the pyramid's volume, is associated with its height x and is the limit of the Riemann sum of that function. The areas of the indivisible sections of the two pyramids being compared are the derivatives of the "volume function"; the "volume functions" of the two pyramids, having the same derivative and the same initial value (zero, for $x = 0$) are therefore equal. Roberval's method can also be translated into the language of limits; the sequence defined can be considered as the sum of the volumes of n small prisms of equal thickness inscribed within the pyramid; its limit is the volume of the pyramid.

This notion of limit, that proceeds to arithmetisation from the idea of the infinite and infinitesimals, is this perhaps the ultimate secret of the pyramid?

Bibliography

Source texts

ARCHIMEDE, *Traité des Spirales*, Les œuvres complètes d'Archimède, trad. P.ver Eecke.Vaillant Carmanne, Liège, 1960.

ARISTOTE, *Physique* , traduction de H. Carteron, Les Belles Lettres, Paris, 1969.

ARISTOTLE, "Physica" in W. D. Ross (ed.), *The Works of Aristotle,* Oxford, 1930.

BRICARD, "Sur une question de géométrie relative aux polyèdres", *Nouvelles Annales de Mathématiques,* **15**, 1896, p. 331-334.

CARNOT, *Réflexions sur la métaphysique du calcul infinitésimal,* Paris, 1797. repr. 1813. repr. Blanchard, Paris, 1970.

CARRE, *Méthode pour la mesure des surfaces, la dimension des solides, leurs centres de pesanteur, percussion et d'oscillation par l'application du calcul intégral,* Paris, J. Boudet, 1700. repr., Durand, Paris, 1750.

CAVALIERI, *Geometria degli indivisibili,* a cura di Lucio Lombardo-Radice, Turin, UTET, 1966.

CAVALIERI, *Exercitationes geometriae sex.* Bononiae, 1647. repr. 1980.

CHILD, *The early mathematical manuscripts of Leibniz,* Open Court, Chicago, 1920.

DEHN, "Ueber raumgleiche Polyeder, Nachrichten der Akademie der Wissenschaft in Göttingen", *Mathematisch physikalische Klasse,* 1900, p. 345-354.

DEHN, *Ueber den Rauminhalt,* Mathematische Annalen, **55**, 1902, p. 465-478.

DIJKSTERHUIS, *Archimedes,* Copenhagen, 1956.

EUCLIDE, *Les Eléments*, trad. Peyrard, repr. Blanchard, Paris, 1966.

GERWIEN, "Zershneidung jeder beliebigen Anzahl von gleichen geradlinigen Figuren in dieselben Stücke", *Journal für die reine und angewandte Mathematk,* 1833, p. 228-234.

HADWIGER & GLUR, "Zerschneidungsgleichheit ebener Polygone", *Element. Math,* 6, 1951, p. 97-106.

HADAMARD, *Leçons de Géométrie,* Paris, 1898. Repr. 1949.

HEATH, *The Thirteen Books of Euclid's Elements,* Cambridge, 1925. Repr. Drols, Dover, New York, 1956.

HILBERT, "Sur les problèmes futurs des mathématiques", trad. Laugel, in DUPORCQ E. *Compte Rendu du Deuxième congrès international des Mathématiciens tenu à Paris du 6 au 12 Août 1900,* Gauthier-Villars, Paris, 1902, trad. angl. voir p. 52.

HILL, *Determination of the volumes of certain species of tetraedra without the employment of the method limits,* Proceedings of the London Math. society. **27**, 1896, p. 39-53.

LAMY, *Elémens de Géométrie ,* Paris, 1685.

LEGENDRE, *Eléments de Géométrie ,* Didot, Paris, 1794, 12ᵉ édition, Paris, 1823.

L'HOSPITAL, *Analyse des infiniments petits pour l'intelligence des lignes courbes,* Paris 1696. Repr. ACL, Paris, 1988.

MAC LAURIN, *A treatise of Fluxions,* Edimbourg, 1742, trad. de Pezenas, Paris, 1749.

NEWTON, *Opuscula,* Jombert, Paris, 1744.

ROBERVAL, *Traité des Indivisibles* in *Divers ouvrages de Mathématiques et de physique* par l'Académie royale des sciences, Paris, 1693. Repr. IREM Paris VII, 1987.

SMITH, *A Source Book in Mathematics,* McGraw-Hill, New York: 1929. Repr. Dover, 1959.

STRUIK, *A Source Book in Mathematics, 1200-1800,* Cambridge, Massachusetts: Harvard University Press, 1969.

TACQUET, *Elementa Geometriae planae ac solidae,* Anvers, 1654. Repr. 1754

TACQUET, *Cylindrica et Annularia in Opera mathematica,* Anvers, 1668.

SYDLER, "Conditions nécessaires et suffisantes pour l'équivalence des polyèdres de l'espace euclidien à trois dimensions", *Commentarii. mathematicii. helvetica,* **40**, 1965 p. 43-80.

WAGNER, "An early derivation of the volume of a pyramid: Lui Hui, third century A.D.", *Historia Mathematica,* **6**, 1979, p. 164- 188.

General works for further reading

BOLTIANSKII, *Hilbert's third problem*, trad. par R.A. Silverman, Winston and Sons, Washington, 1978.

BOYER, *The History of the Calculus, and its conceptual development* 1949. Repr. Dover, New York, 1959.

CLERO, LE REST, "La naissance du calcul infinitésimal au XVIIe siècle", *Cahiers d'histoire et de philosophie des sciences,* n° 16, Centre de documentation Sciences humaines, 1980.

DHOMBRES, *Nombre, mesure et continu*, Cedic Nathan, Paris, 1978.

FESTA, "La querelle de l'atomisme, Galilée, Cavaliéri et les Jésuites", *La Recherche*, **224**, 1990, p. 1038-1047.

MARTZLOFF, *Histoire des Mathématiques chinoises*, Masson, Paris, 1988. English tr. *A History of Chinese Mathematics*, Springer, Berlin, 1995.

How did you get on?

Ex. 1 AHI + IBJ = BEI + IBJ = BEF = JH'D
AHK = KFC = KLC + LFC = KLC +LH'D
Therefore: ABC = BJKC + KLC +LH'D +JH'D = DBC

Ex. 2 1)The two shaded prisms are isometric and therefore equivalent. If one or other of the prisms is added to the solid ACBLNHMO then we obtain one or other of the parallelepipeds ACBLMFDH or ACBLPOQR.
2) All that is required is to use Prop.29 of Book XI twice: The white and shaded parallelepipeds are "on the same straights lines" (PR) and (QO), the white and hatched parallelepipeds are "on the same straight lines" (PO) and (RQ).

Ex. 3 Figure 26 shows the central cube of side a, figure 27 shows this same cube together with four prisms and fig. 28 uses up the remaining 21 pieces. Each "corner" cube, like ABCDA'B'C'D' consists of three pyramids AA'B'C'D', ACDD'C', ABCC'B'. The hatched cubes, like AEFBA'E'F'B', are each composed of two prisms.

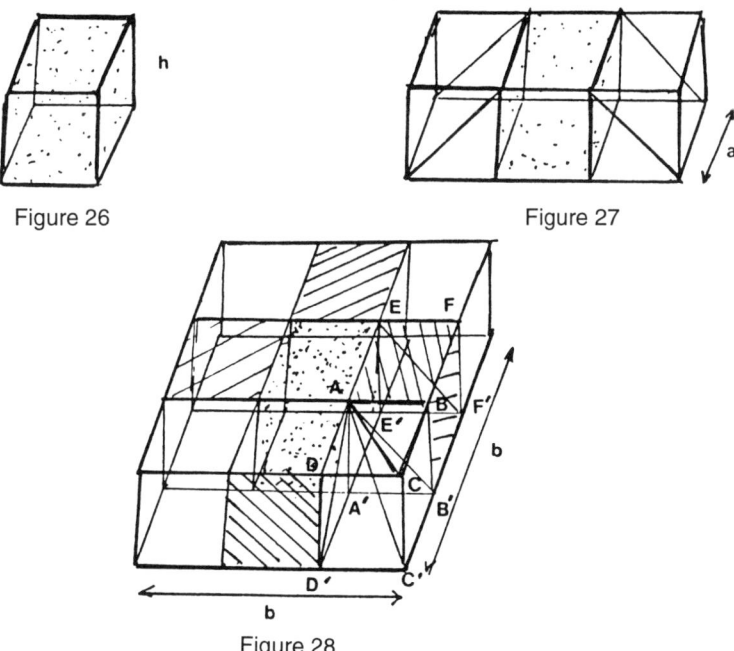

Figure 26   Figure 27

Figure 28

Ex. 4 See figure 29. I and I'are the centres of the inscribed circles of the two triangles.

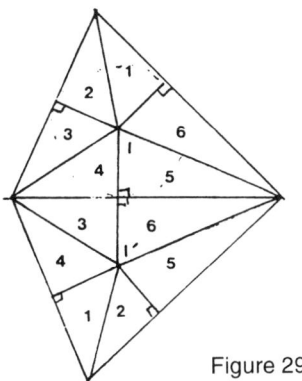

Figure 29

Ex. 6   Use the identity $(p + 1)^3 = p^3 + 3p^2 + 3p + 1$ written out for $p = 0$, $p = 1$, $p = 2$, ..., $p = n$ and adding we get, after cancellation,
$(n + 1)^3 = 3(1^2 + 2^2 + 3^2 + ... + (n - 1)^2 + n^2) + 3(1 + 2 + 3 + ... + n) + n + 1$
The result follows by using $1 + 2 + 3 + ... + n = n(n + 1)/2$.

Ex. 7   Figure 30 shows the way of calculating the area under a parabola according to Roberval's method. Consider rectangles of width H/n with sucessive heights $(H/n)^2$, $(2H/n)^2$, $(3H/n)^2$,etc. The sum of the areas of these rectangles will be $(H^3/n^3)(1^2 + 2^2 + 3^2 + ... + n^2)$. The infinitude of small rectangles is thus equivalent to the infinitude of small solids which form the pyramid of side H and height H considered by Roberval. The integral calculus also shows that the area under the parabola and the volume of the pyramid can be found by use of the same type of integral, namely in taking a primitive of x^2.

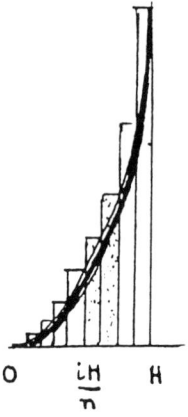

Figure 30

Ex. 8   To make it simpler, suppose that (AP) is perpendicular to the plane (PMm). Let AP = c, PM = a, Pm = b, AD = x, DG = h. From similar triangles APM, ADE and APm, ADe it follows that ED = ax/c and De = bx/c. Let V(x) be the volume of ADEe so that $V(x) = abx^3/6c^2$. Taylor's formula gives:
$V(x + h) - V(x) = (abx^2/2c^2)h + (abx/c^2)(h^2/2) + (ab/c^2)(h^3/6)$.
It can be verified that the prism DEeGIi with base EDe and height DG has a volume $abhx^2/2c^2$. The first fluxion is then the first term in the Taylor expansion of V(x). In the same way, the prism EHIeNI with height De = bx/c and base EHI similar to APM and of area $ah^2/2c$, has a volume $abxh^2/2c^2$, and so the second term in the expansion is half the second fluxion. The pyramid eNih with height Ni = ha/c and with base eih, similar to APm of area $bh^2/2c$, has a volume $abh^3/6c^2$. This is the last term in the Taylor expansion and, being independent of x, its fluxion is zero.

Ex. 9   Each of the prisms of the dissection contains a pyramid isometric to one of the small pyramids and so is greater than a small pyramid. The large pyramid, of base B and height h, is the sum of two prisms and two small pyramids and is less than four prisms. The prism AGHFDE has a base B/4 and height h/2; its volume is (B/4)(h/2). The four prisms have a total volume Bh/2.

Ex. 10   If we let SABC = SO x (ABC/3 – M), we can prove for any pyramid
SDEF = SP x (DEF/3 – M)
and, since there exists a pyramid Sabc for which abc < 12M, we shall have
Sabc = Sp x (abc/3 – M) < Sp x abc/4.
This result contradicts corollary I of Legendre deduced from a dissection of the pyramid (see Ex. 9): the large pyramid is larger than two prisms so its volume, and the volume of any triangular pyramid, is greater than a quarter of the product of its height and its base.

4

Why Ruler and Compass?

Joëlle Delattre and Rudolf Bkouche
IREM Lille

> *Then if [geometry] compels us to behold being, it is suitable; if to behold becoming, it is unsuitable. That is our view.*
>
> Plato, *Republic*, VII 526 e.

Movement was a problem for the Ancients. The philosophers and scientists of antiquity questioned the possibility of its existence, and by a great deal of argument they looked for a definition of the principles which would account for its diverse forms. It is the object, for example, of a true "physical philosophy" like that of Aristotle. The paradoxes of Zeno of Elea continue today to claim the attention of philosophers, logicians and mathematicians. The most puzzling, in fact, is how the fast runner (Achilles) is prevented from catching up with the slow runner (the tortoise) which started before him, because it is always necessary first to reach the point that the other has passed.

Among the significant questions which haunt the history of ancient geometry, the most compelling is perhaps this: *certain curves are impossible to construct without a combination of movements* which are difficult to conceive, or to effect mechanically.

In recounting this history, we shall meet the *Spiral of Archimedes* (3rd century B.C.), the *Quadratrix of Dinostratus* (4th century B.C.) and the *intersection of three conics by Archytas* (*c*.400 B.C.) to find two mean proportionals to two given lengths. The study of these examples will show how mechanical constructions progressed in simplicity and efficiency alongside geometry, the two sciences improving by each drawing upon the other. In fact, only an historical interpretation of metaphysical or religious influence will permit us to understand the evolution of mathematical progress, from a stage which is "mechanical", which some might describe as a sort of DIY approach, to a stage which is geometrical and rational. The progress made in each area took place in parallel and each helped to expand and develop the other.

In the 3rd century B.C., Archimedes wrote to Eratosthenes to tell him about a new method allowing him to *"understand certain mathematical realities using a mechanical device"* In fact, according to him, it is the properties where

> *certain things first became clear to me by a mechanical method, although they had to be demonstrated by geometry afterwards[1].*

For example, before a complex problem can be solved, it is a question of using the mechanical principle of the lever or some other mechanical analogy to discover the principles of an exact solution. It then remains, after the logical demonstration, to secure the geometrical foundation of the hypothesis.

Well aware of the value of his heuristic method (for discovery and invention), the great scientist was still not able to abandon the Platonic principle of a "semi-initiatory" geometry, a geometry of the necessary and the absolute which is sometimes called *"the geometry of ruler and compass"*[2] or *"straight line and circle"*. (In fact, such a change would require a kind of conversion, almost of a religious kind.) Given this, we come to see how the *geometers excluded movement from their demonstrations, and preferred to delegate that study to students of mechanics,* refusing to use "geometric" for any construction or demonstration not abiding by the rules of ruler and compass.

But why are the ruler and compass considered to be the only geometrical instruments? Why are the ingenious tools like the solution of Archytas or Plato's machine or even the procedures of Nichomacus and Eratosthenes not recognised as true geometrical constructions by the geometers, and only seen as "empirical tinkering" or "irrational processes"? What then is at stake in this restrictive choice? Is it only mathematics?

Mechanical "Do It Yourself" processes and disputed geometrical constructions

Spirals and the method of intercalation

It is very difficult to understand why Archimedes is so interested in spirals. Is it for purely geometrical reasons because he studied this curve as a means of calculating the ratio [π] of the circumference of a circle to its diameter, and *"squaring"* the circle? Is it because of his astronomical interests, trying to calculate geometrically the spiral movements of the wandering stars [the planets]? Or is it finally through the interest of a mechanical mind in a curve which results from the combination of two regular and uniform movements, one in a straight line, the other in a circle? These three reasons are evident at one and the same time, but the texts are missing.

Archimedes defines the spiral in the following manner:

> *If a straight line drawn in a plane revolves uniformly any number of times about a fixed extremity until it returns to its original position, **and if, at the same time as the line revolves, a point moves uniformly along the straight line** beginning at the fixed extremity, the point will describe a spiral in the plane.[3]*

---

1.   Heath, *The Method of Archimedes*, p. 13.
2.   Rey, *L'Apogée de la science technique grecque*, p. 248.
3.   Archimedes, *On Spirals*, Heiberg, II, 44, 17-46, 21, and see Thomas, *Greek Mathematical Works*, II, p. 182-183.

And this is how Archimedes sets about demonstrating a fundamental property of this curve: everything works *as if the comparison of distances described by two free points, driven by a uniform movement, allows us to make the geometrical abstraction,* simply as a consequence of the uniformity (or of the regularity) *of the movement itself.*

In fact, Archimedes has already demonstrated, at the beginning of chapter II, that *"two uniform displacements in equal times result in a proportionality of the distances travelled"*; and in particular he established the second property as follows:

> *If each of two points on different lines respectively move along them each at a uniform rate, and if lengths be taken, one on each line, forming pairs, such that each pair are described in equal times, the lengths will be proportionals[1].*

$$- - - A - - - - - - - B - - - - - - - - - - - C - - - - D - - - - - - - -$$
$$- - E - - - - - - - - - - - F - - - - - - - - G - - - - - H - - - - - - -$$

AB/EF = CD/GH. *Thus the equal ratios obtained become independent of the movement of the points.*

So this is why it is then possible to state: the *property of the spiral* (Proposition 14):

> *If, from the origin of the spiral, two straight lines be drawn to meet the first turn of the spiral and produced to meet the circumference of the first circle, the lines drawn to the spiral will have the same ratio one to the other as the arcs of the circle between the extremity of the spiral and the extremities of the straight line produced to meet the circumference, the arcs being measured in a forward direction from the extremity of the spiral.*

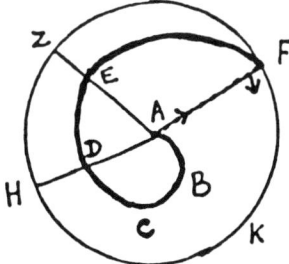

Figure 1

> *Let ABCDEF be the first turn of the spiral, let the point A be the origin of the spiral, let FA be the initial line, let FKH be the first circle, and from the point A let AE, AD be drawn to meet the spiral and be produced to meet the circumference of the circle at Z, H. It is required to prove that AE : AD = arc FKZ : arc FKH.*
>
> *When the line AF revolves it is clear that the point F moves uniformly round the circumference FKH of the circle while the point A, which moves along the straight line, traverses the line AF; the point F which moves round the circumference of the circle traverses the arc FKZ while A traverses the straight line AD in the same time as F traverses the arc FKH, each moving uniformly; it is clear therefore, that AE:AD = arc FKZ:arc FKH[2].*

Exercise 1   Using these two properties prove that this is also true for the second turn of the spiral and generalise for the nth turn of the spiral.

In order to study some other properties of the curve, in particular the problems connected with the construction of its tangents, Archimedes resorted to the method

---

1.   *Ibid.,* 50, 9-52, 15; see Thomas, II, p. 184-185.
2.   *Ibid.,* p. 185.

which is attributed to him by the Arabs, but which in fact, corresponds to a practice current in the 5th century BC. that is, *neusis* or *intercalation*. This method was criticised by Pappus and according to him, raises the issue of a *"geometrical impotence"* for it consists in a *"mechanical construction"*. In effect it becomes quite simple to set in place a segment of given length in an "ad hoc" way, and *by simple empirical adjustment, we can guarantee that "it works"*! It is correct to think that the Sophists in Athens, and in particular Hippias of Elis, must have employed this "magical" artifice to amaze a naive public ignorant of mathematics, and so contributed to making this process famous.[1]

The problem of the trisection of the angle and the need to use conics

Dividing an angle into two parts is easy with ruler and compass, and the geometrical construction of the bisector is considered quite simple. But how do we divide an angle into three equal parts?

This problem interested the geometers of the 5th century B.C., the time of Hippocrates of Chios, as well as the duplication of the cube.

The following solution seems to be the oldest[2], which allows us to have a better understanding of exactly what the *method of neusis* consists of, which relies on an approximate drawing, then successive displacements of a line, until an exact coincidence of it with the position sought.

> Let an angle A be divided in three parts. Through a point B, placed on one of its sides, we draw a line parallel to the other side. From this point B, we drop a perpendicular to meet the other side in C.

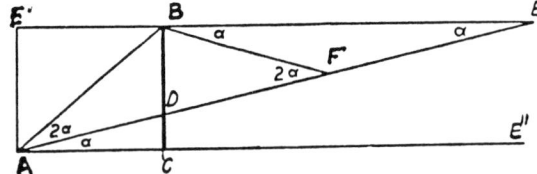

Figure 2

> **Intercalate** *a line AE such that it intersects with BC in D, and with the parallel through B in E, such that DE is twice the length of AB. Bisect DE in F and join BF.*
>
> *Thus we have: angle BAF = angle AFB. The triangle ABF is isosceles, hence angle BAF = angle AFB; similarly, the triangle BFE is isosceles, and consequently the angle AFB = 2BEF, and since angle BEF = angle CAD, we have the angle CAD = 1/3 of angle CAB[3].*

It was proved in the 19th century that the *line DE cannot be constructed by ruler and compass*. We have believed for some time, that Nichomacus was able to produce this construction with his "modern" machine in the 2nd century B.C. We will go back over the geometrical and mechanical interest in this process. To construct DE it is necessary to show that the intercalation is possible, that is to say that the segment DE which is double AB actually passes through A. In the 4th century AD, Pappus[4] constructed the rectangle AE'BC, with diagonal AB, and then the parallelogram ABEE" with diagonal AE without reaching a very convincing geometric generalisation.

---

1.   See a comment on Hippias in *Les Présocratiques*, p. 1554. See also *Les écoles présocratiques*, p. 949.
2.   See Rey, *L'Apogée de la science technique grecque*, p. 111.
3.   Pappus, *Mathematical Collection, IV 43.*
4.   *Ibid.*

On the other hand, let us now consider *"Archimedes' demonstration"*, which appears in the Arab tradition, in the *Book of Lemmas*, attributed to Archimedes.

> *let ABC be the angle at the centre of a circle. The sides AB and CB are radii of the circle. Produce AB to obtain a diameter AD, and intercalate this line in E with the line CF drawn from C, cutting the circle in F such that FE = FB[1].*

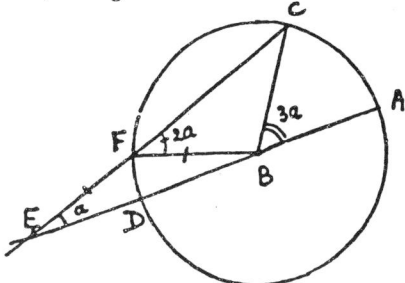

Figure 3

Exercise 2   Show that: "angle DEF = 1/2 BFC = 1/2 FBC = 1/3 ABC."

At the beginning of the 2nd century B.C., Apollonius of Perga proposed another solution, this time using conics. In fact, this problem is one of those which *Pappus proposed to call "solid problems"* because in its construction, it is necessary to use curves which can only be defined on a solid, namely conic sections. Here is the solution using the sort of *"loci of solids"* which Pappus uses.

> *On the line AC construct a triangle ABC such that the angle ACB is double the angle CAB. We will show that the point B is on a hyperbola.*
>
> *Draw BD perpendicular to AC and find the point E such that DE is equal to CD; BE will then be equal to AE.*
> *Make EZ equal to DE; then CZ is equal to 3CD.*
> *Now AC is three times CH, and so the remainder, AZ will be the triple of HD[2].*
> *[...]*

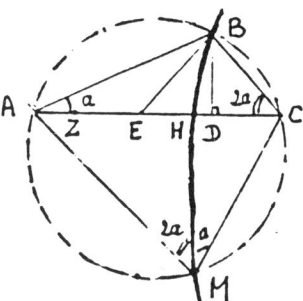

Figure 4

It can be shown that:

$$3AD.DH = BD^2$$

which implies that B lies on the hyperbola with axis AH and second axis $\sqrt{3}AH$. The hyperbola described in this manner cuts the large circle ABC and M "trisects" the large arc AC.

Exercise 3   Let M be any point on the circle ABC, as in the figure, such that the angle AMC = 3BMC. Find a construction for the trisection of the angle using the intersection of a hyperbola and a circle.

---

1.   Pappus, *Mathematical Collection*, IV 43-44 in Thomas I, p. 356.
2.   *Ibid.,* IV, 44. See Thomas I p. 361.

We see, by this demonstration how the mechanical process of intercalation has been entirely abandoned in favour of a more geometric approach where the claim is that it gets away from empiricism and the use of a particular construction in order *"to achieve the necessity and the generality"*[1] However, it was proved in the nineteenth century that the construction problem of the *intersection of a hyperbola and a circle was not possible using ruler and compass* [see chapter 13].

The curve of the quadratrix: geometrical analysis or mechanical construction?

If we believe Proclus, this ancient problem of the trisection of the angle gave rise to the invention of the curve called the quadratrix, as he remarks in his *Commentary on the First Book of the Elements of Euclid* (5th century A.D.). Some think that Hippias of Elis is the original inventor, others believe it was Dinostratus, the brother of Menaechmus, who gave the curve its name.[2]

This is how Pappus states it in his *Mathematical Collection*[3]:

> *Let ABCD be a square, and with centre A let the arc BED be described, and let AB be so moved that the point A remains fixed while B is carried along the arc BED; furthermore let BC, while always remaining parallel to AD, follow the point B in its motion along BA, and in equal times let AB, moving uniformly, pass through the angle BAD (that is, the point B pass along the arc BED), and BC pass by the straight line BA (that is, let the point B traverse the length of BA).*
>
> *Plainly then both AB and BC will coincide simultaneously with the straight line AD. While the motion is in progress the straight lines BC, BA will cut one another in their movement at a certain point which continually changes place with them, and by this point there is described in the space between the straight lines BA, AD and the arc BED a concave curve, such as BZH, which appears to be serviceable for the discovery of a square equal to the given circle.*

[PROPERTY OF THE QUADRATRIX]

> *If any straight line, such as AZE, be drawn to the circumference, the ratio of the whole arc to ED will be the same as the ratio of the straight line BA to ZF; for this is clear from the manner in which the line was generated.*[4]

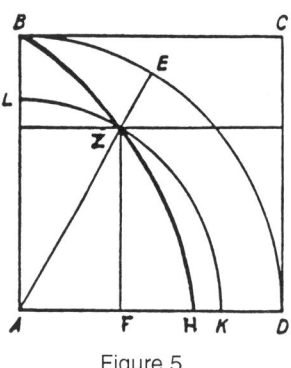

Figure 5

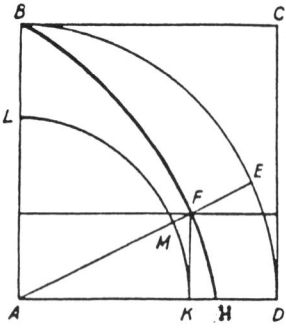

Figure 5 bis

1.   Rey, *L'apogée de la science technique grecque*, p. 131.
2.   Lasserre, *De Léodamas de Thasos à Philippe d'Oponte*, p. 565.
3.   Pappus, *Mathematical Collection*, IV, 30. See Thomas I p. 337-341.
4.   *Mathematical Collection*, IV, 33.

According to Becker, Dinostratus bases his proof on "the **method of exhaustion**, used by his teacher Eudoxus":

> *If the ratio of the quadrant of the circle to the side AB is the same as to the side AH, then there must be a ratio AB/AK such that AK is either greater than AH or less than AH.* [1]

1) If it is greater, then the arc traced by the segment AK, centre A, cuts the quadratix in Z, and AZ cuts BED in E, and,

> *by comparing the chords and the arcs thus drawn, we obtain:*
> *arc BED/AB = arc LZK/AK and AB/AK = arc BED/arc LZK*
> *therefore AB = arc LZK*
> *But from the properties of the quadratrix,*
> *AB/ZF = arc BED/arc ED = arc LZK/arc ZK*
> *in consequence of which ZF will be equal to ZK, which is absurd.*

2) Now suppose AK is less than AH,

> *we draw the quadrant of the circle LK with the segment AK, and a tangent to it at K which cuts the quadratrix in F, and we draw AF which meets the arcs LK and BD in M and E respectively. And so from the properties of the quadratrix,*
> *AB/FK = arc BED/arc DE = arc LMK/arc MK*
> *now since AB = arc LMK as we have demonstrated above, we have*
> *FK = arc MK, which is impossible.*
> *And so it can only be that AK = AH; and so, the length of the quadrant becomes AB^2/AH, provided that $\sin\phi < \phi < \tan\phi$, when $\phi < 90°$!*

Exercise 4   Find the polar equation of this curve if AZ = ρ, the angle ZAD = φ and AB = a.

Suppose that AB = 1, for H, the point of intersection of the curve with AD, calculate AH and deduce that the quadratrix allows us to obtain the quadrature of the circle. (We will meet this later in the construction of certain points of this curve with ruler and compass).

Now, Pappus tells us:

> *Sporus* [end of the 3rd century] *did not have a liking for this curve. In fact, from the outset, what it seems to set out to do, is in fact contained in its statement.*
> *How is it possible to have two points, starting to move simultaneously from B, one towards A on the straight line, and the other on a curved line towards D, and arriving at D in equal time, before we know the relationship between the line AB and the arc BED? For it is also necessary for the velocities of the points which move, to be in the same ratio* [2].

Exercise 5   Calculate the ratio of these velocities.

> *Let us think about the descending construction of the line. When the moving CB and BA reach AD, they will not cut each other any more* [3],

because the cut which has to be at the end of the line when it meets AD, stops before it gets to AD.

---

1.   Becker, *Das mathematische Denken der Antique*, p. 96-97.
2.   Pappus, *Mathematical collection*, IV, 31. See Thomas I, p. 337-341.
3.   *Ibid.*

The point H cannot in fact be found unless one knows beforehand the ratio of the circle to the straight line, but on the other hand, the calculation of π is not possible without a mechanical process that allows us to trace the curve in a continuous movement and to determine the point H. (Lebesgue explained this at the beginning of the 20th century[1], and we show later how the trammel process which he devised does this); and since the ratio is not given, it is impossible.

Also, before the critiques of Sporus became relevant, Pappus investigated the *geometrical analysis of the generation of this "too mechanical" line* and proposed that it should be obtained by means of "loci on a surface", that is to say of geometrical loci produced by the surfaces of the cone, cylinder, sphere or other surfaces of the second degree (the subject of two books of Euclid which are lost today, but which seem to have been in Pappus possession).

[THE GEOMETRICAL CONSTRUCTION OF THE QUADRATRIX]

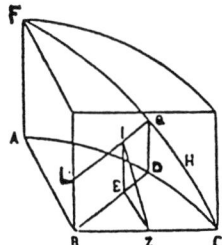

Figure 6

Consider the quadrant ABC. Take any line BD as a transversal, and draw perpendicular to the line BC the line EZ in a given ratio with the arc DC, and I say that E is on the line.
So, let us imagine the surface of a right cylinder [generated] by the arc ADC, and the helix CHQ [a curve in space] described in this surface. Let QD be a side [a generator] of the cylinder; draw the lines EI and BL perpendicular to the plane of the circle, and from the point Q, draw the line QL parallel to the line BD. Since the ratio of the line EI to the arc DC is given by the helix, and the ratio of the line EZ to the arc DC is given, the ratio of the line EZ to the line EI is also given. Moreover, since the lines ZE and EI are in **juxtaposition** *[respectively parallel to AB and QD]; then the position of the line joining ZI is also given [the hypotenuse of the right triangle IEZ]. So, this line is perpendicular to the line BC; since the line ZI is in a secant plane [of the cylinder]; so also is the point I. Since this point is also in a "plectoidal" (twisted) surface [helical, since the line QL moves between the helix QHC and the line LB, whose position is given, being continually parallel to the plane below]; then, the* **point I lies in a line such that the point E is also in that line. In this way, this property is analysed in a general manner**, *and if the ratio of the line EZ to the arc DC is the same as that of the line BA to the arc ADC, one obtains the quadratrix that we referred to above[2].*

To prove that the point E is on the quadratrix, we need to imagine the surface of the cylinder generated by the arc ADC and the curve FHC described in its surface (obviously, we have here the well-known helix of Hero, that is to say that the curve in space CHQ has the point C as origin and the ordinate LB for any point Q proportional to the curved abscissa CD of this point), so now *we can state that EZ/arc DC = AB/arc ADC*. We see clearly the attempt at a *"general analysis"* used by Pappus in order to escape from the rather empirical mechanical manipulation, the DIY approach, used by Hippias and Dinostratus.

1.   Lebesgue, *Leçons sur les constructions géométriques*, p. 12.
2.   Pappus, *Mathematical Collection*, IV 33. See Thomas I, p. 337-341.

The line LQ moves between the helix QHC and the line LB parallel to the plane ABC, and so produces a *"rising helical surface"* which is also called a *"screw surface with a square thread"*; and the quadratrix of Dinostratus is found by the orthogonal projection onto a plane perpendicular to the axis of this surface, of the section made by a plane passing through one of the generating lines.

In his 19th century work *Apercu Historique* Chasles produced a brilliant review of Pappus' geometrical analysis, and achieved a remarkable theoretical synthesis. He gave the name *"conicoids"* to this class of surfaces, defining them as

> *generated by a moving line which depends upon a fixed line and a curve, and which remains always parallel to the same plane[1].*

Here is his way of generalising Pappus' approach[2]: knowing that "*a rising helical surface*" is the surface which is generated in space by a line turning about a fixed line, and always remaining parallel to a horizontal plane (in the figure LQ turns about LB), in his description of the helix he says:

> *Firstly, if we cut the rising helical surface with a plane through one of its generators, the section is projected orthogonally onto a plane perpendicular to the axis of the surface and forms a **quadratrix of Dinostratus**.*
>
> *Secondly, a cone of revolution which has the same axis as a rising helical surface, cuts this surface in a curve of double curvature (in space) which is projected orthogonally onto a plane perpendicular to the axis, and forms **a spiral of Archimedes**.*

Which therefore constructs a spiral from loci on surfaces, analogous to the construction of the quadratrix! The constructions proposed by Chasles have the main advantage of establishing *constant geometric relations* between these curves and those which, in the normal coordinate system, carry the same name: for example, the hyperbolic spiral and the hyperbola; and, in this system, he claims that *"the spiral of Archimedes corresponds to the straight line"*.

So the same *process* which uses certain curves to construct others, gives *a priori* geometrical reasons for their properties and enables us to *establish a relationship* between them which even extends to the same form of equation, and to a simple analogy. The *"geometric reason"* becomes a truly established relationship where *the movement, the actual object of study, has acquired a real geometric status*. But, without anticipating the chapter which expressly considers Chasles' processes [see chapter 6] let us try to understand why the Ancients did not want movement in their geometrical processes.

The explanation for the movement of the stars: geometrical hypotheses and mechanical constructions

Theon of Smyrna expressed the problem of the ancient astronomers very clearly: what is the source of the "anomaly" [irregularity or lack of uniformity] in the movement of the Sun and the planets?

> *The changing aspects of the revolution of the planets is because, being fixed in their own circles or in their own spheres whose movements they follow, they are carried across the zodiac, just as Pythagoras had first understood it, by a regulated, simple and equal*

1.   Chasles, *Aperçu historique sur l'origine et le développement des méthodes en géométrie*, p. 30.
2.   *Ibid.*, note VIII, p. 297.

revolution, but which results by combination in a movement that appears variable and unequal[1].

This untidy and complex movement of the stars is only apparent. The particular purpose of the mathematician here is again *to identify the elementary, that is, the circle and the line, and so to discover a "simple and equal" theory which enables us to give an account for the apparent movement* or, as we sometimes translate, to *"save the phenomena"*. In the 4th century B.C., Eudoxus of Cnidos[2], was the first to suggest a model of concentric spheres as a theoretical model for movement which explained the observable displacements. But *the need* to explain the movements in terms of simple forms was not only theoretical, it was also *metaphysical and religious:*

> *It is natural and necessary [from the point of view of physics] that all the divine creations [the stars] have a uniform and regular movement[3].*

Theon reminds us many times of this Pythagorean and Platonic principle. For how do we explain that the Sun which is at A in the spring, at B in the summer, at C in the autumn and at D in the winter, travels equal arcs in unequal times when seen from the point Q? It is obvious that the cause of this phenomenon is a different movement which is not performed around the centre Q. The point Q can occupy three positions: inside the circle, on the circle or outside the circle. Now, it is impossible for the solar circle to pass through the point Q since the Sun would meet the Earth so that half its inhabitants would always be in daylight, half in darkness; there would be no sunrise or sunset, and the Sun would never be seen to turn about the Earth, which is absurd. Theon writes:

> *It remains therefore to suppose that the point Q is either inside or outside the solar circle.* **For whichever hypothesis we finally decide upon, the phenomenon will be explained**, *it is for that reason, that one has to consider as useless the discussions of the mathematicians who say that the planets are only carried on eccentric circles [where Q is interior, but not at the centre], or on epicycles [where Q is exterior to the solar circle], or around the same centre as the sphere of stars[4].*

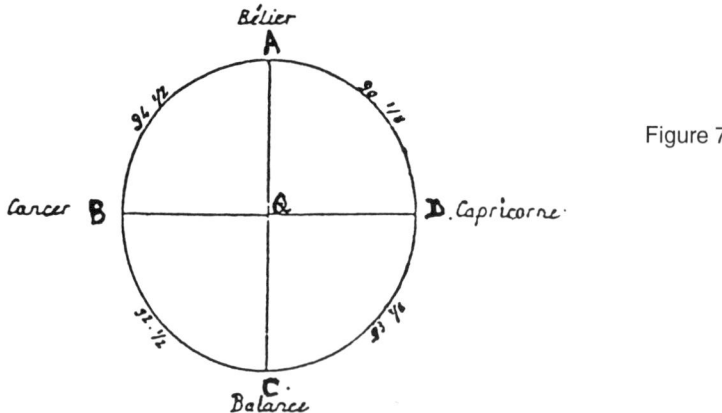

Figure 7

1.   Theon, *Mathematics Useful for Understanding Plato*, p. 244-245.
2.   See Aristotle, *Metaphysics*, Λ 8, 1073 b 16.
3.   Theon, *Mathematical Useful for Understanding Plato*, p. 249.
4.   *Ibid.* p. 251

In this way *the geometric method leads to the formulation* of the hypothesis of a shift of the centre of the celestial spheres in relation to the centre of the universe, *a hypothesis at the same time paradoxical and yet compatible with the phenomena.* Theon of Smyrna shows that:

> the reason why the astronomers adopted epicycles and eccentric circles with centres that were purely geometric[was certainly]their wish to explain the inequality of the seasons[1], and the irregular appearance of the solar motion.

But it would be even better if the apparent movement of the heavens could be reconstructed mechanically with the aid of complicated gear wheels! And in the end, the only geometric hypotheses which were retained were those which allow such a construction. It is for this reason, that Theon, at the end of his work reminds us of the two sorts of *"resulting"* spiral that the planets describe; one, as around a spindle corresponds to the daily rapid movement, combined with the slower movement in an opposite sense to the stars, the displacement in the zodiac being not in a straight line. He compares it to *"the coiled straps on the scytales[2] of Laconia on which the ephores wrote their dispatches"[3]* or to the tendrils of the vine; while *"the other as though on a plane surface"* is a continuous helix and corresponds to the slow and proper movement of the planets which carries them from one tropic to the other. In the end, we find ourselves back with the marvellous spiral of Archimedes, the favourite of astronomers and students of mechanics!

In fact, the apparent movement of the stars did not only attract geometers, but also fascinated physicists and mechanics. Let Aristotle's distinction act as a guide: he asks how the study of mathematics is different from the study of physics:

> When **the mathematician** studies surfaces, the lines and points do not occupy (space) in so much as they are the limit of a natural body and do not consider the properties which may accidentally belong to them and by which they would have real existence; also **it is possible to abstract these ideas so that the understanding effectually separates them without the trouble of movement.**[4]

But physical objects are less susceptible to abstraction than mathematical objects. If the odd and the even, the straight line and the curve, the number, the line, the diagram can exist without movement, things like flesh, bone and man cannot be conceived without movement. And in a certain manner, optics, harmony, astronomy are *"the opposite"* of geometry for, as Aristotle says,

> **geometry** studies the form which is very concrete, **but it does not study it such as it is in nature**, and optics on the other hand studies the mathematical form but not as mathematics, but in such a way as it takes a part in natural reality.[5]

But what is happening when Archimedes constructs his Armillary Sphere (that is his moving model of the Universe)? Let us read the words of Cicero:

> ... And we must admire the creative genius of Archimedes who had found a means of constructing the unequal movements and the different orbits by the rotation of a single object. When Gallus was making the sphere move, we saw the Moon and Sun change position by rotations operating in the metal equal in number to the days in the heavens;

1.   Duhem, *Le Système du monde*, vol. I, p. 448
2.   scytale = staff: a method of Spartan secret writing on a strip wound about a stick, unreadable without a stick of like thickness; ephor = magistrate. [Chambers Dictionary]
3.   Theon, *Mathematics Useful for Understanding Plato,* p. 203-204.
4.   Aristotle, *Physics*, II 2, 4. 193 b 23-194 a 11.
5.   *Republic,* II 2, 4.

and so in the sphere just as in the sky, the Sun would disappear and the Moon appeared in the shadow projected by the Earth, when the Sun moved to a region in the heavens[1]...

Unfortunately the text is broken off by a lacuna. It is clear in this case that *Celestial Mechanics studied the orbits and the trajectories not so much as mathematical objects, but in relation to the role that they played in reality.* It is about understanding eclipses, phases of the Moon, the retrograde movement of the planets, etc.; and *all these phenomena could not have been studied as abstractions of the movement which produces them.*

And this was the stimulus for inventing machines or complex illustrative mechanisms. Whatever the opposition of geometers and 'mechanical engineers' to such projects, if we are to believe the philosophers, they nevertheless devoted all their efforts to producing such models.

Mechanical mysteries, geometrical mysteries, puzzling history

Legend and history

The most famous problem of antiquity, which became almost legendary, was without doubt that of the duplication of the cube. Theon of Smyrna tells how the famous problem was posed to the inhabitants of Delos by the oracle of Apollo. The god Apollo

> *when asked the best way to deliver the citizens from the plague, (...) ordered the construction of an altar double that which already existed.*[2]

The builders, in confusion, consulted Plato: how do we make a solid the double of another? Plato then gave an interpretation of the oracle:

> *If God had made this response, it doesn't mean he needs an altar double in size, but he wished to reproach the Greeks for neglecting mathematics, and rebuked them for neglecting geometry*[3].

This is without doubt the interpretation which Plutarch was echoing when he wrote:

> *Plato reproached the disciples of Eudoxus, Archytas and Menaechmus for resorting to mechanics and instrumental means for resolving the problem of the duplication of a volume; for in their desire to find, in some fashion, two mean proportionals, they resorted to a method that was irrational. In proceeding in this way, did not one lose irredeemably the **best of geometry**, by a regression to the level of the senses, which prevents one from creating and even perceiving the eternal and incorporeal images among which God is eternally god?*[4]

Does the writing of Plutarch faithfully give us the actual solutions of Eudoxus, Archytas and Menaechmus? Is it not surprising that another tradition attributes to Plato the invention of a procedure using a complex machine different from the ruler and compass, while the solutions of Archytas and Menaechmus are as theoretical as possible,[5] as doubtless was that of Eudoxus which is lost?

1.   Cicero, *Republic*, I, ch 14.
2.   Theon, *Mathematics Useful for Understanding Plato*, p. 2.
3.   *Ibid.*
4.   Plutarch, *Propos de table* VIII, quest. 2, ch. 1, 718E.
5.   Tannery, *La Géométrie grecque*, p. 79-80.

Eratosthenes, in the *Letter to Ptolemy* which Eutocius has preserved for us (whether it is a literary fiction, containing simply a demonstration and an epigram of Eratosthenes, who was the private tutor of the son of a king Ptolemy, is another enigma which we leave aside!), proceeds as follows:

> In the course of works undertaken by the geometers of the Academy in order to resolve the Delian problem, and **to find how to construct two means between two given segments,** Archytas of Tarentum found the solution by means of his half-cylinders,[1] and Eudoxus with so-called curved lines. However, **they dealt with the problem by reasoning but did not arrive** at a practical realisation **to make their solution usable**, except perhaps Menaechmus, and that tediously![2]

In fact, the methods of Archytas and Eudoxus provided a difficult demonstration, but it seems they did not provide for a practical construction.

Before studying a few of the different procedures, let us recall how the investigation of the two mean proportionals x and y between a and 2a allows us to solve the duplication of the cube of side a:

$2a/y = y/x = x/a$, therefore $y^2 = 2ax$ and $x^2 = ay$; and, $x^4 = a^2y^2 = a^2(2ax) = 2a^3x$

consequently,

$x^3 = 2a^3$. The volume with side x is the double of the volume of side a.

The solution of Archytas

In his solution Archytas uses a construction in space and the intersection of three surfaces of revolution, namely a right cone, a cylinder and a torus. According to the account of Eutocius:

> This is the solution of Archytas, reported by Eudemus, [to the following problem]: *let the two given lines be AD and G; it is required to find two mean proportionals between AD and G.*

[CONSTRUCTION]

> Draw the circle ABDZ having AD as diameter where AD is the greater [of the two lines]; and inscribe [the line] AB, of length equal to G and produce it to meet at P the tangent to the circle at D. Draw [from B] the line BEZ parallel to PDO, and imagine a half-cylinder which rises perpendicularly on the semi-circle ABD, and that on AD is raised a perpendicular semi-circle standing on the parallelogram [in other words, the rectangular section] of the half-cylinder.
>
> When this semicircle is moved from D to B, the extremity A of the diameter remaining fixed, it will cut the cylindrical surface in [carrying out its] movement, and will trace on it a certain bold curve. Then, if AD remains fixed, and if the triangle APD pivots [about its base AD] with a movement opposite to that of the semicircle, it will produce a conical surface by means of the line AP which, in the course of its movement, will meet the curve [drawn] on the cylinder at a particular point. At the same time, B will describe a semi-circle on the surface of the cone.

1. *Les Présocratiques,* p. 527 and note. See also Baccou, *Histoire de la science grecque,* p. 254.
2. Eutocius, *Commentaire sur la Sphère et le Cylindre d'Archimède,* p. 65.

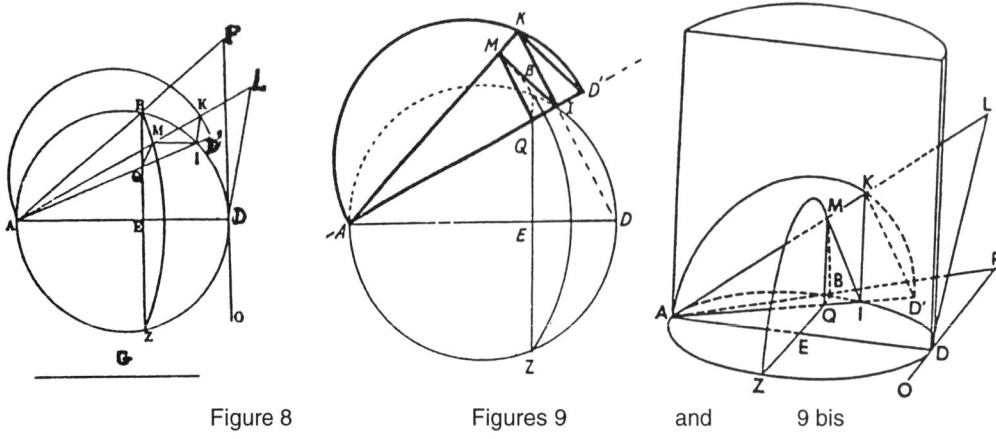

| Figure 8 | Figures 9 | and | 9 bis |

[THE COMBINATION OF MOVEMENTS]

Consider now the correspondence between the meeting point of the curves, the semi-circle moved [from D to B] *to the position D'KA* [the distinction between D and D' is modern; as the translator indicates, "what you gain in clarity you also gain in stability, and lose in movement. **Archytas' imagination is not static**] *and to the triangle moved by an opposite motion to the position DLA; let K be the meeting point already mentioned, and let BMZ be the semi-circle described from B; also let BZ be the section in common with the circle BDZA; if we drop a perpendicular from K to the plane of the semi-circle BA it falls on the circumference of the circle given by the right circular cylinder.* [Call] *the perpendicular KI;* **the straight line from I to A meets BZ in Q,** *while AL* [cuts] *the semicircle BMZ in M, and KD, MI and MQ are thus joined.*

[DEMONSTRATION]

Now since each of the semicircles D'KA and BMZ are perpendicular to the horizontal plane, their common section MQ is perpendicular to the plane of the circle, and in the same way, MQ is also perpendicular to BZ. So the [rectangle produced by the unequal lines BQ.QZ] *formed by BQZ, namely the* [rectangle AQ.QI] *formed by AQI, is equal to* [the square MQ²] *raised on MQ. The triangle AMI is then similar to each of the triangles MIQ and MAQ, and the* [angle] *IMA is a right angle.*

As for [the angle] *D'KA, this is also a right angle. So KD' and MI are parallel and they are proportional in the way that D'A is to AK or also as KA is to AI and IA is to AM, due to the similarity of the triangles.*
So the four lines DA, AK, AI and AM form a continuous proportion [DA/AK = AK/AI = AI/AM]. *And AM is equal to G, since it is equal to AB. And so for the two given lines, AD and G, two mean proportionals AK and AI have been found.*[1]

We see here that Archytas *"used a curve of double curvature, the intersection of the surfaces of a cylinder, a cone, and a torus"*. And Baccou[2] points out that the problem today would be resolved by solving the system of three equations with three unknowns:

$$\begin{cases} x^2 + y^2 = ax & \text{(cylinder)} \\ x^2 + y^2 + z^2 = a\sqrt{x^2+y^2} & \text{(torus)} \\ x^2 + y^2 + z^2 = a^2x^2/b^2 & \text{(cone)} \end{cases}$$

1.   Eutocius, *Les Présocratiques*, p. 525-526.
2.   Baccou, *Histoire de la science grecque*, p. 256.

Exercise 6   1) Explain why solving these equations gives the solution.

2) Try to reconstruct the method used here by using the ratios of segments with the help of the different diagrams, starting from what Eutocius says, as told by Eudemus.

As a final remark, this solution is abstract and geometrical, rather than mechanical. In fact, the combinations of movements that the demonstration uses are fairly complex to carry out mechanically; and if today, with the aid of some software, we can produce an animation corresponding to the movements conceived by Archytas, we would be hard put to it to imagine *what mechanical process was available to help the great Pythagorean scientist, or even if he had one!* For historians of science and technology, this remains an enigma.

On the other hand, many of the other solutions collected by Eutocius[1] were strictly connected to the description of a machine: Hero of Alexandria used a ruler moving about a fixed point B, at the corner of a rectangle with sides a and b; Philo of Byzantium produced a similar technique. We do not have any way of knowing if these machines were constructed or used … As for Plato, this is the one which is attributed to him.

Plato's machine

Let there be two rulers KL and HQ where one moves parallel to the other which is fixed, and slides in the two grooves of the supports MQ and ZH, which are mounted perpendicularly to the fixed ruler.

Let AB = a and BC = b, be the two segments between which we seek the two mean proportionals. The geometric solution used by Plato is as follows: if two straight lines AD and EC are perpendicular at B such that the angles AED and EDC are right angles, we can deduce that

$$AB/EB = EB/DB = DB/BC.$$

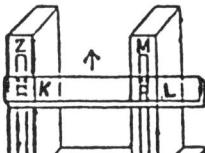

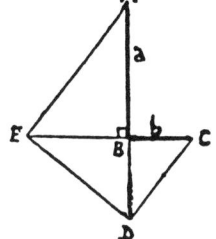

Figure 10

Figure 11

Exercise 7   Show that AB/EB = EB/DB = DB/BC.

The straight lines AB (length a) and BC (length b) being perpendicular at B, we use the instrument in the figure, and placing C on the fixed ruler HQ, and A on the moving ruler KL, and the extension of AB passing through H. Now move the sliding ruler away, and in doing so turn the instrument so that C is displaced along HQ. With A on KL and, the line AB always passing through H, we can manage to make the extension

1.   Eutocius, *Commentaire sur la Sphère et le Cylindre d'Archimède*, p. 47-62.

of BC pass through K. In this manner we obtain the points D and E respectively coincident with H and K.

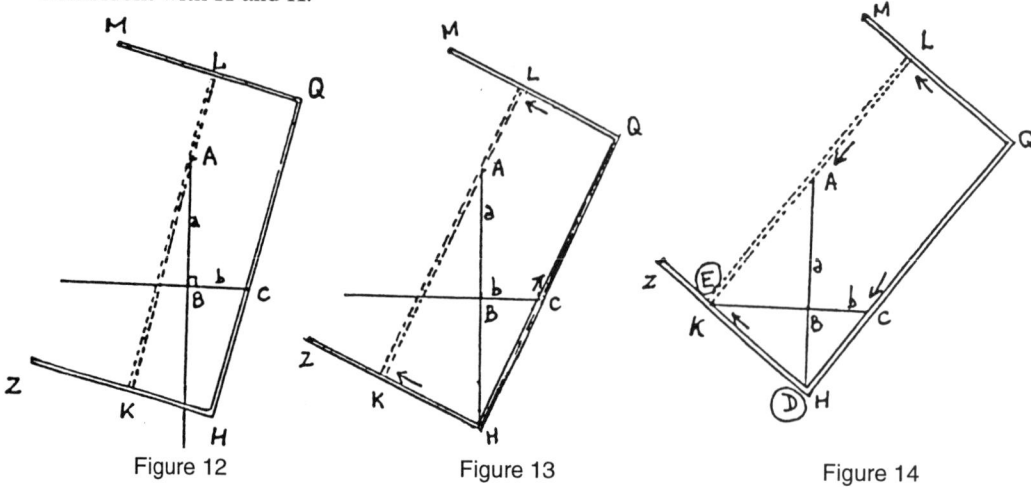

| Figure 12 | Figure 13 | Figure 14 |

Exercise 8   Find the equations defining the positions which give the solution (the point K which is the intersection of BC produced and HZ is the same distance as A from HQ); in other words, find HC and the angle CHA.

The machines of Eratosthenes and Nichomacus

In the *Letter to Ptolemy,* Eratosthenes claimed that he had imagined an instrumental method capable of finding between two given segments, not only two means, but as many as one required, and in this way to allow him to

> *make any solid figure of one form or another similar to a given figure, or to increase it in proportion, whether it is a question of altars or temples.*[1]

And the process is again an example of the desire to move away from the particular case and attain "necessity and generality". Between the segments AE and BQ, we construct a series of parallelograms which by virtue of the parallelism of the diagonals, allows us to compare the intermediate segments between AE and BQ. The operation of this machine, which we call the *mesolabe* [Greek: middle base] is simple[2]: it is a matter of sliding the blocks, each one in relation to the others, in such a manner that the points A, E, F and B are aligned. And so we obtain:

$$2a/y = y/x = x/a$$

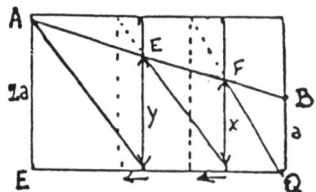

Figure 15

---

1.   *Ibid.,* p. 65.
2.   See Authier, *"Archimède: le canon du savant"*, p. 107-108.

The instrument consists of a small slab of wood, ivory or bronze containing three equal blocks made as thin as possible. The one in the middle is fixed, the others can be moved in the grooves [...] So that the line segments may be constructed more accurately, it is necessary to be careful when moving the blocks so that they remain parallel and the different pieces fall perfectly into place.[1]

If many means are required, we put a greater number of blocks in the instrument, but the demonstration remains the same. And Eutocius, through the voice of Eratosthenes, describes a votive offering of bronze sealed in lead on the gravestone of Ptolemy the king, which, as well as the diagram, included the description of the demonstration and the epigram of Eratosthenes as an "advertisement" which accompanied it:

... Give up looking for the mean with the laborious device of the cylinders of Archytas, or the three lines of Menaechmus obtained by conic sections, or the construction with curved lines invented by the divine Eudoxus. By using our tablets, you will easily be able to produce thousands of mean proportionals starting from a little seed.[2]

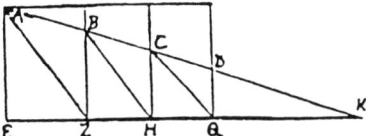

Figure 16

Exercise 9   Show that if the lines AZ,BH and CQ are parallel, then we have
AE/BZ = BZ/CH = CH/DQ

Is this account more reliable than that of Plutarch? We now find in the 3rd century, "modern" methods which are simpler and mechanically more elegant than the rather theoretical and "laborious" methods proposed in the 4th and 5th centuries. In fact, if Archytas had been the first to introduce the movements of instruments in a geometric figure[3], there is no evidence to say that he had really tried to carry out his construction mechanically.

Let us move on to the modern solution by Nichomacus, that Eutocius compares to that of Eratosthenes:

Imagine two rulers AB and CD linked together in the same plane so that they form a right angle. In AB there is an H-shaped groove in which slides a peg carved from tortoiseshell; in CD, where D is on the line which divides the larger ruler [AB] into two equal parts, a small cylinder is attached to the ruler which sticks up slightly from the upper surface; another ruler EZ, at a small distance from Z, has a groove HQ where it can engage with the cylinder attached at D, and from the point E a round hole allows the small solid axis of the peg to travel back and forth in the groove in the ruler AB[4] *...*

1.   Eutocius, *Commentaire sur la Sphère et le Cylindre d'Archimède*, p. 67.
2.   *Ibid.*, p. 68-69.
3.   Diogenes Laërtius, *Lives*, VIII, 83, in *Les Présocratiques*, p. 518; *Les Ecoles présocratiques*, p. 275.
4.   Eutocius, *Commentaire sur la Sphère et le Cylindre d'Archimède*, p. 69-70.

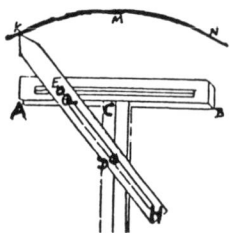

Figure 17

If we imagine a point K such that EK extending beyond the groove, remaining always the same length, and with a marker [at K] touching the page, the movement of the ruler describes a curve called a "first conchoid line" by Nichomacus, with radius EK and pole D (the small cylinder fixed on CD).

Nichomacus showed that this curve has the property of becoming closer and closer to the ruler AB without ever reaching it, and any straight line drawn between this curve and the ruler AB will cut the curve.

But how do we use this to find the mean proportionals? Here is the demonstration given by Eutocius:

> Let there be given a straight line AB and a point C exterior to it, and let the straight line CH be drawn such that the segment KH [intercepted by the line AB] is equal to a given segment. From C, drop the perpendicular CQ on to AB, and extend it to D such that DQ is equal to the given segment. Describe the first conchoid EDZ by taking the point C for the pole, the given segment DQ for the radius, and AB for the ruler. This curve meets the straight line AH at H, the intersection point. Describe the line CH where the segment KH is equal to the given segment.

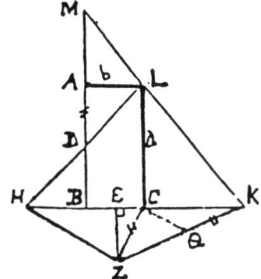

Figure 18

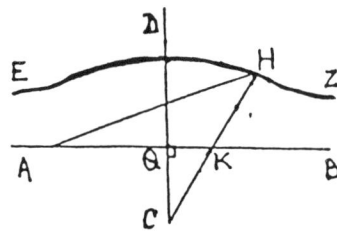

Figure 19

> This having been shown, let there be given two straight line segments CL and LA perpendicular to each other between which we wish to find two mean proportionals in continuous proportion. Complete the parallelogram ABCL, dividing each of the sides AB and BC into two equal parts by the points D and E. Draw the straight line DL and let H be the meeting point of the extension of DL and the extension of CB. Draw EZ perpendicular to BC and join C to the point Z such that CZ is equal to AD. Join ZH and draw CQ parallel to ZH. The angle KCQ being thus given, draw from the given point Z the straight line ZQK in such a manner that the segment [intercepted] QK becomes equal to AD or CZ. We have, in fact shown that this is possible by using the conchoid. Let M be the meeting point between the extension of AB and the extension of KL; and I say that CL is to KC as KL is to MA and as MA is to AL[1].

In other words, CL/KC = KC/MA = MA/AL and the two means are MA and KC. We note here that the machine that constructs the conchoid quickly and surely guarantees the equality of the segments, which would not have been possible with a

---

1.   *Ibid.*, p. 73-74.

simple compass. At the same time, notice that this process appears to be very close to the Archimedean process of intercalation.

> **Exercise 10**   Show how this apparatus also solves the trisection of the angle.

Ruler and compass... and movement

All these scientists of Antiquity, with Plato at their head, competed with each other in producing instruments which they refused to call "geometrical", yet which were used for carrying out what can certainly be called geometrical constructions. This leaves us with the question of why it was that constructions by ruler and compass were considered to be so superior.

A movement of another kind

Through reading Theon of Smyrna[1], we have seen the concern of the mathematician to identify behind movements that appear to be disordered, *a movement which is true and elementary* (that is, reducible to the circle and the straight line by a combination of revolutions and translations), and that this concern was not merely geometrical, but also metaphysical. For Plato, *mathematical objects* have a reality in the intellectual order, compared with which the moving and changing objects of our visible environment are but mere images, shadows, or reflections.They are the true foundations of our intellectual activity, their truth and reality results from their stability and permanence. And we can symbolise this hierarchy with a geometric line divided into segments according to a given relationship, each segment being divided in its turn according to the same relationship, in the following manner:

$$AB/BC = AD/DB = BE/EC$$

VISIBLE DOMAIN		INTELLIGIBLE DOMAIN	
images	objects	objects	ideas, notions
A--------------D-------------------B------------------E---------------------------C			
reflections	visible	mathematical	dialectics
	sensible	geometrical	

In the third century AD, in the commentary on Plato's myth of *Phaedra* Plotinus describes the intelligible domain:

> *There, life is easy; truth is their mother and their nourishment, their substance and their food; [...] each one has everything in himself, and can see everything in the others: [...] the splendour is without limit; [...]* **movement there is pure movement**; *for it has a mover which does not disturb it in its progress, since the mover is not distinct from it;rest is not troubled by movement, because it is not involved with anything unstable; beauty is pure beauty[2] ...*

How do we understand a "pure movement" which contains its essence within itself?

---

1.   See earlier in this chapter.
2.   Plotin, *Ennéades*, V 8, *"De la beauté intelligible"*, ch. 4.

We know that Aristotle identified supreme happiness with, above all, the exercise of intelligence, or with contemplative thought: here we can see what that means; there he is, man in the highest state *"only the concern of himself"*.[1] Pursuing no goal exterior to himself, conforming to a pleasure which is his own and whose perfection continues to increase, this intellectual activity is the actual creation of perfect happiness if it continues throughout his life.

But the act of contemplation, the divine and supremely pleasant life which is given to us as human beings, to live for but a brief moment,[2] *is the necessary and eternal essence of the principle of all movement.* Nevertheless, to be this absolute mover it must be still [i.e. not moved][3], because *"it must stop itself"* in the sequence of movements [the connection of movers and things moved] and since

> *we do not think science and reflections except as a rest and time of suspension*[4]

otherwise, it would be necessary to argue that

> *there is a creation for sight and touch, and that the activity for the things of intelligence is exactly similar to those of sight and touch*[5],

two epistemological a-priori statements that Aristotle leaves to the sophists and empiricists.

The movement of purely intelligible intelligence, that we carry over to the contemplative life, is nothing physical, in the sense that we use the term today, although it is in principle in our nature *(physis)*; and it is by abstraction and reduction or rather by purification *(catharsis)* that one is able to attain it.

Thus Proclus, in the fourth century A.D., in his commentary on the first three postulates of the Elements of Euclid[6], quoted the opinion of the Stoic astronomer Geminus,[7] of the first century, concerning the first postulate,

> *the line is a flowing point,* [and if we imagine] *the point moved by a movement entirely equal and minimal, we come to the other point.*[8]

Likewise, for the two other postulates, according to Geminus, it is a question of *imagining movement which can be described as if it were the manipulation of a point or a straight line.* Only in the empirical and materialist perspective of the Stoics, is scientific knowledge possible by this *"comprehensive representation"*[9] and inscription or imprinting on the "tabula rasa" of the intelligence, and Proclus strongly rejected this argument. His adherence to the Neoplatonic doctrine did not permit him to consider intelligence as *"an empty tablet upon which nothing has been inscribed"*,[10] for he defined the Soul as being always imprinted there, either by itself, or by Intelligence.

He also rejected the empiricist explanation that a geometric movement was simply an abstraction from a bodily movement. He wrote:

> *The indivisible things* [here geometrical concepts] *are free of all corporeal connections and exterior movements; but among them we consider **another type of movement***

---

1.   Aristotle, *Ethique à Nicomaque*, X 7.
2.   Aristotle, *Metaphysics,* Λ 7, 1072 b 14.
3.   As Aristotle demonstrates in *Physics*, VIII, 5.
4.   Aristotle *Physics*, VII 3, 247 b.
5.   *Ibid.*
6.   Proclus, *Les Commentaires sur le premier livre des Elèments d'Euclide,* p. 163-164.
7.   See Aujac, *Introduction aux Phénomènes*, Introduction.
8.   Proclus, *Les Commentaires sur le premier livre des Elèments d'Euclide*, 162-163.
9.   See *Géométrie ou géométries antiques?*, second part.
10.   See Proclus, *Les Commentaires sur le premier livre des Elèments d'Euclide*, p. 12.

and another domain for those movements, [...]: *the forms of geometers are therefore properly differentiated from those which they establish* [namely, their constructions]; *the movement of bodies is different from the movements conceived in the imagination; ... and they must not be confused!*[1].

In fact, although it is an essential point of doctrine for Proclus, to recognise the intermediate role of mathematical science which, he says:

begins externally by reminiscence and finishes internally in the reasonings, [or again] *arrives at perfection coming from imperfection*[2],

it is also quite clear in the first pages of the Prologue that

its operation is not stationary like that which is [purely] *intellectual, it takes place in a movement which is neither local nor modifying, [...] but living*[3].

And so we can come to recognise here, on one hand the *boldness of the movement of the soul* projecting itself outside the matter which enfolds it towards pure and perfect spirit, and on the other hand the *movement of the generous overflowing of intelligence*, one and indestructible, which allows the principles and figures of geometry, astronomy and harmony to emanate from the soul.

This *"living"* movement goes *in two directions rather like a "circulation"* of life like our model of the pulsing of the blood, contracting to restore to order and unity a complex system of propositions, and then dilating to multiply and develop theorems and demonstrations. One can extend the metaphor to see the use of constructions, of figures and the written text in geometry, as a source of filfth and pollution, and yet to see the technical, mechanical and strategical applications, as a source of oxygen!

Two sorts of geometric constructions

But let us return to the confusion of which Proclus warned us, between the kinds of geometers and their constructions ...

The error, he points out, is to take the line drawn on the tablet by the stylus for the geometric line. We have seen above, with Aristotle[4], that *geometric abstraction* consists *in separating its forms and notions from natural motion*. But what mechanics and philosophy shows us, is that that operation does not at all signify the suppression of movement, but rather a *reduction of this movement to its simple geometric expression*.

Thus in Archytas' solution, the movement which has to be imagined would be extremely complicated to carry out mechanically. In the same way, as we saw, the mechanical construction of the quadratrix poses serious problems about the determination of the ratio for π!

"The Greeks were not able to envisage", writes Lebesgue, in drawing the quadratrix in a continuous manner; that:

"the quadratrix as a mechanical curve that one cannot actually draw mechanically"[5].

---

1. *Ibid.*, p. 164.
2. *Voir page 108*, p. 15.
3. *Ibid.*
4. Aristotle, *Physics*, II 2, 4. See note 28.
5. Lebesgue, *Leçons sur les constructions géométriques*, p. 12.

In fact, to draw this curve in a continuous manner, it is necessary to construct a trammel where "*the plan for the cogs would require precisely the construction of points entirely comparable with those of the quadratrix itself*"[1].

And this mathematician proposes a *construction by dichotomy with ruler and compass*, to respond to Sporus.

> *Let us imagine two straight lines displaced by uniform movements where the first, Hd, is parallel to Ox, and the other, Oδ, turns around O and they leave their starting positions AT and Oy respectively, at the same instant, and arrive at their common final position Ox simultaneously. The locus of their meeting point M is the **quadratrix of Dinostratus**.*

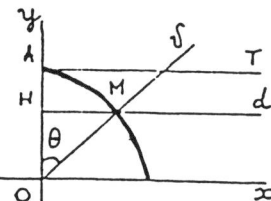

Figure 20

> *To construct it by plotting points, divide both OA, and the angle xOy into 2^n equal parts, to obtain the required points*[2].

Exercise 11   Suppose OA = 1, show that the points of the quadratrix are given by dyadic numbers (a dyadic number is written in the form $\sum_{i=1}^{i=p} 1/2^{n_i}$ where $n_1 < n_2 < \ldots < n_p$) and are constructible with ruler and compass.

You might try to find the position of the points such as M using suitable software such as LOGO or Cabri-Geometry, for example.

In effect, by this means we obtain the point by point construction of the curve; but as a result the movement is interrupted, as if, in a sense, the reduction of the lines and the circle freezes the problem, aventhough the large number of points obtained by dichotomy can then be used to generate the curve as an "animation" (especially if we use the computer).

We refer to "*a graphical solution*" each time we resort to a technical procedure which gives a solution which approaches defining the locus of the points M without giving us a scientific solution to that approximation.

On the other hand, when a curve is not constructible point by point, but only in a continuous manner, as in the case of the conchoid for example, then the nature of the curve drawn with a sliding mechanism, like that of Nichomacus, results only in an "*imperfection of material achievement.*"

One can imagine:

> *an ideal apparatus constructed on the same principle as the real apparatus which exactly traces the breadthless curves of geometry*[3].

---

1.   *Ibid.*, p. 13.
2.   Lebesgue, *Leçons sur les constructions géométriques*, p. 12.
3.   *Ibid.*, p. 15.

It is for that reason that we talk here of a *"mechanical solution"*; we obtain a figure or a representation as far removed from the geometric curve being studied as is the drawn triangle from the geometric triangle that is being studied. On the other hand, it is necessary to remember that the movement realised in the construction of the curve is a physical movement, and we are not trying to suppress it since it is a component of the very generation of the curve. Simply, *we have isolated this movement from time and speed for no other reason than to control the simple "geometric expression"*. We recognise here an algebraic curve that the Ancients did not wish to distinguish from a mechanical curve, even though, paradoxically, it was precisely the mechanical curve (the quadratrix) which they were not able to construct mechanically!

So, why deprive the mechanical constructions of the label *"geometric"*? In spite of the construction today of a large number of models of instruments whose practical utility is controversial, in fact, we still continue to use only the ruler and compass. It is as if the technical and electronic feats of our century were unable to unseat traditional geometrical instruments!

Should only these two tools be the special instruments with which to construct geometric figures? Is it not, rather, that *their very simple and rudimentary character is a true guarantee that the drawn figure and the figure being studied theoretically will not be confused?* Moreover, what is more mechanical and approximate than drawing with the compass, and can one say that movement and time do not intervene in such a construction?

Certainly, Euclid, as represented by Raphael, with compass in hand on the pavement before *"The School of Athens"*, contemplating the complex figure before him, clearly seems to escape from time and change. Surely, he finds himself in this *"time of suspension"* by which Aristotle defined the reflection of intelligence, unless on the contrary one prefers to imagine him in the middle of a *"brain storm"*, in the process of discovering and giving birth to a new theorem.

However, the long history of attempts by men of philosophy, mechanics and geometry to understand movement continues up to our own time, and Newton who is also represented compass in hand[1] takes up the baton again in the preface of the *Principia*, when he remarks that geometry assumes the circle and the straight line already traced, without us having to learn to draw them.

> *To describe right lines and circles are problems, but not geometrical problems. The solution of these problems is required from mechanics, and by geometry the use of them, when so solved, is shown* ...[2]

This was, of course, before the birth of the kinematics!

---

1.   Blake, *Space and Time*, Portrait of Newton, Tate Gallery, London.
2.   Newton, *Principia*, p. XVII.

Bibliography

Sources texts

ARCHIMEDES

    Opera omnia cum Commentariis Eutocii, ed. J.L.Heiberg, Teubner, Leipzig, 1910-1915. Repr. Stuttgart, 1972.

    Des corps flottants, ed. J.L.Heiberg.

    Sur les spirales, ed. J.L.Heiberg.

    La méthode d'Archimède relative aux propositions mécaniques, à Eratosthène, Les Belles Lettres, Œuvres complètes d'Archimède, Paris 1972.

    T.L. Heath, *The Works of Archimedes* and *The Method of Archimedes*. Repr. Dover, New York, (1953).

ARISTOTLE

    Du ciel, tr. P. Moraux, Les Belles Lettres, Paris, 1965.

    Metaphysics and *Physics* Ross, W.D., Revised Text with Introduction and Commentary, Clarendon Press, Oxford, 1908 on

    Ethique à Nicomaque, tr. Voilquin, Paris, 1965.

CICERO

    De Republica, tr. C.W. Keyes, Loeb Classical Library, Harvard University Press, 1972.

    Republic, French tr. Ch. Appuhn, Paris, 1965.

DIOGENES LAËRTIUS, *Lives of Eminent Philosophers*, 2 vol., Hicks, R.D., Loeb Classical Library, Harvard University Press. Repr. 1991.

EUCLID, T.L.Heath, *The Thirteen Books of Euclid's Elements*, 2nd edition, Cambridge: Cambridge University Press, 1926. Repr. Dover, New York, 1956.

EUTOCIUS, *Commentaire sur la sphère et le cylindre d'Archimède,* French tr. Ch. Mugler, t. IV of *Œuvres complètes* d'Archimèdes, Paris, 1972.

GEMINUS, *Introduction aux phénomènes,* Introduction and French tr. by G. Aujac, Les Belles Lettres, Paris, 1975.

NEWTON, *Mathematical principles of Natural Philosophy*, tr. Motte, ed. Cajori, University of California Press, Berkeley, 1934.

PAPPUS, *Collections mathématiques,* French tr. Ver Eecke, repub. Repr. Blanchard, Paris, 1982.

PLATO

    Platonis Opera, 5 vol., Burnet, J., Oxford.

    Plarménide, French tr. A. Diès, Les Belles Lettres, Paris, 1965.

    Plato's Parmenides, ed. Allen, Blackwell, Oxford, 1983.

    La République, French tr. R. Baccou Garnier, Paris, 1963.

    Republic, tr. A.D. Lindsay, Everyman, Dent & Sons, London, 1935.

PLOTINUS

    Plotini Opera, 3 vol., ed. P. Henry and H.R. Schwyzer, Bruxelles-Parid-Leyden, 1951-1973.

    Plotinus, 7 vol., tr. A.H. Armstrong, Loeb Classical Library, Harvard University Press, 1966-1988.

    Ennéades, French tr. E. Bréhier, Les Belles Lettres, Paris, 1967.

PLUTARCH

    Plutarch's Parallel Lives, 16 vol., tr. B. Perrin, Loeb Classical Library, Harvard University Press, 19...

    Plutarch's Moralia, vol. 8, tr. P.A. Clement and H.B. Hoffleit, Loeb Classical Library, Harvard University Press, 1969; vol. 9, tr. E.L. Minar, Jr., F.H. Sandbach and W.C. Helmbold, Loeb Classical Library, Harvard University Press, 1961.

    Propos de table

    Vie de Marcellus, French tr. Amyot, Paris, 1587.

PRESOCRATICS, The (*Archytas, Hippias*, ...) in Bibliothèque de La Pléiade, French tr. J.P. Dumont, J.L. Poirier and D. Delattre, Paris, 1988. See also *Les Ecoles présocratiques* collection "Folio", Paris, 1991 and Freeman, K., *Ancilla to the Pre-Socratic Philosophers*, Oxford, 1948-1962.

PROCLUS

    Commentaire au premier livre des Eléments *d'Euclide,* French tr. P. Ver Eecke, Bruges, 1948.

    A Commentary on the first Book of Euclid's Elements, tr. Morow, Princeton University Press, 1970

THEON OF SMYRNA

    Mathematical knowledge serviceable to read Plato, III. Astronomy, ed. Hiller, E., Leipzig 1878.

    Exposition des connaissances mathématiques utiles pour la lecture de Platon, III. Astronomie, French tr. J. Dupuis, Paris, 1892; reprint Culture et Civilisation, Brussels, 1966.

    Mathematics Useful for Understanding Plato, tr. of French by Lawlor, Wizards Bookshelf, San Diego, 1979.

THOMAS, Ivor, *Greek Mathematical Works,* Loeb Classical Library, Harvard University Press, Cambridge, Mass. and London, 1939.

General works for further reading

BACCOU, *Histoire de la science grecque,* Aubier, Paris, 1951.

BECKER, *Das mathematische Denken der, Antique* Göttingen, 1966.

CHASLES, *Aperçu historique sur l'origine et le développement des méthodes en géométrie,* Paris, 1889. Repr. Gabay, Paris, 1989.

CROWE, *Theories of the World from Antiquity to the Copernican Revolution,* New York, 1990.

DUHEM, *Le système du monde,* repub. Paris, 1979.

HEATH, *Greek Astronomy.* Repr. Dover, Toronto, 1991.

HEATH, *A Manual of Greek Mathematics,* Oxford, 1931.

LASERRE, *De Léodamas de Thasos à Philippe d'Oponte,* Naples, 1986.

LEBESGUE, *Leçons sur les constructions géométriques.* Repr. Paris, 1987.

NEUGEBAUER, *A History of ancient mathematical astronomy,* Berlin, Heidelberg, New York, 1975.

REY, *L'Apogée de la science technique grecque. Essor de la mathématique,* Paris, 1948.

SERRES, *Eléments d'histoire des sciences,* Bordas, Paris, 1988.

TANNERY, *Mémoires scientifiques,* ed. Heiberg et Zeuthen, 17 vols., Privat, Toulouse, 1912-1950.

TANNERY, *Géométrie grecque,* Gauthier-Villars, Paris 1887. Repr. Gabay, Paris, 1988.

How did you get on?

Ex. 1   Generalisation of the property of the spiral. (prop. 15). See note by Ivor Thomas *op. cit.* p. 187:

"... if AE, AD are drawn to meet the second turn of the spiral, while AZ, AH are drawn, as before, to meet the circumference of the first circle, then
AE : AD = arc QKZ + circumference of the first circle : arc QKH + circumference of the second circle and so on for the higher turns. In general, if E, D lie on the nth turn of the spiral, and the circumference of the first circle is c, then
AE : AD = arc QKZ + (n – 1)c : arc QKH + (n – 1)c
These theorems correspond to the equation of the curve r = aσ in polar co-ordinates."

Ex. 2   Trisection of the angle "a"

If EF = R, angle FEB = FBE; in the isosceles triangle EFB, angle BFE = $180° – 2a$; so angle BFC = $180° – BFE = 2a$; in the same way, in the isosceles triangle FBC, the angle FBC = $180° – 2(2a)$; so, since angle FBE = a, angle ABC = $(180° – $ angle FBC$) – a = 3a$.

Ex. 3   It is sufficient to notice that the arc AB is double the arc BC
AC = 3CH and CZ = 3CD so, AC – CZ = 3(CH – CD), or AZ = 3HD.
$BD^2 = BE^2 – EZ^2$, now DA.AZ = BD^2, since AZ = 3HD
so 3DA.HD = BD^2.

Ex. 4   This amounts to calculating the limit of ρ as φ → 0.

Ex. 5   Let v be the velocity of the translation of the line BC and w the angular velocity of the straight line AB. Then when $v/w = 2a/\pi$, the two lines arrive at AD at the same time.

Ex. 6   If you use $u = \sqrt{x^2 + y^2}$ and $v = \sqrt{x^2 + y^2 + z^2}$, then you can show that $a/v = v/u = u/b$.

Ex. 7   The two rulers of the machine define the similar triangles AEB, BEH and HBC; so let x = EB, a = AB, y = DB, b = BC. Then we have $a/x = x/y = y/b$.

Ex. 8   Express the tangent of angle CHA as a function of a and b,
[tan CHA = $\sqrt[3]{b/a}$] and you can deduce HC.

Ex. 9   Use Thales' theorem! [A line drawn parallel to the base of a triangle divides the sides in the same ratio.]

Ex. 10   See O. Becker, *op. cit.* p. 87, after Pappus (refer to figures 2 and 18):

"*Place the 'pole' of the machine to trace the conchoid at B, its ruler [AB] at AC, while its double ray extends to 2AB. The intersection with the parallel drawn from A to BC produces the point E such that the intersection with B gives the angle EBC which is a third of ABC.*"

5

The Straight Line and The Curve

Evelyne BARBIN and Gilles ITARD
IREM Pays de Loire
Centre du Mans

Two friends, called Kromme and Recht, were leaving a reception having drunk perhaps a little too much wine. Each thought himself sober and steady but considered the other to stagger. However, a policeman advised them to take care: according to him, both were a little the worse for wear. Figure 1 illustrates the three cases:

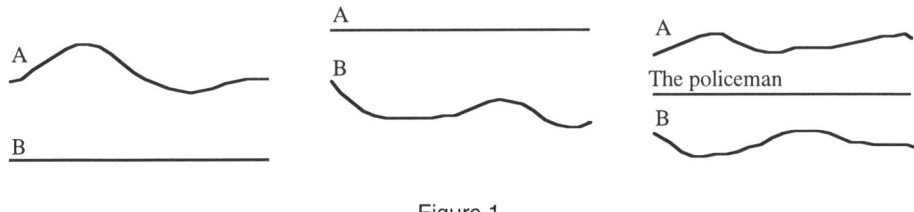

Figure 1

The following day, their discussion was lively. How could they distinguish between a curve and a straight line? Just like the Eleates, our two friends enjoyed debating. Recht held that the line held prior place, for to be curved meant being not straight. While Kromme claimed that a straight line was simply a special case of the set of curves.

Recht: Geometry concerns itself initially with the straight line. Don't we talk about 'rectifying' a curve, in much the same way as we would talk about rectifying an error? And our wandering walk yesterday evening only goes to prove that we were not going straight.

Kromme: Yes, geometry does study the line, but Euclid first defined a line and only then defined a straight line as a special type of line. Let's go and check in the library!

The first day: To define the curved line and the straight line

After having asked for a number of scholarly works, our two friends settled down near a painting in an academic style. A young woman seated on a throne, crowned with laurels and holding an open book in her hand which she was examining with a lofty air. "Clio, the Muse of History", observed Kromme with a shrug of the shoulders, before rushing to a reading of Euclid.

Kromme: In Book I of Euclid's *Elements*, definition 2 states that *"a **line** is breadthless length"*, but definition 4 specifies that *"a **straight line** is a line which lies evenly with the points on itself."*[1]

Recht: Yet I read in the 1988 edition of *Petit Robert* dictionary that a curve is that *"which changes direction without forming an angle"* and that a straight line is that *"which does not deviate from one end to the other [and] is like a ruler."* But the words 'direction', 'straight' and 'ruler' all come from the same root![2] You can see well enough that 'straight line' is self defining and that 'curve' is merely its opposite.

Kromme: Not at all. "Without deviating" means "without curving", and "like a ruler" is simply a conventional expression. On the other hand, according to that dictionary, a polygonal line is neither a straight line nor a curved line. That shows that the definition of "curve" in the dictionary is not satisfactory. The concept of line is primary, as in accordance with Euclid, and the rest are special cases. This reminds me that Euclid defines angle as between any two lines, whether straight or curved. He wrote, in definition 8, that *"a **plane angle** is the inclination to one another of two lines in a plane which meet one another and do not lie in a straight line[3]"*, and he continues, in the following definition with *"and when the lines containing the angle are straight, the angle is called **rectilineal**."*[4]

Exercise 1   Draw the line given by $y = |x^2 - 1|$. Is this a curve according to the definition given in *Petit Robert*?

Kromme: This dictionary teaches us nothing ! We would do better to consult the *Encyclopédie Méthodique*.

Clio: The *Encyclopédie méthodique* presents, in its entirety, the articles of the *Grande Encyclopédie* written by Diderot and d'Alembert. The three volumes dedicated to mathematics were among the first to be published between 1784 and 1789. They were written by the mathematicians d'Alembert, l'Abbé Bossut, le Marquis de Condorcet and the astronomer Lalande. They constitute a complete and exhaustive document on the state of the exact sciences on the eve of the French Revolution.

Kromme: We learn from the Encyclopaedia that *"a line can be considered as formed by the flux or movement of a point"* and then that the straight line is *"the shortest distance between two points."*[5] Now, there's something that I'd missed in the *Robert* dictionary: "changes direction" implies the idea of time. And then, what is the direction, the "straight line" [droit], of a curve? If the curve changes its "direction" [droit chemin] at every instant then it means that it has the potential of "straightness" [droit], it carries within itself the seed of being "straight" [droit] without it ever being realised.

Recht: Your remark leaves me not knowing what to think. I find it very unsatisfactory to make an appeal to the idea of time in the context of geometry. Geometry is timeless ! Furthermore, to talk about the straightness of a curve as something that never realises itself, which never has the time to realise itself, is just so much verbiage. The mariner and the astronomer who point their telescopes at the stars do so along straight lines that

---

1.   Heath, *The Thirteen Books of Euclid's Elements*, vol. 1, p. 153.
2.   The French words *direction, droit,* and *règle* have the same Latin root *regere* [tr.].
3.   Heath, *The Thirteen Books of Euclid's Elements*, vol. 1, p. 153.
4.   *Ibid.*
5.   Diderot, *Encyclopédie méthodique*, vol. 2, p. 307.

are immaterial, timeless and purely geometric. Neither time nor distance are in question with such ideal straight lines. The path of a stone that has been thrown cannot produce the ideal geometric parabola since that would presuppose that our memory already has a record of the set of points that make up the parabola and that we have a clear image of it. The evidence for this can be seen in the difficulties encountered by the 16th century scientists in trying to determine the trajectory of a cannon ball. Tartaglia managed to establish that at no point was it a straight line, but it was not until Galileo that its geometry was proved. Now, engravers of the period were quite capable of showing jets of water perfectly and so perhaps, by imitation, the path of cannon balls. But a jet of water produces something material to draw as it lasts, but it does not consist of an ideal, geometric parabola.

Clio: In 1537, the Italian Tartaglia published *Nova Scientia*, a work in which he invented the "new science" of ballistics, concerned with solving artillery problems. It considered, in particular, how to find the range of a cannon ball given its firing angle. Tartaglia therefore investigated the trajectory of the cannon ball. Himself a prisoner of scholastic thought, he argued that the trajectory was made up of three parts: motion along a straight line, then part of a circle and finally a straight line fall.

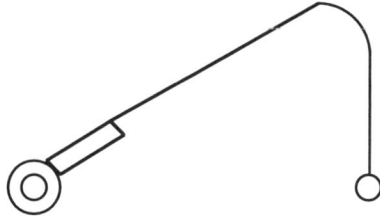

Figure 2

But in *Quesiti et inventioni diverse* published in 1546, Tartaglia returned to this idea in order to prove that the trajectory is not rectilineal at any of its points. He wrote:

> *the cannon ball in such a situation does not travel along a perfectly straight line for any part of its motion (even if it comes out at a speed as fast as one wishes) because the speed (as great as it may be) is never enough, in such a situation, to make it travel in a straight line. It is true that the faster it travels, the more it approaches rectilineal motion, that is to say, straight line motion, however it never quite gets there. It would be better to say, in such a case, that the faster the cannon ball, the less is its movement curved.* [1]

In 1638, Galileo published his *Discourses on two new Sciences,* where he proved that the trajectory of a cannon ball in space was part of a parabola. To do this, he found the composition of the two motions to which the ball is subject: a vertically downwards motion of uniform acceleration and a uniform rectilineal motion. The parabola was already known to the Ancient Greeks: Appolonius had written a work on conics in the second century B.C. For the Greek geometers, the parabola was a static object, defined geometrically by the intersection of a cone with a plane. With Galileo, the parabola became a dynamic idea, conceived of cinematically as the trajectory of a moving point. The study of motion in the 17th century would become a powerful instrument of invention leading, at the end of the century, to the infinitesimal calculus, that is the calculus of derivatives and integrals.

1.   Tartaglia, *Quesiti et inventioni diverse*, p. 12, Fr. tr. by Annie Gele.

Exercise 2	How do we prove nowadays that the trajectory of a cannon ball moving in space, without resistance, is part of a parabola?

Kromme: We cannot get waylaid into archaeology, we do not have the facts. Moreover, I am happy to allow that the "visual straight line" ["la droite optique"] has meant a lot in the creation of a geometry of ideal objects, but I hold that it is the idea of line itself that is paramount, not in chronology but in universality. Also, we say curved line, straight line, polygonal line in order to distinguish between different types. The *Encyclopédie Méthodique* invites us to consider the problem differently. I read, for example, in the article *Curve:*

> *A CURVE, it is said, is a line of which the different points have different directions, or are differently placed one with another. That at least is the definition given by Chambers, following a host of authors. See LINE.*
> *A* curve, *it may be added, considered in this way, is the opposite of a* straight line, *whose points are all placed in the same way one with another.*
>
> *Perhaps each of these definitions will be found imprecise: which is the case. However, they agree well enough with the idea of curved line and straight line that everybody holds: furthermore it is very difficult to give to these lines a notion that is more clear to the mind than the simple notion that we gain from the words straight and curve themselves. The most exact definition that one might be able to give to both of them is as follows: the straight line is the shortest path between two points, and the curved line is a line leading from one point to another which is not the shortest path. But the first of these definitions embodies in itself a secondary property but not the essence of a straight line: and the second, although it only embodies a negative property, applies just as much to a set of straight lines with angles as to what are properly called curves, and could be regarded as an infinite collection of small lines joined one to the other by infinitely obtuse angles. See below POLYGONAL CURVE; see also CONVEX. Perhaps it might be better not to define curved line and straight line on account of the difficulty, and perhaps the impossibility, of reducing these words to a more elementary idea than that which they themselves present.* [1]

Recht: Let us accept, then, the notion of straight line and curve; let us continue as if the nature of the straight line exists. Non-Euclidean, axiomatic and formalist geometries have accustomed us to reject such a notion, to go beyond it. However, without it, without this link between matter and ideas, there would perhaps be no mathematics other than for it to be reduced to an abstract language. Let us, rather, find out how the straight line and curve are linked together since we are agreed on the existence of a link between these notions.

Clio: In 1733, Saccheri thought to prove Euclid's famous axiom on parallel lines in his work *Euclid cleansed of all stain* [Euclides ab omni naevo vindicatus]. To do this he proceeded by *reductio ad absurdum* and concluded that the contradiction of Euclid's axiom *"is repugnant to the nature of the straight line."* With the invention of non-Euclidean geometries by Bolyai and Lobachevsky, at the beginning of the 19th century (see chapter 11), the status of geometrical objects underwent a change: a reference to the "nature" of the straight line was no longer admissible. This idea would be reinforced with the formalist school. Hilbert, in his *Foundations of Geometry* in 1899, did not admit any intuitive idea of geometric objects: points, lines and planes are only different

---

1.   *Encyclopédie ...* vol. I, p. 449-450.

classes of objects whose nature is not specified – *"schematic concepts devoid of content"*, as Einstein wrote.

The second day: the curve generated by the line

Recht: A rather attractive curve occurs to me. It easily shows that the straight lines can produce curves. Take line segments, each the same length, and put one end on a line (x'ox) and the other end on a line (y'oy), perpendicular to (x'ox).

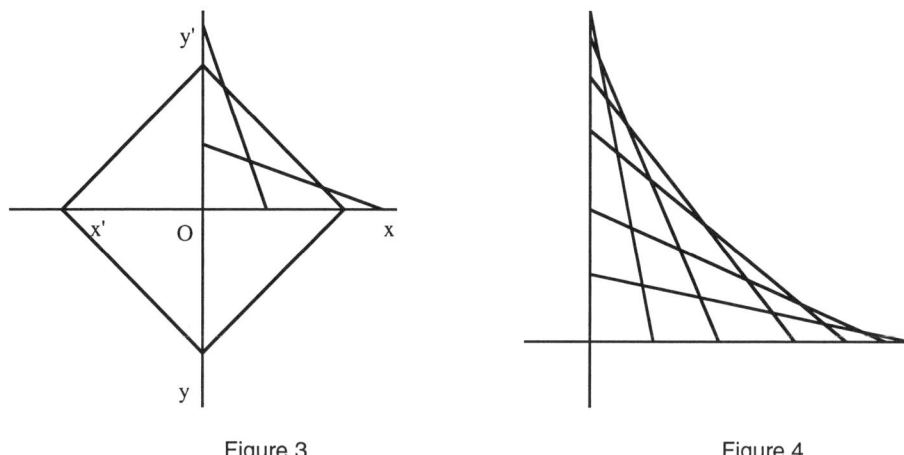

Figure 3   Figure 4

The evident symmetries follow from the construction: let's look at one quadrant. The line segments fill a space apparently bounded by a curve "which changes direction without forming an angle," and yet we have only constructed a polygonal line. It is as if our eye is unable to notice very small changes in direction. Galileo thought of the circle as a regular polygon having an infinite number of sides in his *Discours on two new sciences.* [1]

Kromme: Yes, that's certainly interesting, and we can sense the "straightness" that is potentially inherent in the curve. However, I am not convinced. My eyes see the appearance of a curve, but neither sight nor appearances have a place in geometry. In order to determine the curve suggested by your drawing, I need to be able to determine each point. I think I have a way of going about it. The diagram is so convincing that I am unable to reject the result, but for me the diagram itself is not enough. Take two of the line segments, only slightly inclined one to the other and, having found their common point, let's find a way of removing the inclination of one to the other. Perhaps, then, we shall be able to determine the very instant when the curve is being created from the line. Please excuse my use of "instant". I don't mean to be provocative, it is merely a figure of speech.

Recht: Do carry on.

Kromme: The calculations needed for your curve are rather long and I would like to try out the method in a rather easier case. Take a straight line (D) and a point F, not on (D). For each point H on (D), draw the mediator [perpendicular bisector] of FH. Just as in

---

1.   *Discorsi e dimostrazioni matematiche inforno a due nuove scienze,* 1638, The two new sciences being "mechanics and local motions." [tr.].

your example, we see a curve and an apparently familiar one. To make the calculations easier take the mediator of FH_0 as abscissa [the x-axis], where H_0 is the foot of the perpendicular from F onto (D), and take FH_0 as ordinate [the y-axis].

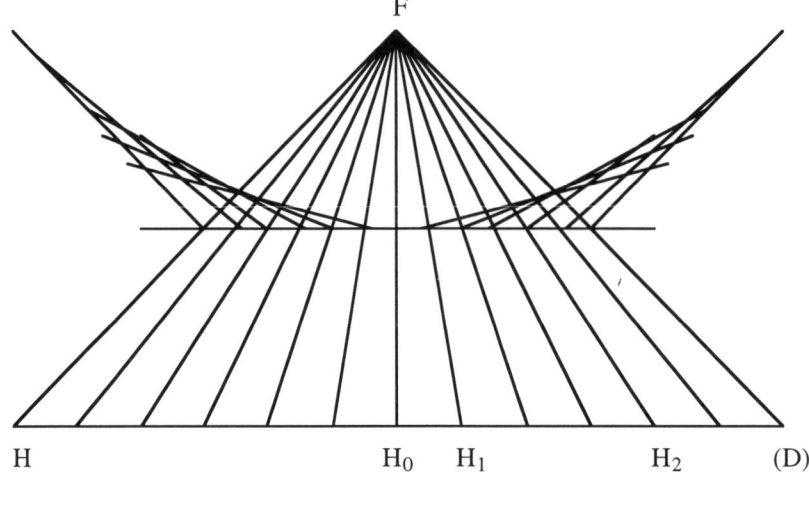

Figure 5

Recht: You're going to use calculus. Is that satisfactory? Aren't you simply introducing the idea of time under the guise of the abscissa?

Kromme: I don't know, but it is worth exploring the method. Don't let us mix in everything, or else we'll never get anywhere. Take two points H_1, H_2 on (D) with abscissae x_1, $x_1 + h$ respectively, h being non zero. Write down the equations of the mediators of FH_1 and FH_2 find their common point. I get $x = x_1 + h/2$. Having done this with h non zero, I now assume that the two mediators become coincident and I put $h = 0$, from which I get $x = x_1$ and it is easy to calculate y which has the form ax^2. The curve which we see here, called the envelope of the straight lines, is a parabola.

Exercise 3   Carry out the calculations that Kromme suggests. Do the same for Recht's first example: the curve is an astroid.

Recht: Very ingenious! But your calculations are hardly surprising since in constructing the parabola with focus F and directrix (D), we know that the mediators will be tangents to the curve. Furthermore, your method is strongly reminiscent of Fermat's procedure for finding tangents to curves.

Clio: In 1637, Fermat published his *Method of Finding Maxima and Minima*. In it he describes the method he had invented for finding tangents. He wrote:

> *Consider, for example, the parabola BDN with apex D and diameter DC; let B be a point on the parabola from which it is proposed to draw a straight line BE, tangent to the parabola, and meeting the diameter at E.*
>
> *Take some point O on BE and draw the ordinate OI and also the ordinate BC. We have: $CD / DI > BC^2 / OI^2$, since the point O lies outside the parabola.*
>
> *But, by similar triangles, $BC^2 / OI^2 = CE^2 / IE^2$.*
>
> *Hence, $CD / DI > CE^2 / IE^2$.*
>
> *Now, given the point B, the ordinate BC, the point C and the length CD are defined..*
>
> *Hence, let $CD = d$ be given. Put $CE = a$ and $CI = e$; whence $d / d - e > a^2 / a^2 + e^2 - 2ae$.*

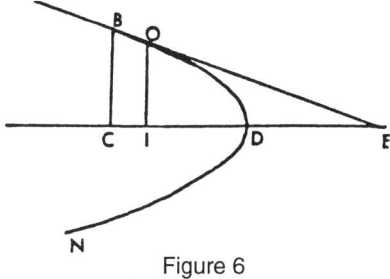

Figure 6

Multiplying means and extremes: $da^2 + de^2 - 2dae > da^2 + a^2e$.
Making these become more equal [adegalons], *following the previous method, we get,*
after subtracting common terms: $de^2 - 2dae \approx - a^2e$,
or, which is the same thing: $de^2 + a^2e \approx 2dae$.
Dividing all terms by e: $de + a^2 \approx 2da$.
Suppressing the term de: there remains $a^2 = 2da$, and hence $a = 2d$."[1]

Kromme: You have inverted the problem! You start from the parabola and you verify that it agrees. That's not fair, it was my example!

Clio: The inverse problem of tangents, that is to say to find the nature of a curve given some property of its tangents (or normals), was posed by Kepler, in a particular case, in 1604 in his *Paralipomènes à Vitellion*. The origin of the question came from optics: to find the glass shape that refracts a parallel beam of light into a conical beam. The inverse tangent problem was also proposed to Descartes by Florimond de Beaune in 1639.

Recht: Very well, then! But your method employs calculus. The latter is not a simple intermediary tool, it is necessary, it is instrumental, and it that what bothers me. And besides, when at the end of your calculation you let h become zero, your calculation loses all meaning: you claim to have determined a unique point whereas its sole characteristic is that it is the common point of two coincident lines!

Kromme: We shall have to return to the fundamental role of the calculus. That was a fruitful mutation in the evolution of geometry. Our discussion may clarify its necessity. But I think you have not understood my method. On the mediator of FH_1, the family of the other mediators determines a family of points with abscissae $x_1 + h/2$ which stretch out continuously except at the point where $x = x_1$. And I say that this particular point is the point of the ideal curve at the instant where real intersection points vanish and appear.

Recht: So you claim to grasp the ungraspable, the notion of the instant returns and the notion of "straightness" being potential in the curve. Perhaps we would find it more interesting to start our discussion from the other end. According to the geometers, as I have already said, your straight lines are nothing more than tangents to a parabola. Now, as far as I am aware, the notion of tangent does not require your acrobatics which are far too metaphysical for my taste.

---

1.   Fermat, *Œuvres*, vol. III, p. 122-123.

The third day: the direction of a curve at a point

Kromme: Euclid talks about the tangent to the circle in his *Elements* but he does not consider other curves. The definition 2 in Book III is satisfactory for the circle: *"A straight line is said to touch a circle which, meeting the circle and being produced, does not cut the circle."* [1] But it is easy to imagine twisting curves which recut their tangents. For my part, I would prefer to see whether proposition 16 in the same Book III might not serve as a definition. In fact Euclid proves that *"The straight line drawn at right angles to the diameter of a circle from its extremity will fall outside the circle, and into the space between the straight line and the circumference another straight line cannot be interposed; further the angle of the semicircle is greater, and the remaining angle less, than any acute rectilineal angle."* [2]

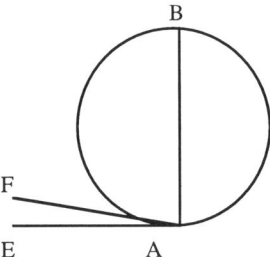

Figure 7

Recht: Don't let's get side-tracked. You can come back to this if you wish but I have a feeling that it's this angle "smaller than any acute rectilineal" angle that you find so tempting – that which the geometers call a *"horned angle"*. You hope to come back to the "instant of appearing" of something from nothing. However, if Euclid only talked about tangents to the circle, Archimedes was interested in difficult curves and Appolonius was acquainted with tangents to conics.

Clio: In the third century B.C., Archimedes proved, in his treatise *On Spirals,* that *"If a straight line touch a spiral it will touch in one point. only"*[3] Then he determined the exact position of the tangent at a point on the spiral.

Kromme: However, in the 17th century, mathematicians became critical. The Ancients knew the tangents to some curves but lacked the means of discovering tangents to new curves. The Ancients' enquiries did not carry them to the general idea of the tangent but only to the particular behaviour of certain straight lines relative to certain defined curves, mostly in an entirely geometric way. Their proofs were geometric and mostly used the method of *reductio ad absurdum*. Our enquiry is different and perhaps more universal: we need answers that are direct and to the point.

Recht: And, for sure, you are going to tell me, just like Roberval, that the tangent to a curve is a direction of motion.

Clio: The name of the geometer Gilles Personne de Roberval remains associated with the name of a balance which he proposed to the Academy of Sciences in 1669. Roberval's method for finding tangents to curved lines dates from the 1630s, a time

1.   Heath, *The Thirteen Books of Euclid 's Elements*, vol. 2, p. 1.
2.   *Ibid.*, vol. 2, p. 37.
3.   Archimedes, *On Spirals*, Prop. 13. in Dijksterhuis, *Archimedes*, p. 265.

when geometers sought to determine the nature of the cycloid. His work was not published until 1693 under the title *"Observations on the composition of movements and on a way of finding tangents to curves."* To use Pascal's definition, the cycloid is *"nothing more than the path through the air made by a stud on a wheel in ordinary motion, since the stud begins to rise from the ground at the very moment when the continuous motion of the wheel returns it to the ground after a complete revolution: supposing the wheel to be a perfect circle, the stud to be a point on its circumference and that the ground to be a perfect plane."* [1] This curve had been proposed by Mersenne in the 1620s. It became the object of numerous studies in the 17th century and of numerous paternity disputes, to the extent that it became known as *"the geometer's Helen."*

Roberval's method is based upon the following *"principle of invention"*: *"The direction of the motion of a point which describes a curve is the tangent to the curve at each position of the said point."* Hence, the general rule for finding the tangent at a point is as follows: *"by the specific properties of the curve (which you are given), examine the different motions of the point which describe it in the place where you wish to draw the tangent; find the single motion of which these motions are the components and you will have the tangent to the curve."* [2]

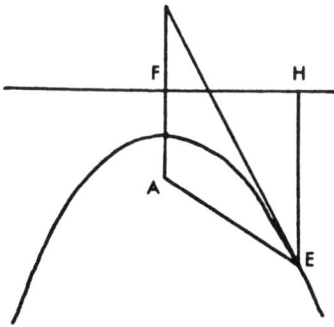

Figure 8

Roberval explains how to find the tangent at the point E of a parabola with focus A, apex F and whose directrix passes through the point B where AF = BF. Since the point E is equidistant from the focus and the directrix, Roberval deduces that *"the motion of the point E describing the parabola consists of two straight line equal motions, of which one is [in the direction] AE and the other is the line HE along which it moves with equal speed."* [3] And so, in combining these two motions, Roberval found that the direction of the tangent at the point E is that of the bisector of the angle AEH.

Exercise 4	Let F and F ' be the two foci of an ellipse. Let the point M be such that MF + MF' = 2a and find the tangent to the ellipse at M by Roberval's method. Verify that the result agrees with present-day methods. Repeat the exercise for the cycloid.

1. Pascal, *Œuvres complètes*, p. 117.
2. Roberval, *Observations sur la composition des mouvements...*, p. 24-25.
3. *Ibid.*, p. 26.

Kromme: I know that you dislike the idea of motion, but I find this notion of speed just as tricky as that of tangent and together I fear that they may spin round uncontrollably in my head. But since you love geometry, let's go back to Euclid III 16 and use this as the starting point for our analysis. Let A be a point on a curve and consider a pencil of ·lines through A. We wish to identify that line T which, according to my reading of Euclid, has the property that no other line of the pencil may lie between T and the curve.

Recht: You are taking the property that Euclid established for the tangent to a circle as your starting point, so his definition of the tangent to a circle becomes a consequence, a special case. Before going further, I'd like to remark that you appear to be assuming the existence and the uniqueness of T. The analysis is fine, but let us not lose sight of this.

Kromme: Your are right and, in order to allow me to continue with my calculations, I am also going to assume that the curve can be expressed in the form $y = f(x)$ in some coordinate system, where f is a function.

Recht: And you talk of universality!

Kromme: I don't claim complete facility with the universality of the concept. To follow Roberval, let us know our limits but we should avoid the inertia that comes from scepticism.

Clio: In 1637 Descartes explains how to associate an equation with a curve in his famous work *La Géométrie*. Each point of a curve could be fixed by using axes, an origin and direction – which will become our Cartesian coordinate system. Descartes restricted himself to those curves whose equations were polynomials. At that time, Fermat introduced the idea of the "characteristic property" of a curve. In 1655, in his *Arithmetica Infinitorum* Wallis extended Descartes' idea to include curves represented by infinite algebraic expressions.

The concept of function was introduced towards the end of the 17th century by Leibniz and Newton, the inventors of the infinitesimal calculus. In his authoritative *Introductio in analysis infinitorum,* published in 1748, Euler begins by defining the concept of function: *"A function of a variable quantity is any analytic expression whatsoever composed of that variable quantity and numbers or constant quantities."*[1] From then on the study of curves became a part of what we now call analysis.

Kromme: Let us note firstly, that for two straight lines issuing from A, and no parallels to (Oy), the difference in the ordinates is proportional to the difference in the abscissae, something that is not true in case of a straight line and the curve through A. If no straight line can be interposed between the curve and T, at least in the neighbourhood of A, then the difference of their ordinates is less than all magnitudes proportional to x.

Recht: I agree with you. However, so that we do not get trapped by words, let us take A as the origin, $y = f(x)$ as the equation of the curve and $y = mx$ as the equation of the line T.

---

1.   Euler, *Introduction à l'analyse infinitésimale*, p. 2.

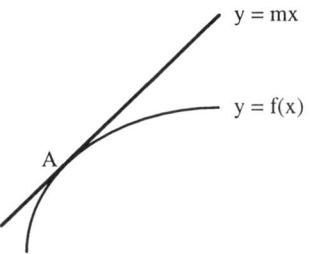

Figure 9

Kromme: You anticipate me! For we want, near x = 0, the difference f(x) – mx to be less, absolutely, than any kx where k is a non zero real number. Moving away from A, we therefore impose the condition that |f(x)/x – m| should be less than any positive real number, and yet we know that this quantity is not zero.

Recht: I have fallen into your trap! By another route you have come back, more or less, to that Euclidean "horned angle", less than any rectilineal angle. The magnitude of |f(x)/x – m| is not zero and yet is less than any given magnitude. Despite my reluctance, I am close to thinking that we can see here the tangent born out of nothing. But I recall that the great Euler in his *Elements of Algebra,* published in 1774, talked about complex numbers like i and 3 + 2i as being neither nothing, nor less than nothing, nor greater than nothing.

Kromme: We are concerned here with 'imaginary' quantities, but not in the sense of complex numbers, which cannot be put on a straight line, and which were represented by line segments or by points of a plane at the beginning of the 19th century (see chapter 13). Our quantity |f(x)/x – m| certainly belongs to the domain of the reals. It is not, however, zero and it always exists in an interval containing 0, on which it is less than all fixed reals.

Recht: You have fallen into your own trap! I admit to having given in to the temptation of believing in your victory, but we cannot accept these fantasies if we wish ourselves to be geometers.

Kromme: Just a moment! The idea seemed attractive to you, you admitted as much: would it not be better to go into it more deeply than to reject it out of hand? I can well understand that this "vanishing" real number, which is neither zero nor more than zero, poses a problem. It requires one to accept an idea of flux or variation, that a number is variable in a continuous way, and I understand your reluctance. For you, a number is fixed: there exist many small than it (and greater than it), but a number is a number.

Recht: There you go again with time, variation, flux, … Geometer you enquire after the eternal, but in your research you always come back to time. It remains the case that, for each x, |f(x)/x – m| has a precise and assignable value.

Kromme: Certainly, but your over static-vision is too restricting. You have a stroboscopic view of what I am describing globally. The truth is that if I take a fixed point b (figure 10), then (BF/AB) – (BE/AB), that is EF/AB can be found to have a value k, say. If I increment b to b', then E'F'/AB' can be found to have some value k'. And the ratio k to k' is itself determined. And also, and this is the point, it is impossible to find a number r such that for all b' the value of E'F'/AB' ≥ r. To be more precise I say that if, every time I fix r all the E'F'/AB' are less than r, for those b' sufficiently close to a, I can be sure that no straight line can be interposed between (AT) and the curve. Hence (AT) is the tangent.

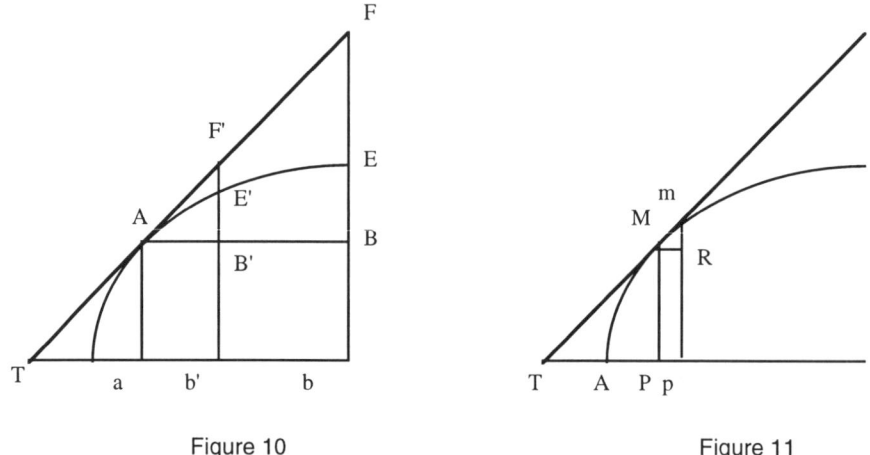

Figure 10   Figure 11

Recht: It's a bit complicated but you very ingeniously avoid the idea of time by this "overall view". You have also avoided a negative definition, although it seemed to be conveyed by your first idea. In listening to you I was thinking that the non-rectilineal triangle AB'E' which is certainly not similar to the rectilineal triangle TaA is, perhaps, quasi-similar to it. It brings to mind Leibniz' characteristic triangle which formed the basis of his invention of the infinitesimal calculus. Furthermore, your construction caused me to think of the Marquis de l'Hospital's construction for finding tangents.

Clio: In a memoire dated 1713 entitled *"The History and the Origin of the Differential Calculus"* Leibniz explains that the idea of the differential calculus came to him when he was thinking about the triangle MmR which he called "characteristic" (figure 11): Two sides are parallel to the axes and the third is part of the tangent. He wrote: *"It seemed always possible to identify triangles similar to this characteristic triangle, no matter that the triangle was itself unidentifiable or infinitely small."* [1] This triangle led him to see that in problems of calculating area and finding tangents each was the inverse of the other.

The differential calculus invented by Leibniz was popularised by the Marquis de l'Hopital in 1696 in his work *Analysis of infinitesimals*. This calculus is based upon considering the *"infinitely small portion by which a variable quantity may be continually increased or diminished, called the Difference."* It requires that *"a curve may be considered as a collection of an infinite number of straight lines, each one infinitely small: or as a polygon with an infinite number of sides."* [2] The difference of a quantity x is denoted by dx. In his treatise, l'Hospital explains how to use the differential calculus to find tangents to curves (figure 11):

> Let AM be a curve such that the relation between the coupée [abscissa] AP and the appliquée [ordinate] PM may be expressed by any equation whatsoever, and that it is required to find the tangent MT at M.
> Having drawn the appliquée MP, and supposing that the straight line MT that meets the diameter at T is the required tangent, imagine another appliquée mp infinitely close to the first and a small straight line MR parallel to AP. And by letting the given AP, PM be x

1.   Leibniz, *Histoire et origine du calcul différentiel*, M.S., vol. V, p. 392.
2.   L'Hospital, *Analyse des infiniments petits pour l'intelligence des lignes courbes*, p. 2-3.

and y; (whence Pp or MR = dx, and Rm = dy) the similar triangles mRM and MPT imply mR / RM = MP / PT (whence dy/dx = y / PT). Now, by use of differences from the given equation, the value of dx can be found in terms of dy, which being multiplied by y and divided by dy, will give the value of the sub-tangent PT in terms which are entirely known and free of differences, by which means the desired tangent MT may be found.[1]

Recht: However, your wish for universality is hardly satisfactory. Without having to look very far, your method founders with y = √x if A is taken as origin.

Kromme: No, my definition fits Euclid's proposition well enough, even if its use in that example founders. You can show without too much difficulty that no line can be interposed between the curve and the y-axis. There are doubtless more difficult cases, even much more difficult. But I believe we have identified here the concept of tangent, its essence, and found a process which deals with a very large number of cases.

Recht: We have really come a long way and I would like to lay to one side some difficulties which I can see, for example the need for an equation y = f(x), at least in the neighbourhood of A. I would prefer to put forward some ideas which came to me while I was listening to you which we might well think about. This is not very well thought out but I'm afraid I might forget about it in discussion. We cannot interpose a straight line between a tangent and a curve, but it is obvious that we can interpose a curve, be it as flat as we wish. While a straight line is always straight, a curve seems able to be more or less curved – or, if you prefer, more or less flat.

Kromme: Yes indeed, and if we do not want the curve and the straight line to be two separate Aristotelian qualities, we must find some way of finding a linear measure, a quantity.

Recht: Exactly so, but I can go further. May not my earlier remark about the property of quasi-similarity of triangles ABE and TaA allow us to "nearly calculate" the length of the arc AE?

Kromme: That is very promising, though not without some difficulty. First of all, I'd like to examine this question of being more or less curved which appears rather easier. It's a question of introducing some quantitative measure and going beyond the qualitative aspect.

The fourth day: using a circle to measure curvature

Kromme: At the point A of the curve we know the tangent (T), the "potential straightness" of the curve. Let us construct a family of circles with common tangent (T) at A. This family of circles offers a sort of graduation of curves, a scale of curvature.

---

1.   *Ibid.*, p. 11.

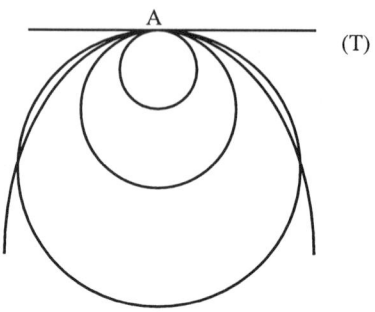

Figure 12

Recht: It appears easy for us to pick out the greatest and the least curve, but your idea which appears simple shows just how fragile our intuition is.

Kromme: I don't understand your concern!

Recht: The circle appears always similar to itself and we can talk of its "constant curvature": it can slide upon itself, just as a straight line can, and it seems entirely appropriate to talk about its curvature. However, a large circle is nothing more than a small circle enlarged without deformation. It is similar and homothetic[1] to the small one. It follows that it can be neither more curved nor less curved. No straight line is straighter than any other and no circle can be more curved than another: and that contradicts our intuitive idea about more curved and less curved.

Clio: In the 14th century, Nicole Oresme touched upon this contradiction in his *Tractatus de configurationibus qualitatum et motum.* He distinguished between the "intensity of curvature" and the "amount of curvature":

> In a circle, the "intensity of curvature" is equal, uniform, constant and is inversely proportional to the diameter. On the other hand, the extensio, that is the length of the circumference, is directly proportional to the diameter, the "total curvature", that is the amount of curving is the same for all circles[2]

Kromme: Your remark embarrasses me. I understand well enough what you are saying and there is no way that I can rebut it. The family of circles shows the situation clearly enough and it convinces me even though I have to defer to your argument. Roberval said, in effect, that logic can be a deceiver, and I should like better it if your correct reasoning and its conclusion concerned premises other than those that relate to our question.

Recht: Relax! The *Encyclopédie Méthodique* has a ruling on this question. I'll read the entry on curvature:

> CURVATURE. This name is given to the amount by which an infinitely small arc of any curve deviates from a straight line: now, an infinitely small arc may be considered as an arc of a circle [...]; and so one may determine the curvature of a curve from the infinitely small arc of a circle. Imagine, on the same infinitely small chord, two circular arcs with different radii; the lesser will deviate further from the chord than the greater, and it can be proved geometrically that these deviations will be in the inverse ratio of the radii of the circles: hence, in general, the curvature of a circle is in the inverse ratio of its radius, and the curvature of a curve, at each point, is in the inverse ratio of the radius of its circle of osculation. However, this definition is arbitrary: for if, on the one

1.   Mathematical enlargement [tr.].
2.   Quoted by Clagett, *Nicole Oresme and tthe Medieval Geometry of Qualities and Motions.*

hand, one may say that a small circle is more curved than a large circle with reference to the same chord, one may say, on the other hand, that the arcs are equally curved with reference to different chords, each proportional to its radius; and, talking in this way, it has further to be admitted that all circles are similar curves. [1]

Kromme: That's good enough for me. However, some day we must talk more about this apparent paradox. How can it be that the very definitions for the circle, which refer to all circles, prove to be an obstacle to identifying their different curvatures? I am sure I pose the question badly, but no matter for the present. The family of circles under consideration can certainly serve as a scale of curvatures.

Recht: A comparative scale, yes, and the benefit is important; but we wanted to define the curvature at A. Among the pencil of straight lines through A we have picked out the tangent, and now from among the family of circles through A we must pick out the best.

Kromme: But we also need to look for a meaning for this word "best". Perhaps we ought to identify that particular circle among the family of circles that, in the study of equations, is called the circle of osculation. Take, for example, the parabola with equation $y = x^2$. The family of circles through its vertex is given by $x^2 + (y - b)^2 = b^2$. From these two equations I get $x^2(x^2 + 1 - 2b) = 0$. The value $x = 0$ at the vertex is the solution twice for all cercles but four times for the circle given by $b = 1/2$. Any larger circle will recut the parabola in two distinct points. I do not dare to generalise too much from this one example, but I just want to hold on to the fact that the value $x = 0$ is a multiple solution and is so more than twice.

Exercise 5 Carry out similar calculations for a general point $A(x_0, y_0)$ of the parabola. Taking $\Omega(a,b)$ as the centre of the circle, verify that $(x - x_0)^2 (x^2 + 2x_0x + 1 + 3x_0^2 - 2b) = 0$ and find the circle of osculation. "How many times" is x_0 the solution in this case?

Recht: You surprise me! Your method, which you suppose to be general, relies upon an algebraic singularity and neglects the essence of the problem. Your method of tangents has so convinced me that I would like, to borrow your expression, to seize the instant when the curve, unable to realise its straightness, endeavours to make itself like a circle.

Kromme: I have never made any claims on the curve and I have introduced circles only as a way of measuring curvature.

Recht: We are at one, but my criticism remains. So far you have avoided this type of method in the study of tangents. The family of straight lines at the vertex A is given by $y = m(x - x_0) + x_0^2$. From this we get the equation $(x - x_0)(x + x_0 - m) = 0$ and the algebraic singularity is given by $m = 2x_0$ which makes x_0 a double solution. We have certainly found the tangent but we are no more enlightened.

Kromme: There are two solutions, x_0 and $m - x_0$. I can let $m = 2x_0 + h$. Disregard the family of straight lines through A and imagine a single moveable straight line through A, then allow h to be a real variable; in letting $h = 0$, the two intersection points coalesce. This remark brings to mind Descartes' method for finding tangents to curves.

Clio: Descartes' method, as given in his *Geometry,* consists in considering a circle to pass through two points C and E of a curve, the circle to have its centre on a coordinate axis; then let the intersection points coincide. Descartes considered the case where the

---

1.   Diderot, *Encyclopédie...*, vol. I, p. 465.

curve was an ellipse: the coordinates of the intersection points were thus the solutions of a second degree equation. The point P, where the circle "touches" the curve at one point, was found where the equation produces a double solution. As Descartes wrote: *"the nearer together the points C and E are taken however, the less difference there is between the roots; and when the points coincide, the roots are exactly equal."*[1] The tangent to the circle is already known: it will also be the tangent to the curve at the point where circle and curve "touch".

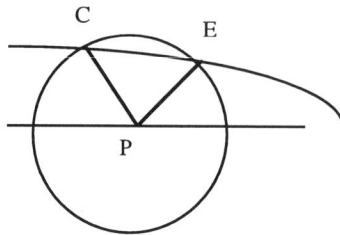

Figure 13

Kromme: We could proceed in the same way with circles by considering the continuous transformation of a circle.

Recht: The idea of time, through flux, keeps returning, but I am beginning to think that it is difficult to avoid continuous flux. Wasn't that the basis of Newton's differential calculus? We must look out a copy of his *Method of Fluxions and Infinite Series.*

Clio: In this work, written in 1671 and published in 1736, Newton considered mathematical quantities to be generated by *"continual Increase, after the manner of a Space, which a thing or point in motion describes"*:

> *Now those quantities which I consider as gradually and indefinitely increasing, I shall hereafter call Fluents, or flowing Quantities, and shall represent them by the final letters of the alphabet v, x, y, and z; that I may distinguish them from other quantities, which in equations may be considered as known and determinate, and which therefore are represented by the initial letters a, b, c, etc And the velocities by which every Fluent is increased by its generating motion (which I may call Fluxions, or simply Velocities or Celerities,) I shall represent by the same letters pointed [...]*[2]

Recht: The idea of fluxions plays the same role for Newton as Differences does for l'Hospital. Let's read on further, for I think that Newton's solution will commend itself to us:

> *If a Circle touches any Curve on its concave side in a given point, and its magnitude be such that no other Tangent Circle can be interscribed in the Angle of contact nearer that point, that Circle will be of the same Curvature as the Curve is of in that point of contact. For that circle which comes between the curve and another Circle at the point of contact, varies less from the Curve and makes a nearer approach to its Curvature, than that other Circle does; and therefore that Circle approaches nearest to its Curvature, between which and the Curve no other Circle can intervene.*
> *Therefore the Center of Curvature at any point of a curve, is the Center of a Circle equally curved, and thus the Radius, or Semi-diameter of Curvature is part of the perpendicular which is terminated at that Center. [...]*
> *The Problem then is reduced to this, viz. To find the Radius or Center of Curvature.*

1.   Descartes, *The Geometry*, p. 103-104.
2.   Newton, *The mathematical works of*, vol. I, p. 49.

> *Imagine therefore that at three points of the Curve ∂, D, d, [figure 14] perpendiculars*
> *are drawn, of which those that are at D, ∂ meet in H, and those that are in D, d, meet in*
> *h, and the point D being in the middle, if there be a greater curvature at that part D∂*
> *than at Dd, then DH will be less than dh; but by how much the perpendiculars ∂H and*
> *dh are nearer to the intermediate perpendicular, so much the less will the distance be of*
> *the points H and h, and at last, when the perpendiculars meet, the points will coincide.*
> *Let them coincide in the point C, and C will be the Center of Curvature.*[1]

Kromme: So Newton does with this circle what we were doing with the tangent. No circle can be interposed between the curve and this singular circle. This brings us back again to Euclid.

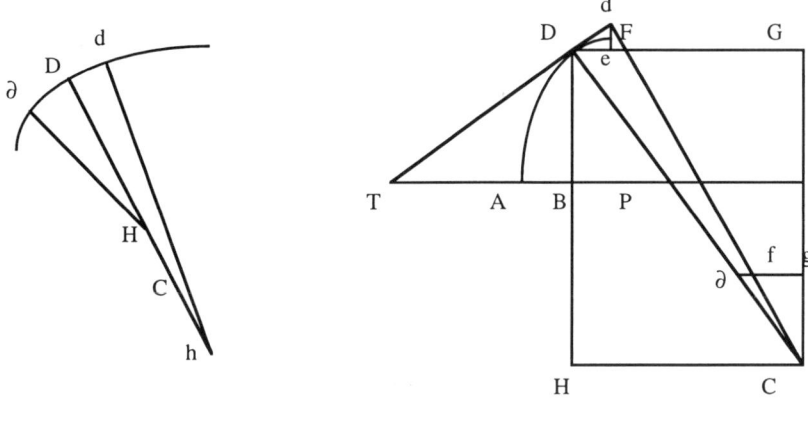

Figure 14   Figure 15

Recht: In order to find the centre of curvature C of any point D of a curve, Newton chose DB as ordinate, AB as abscissa and DT as the tangent at D. Then he drew CG perpendicular to AB and DG and ∂g parallel to AB. (fig. 15) Then he imagined the point D to *"move along the curve an infinitesimally small distance Dd"* and he drew a perpendicular to DG and made Cd perpendicular to the curve. Assuming d to be a point on the tangent DT and using similar triangles Cg∂ and Ded, Newton obtained:

Cg/g∂ = De/ed = Fluxion of the abscissa AB / Fluxion of the ordinate DB.

Since DG is parallel to ∂g, we also have: CG/Cg = CD/C∂ = DF/∂f. The value of DF can be found from the right angled triangle DdF: DF = De + (de x de/De). From this Newton found the value of CG and hence CD.

Kromme: I can see another advantage in Newton's method: it uses perpendiculars to curves, what mathematicians call normals, in much the same way as we used tangents to determine the envelope curve. We found their intersection when they were inclined one to the other and then we removed their inclination. In a way, the centre of curvature for D is the point where the normal at D is a tangent to the envelope of normals.

Exercise 6   Find the envelope of the normals to the parabola.

---

1.   *Ibid.*, vol. I, p. 76-77.

Recht: The foundation of the method is certainly there but Fermat's approach is not the same as the Marquis de l'Hospital's or Newton's. These latter use quasi-similar triangles or, if you prefer, they take an infinitely small part of the curve to be a straight line.

Kromme: Yes, and it seems curious that one gets an answer that is consistent when taking things to be alike that are so only approximately. The way of proceeding is clear enough but it is much more difficult to establish its validity and there is an upsetting instability.

Recht: There is no actual similarity except at the very instant when the triangle disappears and with it the possibility of similarity. In fact, an error is neglected and this 'neglecting' is used to produce a reliable result which, although I feel convinced, I hardly dare call certain. But it lacks the rigour to which I am accustomed ... and so I distrust it and want to keep my distance. But you, having brought me thus far, even you seem to hesitate.

Kromme: Yes, we are absolutely sure that these methods are well founded yet we do not know on what foundations to base them.

Clio: The question of the foundations of analysis which so worry our two friends took its place in the history of mathematics at the beginning of the 19th century. At the end of the 17th century, different methods invented by Descartes, Roberval and Pascal overshadowed the creation of the infinitesimal calculus by Newton and Leibniz. The mathematicians of the 18th century were keen to use the marvellous power of the calculus to solve successfully many difficult problems. They set about producing new results disregarding established forms. Their confidence in their results derived from their manipulating symbols according to rules and to the exactness of the physical results. The 18th century belonged to Euler and Lagrange and produced a wealth of creative results.

At the beginning of the 19th century mathematicians began to question the concepts and proofs of analysis: the concept of function was not clear, the use of series without regard to their convergence produced paradoxes and the ideas of differentiation and integration were not properly defined. The main leaders of the critical movement wanted to restore order to chaos and to construct the foundations of analysis upon the concepts of arithmetic. The leading mathematicians in this work were Bolzano, Cauchy, Abel, Dirichlet and Weierstrass. The use of rigour is explicit in Cauchy's work:

> *My principal aim is to reconcile rigour [...] with the simplicity which comes from considering infinitely small quantities.*[1]

In the 1820s Bolzano and Cauchy introduced the idea of a continuous function. In doing so, Cauchy used such terms as 'limit', 'infinitely small increase' and 'decreasing indefinitely'. It was in order to avoid the vagueness of these expressions that Weierstrass, in the middle of the 19th century, introduced the definition of limit we use today. The efforts to establish a rigourous basis for analysis were motivated by the wish to prove several results for continuous functions that had been accepted intuitively.

---

1.   Cauchy, *Résumé des leçons ...*, p. I.

The fifth day: rectification of the curve

Recht: I propose to examine how we might imagine calculating the length of an arc of a curve. It is a question of deciding how we might establish an equality between the length of an arc and a segment of a straight line, what mathematicians call rectification. The question makes me feel ill at ease.

Kromme: However, the idea is tempting.

Clio: In the *Elements,* Euclid ventured to compare the area of a circle with that of a square and the volume of a sphere with that of a cube, albeit in an indirect way. But he never compared the circumference of the circle with a straight line. To use the language of the Ancients, Euclid strove to square the circle and to cube the sphere but never to rectify the circumference of a circle. Archimedes, on the other hand, in his *Measurement of a Circle,* did not hesitate to suppose that a straight line might be made equal to the circumference of a circle. Throughout history there has been this dialectic between a quasi-mystical rigour, which is necessary even if it sometimes paralyses thought, and a boldness which assumes the risk of error.

Recht: I have before me a text by Descartes which is hardly encouraging: *"the ratios between straight and curved lines are not known, and I believe cannot be discovered by human minds."*[1] Just look where all our research has come to.

Clio: Descartes' assertion was contradicted some twenty years after the publication of his famous *Geometry.* At the beginning of 1658, Pascal threw down a challenge to all geometers: he offered 40 pistoles [an old Spanish coin] to anyone able to find the centres of gravity of all volumes of revolution generated by an arch of the cycloid. Not a single geometer was able to furnish the result and Pascal produced his own solution to the admiration of his colleagues. But in August he received a letter from the English mathematician Wren which contained the rectification of the curve. Pascal wrote: *"there is nothing more beautiful than what has been sent by W.Wren; for besides the beautiful way in which he has shown how to measure the area of the roulette, he has shown how to compare the curve itself and its parts with a straight line. His assertion is that the length of the roulette is four times that of its axis, though he has sent the result without a proof."*[2]

Wren's proof can be found in Wallis's work *Tractatus Duo De cycloïde de cissoïde* and uses a double *reductio ad absurdum*, in the manner of the Ancients. It uses the property that the length of the arc OP of a cycloid lies between the length of the tangent OV and the length OE, drawn parallel to the tangent at P, (figure 16). If the length of the semi-cycloid is supposed greater or less than 2AD, and if X is the difference, Wren shows that there are circumscribed and inscribed polygons for the curve of lengths p_1, p_2 respectively and that $p_1 - p_2 = X$. He then proves that both the assumptions, that $p_1 - 2AD$ is less than X and that $2AD - p_2$ is greater than X, result in contradictions.[3]

1.   Descartes, *The Geometry,* p. 91.
2.   Pascal, *Œuvres,* p. 121.
3.   Wallis, *Tractatus Duo De cycloïde de cissoïde,* p. 68.

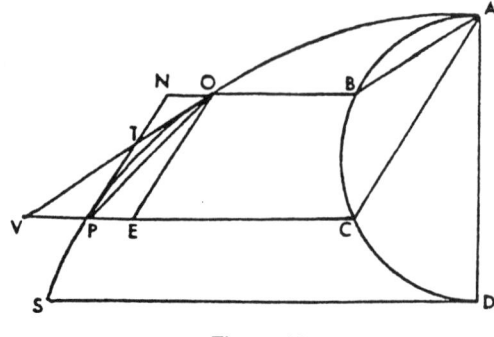

Figure 16

Wren's discovery seems to have astonished his contemporaries to judge by the enthusiasm of the Dutchman Huygens who wrote, in 1659, to one of correspondents: *"this invention will make this a famous curve because it is the first, and perhaps the only, curve capable of being rectified."*[1] At his request Pascal sent Huygens a rectification of the cycloid in February 1659 obtained by the summation of an "infinite number" of arcs of the cycloid and arcs of circles. He concluded his letter in writing that *"this admirable equality of the roulette curve with a straight line was nothing more than an equality by accident."*[2] Thus Descartes' prejudices retained their strong hold. Being able to rectify the cycloid did not imply that it could be done with any curve whatsoever: in order to consider the cycloid one had to assume that its base was the length of the circumference of its generating circle!

Kromme: Here's what I suggest. Over a small interval the arc of a curve differs by a small amount from the tangent and I may find it by using Pythagoras' theorem: in this way I can find a good approximation to the arc over a small interval. For larger arcs all I need to do is to add up a number of small pieces.

Recht: The errors will accumulate so your result will hardly be of value.

Kromme: What encourages me is our study of curves as envelopes of their tangents: the envelope has to be thought of as if its arc is almost the same as the polygonal line made by the pieces of tangent between their intersection.

Recht: We could make the interval more and more small. But for a given arc this would increase the number of small arcs to be summed. I concede, intuitively, that the error will diminish for each small interval, but with an increasing number of intervals it is hardly likely that the total error will 'vanish', the 'derived values' will accumulate and we shall probably end up with goodness knows what.

Kromme: Perhaps that is so but for any given curve the problem is to show that it is not so.

Recht: If we have two polygonal lines, the one greater and the other less than the arc in question and if, as we take increasingly smaller intervals, we can show that both lines yield the same result, then we should be satisfied. Isn't that how Fermat proceeded in the 17th century? Let's find his geometrical treatise on this subject.

Clio: In 1660, following Wren's sensational discovery, Fermat wrote his treatise *On the comparison of curves and lines (De la comparaison des lignes courbes avec des lignes droites)*, where he inveighed against the prevailing prejudice.

---

1.   Huygens, *Œuvres complètes,* vol. II, p. 315.

He wrote:

> *Never before, to my knowledge, has a purely geometric curve been shown by the geometers to be equal to a given straight line. That which, in essence, a subtle English mathematician has recently discovered and proved, namely that the primary cycloid is four times the diameter of its generating circle must, according to the wisest geometers, be considered as a special case. They think, in effect, that it is the law and order of nature that it is not possible to find a straight line equal to a curve, at least unless one supposes first of all that there is another straight line equal to a curve, and taking this example of the cycloid, they show that this is the case here. I do not deny it: it is clear, in fact, that the graph of the cycloid presupposes the equality of another curve with a straight line, namely that of the circumference of the generating circle with the straight line that forms the base of the cycloid. But we shall see below what is the nature of this base that they seek to establish, and how it is dangerous to deduce an axiom directly from the examples of one or two results. I am going to show, in essence, the "equality between a straight line and a purely geometric curve" and for its construction one does not have to presuppose any other equality of another curve with a straight line, and I shall deal with the whole question as succinctly as possible.*[1]

Kromme: Fermat considered a concave curve with base AF and axis FG and from a point H of the curve he drew a tangent IHK. (figure 17) From the two points I and K of this tangent he drew perpendiculars IB and KD to the base AF, cutting the curve at R and M. Then he drew IX, KY perpendicular to the axis FG, cutting the curve at O and P. Fermat compared segments of the tangent with parts of the curve and proved that:

$$\text{arc HO} < \text{IH} < \text{arc HR} \qquad \text{and} \qquad \text{arc HM} < \text{HK} < \text{arc HP}$$

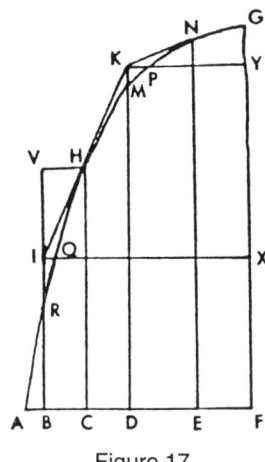

Figure 17

Fermat's method consisted in constructing two figures formed by segments of tangents circumscribing the curve such that, by the preceding result, the length of the first circumscribed figure is greater than that of the curve, which is itself greater than the second circumscribed figure. He then went on to show that it is possible for each of these greater amounts to be made less than and given length. In this way he obtained the length of the curve itself by a 'summation' of segments of tangents.

2.   Pascal, *Œuvres*, p. 184.
1.   Fermat, *Œuvres,* vol. III, p. 181.

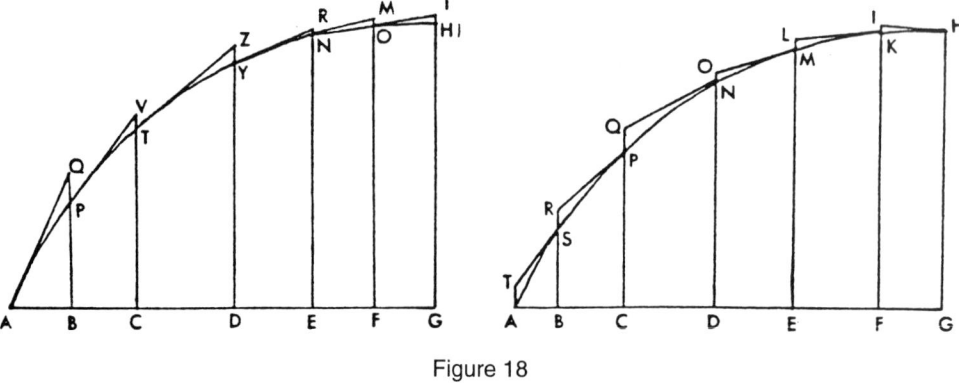

Figure 18

Clio: Having explained his method, Fermat wrote: *"I dare to claim that it is possible to find an equality between a truly geometric curve and a straight line."* [1] In order to convince his contemporaries, he followed this by finding the length of the semi-cubical parabola.

Kromme: However, I return to my first idea. Could we not, in the same way as we did in trying to find the curvature at a point of the curve, simply take an 'infinitely small' part of the curve to be the same as a bit of the tangent? Would not calculating the length of the segment of tangent, using Pythagoras, give the flux of the length of the curve, that is the speed with which it is increasing? I believe we shall gain something from another reading of Newton.

Recht: In fact, Newton obtained the fluxion of the length of a curve as the square root of the sum of the squares of the fluxions of the abscissa and the ordinate. He wrote:

> *For let RN be the perpendicular Ordinate moving upon the Absciss[2] MN. And let QR be the proposed Curve, at which RN is terminated. Then calling MN = s, NR = t, and QR = v, and their Fluxions, and, respectively; conceive the line NR to move into the place nr infinitely near the former, and letting fall RS perpendicularly to nr; then RS, Sr, and Rr, will be the contemporaneous moments of the lines MN, NR, and QR, by the accession of which they become Mn, nr, and Qr; but as these are to each other as the Fluxions of the same lines, and because of the Rectangle RSr,[3] it will be $RS^2 + Sr^2 = Rr^2$ [4]*

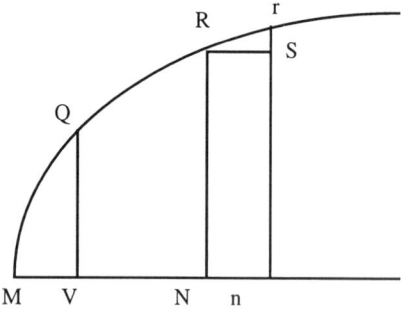

Figure 19

---

1.   *Ibid.*, p. 185.
2.   abscissa [tr.].
3.   Right angle RSr [tr.].
4.   Newton, *The Mathematical Works of,* vol. 1, p. 128-129.

Kromme: There remains the question of how we may determine a fluent starting with its fluxion! That is to say, how can we find a curve when we know its tangents? And we can revisit one our earlier problems.

Exercise 7	Find the length of an arch of a cycloid by Newton's method. Express x and y as functions of the angle θ through which the generating circle turns.

Recht: Well, that's enough! As the policeman said the other night, there's both right and wrong on each side and this dialectic discussion has brought us closer together.

Kromme: I am delighted as well. All is not clear, far from it. But it was a pleasure to find in the past echos of our questions and to discover the course of mathematical research.

This lively exchange between Kromme and Recht had occupied five of their afternoons without them ever being able to conclude which of them was weaving about and which was going in a straight line. As for us, we have been able to spell out the curious difficulty that exists in defining a straight line and a curved line and, more surprising, to understand the difficulty in constructing the one in relation to the other, particularly with reference to tangents. We have also discovered, along with the notions of 'circle of curvature' and 'rectification of a curve', that geometrical thought developed and deepened, going forward and turning round on itself at the same time, in a complex movement that ebbed and flowed between the need for rectitude and undeviating rigour on the one hand and, on the other, the irresistible attraction of multiple 'curves' which characterised the deviant, the non conforming, the uncalibrated, the unmeasured, ... and, even, the unimaginable!

Bibliography

Sources texts

ARCHIMEDE, *Œuvres complètes*, tr. Mugler, Les Belles lettres, Paris, 1971.

CAUCHY, *Résumé des leçons données à l'Ecole Royale Polytechnique sur le calcul infinitésimal*, Debure, Paris, 1823, réédition ACL-éditions, 1987.

DESCARTES, *La géométrie*, 1636, ed. Christian David, Paris, 1705.

    Encyclopédie Méthodique, Mathématiques, Tomes I, II et III, édition Panckoucke, 1784, réédition ACL-éditions, 1987.

    The Geometry, (facsimile edition of original and English translation by D.E.Smith and M.L.Latham), Open Court Publishing Co., 1925: Dover, New York, 1954.

DIJKSTERHUIS, *Archimedes,* Copenhagen, 1956.

EUCLIDE, *Les Eléments*, édition P.U.F., Paris, 1990.

EULER, *Introduction à l'analyse infinitésimale*, Barrois, 1796, reédition ACL-éditions, Paris, 1987.

FERMAT, *Œuvres*, tome III, ed. Tannery et Henry, Gauthier Villars, 1896.

HEATH, *The Thirteen Books of Euclid's Elements,* Cambridge, 1925: Dover, New York, 1956.

HUYGENS, *Œuvres complètes*, pub. Société Hollandaise des Sciences, Nijhoff, La Haye, 1888-1950.

LEIBNIZ, Histoire et origine du calcul différentiel, 1713, M.S., tome V, trad. Szeftel-Zylberbaum, *Cahiers de Fontenay*, n° 1, 1975.

L'HOSPITAL, *Analyse des infiniments petits pour l'intelligence des lignes courbes*, 1696, Montalant, Paris, 1716.

NEWTON, *The Mathematical Works of Isaac Newton,* New York, Johnson Reprint Company, 1964.

NEWTON, *La méthode des Fluxions et des suites infinies*, 1736, tr. Buffon, Blanchard, Paris, 1966.

ORESME, *Tractatus de configurationibus qualitatum et motum*, in CLAGETT, *Nicole Oresme and the Medieval Geometry of Qualities and Motions,* University of Wisconsin Press, 1968.

PASCAL, *Œuvres complètes*, Seuil, Paris, 1963.

ROBERT, *Dictionnaire alphabétique et analogique de la langue française*, Robert, Paris, 1988.

ROBERVAL, *Observations sur la composition des mouvements et sur le moyen de trouver les touchantes des lignes courbes*, Recueil de l'Académie, tome VI.

TARTAGLIA, *Quesiti et inventioni diverse*, Brisciano, Venise, 1546.

WALLIS, *Tractatus Duo De cycloïde de cissoïde*, Typis Academicius Liechfieldianis anno Dom., 1659.

General works for further reading

BOYER, *The history of the calculus and its conceptual development*, Dover, New York, 1959.

CLERO & LE REST, La naissance du calcul infinitésimal au XVIIᵉ siècle, *Cahiers d'histoire et de philosophie des sciences*, n° 16, 1980.

KLINE, *Mathematical Thought from Ancient to Modern Times*, Oxford University Press, New York, 1972.

LEIBNIZ, *Naissance du calcul différentiel*, trad. Parmentier, Vrin, Paris, 1989.

How did you get on?

Ex. 1   Imagine a point moving along the graph. As it passes the points with abscissa -1 or 1 it changes direction "making an angle". According to the dictionary, neither the word straight line [droit] nor curve can be used for this graph.

Ex. 2   The proof rests on the concept of an acceleration vector and its invariance when the only acting force is the force of gravity, assumed constant within small variations of altitude. In a rectangular coordinate system take the origin as the point of projection at time $t = 0$, the acceleration vector as $(0, 0, a)$ and the vector of initial velocity as $(m, 0, p)$. Write down the velocity vector at some time t and hence find the position vector. In eliminating t the value of z can be found as a polynomial function of the second degree in x, and $y = 0$.

Ex. 3   Part 1: Using the suggested coordinate system, F is at $(0, p)$. Consider $H_0(0, -p)$ and $H_1(x_1, -p)$. The points $M(x, y)$ of the mediator of $[FH_1]$ satisfy the equality $FM^2 = MH_1^2$ giving $4py = 2x_1 x - x_1^2$. Putting $x_1 + h$ for x_1 gives the mediator of $[FH_2]$. The intersection is at $x = x_1 + h/2$ and by letting h become zero $x = x_1$, whence $y = x_1^2 / 4p$.

    Part 2: The calculation is rather more difficult. Let L be the fixed length. The straight line passes through $(x_1, 0)$ and $(0, \sqrt{(L^2 - x_1^2)})$. Write out its equation in the form $y = ax + b$, assuming $x_1 \neq 0$. In this equation replace x_1 by $x_1 + h$. The symmetry of the curve allow us to square these expressions. Equating the expressions for y^2 and simplifying, including cancelling through by h (non zero), then putting $h = 0$ in the simplified expression, we get $x = x_1^3 / L^2$ giving $y = (L^2 - x_1^2)^{3/2} / L^2$ and hence $x^{2/3} + y^{2/3} = L^{2/3}$.

Ex. 4   For the ellipse: $MF + MF'$ is constant so M moves towards F as it moves away from F' by the same amount. The point M is thus subject to two "equal motions", the one in the direction FM towards M the other in the direction MF'. The composition of these two motions is the diagonal of a rhombus based on [FM] produced and [F'M]. So the tangent is the external bisector of the angle FMF'.

    For the cycloid: The point M is subject to two 'equal motions': a circular motion whose direction is the tangent of a circle, and motion in a horizontal direction. The composition of these two motions gives the diagonal of a rhombus with one side horizontal and the other the tangent to the circle. Hence the tangent to the cycloid is the chord joining M and the 'highest point' of the circle.

Ex. 5   Let $A(x_0, y_0)$ be a point on the parabola $y = x^2$. The tangent at A is given by the vector $(1, 2x_0)$. Let θ (a, b) be the centre of a circle with a common tangent at A with the parabola. We have $(x_0 - a) + 2x_0(y_0 - b) = 0$ whence $a = x_0 + 2 x_0(y_0 - b)$. The equation of the circle is given by $(x - a)^2 + (y - b)^2 = (x_0 - a)^2 + (y_0 - b)^2$ and its intersection with the parabola is found by putting $y = x^2$ and $y_0 = x_0^2$. Having done this, and using the expression for a, we get the required equation. x_0 is always a double solution: in order to obtain more than two solutions we need to put $b = (6 x_0^2 + 1) / 2$ which gives $a = -4x_0^3$. The equation then becomes $(x - x_0)^3(x + 3x_0) = 0$. x_0 is a triple solution if $x_0 \neq 0$ and is only a solution four times when $x_0 = 0$, the case considered by Kromme.

Ex. 6   Let $A(x_0, y_0)$ be a point on the parabola $y = x^2$ and let $M(x, y)$ be a point on the normal. The vector AM is orthogonal to the tangent vector $(1, 2x_0)$. The equation of the normal becomes, after simplification, $2x_0 y = -x + x_0 + 2x_0^3$. In the same way, at the point on the parabola with abscissa $x_0 + h$ we get

$$2(x_0 + h)y = -x + (x_0 + h) + 2(x_0 + h)^3$$

and hence

$$2hy = h + 2(3x_0^2 h + 3x_0 h^2 + h^3)$$

and, for $h \neq 0$,

$y = 1/2 + 3x_0{}^2 + 3x_0 h + h^2$.

Then putting $h = 0$ we get

$y = \dfrac{1}{2} + 3x_0{}^2 = (6x_0{}^2 + 1) / 2$

see. preceding example) and

$x = x_0 + 2x_0{}^3 - x_0 - 6x_0{}^3 = - 4x_0{}^3$,

giving the equation of the envelope of the normals as

$x^2/16 - (y - 1/2)^3 / 27 = 0$.

This curve can be visualised by drawing normals to the parabola or by plotting the centres of circles of osculation.

Ex. 7   If the origin is taken as the starting point then the vector OM is equal to $[\, R(\theta - \sin \theta), R(1 - \cos \theta)\,]$. So,

$dx = R(1 - \cos \theta)\, d\theta$

and

$dy = R \sin \theta\, d\theta$,

and from this we get

$ds^2 = dx^2 + dy^2 = 2\, R^2 (1 - \cos\theta)\, d\theta^2 = 4\, R^2 \sin^2 (\theta/2)\, d\theta^2$.

By integrating

$ds = 2\, R \sin (\theta/2)\, d\theta$

over the interval $[0, 2\pi]$ we obtain 8R for the length of the arch of the cycloid.

6

Movement and Geometry

Rudolf BKOUCHE and Joëlle DELATTRE
IREM Lille

Introduction

It is as a result of a confusion of ideas that many geometers wish to ban considerations of movement *from elements of geometry.*
The idea of movement as abstracted from the time used to accomplish it, that is the idea of geometric movement, *is an idea no more complex than that of size or distance. It could even be said, entirely rigorously, that this idea is identical to that of size, since it is precisely by means of movement that we come to appreciate size.*
This geometric movement, that equivocal language confuses with movement *in time which is the subject of kinematics, does not depend on any other science than pure geometry.*
It is advantageous to introduce this idea of geometric movement as soon as possible and as explicitly as possible. Much will be gained from the clarity and precision of the language, and it offers the opportunity to the learner of being better prepared for the later introduction in the movement the ideas of time and speed.
Moreover, all authors use the idea of movement unconsciously; it is difficult to find a single proof of a fundamental theorem in geometry, in which the idea of geometric movement, more or less disguised, does not have a part.[1]

It is in these terms, that Jules Houël makes the place of movement in geometry explicit, in his *Essai critique sur les Principes de la Géométrie élémentaire.* We find here the double idea of movement that we met in Chapter 5. Alongside movement *in time*, which received mathematical treatment with the birth of mathematical physics in the eighteenth century, we have the notion of movement *outside of time* which takes place only in geometry.

In order to understand this novel introduction of time into geometry, we should recall the mathematisation of time that occurred at the beginning of the Classical Age, particularly with Galileo, a mathematisation that led to the development of the mechanics that he knew. In this context, the study of the notion of velocity led to the problem of determining tangents to curves as trajectories of moving points (see Chapter 5), a problem which, in its turn, led to the use of movement as a means of constructing

---

1.   Houel, *Essai critique...*, p. 59-60.

tangents. These constructions were to give a new legitimacy to the use of movement in geometric argument, a use that had been rejected by the Greeks.[1]

Later, at the beginning of the nineteenth century, Ampère would identify, within mechanics, a study of movement *"independent of the forces that produce it"*. This was that part of mechanics that he called kinematics (from the Greek *kinesis*: movement). Ampère went on to explain the pedagogical arguments in favour of this approach:

> *A work which in this way considered all movements independently of the forces which produced them, would be extremely useful for teaching, in presenting the difficulties that arise with a variety of different machines, without the mind of the pupil having at the same time to cope with resolving the equilibrium of forces.*[2]

The result for works on mechanics was to separate the treatment of kinematics from dynamics (which dealt with the causes of movement), whereas in the eighteenth century, movement and the forces which caused it, were treated simultaneously: examples are D'Alembert's *Traité de Dynamique* and Lagrange's *Mécanique Analytique*.

Kinematics rests on geometry, and the connection between kinematics and geometry was developed throughout the nineteenth century. This led to the creation of a purely geometric study of movement, a study which Mannheim defined in his course at the Ecole Polytechnique, in the following way:

> *The object of Kinematics is the study of movement independent of the forces which cause it; the object of Kinematic Geometry is the study of movement independent of forces and of time, that is it concerns the study of displacements. We shall use the term displacement for a movement that is considered without taking account of speed.*[3]

And so movement was studied in the nineteenth century, on the one hand, independently of its causes, and on the other, independently of time.

Constructing tangents using composition of movements

The use of movements in finding tangents rests on the composition of movements. Given a curve defined by the movement of a point subject to a number of conditions, the tangent to the curve (determined by the direction of its velocity) will be found by the composition of the velocities of the component movements.

The composition of movements had already appeared in Aristotle, when he defined what was called the parallelogram of movements (the notion of velocity had not yet been defined). He wrote:

> *When a body is moved in a certain ratio, it must move in a straight line, and this straight line is the diagonal of the figure formed from the two straight lines which have the given ratio.*

1.   Bkouche, *Variations autour de la réforme 1902/1905*, p. 187.
2.   Ampère, *Essai sur la philosophie des sciences*, p. 52.
3.   Mannheim, *Cours de géométrie descriptive*, third lesson.

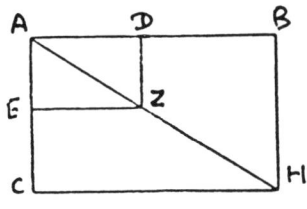

Figure 1

For, let the ratio according to which the body moves be that of AB to AC; let AC be moved towards B while AB be moved towards HC; and let A travel to D, while AB travels to a position marked by E. If the ratio of the movement is that of AB to AC, then AD must needs have the same ratio to AE. Therefore the small quadrilateral is similar to the larger, so that they have the same diagonal, and the point A will be at Z.[1]

It was by using the composition of uniform movements and uniformly accelerated movements, that Galileo, in his *Discourse on Two New Sciences*, was able to show that the trajectory of a projectile was parabolic. We may recall that the work was in the form of a dialogue between three persons, which took place over four days. After studying uniform movement and uniformly accelerated (rectilinear) movement during the first three days, Galileo wrote this at the beginning of the fourth day:

> *The properties of uniform movement, such as those of naturally accelerated movement on inclined planes, have been the object of our earlier considerations. In the study which I am about to embark on, I shall do my best to throw light upon, and to establish on firm proofs, certain consequences which are particularly important and worth knowing, concerning a particle whose behaviour is the subject of a double movement, namely a uniform movement and uniformly accelerated movement; for this type certainly appears to be the movement which we ascribe to projectiles. As for how this movement is produced, I shall explain in what follows.*
>
> *I imagine that a particle is projected on a horizontal plane from which all obstacles have been removed; it is certain, following what has been already said elsewhere a long time ago, that its movement will proceed uniformly and eternally on this same plane, provided that it is extended to infinity. Suppose, on the other hand, that the plane is limited and at a certain height; the particle which I imagine to be endowed with gravity, reaching the end of the plane, and continuing its course, will add to its preceding uniform and indelible movement, the tendency downwards which gravity gives to it; the result will be that movement, composed of a horizontal uniform movement and a naturally accelerated downwards movement, which I call projection.[2]*

By the composition of the movements that he described, Galileo was then able to determine the trajectory of the projectile. The composition of movements is, in fact, the basis of Roberval's method for finding tangents to curves (see Chapter 5).

The tangent to the cycloid (Descartes' approach)

In 1638, Descartes' approach to finding the tangent at a point of the cycloid used only the direction of motion, thus eliminating all reference to time in his argument. He describes his method in a letter to Père Mersenne (23 August 1638), a part of which we reproduce below.

---

1.   Aristotle, *Mechanics* 1, 848 b, in Thomas I, p. 432-433.
2.   Galileo, *Discours concernant deux sciences nouvelles*, p. 203.

The cycloid is the curve generated by a point of a circle (the roulette) which rolls without slipping along a straight line (the base). Descartes' method consists in considering what happens when a regular polygon is rolled along a line, and then considering the circle as a regular polygon composed of an infinite number of sides.

The first of these questions is to find the tangents to curves described by the movement of a roulette. To which I reply that the straight line which passes through the point of the curve where one wishes to find the tangent, and the point of the base which the roulette touches when it describes that point, always cuts that tangent at right angles. It follows that if, for example, one wishes to find the straight line at B which touches the curve ABC, described on the base AD by one of the points of the circumference of the roulette DNC, one must draw through this point B, the line BN parallel to the base AD, then draw another line from the point N, where this parallel meets the circle, this line being towards the point D where the roulette touches the base, and afterwards draw BO parallel to ND, and finally draw BL which is at right angles to ND; now this line is the required tangent.

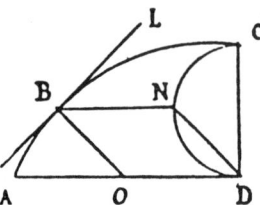

Figure 2

For this I shall only offer a proof which is extremely short and extremely simple. If a rectilinear polygon, of any sort, is made to roll along a straight line, the curve described by one of its points, whatever it might be, will be composed of many portions of circles, and the tangents at all the points of each of these portions of circles, will cut at right angles the lines drawn from these points towards the point at which the polygon touches the base when it describes that portion of a circle. Following from which, taking the circular roulette to be a polygon with an infinite number of sides, it can be clearly seen that it must have this same property, that is that the tangents at each of the points which are on the curve which it describes, must cut at right angles the lines drawn from those points towards the points of the base which they meet [sont touchés par elles], at the same moment when it describes them.

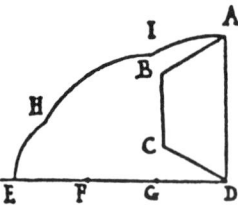

Figure 3

Thus, when the regular hexagon ABCD is made to roll along the straight line EFGD, its point A will describe the curve EHIA, composed of the arc EH, which is described while this hexagon touches the base at the point F which is the centre of that arc, the arc HI whose centre is G, the arc IA whose centre is D etc., through which centres pass all the lines which meet the tangents at right angles. Now the same will happen with a polygon composed of a hundred thousand million sides, and consequently also for a circle. I could demonstrate this tangent in another way, which is more beautiful in my view and more geometric; but I omit it in order to save me the trouble of writing it out because it would be a little longer[1]

Descartes thus shows, that the movement of the plane containing the circle, on the fixed plane, is at each instant, a rotation about the point of contact between the circle and the base. In the later part of his letter, Descartes studies the movement of points of the moving plane, which lie both inside and outside of the circle.

Exercise 1	Determine the path of the trajectory of a point in the moving plane and construct the tangent at a point of that trajectory.

Exercise 2	Compare Descartes' method with that of Roberval (see Chapter 5).

Following Descartes, Johann Bernouilli studied displacements of figures and defined a spontaneous centre of rotation, and then Euler and D'Alembert used methods of analytical geometry. This led to the development of analogous methods for the treatment of the movement of solid bodies in space.[2]

This area of study was developed geometrically by Chasles and Poinsot in the nineteenth century. In his memoir *Théorie Nouvelle de la Rotation des Corps*, presented to the Academy of Sciences in 1834 but not published until 1851, Poinsot, in comparing the methods of Euler and Lagrange, makes a case for the study of geometry in its own right:

> *And so we are led by reasoning alone to the clear idea that the geometers did not obtain their formulas from analysis. This is a new example which shows the advantage of this simple and natural way of considering things in themselves, and without them being lost to in the course of the argument.*[3]

In what now follows we shall develop Chasles' methods.

The instantaneous centre of rotation

In 1830, Michel Chasles presented a memoir to the Philomathematical Society with the title *Notes sur les propriétés générales du système de deux corps semblables placés d'une manière quelconque dans l'espace et sur le déplacement infiniment petit d'un corps solide libre* (*Notes on the general properties of a system of two similar bodies at any position in space and on the infinitely small displacement of a free solid body*), of which one part, dedicated to the construction of tangents to a curve, was published in 1878 in *Bulletin de la Société mathématique de France* (*Bulletin of the French Mathematical Society*). In that article, Chasles shows how the idea of movement can be used to construct tangents to curves, as well as second order elements, like curvature and centres of curvature (see Chapter 5 and below). Movement here appears free of all temporal considerations, and in this sense we have here an article that is representative of that species of geometry of movement of which we spoke earlier.

---

1.   Descartes, *Œuvres*, t. II, p. 307-313.
2.   Chasles, *Notice historique*, p. 491-495.
3.   Poinsot, *Théorie nouvelle de la rotation des corps*, p. 86.

Chasles based his method on that of Descartes, which he intended to generalise as he explains in his *Aperçu Historique* (1837) in a resume of his memoir concerning the construction of tangents:

> *The theorem which follows appears to us to offer a generalisation of Descartes' method:*
> When a plane figure is subject to an infinitely small movement in its plane, there is always one point which remains fixed during this movement.
> The straight lines drawn through the different points of the figure moving in its own plane, perpendicularly to the trajectories that they follow during this infinitely small movement, will always pass through this fixed point.
> *From this theorem, when a curve is described by a point of a figure moving in its own plane, then to find the normal through the point, it is sufficient to determine the point that remains fixed at the moment when the point under consideration has the position that is being considered. This point will be determined by the different conditions of the movement of the figure.*
> *For example, if the movements of two points of a figure are known, the normals to the curves that they follow can be drawn through those points, and the point of intersection of these two normals will be the required point.*[1]

After recalling several examples which he had explained in 1830, Chasles continues:

> *What has been said is sufficient to show that the theorem that we have stated is a generalisation of Descartes' idea on the subject of the tangent to the cycloid, and that it constitutes a true method for [finding] tangents, a method that is different from all the others, and even different from Roberval's method, even though that method depends like this one on considerations of movement. But it can be seen that this method, which is so simple, is, like Roberval's, limited in its applications, since it presupposes that the geometrical conditions of the movement of the figure containing the point, are known. However, it can be applied to a large number of particular curves, and to whole families of curves.*[2]

In his memoir of 1830, reproduced in 1878, Chasles began by proving the following lemma:

> *If there are in a plane two equal polygons placed in any manner, and their corresponding vertices are joined by straight lines, the perpendiculars through the centres of these lines will all pass through the same point.*[3]

Exercise 3   Prove the lemma just stated (it is sufficient to prove it for triangles).

This lemma now allows a general property to be stated:

> *Whenever, given a plane figure, it is moved in any manner whatever, to a different position in its plane, there will always be one point of that figure which, following the movement, finds itself in the same place.*[4]

in other words:

> *all displacements of a figure in its own plane can be effected by a rotation about a fixed point.*[5]

We note that Chasles does not only consider the points of a given figure, but also all the points of the plane which are connect to that figure. We also note that Chasles is

1.   Chasles, *Aperçu historique*, p. 548.
2.   *Ibid.*, p. 549.
3.   Chasles, *Mémoire de géométrie...*, p. 210.
4.   *Ibid.*, p. 211.
5.   *Ibid.*

silent about the case where the movement is a translation; the reader is invited to complete the statements made by Chasles where necessary.

By consideration of infinitely small movements, Chasles deduces the following theorem:

> *If a plane figure, of whatever shape, is subject to an infinitely small displacement in its own plane, the normals drawn through its different points, to the directions of the infinitely small lines followed by these points, will all pass through a single point.[1]*

which can be also be stated as:

> *If any plane figure moves in its own plane, the normals, drawn through the different points of the figure, to their respective trajectories, at any particular instant, will all pass through a single point.[2]*

This point, considered as belonging to the figure in movement, remains fixed during an infinitely small movement of the figure. Consequently, the movement of a figure in its plane

> *always reduces, at each instant, to a rotation about a point that remains fixed during that instant.[3]*

Although in this memoir, the term *instantaneous centre of rotation* does not make an appearance, we would like to point out that Chasles did use the term in other articles. We also recall that D'Alembert used the term *instantaneous axis of rotation* in his works on infinitely small displacements of solid bodies. And we further remark that, in a note, Chasles stated an analogous property for the case of a sphere sliding upon itself.

Exercise 4   Prove the result stated by Chasles in the case of a sphere sliding upon itself, by deducing that the movement of a solid about a fixed point is, at each instant, a rotation about an axis passing through the fixed point.

Chasles is thus able to state the first principle of his method:

> ***First Principle.***
> *Whenever a curve is described by a point of a figure moving in its own plane, then to draw the normals to this curve it is only necessary to know the movements of two points of the figure.*
> *Draw normals through these points to the curves they describe, and join their intersection point to the point being described; this line will be the normal to the curve described by this point.[4]*

---

1.   *Ibid.* p. 211-212.
2.   *Ibid.*, p. 212.
3.   *Ibid.*
4.   *Ibid.*, p. 213.

Exercise 5   Consider a moving ruler whose ends A and B are constrained to slide along two fixed axes.

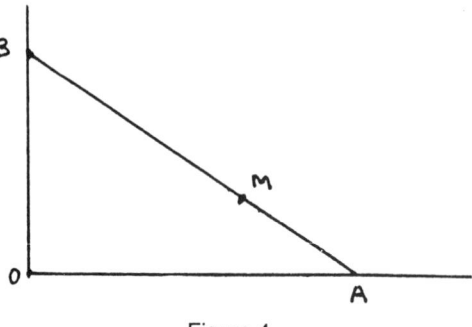

Figure 4

1) First let the axes be rectangular. Find the instantaneous centre of rotation and its geometrical locus in the fixed plane, then in the moving plane. Find the locus of a point M of the ruler. It can be shown that it is an ellipse with auxiliary circle centre O (the intersection of the axes) and radius equal to MA or MB. The solution can also be found analytically. Consider in the moving plane containing the ruler, the circle of diameter AB, and find the loci of the points lying on this circle. Hence find the locus of any point in the moving plane.

2) Show that whatever the axes a line segment can be found in the moving plane, the ends of which are constrained to lie on two rectangular axes.

3) Deduce from this, that if a plane slides upon itself in such a way that two of its points always lie on two fixed axes, then there are an infinite number of points of that plane which are constrained to lie on fixed straight lines which pass through the intersection of the two axes. All these points lie on the circumference of a circle in the moving plane; all other points describe ellipses.

After studying the movement of the moving plane, starting with two of its points, Chasles considered the movement of a curve in the moving plane, set himself the task of finding the envelope of this curve, that is the curve of the fixed plane for which the moving curve is *"perpetually a tangent"*.

Chasles approach to the problem was to state a second principle.

Second Principle.

Whenever a figure, of whatever shape, moves in its own plane, the normals at the points where its perimeter touches the envelope curve of the space through which it passes, at any instant of the movement, all pass through a single point; this point is the point through which the normals to the trajectory of the different points of the figure pass.[1]

In effect, according to Chasles, to find the points where a moving curve touches its envelope, it is necessary to consider that curve in an infinitely close neighbouring position, the required points being the intersections of these two infinitely near curves.

---

1.   *Ibid.*, p. 217.

Here is the proof offered by Chasles:

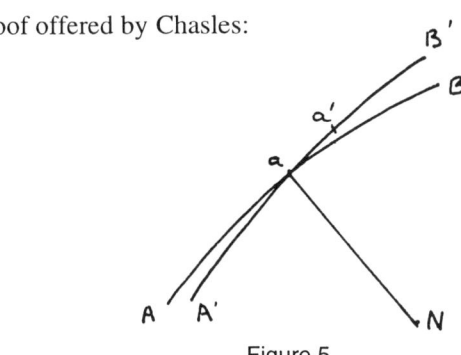

Figure 5

The curve AB, during its movement to take up the position A'B' has, as we have shown, turned about a certain point N, through which pass the normals to the trajectories of its different points. If we drop a normal from the point N onto that curve, then its foot a will, during its infinitely small movement, describe the arc aa' of this curve about the point N, and the point a' where will take one's place the point a will belong to the second position of the curve, A'B'; this point a' will therefore be the intersection of the two infinitely close neighbouring curves AB and A'B'. It is thus proved that the normals to the curve AB at the points of intersection of that curve with its infinitely closely neighbouring position, that is the normals to the points of the curve where it touches the envelope of the space through which it moves, all pass through a single point N, which is the point through which pass the normals to the trajectories of the different points of the moving figure. And so is the theorem proved.[1]

Question: What do you think of the proof given above?

The two principles stated above, then allow Chasles to solve the following problems:

Exercise 6	Let there be an angle of fixed size which moves in such a way that its sides always touch two given curves. Construct the normal to the curve described by the vertex of the angle; what happens when it is a right-angle?

Exercise 7	The vertices of all right-angles circumscribing an ellipse or a hyperbola lie on a circle concentric to that curve.

Exercise 8	The vertices of all right-angles circumscribing a parabola lie on a straight line which is perpendicular to its axis.

The properties stated in Exercises 7 and 8 were well known at the time Chasles was working. He himself, simply presented them as exercises in order to show how movement could be used to prove them.

Chasles then stated a third principle:

Third Principle.

Whenever a figure, of fixed shape, moves in its own plane, in such a way that its perimeter always passes through a fixed point, the normals to the trajectories of the

---

1.   *Ibid.*, p. 218.

different parts of the figure, and the normals at the points where its perimeter touches the envelope curve, pass through a single point situated on the normal to the perimeter, and its point is coincident with the fixed point. [1]

In effect, Chasles is saying that if a moving curve passes through a fixed point, that point can be regarded as *"an infinitely small branch"* of the envelope of the moving curve.

Exercise 9   Construct the tangent to a point on the conchoid of Nicomedes (see Chapter 4).

Exercise 10   The feet of the perpendiculars dropped from a fixed point F onto the tangents of any curve produce a second curve whose normals pass, respectively, through the centre of the straight line drawn from the fixed point to the point of contact of the given curve and its tangents.

If r is the point symmetric to F with respect to the tangent to the given curve, what is the normal to the locus of r? Hence deduce the following proposition:

If rays of light coming from a fixed point are reflected by a curve, and if on the reflected rays, lines are taken from the points of incidence, equal in length to the incident rays, their extremities will produce a curve which is a trajectory orthogonal to the reflected rays. [2]

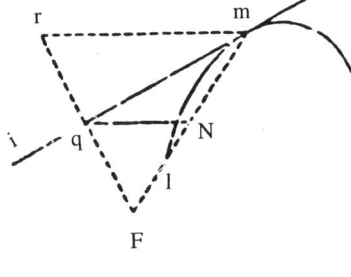

Figure 6

Exercise 11   Consider a right-angle whose vertex describes a given curve and one side of which passes through a fixed point.

Construct the point where the second side touches its envelope. If the vertex of the right-angle describes a straight line, the envelope of the second side will be a parabola.

If the vertex of the right-angle describes a circle, the envelope of the second side will be an ellipse or a hyperbola.

---

1.   *Ibid.*, p. 223.
2.   *Ibid.*, p. 229.

We note that the third principle implies a fourth:

Fourth principle.

When a curve of constant shape moves in its own plane, always passing through two fixed points, each point of the curve describes a trajectory whose normal passes through the intersection point of the normals to the moving curve through the two fixed points. The feet of the other normals from this intersection point to the curve are the points where it touches the envelope curve of the space through which it moves. [1]

Exercise 12 Consider an angle of constant size whose two sides pass through fixed points. Find the instantaneous centre of rotation and its geometrical locus:

I) in the fixed plane,

II) in the moving plane.

The T-square theorem

The study of infinitely small movements of the plane sliding upon itself can be used as a method for constructing tangents. It can also be shown that it can be used as a way of determining second order elements, viz. the centre of curvature and the radius of curvature. We may recall that for a curve defined parametrically (in the case of the trajectory of a moving point, for example, the parameter would be time), the tangent to the curve at a point (a first order element) is defined by the derivative of the position of the moving point with respect to the parameter, while the centre of curvature and the radius of curvature are determined by the second derivative, and so are second order elements (see Chapter 5).

The geometers of the nineteenth century tried to give geometrical interpretations to definitions of an analytic character, like those above, so that they could reason directly form those interpretations, without having to make a detour into calculations using analytic methods. This is the essence of Poinsot's text quoted above.

The centre of curvature at a given point of a curve, can be defined as the point on the normal to the curve, at that point where the normal touches its envelope, that is the point where the normal meets the normal of an infinitely close neighbouring point. The radius of curvature is then the radius of the circle, with centre the centre of curvature, and passing through the given point of the curve. It can be shown that this circle meets the curve in three points, coincident with the given point.

Let us consider a plane curve, moving in its own plane, of which two of its points slide along a fixed curve, then the instantaneous centre of rotation is the intersection of the normals at A and B of the fixed curve; the points where the moving curve touches its envelope are the feet of the normals to the curve drawn through the instantaneous centre of rotation. Suppose now that the point B is infinitely close to the point A, then the moving curve rests tangential to the fixed curve and the instantaneous centre of rotation coincides with the centre of curvature of the fixed curve at the point where it touches the moving curve.

---

1. *Ibid.*, p. 231.

Hence the fifth principle can be stated:

Fifth principle.

If a curve of any shape slides upon another fixed curve in such a way that the point of contact is always at a single point of the moving line, the normal to the curve described by any point of the moving curve will pass through the centre of curvature of the fixed curve, at the point of contact of the two curves; and the feet of the normals dropped from the centre of curvature onto the moving curve are the points where it touches the envelope curve of the space through which it moves.[1]

Exercise 13   Let a right-angle move in its own plane in such a way that one of its sides remains a tangent to the locus of the vertex of the right-angle, then the instantaneous centre of rotation is no other than the centre of curvature of the locus of the vertex of the right-angle (the T-square theorem).

Exercise 14   The tractrice (or tractoid in Chasles' text) is a curve for which the segment of the tangent from its point of contact to a fixed straight line is of constant length.
Construct the centre of curvature at a point of such a curve. Show that the locus of the centre of curvature of a point on the curve is a catenary (the differential equation of the locus can be obtained by taking the fixed line to be the *x*-axis).

Exercise 15   The logarithmic spiral can be considered as the locus of the vertex of an angle of fixed size, one of whose sides passes through a fixed point and the other remains a tangent to the locus of the vertex.

Construct the centre of curvature at a point of the logarithmic spiral and show that it describes a curve that is similar to the given one.

One curve rolling along another

Among those movements of a plane sliding upon itself are those which are defined by a curve rolling along a fixed curve (today we would be careful to say rolling without sliding). Examples are a circle rolling along a straight line, which generates a cycloid, and a circle rolling around another circle, which is the movement which occurs in gear systems. This last was studied by Philippe de La Hire (1694), who considered the locus of a point on a moving circle, today called the epicycloid or hypocycloid, depending on whether the moving circle lies inside or outside of the fixed circle. In particular, Philippe de La Hire found the tangents to this locus.

---

1.   *Ibid.*, p. 233.

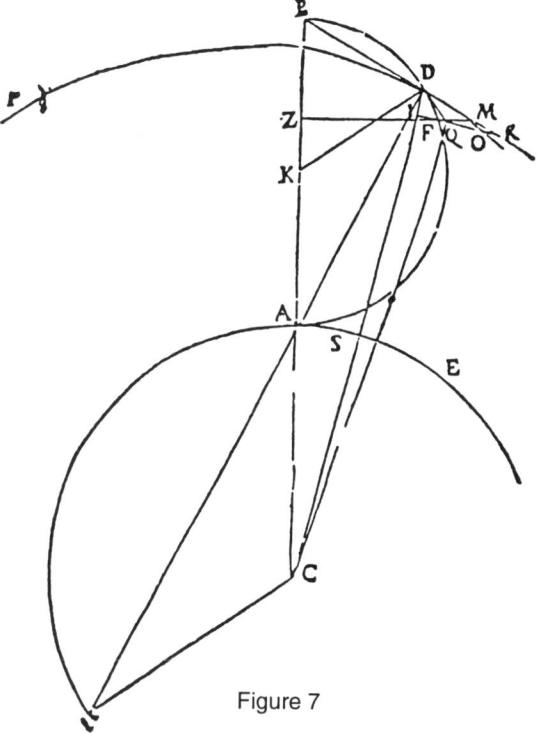

Figure 7

Using the constructions of Descartes and de La Hire, Chasles noticed that when a moving curve rolls along a fixed curve, the instantaneous centre of rotation is none other than the point of contact of the moving curve with the fixed one: he then showed that all movements of a plane sliding upon itself are of this same type. In particular, he showed that the locus of the instantaneous centre of rotation in the moving plane (the roulette) and the locus of this same point in the fixed plane (the base) are tangents at that point and that the locus in the moving plane rolls upon the locus in the fixed plane.

Here is Chasles' proof:

> *In effect, the moving figure turns at each instant about a point that is fixed during this infinitely small instant. Let N, N', N'', ... be the points about which the figure successively turns, these points being considered as belonging to the fixed plane on which the figure moves, and which points form an immobile curve.*
>
> *Let N be one of these points about which the figure is turning at the instant when we are considering it. Take v, v', v'', ... to be the points in the plane of that figure which, during the movement, coincide with the points N, N', N'' ... supposed fixed and belonging to the fixed plane over which the figure moves. The points v, v', v'', ... will form a curve which is part of the figure and moves with it. This curve passes through the point N about which the figure turns at the moment when we are considering it.*
>
> *After an infinitely small movement about the point N, the point v' will come into contact with the point N'; after an infinitely small movement about N', the point n'' will come into contact with the point N'', and so on. The parts Nv', v'v'', v'',v''', ... of the moving curve Nv'v''... will then successively come to coincide with the parts NN', N'N'', N''N''', ... of the fixed curve NN'N'' ... which proves that the curve Nv'v'' ... rolls along the curve NN'N'' ...; hence:*

No matter what the nature or duration of the movement of a figure in its plane, that movement is no other than that which is produced by the rolling of a certain moving curve along another fixed one.[1]

Exercise 16   By using an analogous argument for the movement of a sphere sliding upon itself, show that such a movement is no other than that which is produced by a moving curve on the sphere rolling along a fixed curve on that sphere. Hence deduce the theorem:

Whatever may be the movement of a solid body, subject to the restraint of a fixed point, it is no other than that which is produced by the rolling of a certain conic surface, having that point as its vertex, upon another conic surface having the same vertex.[2]

Remark. Another proof of this last property may be found in the memoir by Poinsot quoted above.

We leave it to the reader to think about the proof offered by Chasles. What appears in the proof is the idea of a pure differential geometry. There is no appeal to the differential calculus: the proof itself making reference to infinitely small movements and to a purely geometric argument about infinitely small parts. Such arguments, which were often used in classical times (like Descartes in the text we quoted dealing with tangents to the cycloid), pose the problem of legitimacy and the tendency today is to use the differential calculus as is the case in works on kinematics. However, the calculus obscures the visual aspect of problems, which is an essential aspect of geometry and elementary mechanics. In this sense Chasles' proof allows us to see how the locus of the instantaneous centre of rotation in the moving plane rolls along the locus of the same point in the fixed plane. We would also like to point out that the use of the differential calculus brings back time into the argument, at least in respect of the use of a parameter which is differentiated. On the other hand, it has the advantage of making it clear that the roulette rolls without slipping on the base, in that the relative speed of the instantaneous centre of rotation on the roulette is shown to be equal to its speed on the base.

Exercise 17   Find the base and the roulette for the movement defined by a ruler whose ends slide along two axes.

---

1.   *Ibid*. p. 235-236.
2.   *Ibid.*, p. 236.

Exercise 18   A circle rolls without slipping on a circle of twice its radius, the moving circle being inside the fixed circle (de La Hire's gearing).

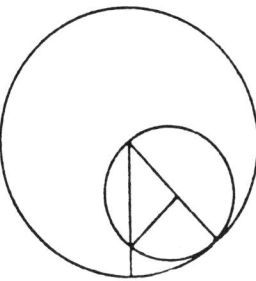

Figure 8

Show that the path of a point on the moving circle is a straight line. Show that the path of a point on the moving plane, not on the circle, is an ellipse.

Exercise 19   Consider a straight line passing through a fixed point and let a point on the line describe a given curve. Find the instantaneous centre of rotation.

I) The given curve being defined in polar co-ordinates with the origin as the fixed point, write down the polar equation of the base.

II) Find the base and the roulette when the given point describes a straight line (this is the movement made by an apparatus that draws concoids).

III) Find the base and the roulette when the given point describes an Archimedean spiral, with the origin as the fixed point. From this deduce a mechanical procedure for constructing an Archimedean spiral.

IV) Find the base and the roulette when the given point describes a logarithmic spiral (see exercise 15).

The duality of the turner

Whenever a moving plane slides upon a fixed plane, we can consider the inverse movement, the moving plane being thought of as fixed and the fixed plane as sliding on the moving plane: this amounts to saying that the base rolls along the roulette. It is this that Chasles calls the duality of the turner:

> *For each object that is worked on by a turner, there are two ways of making it; the first by fixing the object and moving the tool; the second way used by the turner is to fix the tool and make the work move.* [1]

Chasles makes this duality, employed by the turner in his work, explicit in writing:

> *Whenever a plane figure is moving in its own plane, one of its points describes a curve;*

---

1.   Chasles, *Aperçu historique*, p. 409.

The movement of that figure is determined by the constant relations which must exist between it and the points or fixed lines drawn out in its plane;

These points and these lines form together a second figure which remains fixed during the movement of the first figure;

The first figure may now be considered at one of its positions and supposed fixed; then the second figure can be made to move in such a way that it always maintains its same relative position to the first figure;

A fixed pen, placed at a point which describes the first figure, will trace out on the moving plane of the second figure, a moving curve in that plane, and it will be identical (except for position), to the curve that the point originally traced out when drawing the first figure, when that figure was moving.[1]

In other words, there are two ways of drawing a curve on a sheet of paper. The pen can be moved across the paper, or the pen can remain fixed, and the paper can be made to move under it, maintaining the same relations between the pen and the sheet of paper.

This allows Chasles to prove the following property:

When the sides of an invariable angle slide on two fixed points, a fixed pen, placed at any point whatsoever of the moving plane of the angle, will trace out an ellipse.[2]

Exercise 20   Prove the property stated above.

Exercise 21   Determine the inverse of the movement defined by a straight line which passes through a fixed point, and which contains a point which describes an Archimedean spiral with the fixed point as origin (see Exercise 19 III).

Exercise 22   Determine the inverse of the movement defined by a straight line which passes through a fixed point, and which contains a point which describes a logarithmic spiral around that point (see Exercise 19 IV).

---

1.   *Ibid.*, p. 409-410.
2.   *Ibid.*, p. 410.

Bibliography

D'ALEMBERT, *Traité de Dynamique*, 2nd ed., 1758. Repr. Jacques Gabay, Paris, 1990.

AMPÈRE, *Essai sur la Philosophie des Sciences*, 1834. Repr. Culture et Civilisation, Brussels, 1966.

BKOUCHE, "Variations autour de la réforme de 1902/1905" in Gispert, "La France Mathématique", *Cahiers d'Histoire et de Philosophie des Sciences*, Paris, 1991.

BRICARD, *Cinématique et Mécanisme*, Armand Colin, Paris, 1947.

CHASLES, *Aperçu historique sur l'origine et le développement des méthodes en géométrie*, 1837. Repr. Jacques Gabay, Paris, 1989.

CHASLES, "Notice historique sur la question du déplacement d'une figure de forme invariable", *Comptes-Rendus de l'Académie des Sciences de Paris*, **52**, 1861, 489-501.

CHASLES, "Mémoire de Géométrie sur la construction des normales à plusieurs courbes mécaniques", *Bulletin de la Société Mathématique de France*, **6**, 1878, 208-251.

DELTHEIL & CAIRE, *Compléments de Géométrie*, Baillères et Fils Editeurs, Paris, 1951. Repr. with the geometry course "Mathématiques élémentaires" by the same authors in *Géométrie et Compléments*, Gabay, Paris, 1989.

DESCARTES, *Œuvres*, ed. Adam & Tannery, t. II. Repr. Vrin, Paris, 1988.

GALILEO, *Dialogue concerning two new sciences,* Giorgio de Santillana, University of Chicago Press, Chicago, 1953. Repr. Dover, New York.

HOÜEL, *Essai critique sur les principes fondamentaux de la géométrie élémentaire*, Gauthier-Villars, Paris, 1867.

LAGRANGE, *Mécanique Analytique*, 1788. Repr. Jacques Gabay, Paris, 1900.

DE LA HIRE, *Mémoires de Mathématiques et de la Physique*, Paris, 1694.

MANNHEIM, *Cours de Géométrie descriptive de l'Ecole Polytechnique*, Gauthier-Villars, Paris, 1880; the second part of this work is dedicated to geometry of movement.

POINSOT, "Théorie nouvelle de la rotation des corps", *Journal de Mathématiques pures et appliquées*, **16**, 1851, 9-72, 73-129, 289-336.

THOMAS, *Greek Mathematical Works*, 2 vol., Loeb Classical Library, London, 1941. Repr. 1980.

7

Line and Sign
A problem in geometry: use of different figures; Chasles' position

Henry PLANE
IREM Dijon

This chapter considers a problem of a particular type: it is not a question that others might ask of mathematicians but is instead, the sort of question that mathematicians ask themselves. It is not unusual, particularly with geometry, that the study of a problem requires the consideration of a number of different cases, each needing different calculations and different arguments. Faced with this complication, the mathematician naturally asks whether there might not be a method of solution that could deal with all possible cases at the same time. At the beginning of the 19th century such a hope was expressed by J-B Biot, a professor of the Collège de France, in his *Essai de géométrie analytique*, in the following terms:

> *What I mean by this term is the way algebra may be applied to geometry, not by using constructions which have to be varied for each case, but in a completely general way ... and being able to do this would be one of the most important services that could ever be offered to learning.*[1]

Biot went on to consider the problem of using negative numbers and their meaning as applied to the length of line segments. What we shall consider here are other examples, set in a historical context, which also bear witness to this quest for a unified method. The question is an internal one for mathematics but has consequences that go far beyond mathematics.

Discuss!

We can all remember, starting with the first problems in geometry, such as "*given a triangle ABC of height AH and median AM...*", that we were told to draw a number of figures to show that the points B, C, M, H may not be in the same order (figure 1).

---

1.   Biot, *Essai de géométrie analytique*, préface.

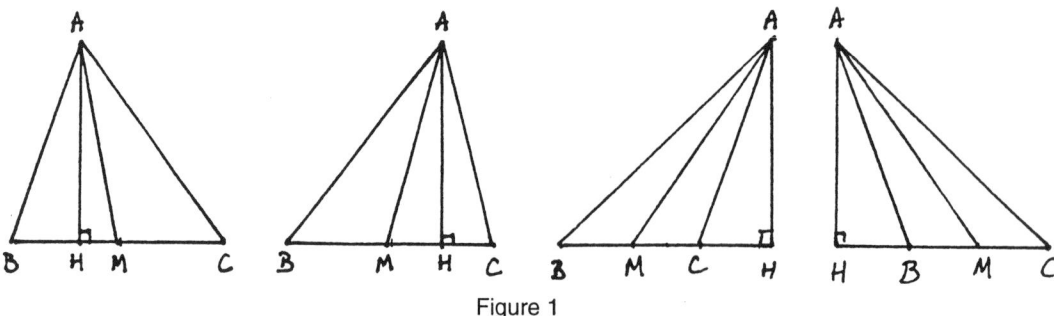

Figure 1

Here we were told: consider the problem according to the different cases. For, to draw conclusions based upon a particular figure, might well produce surprises when applied to other cases. A proof that depended upon a relation such as $BM = BH + HM$ "seen" in one figure can lead to error. It is true that a student at the end of the 20th century might write, whatever the figure,

$$BM = \left|\overline{HM} - \overline{HB}\right| \text{ (or even } \left\|\overrightarrow{HM} - \overrightarrow{HB}\right\|)$$

but a reading of works of geometry in the past shows that such a relation could only come to be stated after a number of detours. What precisely is it that is implied in a geometrical figure? or many figures.

Could there be a general method that can be applied to all cases? The connection between directed line segments and directed numbers only came about after a long and troublesome period of development, a development that we can only touch on here. If the answer lies in linking directed numbers to directed line segments, they are by no means one and the same thing.

For the Greeks, the Arabs and in the Middle Ages, number was thought of as 'magnitude' and usually, as with Euclid, the segment of a straight line. Descartes, right from the start of his *Géométrie*, links arithmetic to geometry. That is why he would not admit negative roots of equations except by transformations of these equations. Starting with a segment length 7, he would not add a segment to obtain a segment length 5. If $7 + x = 5$ is written down, it is only an artifice: an expression like $[x^2 + 5x + 6]$ cannot be put in the form $[(x + 2).(x + 3)]$ except to show the general rule about the number of roots of a polynomial.[1] In this context, it was hardly possible to think of a line segment being "less than zero". Indeed, Carnot, at the brink of the 19th century, struggled with the question in the same way for a long time.

Certainly the problem posed above can be resolved by saying that one of three segments BH, MB, HM will always be equal to the sum of the other two, but that does not avoid us in having to carry out three different calculations!

Before looking at some texts I would like to make some historical remarks about the use of certain terms.

The new line

Up until the 20th century "the straight line AB" was used for what we would now write as $[AB]$, that is a segment of the line AB. The notion that a geometric object, here

---

1.   Descartes, *La Géométrie*, p. 372-373.

a line, could have an infinite property troubled those whose thinking derived essentially from Greek mathematics. Desargues and Pascal, in the 17th century, were responsible for extending the meaning of "line". For this new line, which is not a finite object, there can be no question of associating it with a number or measure.[1]

Calculations

Plane analytic geometry, carried out in the 17th and 18th centuries, is not exactly comparable to our method using two axes. Each point of a curve was "ordered" with respect to a "line" of suitable length and position. The method allowed the conics to be studied other than as sections of a cone, and also permitted the study of other curves suggested by mechanics (see Chapter 6). Examples of curves were chosen so as to avoid difficulties with measures. What we call the origin was carefully chosen, as at the vertex of a conic, and the two branches of the hyperbola were studied separately.

Function or equation?

Nowadays we might calculate the magnitude of a variable as a function of other magnitudes. But the word of function did not appear until the end of the 17th century, with Leibniz, and then in a restricted sense. Geometers, on the other hand, only looked for "equations" between magnitudes equal to the length of line segments, and we still use the expression "equation of a line". For them, it was not a question of, for example, "representing" [$y = 2x - 3$] but, by using a number of figures, establishing an equation between "lines" (that is the lengths of line segments): by extending one of the "lines" ("y" becomes "y + 3") we get twice the other "line" ("$2x$").

Bring forward the witnesses!

In our study here, we shall recall the beginnings of the idea of a directed line segment, in which the search for a unified method of treating different cases of a geometrical figure, plays a not insignificant part. To illustrate the progress towards a unified method, we have chosen three pieces of evidence which are representative, both of the work of the witness, and of the questions that were addressed. Less than a century separates the three, and it was during those years that ideas about representing negative and imaginary numbers were established.

Our first piece of evidence is the article *lieu géométrique* (geometrical locus) found in the 1795 edition of the *Encyclopédie méthodique*[2]. The article is signed by D'Alembert and may have been revised by the Abbé Bossut. The Encyclopaedia appeared in parts, from 1751 onwards, under the title *Dictionnaire raisonné des sciences, des arts et des métiers*. The work was undertaken by Denis Diderot and Jean le Rond D'Alembert and the *Discours préliminaire* of this collection of all theoretical and practical knowledge was a product of the century of the Enlightenment. From 1784 onwards, a new edition of the Encyclopaedia appeared in which the articles were

---

1. In French schools students are taught to distinguish between line, line segment, the length of a line segment and the directed measure of a line segment for which they use (AB), $[AB]$, AB and $\overline{AB}$ respectively.[tr.].

2. D'Alembert, *Encyclopédie Méthodique*, t. 2, p. 304-307.

grouped together in themes. Their were more than 130 volumes of which three concerned mathematics. The Abbé Bossut wrote the introduction and also completed a number of the articles, in collaboration with Condorcet.

The evidence presented here demands some commentary because certain of the expressions used are no longer used. This commentary has been placed alongside the original text. The article offers a good example of how the case of a particular figure affects the problem. It also shows that negative numbers had not yet been given a meaning in terms of the magnitudes of lengths. Indeed, more than a century after the death of Descartes, analytic geometry is still a long way from our present practice

GEOMETRICAL LOCUS is used for a line or curve which is the solution of a geometrical problem. See PROBLEM & GEOMETRICAL.

A locus is a line or curve, each point of which can be the solution of some indeterminated problem. If only a straight line is needed to construct the equation of the problem, then the locus will be called a straight line locus; *if a circle, it will be called* the locus of a circle; *if a parabola, it will be called* the locus of a parabola; *if an ellipse it will be called* the locus of an ellipse; *and the same for other cases, &c.[…]*

The use of the expression "geometrical locus", or "locus" for short, has largely disappeared from the teaching of geometry with the use of the language of sets.

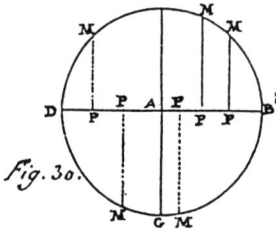

Figure 2
(Figure 30 from the *Encyclopédie*)

In order to get a better idea of what is meant by geometrical locus *imagine two unknown and variable lines* AP, PM *making between them any given angle* APM *(fig.30) one of which,* AP, *say, which has its origin fixed at A and can be infinitely extended in a given direction, we shall call* x, *& the other,* P M, *which changes position and magnitude continually, but always remains parallel to itself, we shall call* y.

The context makes clear what is meant by "unknown and variable line". It is variable in this sense, that its unknown length, x, varies: (x, the abscissa of P, has not yet made its appearance, anymore than the idea of a unit length along *AP*, and certainly not a second co-ordinate axis.) As for the line *PM*, it has no fixed position.

Suppose further that there is an equation containing only the two unknowns x *and* y, *together with other known quantities, & that it expresses a relation between the variable* AP, x, *and the value of* PM, *or the corresponding* y; *finally, imagine that at the extremity of each possible value of* x *has been traced out the corresponding* y *determined by the equation; the straight line or curve passing through the extremities of all the* y *thus traced out, or by all the points* M, *is called, in general, a* geometrical locus *and, in particular, the* locus *of the given equation.*

All equations whose loci *are of the first order can be reduced to one of the four following equations:*

1°. $y = \dfrac{bx}{a}$:   *2°.* $y = \dfrac{bx}{a} + c$:

3°. $y = \dfrac{bx}{a} - c$:   *3°.* $y = c - \dfrac{bx}{a}$,

in which the unknown y *is assumed always to be free of fractions and the fraction which multiplies the other unknown* x *is supposed to have been reduced to the form* $\dfrac{b}{a}$; *& all other terms are supposed combined to form the* + c.

The locus *of the first formula is determined first, since it is evident that it is a straight line which cuts the axis at its origin* A, *& which makes with it an angle such that the two unknowns* x, y *are always to each other as* a *is to* b.

With the idea of a "corresponding y" for each quantity "x" we are close to the idea of function. But function was the subject of only a very small entry in the *Encyclopédie*.

This classification clearly shows that lengths may be added or subtracted but that there can be no question of negative quantities.

An "axis with origin A" is what we would now call a "half-line with origin A" from which we can set out distances.

"x and y are to each other as a is to b" used sometimes to be written: $x: y :: a: b$ and is now written:

$$\frac{x}{y} = \frac{a}{b}$$

Now, supposing this first locus *to be known, then to find that of the second formula* $y = \dfrac{bx}{a} + c$, *take first on the line* AP *(Figure 31) a part* AB = a, *and draw* BE = b *&* AD = c *parallel to* PM. *You must then draw, on the same side as* AP *and towards* E, *the line* AE *of an indefinite length, and the indefinite straight line* DM *parallel to* AE; *I say that the line* DM *is the locus of the equation or of the formula that we wish to construct. For if, from any point* M *of this line,* MP *is drawn parallel to* AD, *the triangles* ABE, APF *will be similar; this gives* AB, a: BE, b::

AP, x: PF $= \dfrac{bx}{a}$, *& consequently*

PM(y) $= $ PF $\left(\dfrac{bx}{a}\right) + $ FM(c).

If we let c = 0, *that is the points* D, A *are coincident and* D M *coincides with* AF, *then the line* AF *will be the* locus *of the equation* $y = \dfrac{bx}{a}$.

To find the locus *of the third formula, proceed as follows: make* AB = a *(figure 32), & you draw the lines* BE = b, AD = c *parallel to* PM, *the one on one of the sides of* AP, *the other on the other side; through the points* A, E *you draw the straight line* AE *which you extend indefinitely towards* E, *& through the point* D *draw the line* DM *parallel to* AE. *I say that the indefinite straight line* GM *will be the desired* locus. *For we shall always have:*

PM(y) $= $ PF, $\left(\dfrac{bx}{a}\right) - $ FM(c).

Figure 3 (Figure 31 from the *Encyclopédie*)

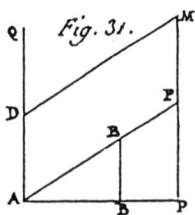

We would write: $\dfrac{AB}{BE} = \dfrac{AP}{PF}$, and hence $\dfrac{a}{b} = \dfrac{x}{PF}$, and so PF $= \dfrac{b}{a}$ x, and y $= \dfrac{b}{a}$x + c.

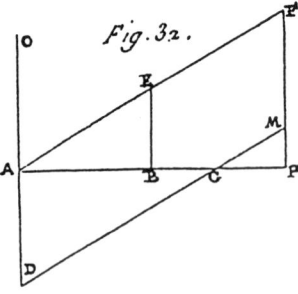

Figure 4 (Figure 32 of the *Encyclopédie*)

In the third case the locus can only begin at *G* since there can only be positive ordinates ...

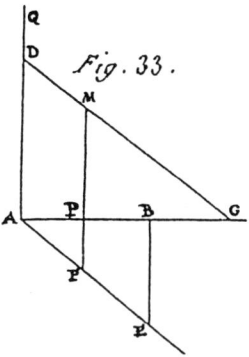

Figure 5 (Figure 33 of the *Encyclopédie*)

Finally, to find the locus *of the 4°*
formula, on AP *(fig. 33) you must*
take AB = a, *and draw* BE = b, &
AD = c, *the one on one of the sides*
of AP, & *the other on the other side.*
Further, through the points A, E
you draw the straight line AE
which you extend indefinitely
towards E, & *through the point* D
draw the line DM *parallel to* AE. *I*
say that DG *is the desired* locus.
For, if from any one of the points
M, *the line* MP *is drawn parallel to*
AQ, *we shall always have*

$$PM(y) = FM(c) - FP\left(\frac{bx}{a}\right)$$

The locus similarly terminates in *G* in the fourth case. In these two cases, *PM* is not allowed to change sense (i.e. its sign …). From the figures it appears that addition and subtraction correspond to "the one and the other side of *AP*". The equations are read as:

$$y = \frac{b}{a}x - c, \quad \text{or} \quad y = c - \frac{b}{a}x.$$

Here we have the equation or the equations of the line, which realte specificaaly to the lengths of line segments, and always in the context of figures. Further, we note the absence of the form $y = -\frac{b}{a}x - c$ since no magnitude can be equal to a negative quantity.

Second witness: *Lazare Nicolas Marguerite Carnot*

Lazare Nicolas Marguerite Carnot appear before us as an expert in organisation [Carnot was also organisator of the victory]. He presents the problem in the context of an example.[1]

§ 11. Let us take, for example, a triangle ABC (fig. 1st) [figure 6], on the base $\overline{BC}$ of which falls a perpendicular $\overline{AD}$ from the angle A, which I suppose to fall between B and C. Let us consider the base $\overline{BC}$ with its two segments $\overline{BD}, \overline{CD}$ and imagine that the point C moves towards the point B, until it has passed the point D. The base $\overline{BC}$ is therefore variable, as well as the segment $\overline{CD}$, while the other segment $\overline{BD}$ remains constant.

This being proposed, let us take for comparison the primitive figure, composed of the triangle ABC with perpendicular $\overline{AD}$; and for the transformed figure, this same figure, after the point C has approached the point D but without yet reaching D. It is evident that in the transformed system, just as in the primitive case, we will always have $\overline{BC} > \overline{BD}$; hence their difference $\overline{CD}$ will always be of the type that I have called direct or in a direct sense; and those quantities themselves $\overline{BC}$, $\overline{BD}$, are what I have called quantities of a direct order.

Now let us take for the transformed system, the same figure considered when the point C has passed beyond the point D; the primitive system staying the same, it is clear that we shall now have $\overline{BD} > \overline{BC}$: hence their difference $\overline{CD}$ has become of the type I have called inverse or in an inverse sense; and the quantities $\overline{BC}$, $\overline{BD}$, are what I have called quantities of an inverse order. [...]

---

1.   Carnot, *De la corrélation des figures de géométrie*, p. 5-23.

§ 17. [...] in the primitive system ABCD already considered, I am able to reason as follows:

The right angled triangle ABD gives: $\overline{AB}^2 = \overline{BD}^2 + \overline{AD}^2,$

and the triangle ACD gives $\overline{AC}^2 = \overline{CD}^2 + \overline{AD}^2.$

Taking the second equation from the first we obtain:
$$\overline{AB}^2 - \overline{AC}^2 = \overline{BD}^2 - \overline{CD}^2.$$

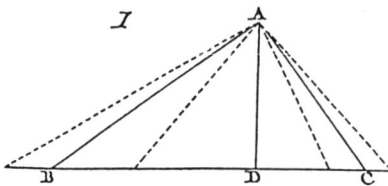

Figure 6

(Figure 1st from *Corrélation des Figures de géométrie*)

Moreover, we have $\overline{BD} = \overline{BC} - \overline{CD}.$
Substituting this value of $\overline{BD}$ in the preceding equation, we obtain
$$\overline{AB}^2 - \overline{AC}^2 = \overline{BC}^2 - 2\,\overline{BC}.\,\overline{CD}.$$

It is clear that this argument can be applied, literally and completely, to the transformed system inasmuch as the point D is placed between B and C; thus the preceding formula is immediately applicable to all cases where $\overline{CD}$ is in a direct sense.

When C has passed beyond the point D, the first part of the argument will still hold, but not the second, since we shall not have, as before, $\overline{CD} = \overline{BC} - \overline{BD}$, but, on the contrary,

$\overline{CD} = \overline{BD} - \overline{BC}$; that is that $\overline{CD}$ has become inverse. Also the final equation will become in this case $\overline{AB}^2 - \overline{AC}^2 = \overline{BC}^2 + 2\,\overline{BC}.\,\overline{CD}$, which differs from the one found in the first case in the sign that governs the second term of the second part, this sign being – in the first case, and + in the second. [...]

– "Is it, then, simply a matter of the sign?"

*§ 40. Such is, it seems to me, the true theory of those quantities which, in Analysis, are improperly called positives and negatives; I say improperly because there exist truly, neither positive quantities, nor negative quantities, of themselves, but only absolute quantities, capable of being added to others, or being subtracted from others when they are the smaller than them, for **there is nothing below 0** [our emphasis]. When, therefore, one speaks of a quantity that has become negative, one can only understand by that, that the system to which it belongs has changed, and that in the system the sign of correlation of the value of that quantity is –.*

– "Please be more specific..."

§ 37. [...] These two formulas

$[\overline{AB}^2 - \overline{AC}^2 = \overline{BC}^2 - 2\,\overline{BC}.\,\overline{CD}\ \&\ \overline{AB}^2 - \overline{AC}^2 = \overline{BC}^2 + 2\,\overline{BC}.\,\overline{CD}]$

are in not alike, since in the second case, $\overline{CD}$ is inverse, while we have $\overline{CD} = \overline{BD} - \overline{BC}$, whereas, on the contrary, in the first case we have $\overline{CD} = \overline{BC} - \overline{BD}$.

But if it is desired to find a formula immediately applicable to both systems, we have only to eliminate the inverse quantity $\overline{CD}$ by substituting its value for direct quantities, namely $\overline{BD} - \overline{BC}$, which gives

$$\overline{AB}^2 - \overline{AC}^2 = 2\,\overline{BC}.\,\overline{BD} - \overline{BC}^2,$$

an equation which is valid whether the point C falls between B and D or the point D falls between B and C.

– "Could the witness now tell us what he understands by the expression '*sign of correlation*"?

– "Certainly, but a development of my deposition will be found in the work that I published in Year IX of the Republic (1801) under the title *De la Correlation des figures de Géométrie*. Starting from the example given here, I explain what it is that I mean by a correlative system, and thereby to have a unified method of dealing with figures according to the relative disposition of the points within them. I also assert that *a quantity can become inverse in two ways: either as it passes beyond 0, or as it passes beyond ∞.*[1]

– "The court begs the witness not to digress too far..."

– "I am able to deposit with the court some extracts from my book which will show my procedure in this matter."

– "The court is indebted to you. Next witness!"

Third witness: Michel Chasles

The witness, who was a member or correspondent of some twenty academies from St Petersburg to Boston, will make two depositions (these will be interrupted in order to provide examples).

First exhibit:

In 1837 we find the following text. Chasles refers to a result due to the 18th century Scottish mathematician Matthew Stewart.

When the point D is taken anywhere on the same straight line as any other three points, Stewart's theorem expresses a relation between them. We have noted that this relation, as well as others between four points on a straight line, is a special case of a more general result concerning five points on a straight line.
Let A, B, C, D, E be these five points. We state that:
$$\overline{EA}^2.BC.CD.DB + \overline{EB}^2.CD.DA.AC - \overline{EC}^2.DA.AB.BD - \overline{ED}^2.AB.BC.CA = 0^{[2]}$$

The text contains no figure. Let us draw one, with five equidistant points, in order to have a simple example of the theorem (Figure 7),

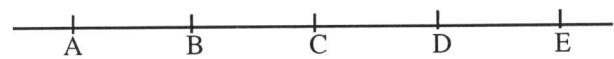

$$\begin{array}{ccccc} A & B & C & D & E \end{array}$$

Figure 7

where $AB = BC = CD = DE = 1$. Taking $\overline{EA}^2$ to have the usual meaning of the square of EA, we get the following values:

$$\overline{EA}^2.BC.CD.DB = 16.1.1.2 = 32,$$

$$\overline{EB}^2.CD.DA.AC = 9.1.3.2 = 54,$$

$$\overline{EC}^2.DA.AB.BD = 4.3.1.2 = 24,$$

$$\overline{ED}^2.AB.BC.CA = 1.1.1.2 = 2.$$

1.   *Ibid.*, § 53, p. 32.
2.   Chasles, *Aperçu historique des méthodes en géométrie...*, p. 176.

and the algebraic sum of the left hand side is, clearly, $32 + 54 - 24 - 2 = 60 \dots$! and not 0.

However, we do obtain 0 for the sum if we draw a different figure with a different arrangement of the positions of the points, as for example:

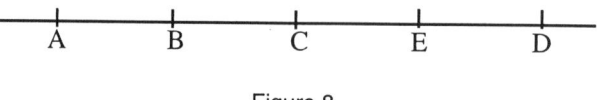

Figure 8

So, where do we go from here?

Exercise 1   Given five points, how many different ways can they be arranged on a line?
... and how many different results for the formula will be obtained?

– "Does that mean that we cannot proceed without first drawing a figure?"

– "Is there some error in the calculation or in the simplification of the formula, that we must rely so much on a diagram?"

Exercise 2   Is it possible to make a perfectly general statement?

– "Let us proceed with the witness!"

It is clear how the terms of the equation are formed. In order to determine their sign, divide all the terms by the product AB.BC.CA. The equation then becomes:

$$\overline{EA}^2\frac{DB.DC}{AB.AC} + \overline{EB}^2\frac{DA.DC}{BA.BC} - \overline{EC}^2\frac{DA.DB}{CA.CB} = \overline{ED}^2$$

In this equation the + sign is given to a product of two segments taken in the same sense starting from their common point, and the − sign where their sense differs.[1]

In the case in question the rule can be applied simply because all the products have a point in common. Chasles does not tell us what we should do with a product such as *AB.CD* or in the case of a product of three factors like the one he divided by.

Let us test his claim with the figures we had earlier.

For the first figure, on the left hand side we have $16 - 27 + 12 = 1$ and we also have $\overline{ED}^2 = 1$, which is fine. But in the case of the second figure, the value of the left hand side is $26 + 6 + 32 = 65$ while the right hand side remains $\overline{ED}^2 = 1$. So there is still a problem!

We'll continue with our reading.

We can now state the relation that exists between four points which can be deduced from the above general result:

1°) If the point E is taken to be at infinity then, on dividing by $\overline{ED}^2$ we obtain

$$BC.CD.DB + CD.DA.AC - DA.AB.BD - AB.BC.CA = 0,$$

each term of which equation is the product of three segments formed by taking three points two at a time.[2]

---

1.   *Ibid.*
2.   *Ibid.*

It is understood here that, given E at infinity, the quotients $\dfrac{EA}{ED}$, $\dfrac{EB}{ED}$ and $\dfrac{EC}{ED}$ are each equal to 1. The sign is not important since we are dealing with their squares.

If we return now to our two figures, we find that the formula is verified in both cases! Is that sufficient?

Exercise 3	Given just four points there are fewer cases to consider. Test the result for, say, the order A, C, D, B. Is it possible to state a general conclusion?

– "But let the witness continue with his deposition:"

2°) If the two points E, D coincide, the result becomes:

$$DA.BC + DB.AC - DC.AB = 0.$$

This equation expresses the simplest relation between four points A, B, C, D placed on a straight line.[1]

Look again at the example we started with and let $ED = 0$, then divide by $DA.DB.DC$. We get:

For the first figure: $(3.1) + (2.2) - (1.1) = 3 + 4 - 1 = 6,$

for the second figure: $(4.1) + (2.1) - (3.2) = 4 + 2 - 6 = 0 \ldots!$

Exercise 4	Repeat Exercise 3 for this relation.

– "Would the witness please draw his remarks to close."

3°) Finally, when the point D is taken at infinity, the general equation becomes:

$$\overline{EA}^2.BC + \overline{EB}^2.AC - \overline{EC}^2.AB = AB.BC.CA,$$

which is Stewart's equation.[2]

Exercise 5	How has Chasles arrived at this result?

The reader will find Stewart's original text as an appendix.

Preliminary and rash conclusion

Here is a rather tricky page to offer to someone new to the subject. Why, in fact, has Chasles decided to begin in this way? Does it not appear to be as much a development of his ideas concerning the orientation of a straight line, as it does his thoughts on the connection between sign and relative sense of a line segment?

---

1.   *Ibid.*
2.   *Ibid.*

Silence: review of the evidence

We examine the second deposition by the same witness. It is contained in his *Traité de géométrie supérieure* of 1860 but we know that Chasles had accepted the main ideas it contains since 1845/50. We also note that this text also contains no figures.

CHAPTER ONE

NOTES REGARDING THE USE OF THE SIGNS + AND –
IN ORDER TO DETERMINE THE DIRECTION OF LINE SEGMENTS OR ANGLES

---

1. DEFINITION. – When the segment between the points a and b will be written ab we shall say that the point a is its origin of the segment. If it is written ba then b will be considered to be its origin.

THE WAY TO INDICATE THE DIRECTION OF LINE SEGMENTS – Whenever we have to consider a number of segments on the same straight line, we shall take as positive *direction all those segments which are in the same sense starting from their origins, and all those in the opposite direction as* negative; *that is, we shall give sign + to the former and the sign – to the latter in all calculations.*
It follows that if the segment between the points a, b when written ab is positive, then it will be negative when written ba; and therefore we can say that ab = –ba.[1]

It seems here that it is a matter of the orientation of the segments and not their measure, since he writes that it is the segments that are to be considered as *positive* or *negative*. Furthermore, no distinction is made here between "direction" and "sense".

1. In making use of this convention we shall make frequent use of the following proposition concerning three points taken on a straight line.

2. FUNDAMENTAL THEOREM. – Given any three points a, b, c taken in any order on the same straight line, the sum of the three consecutive segments ab, bc, ca is always zero.

That is to say that we always have:

$$ab + bc + ca = 0,$$

applying signs to the segments as appropriate.[2]

We see here the germ of the idea of a "zero" segment since the sum of three segments is a segment…

In actual fact, the three points only give rise to three different cases; for, if the points a, b are fixed, then the point c can have just three positions, namely outside the segment ab, to its left or to its right or, of course, within the segment itself, from which there are three possible sequences: a, b, c; c, a, b and a, c, b. Now, we can change from one sequence to another by the permutation of two letters, and the equation

$$ab + bc + ca = 0$$

is unchanged by the permutation; it therefore follows that the theorem is true for all three cases if it is true for one. Consider the points in the order a, b, c as in the first sequence; the three segments ab, bc and ac are of the same sign and the sum of the first two equals the third, namely:

$$ab + bc = ac.$$

---

1.   Chasles, *Traité de géométrie supérieure*, p. 1.
2.   *Ibid.*, p. 2.

But, ac = − ca; *and so*

$$ab + bc + ca = 0$$

which was what had to be proved. Hence, etc.

COROLLARY I Whenever the position of a point a is determined by its distance from some origin O, if it is wished to relate to some other origin O', we shall always have, whatever the relative positions of the two origins with respect to the point a,

$$Oa = O'a - O'O.$$

In fact, since O' a = − a O', this relation can be derived from

$$Oa + aO' + O'O = 0,$$

which is the relation between the three points O, O' and a whatever their respective positions.

Substituting an origin O' in place of the other origin is called changing the origin of the segments.[1]

We have here a situation where three cases of a figure are treated in the same way, and this gives a result which can written in a single unified manner. It will be seen that the first corollary is only concerned with a change of origin on a straight line, the origin being at any point of the line. Further, the notion of distance arises here without Chasles making it explicit, whether or not it is the measure of a length. Finally, as concerns the use of the sign −, he uses it both for subtracting a segment (O' O) from another segment, as well as to show a change of sense, and so of sign, of a segment (O' a = − a O'). Of course, unlike the previously quoted text of 1838, the sign is no longer used for showing the nature of the product of the lengths of two segments.

The second corollary concerns the relation that two points have to a third.

COROLLARY II. The difference of two line segments Oa, Ob having a common origin, namely (Oa − Ob), is always equal to ba whatever the size and direction of the two segments.

For, the equation

$$Oa - Ob = ba$$

implies

$$Oa + ab + bO = 0$$

which is the relation between any three points O, a, b on a line.

Furthermore, one can say that the distance between two points a, b can be expressed in terms of the distances of those points from a common origin O by the relation

$$ab = Ob - Oa.$$

This relation has been largely responsible for making its author so well known to all *lycéens*![2] Yet what teacher would accept it being stated in this way?

Does it really have the importance that everyone accords it? Or is it not just a special case – for points in straight line – of the more general result stated by Argand at the beginning of the 19th century in his *Essai sur une manière de représenter les quantités imaginaires*? Argand wrote:

We conclude that for any points K, P, R, the following relation always holds:

$$\overline{KP} + \overline{PR} = \overline{KR}[3]$$

---

1. *Ibid.*, p. 2-3.
2. In French *lycées*, the property $\overline{AB} + \overline{BC} = \overline{AC}$ between three points of a straight line is frequently called Chasles' relation.
3. Argand, *Essai sur une manière de représenter les quantités imaginaires*, § 9.

$\overline{KR}$ is like a brother of our $\overrightarrow{KR}$. It has to be said that little is known of Jean-Robert Argand other than that he came from Geneva where he was a bookseller. His dates are 1763? – 1818 and he published his *Essai ...* in 1806 but Chasles did not acknowledge Argand's work in his 1837 *Aperçu historique sur l'origine et le développement des méthodes en Géométrie.*

Likewise, we find in Wessel, in his 1797 *Om directionens analytiske Betegning* the statement:

> *When one side of a triangle goes from a to b and the second from* b *to* c, *we can call the third side, which goes from* a *to* c, *the sum of the other two such that ac and ab + bc mean the same thing, or* ac = ab + bc = – ba + bc, *taking* – ba *to mean the segment opposed to* ab.[1]

However, Kaspar Wessel (1745 – 1818) was a Norwegian, and it was as a surveyor of the Copenhagen Academy of Sciences that he published his work. The work only became more widely known when it was published a hundred years later in 1897 in a French translation, under the title *Essai sur la représentation analytique de la direction.* Chasles had published his *Aperçu ...* 60 years earlier and its third edition appeared in 1889.

Last minute witnesses

The civil party wishes to bring some supplementary witnesses before the court to show the way in which directed line segments are used in the proof of a theorem in plane geometry concerning a triangle, its medians and its altitudes. The tribunal agrees.

– "The evidence of Messieurs Lebossé and Hémery, professors of mathematics and authors of several works which were held in high regard in their time."[2]

THE SUM AND THE DIFFERENCE OF THE SQUARES OF THE SIDES OF A TRIANGLE

331. Preliminary relations*. Let A and B be two points on an axis whose origin O is the mid-point of AB, and let H be any other point on the axis.*

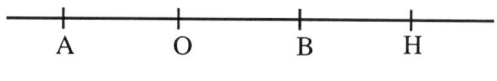

Figure 9
(Figure 282 of Lebossé & Hémery)

We note:

$$\overline{AO} = \overline{OB} = \frac{\overline{AB}}{2} \quad \text{and} \quad \overline{BO} = -\overline{AO}$$

Using Chasles' relation (n° 234), we have:

$$\overline{AH} = \overline{AO} + \overline{OH} \quad \text{or} \quad \overline{AH} = \overline{OH} + \overline{AO}.$$
$$\overline{BH} = \overline{BO} + \overline{OH} \quad \text{or} \quad \overline{BH} = \overline{OH} - \overline{AO}.$$

1.   Wessel, *Essai sur la représentation analytique de la direction*, Part 1, § 1.
2.   Lebossé & Hémery, *Géométrie plane*, p. 179-180.

and on squaring, we have:

$$\overline{AH}^2 = \overline{OH}^2 + \overline{AO}^2 + 2.\overline{OH}.\overline{AO}.$$
$$\overline{BH}^2 = \overline{OH}^2 + \overline{AO}^2 - 2.\overline{OH}.\overline{AO}.$$

Adding and subtracting gives:

$$\overline{AH}^2 + \overline{BH}^2 = 2.\overline{OH}^2 + 2.\overline{AO}^2.$$
$$\overline{AH}^2 - \overline{BH}^2 = 4.\overline{OH}.\overline{AO}.$$

Now, $\overline{AO} = \dfrac{1}{2}.\overline{AB}$ and so $\overline{AO}^2 = \dfrac{1}{4}.\overline{AB}^2$

The two relations now become:

$$\overline{AH}^2 + \overline{BH}^2 = 2.\overline{OH}^2 + \frac{1}{2}.\overline{AB}^2. \tag{1}$$

$$\overline{AH}^2 - \overline{BH}^2 = 2.\overline{AB}.\overline{OH}. \tag{2}$$

332. Problem. *Let A and B be two fixed points and M any other point. Evaluate the two expressions $\overline{MA}^2 + \overline{MB}^2$ and $\overline{MA}^2 - \overline{MB}^2$.*

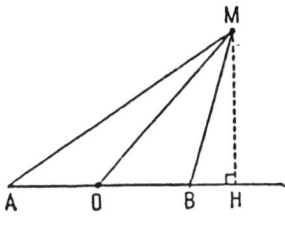

Fig. 283.

Figure 10

(Figure 283 from L. & H.)

Let O be the mid-point of AB and let H be the projection of M upon AB (fig. 283). In the right-angled triangles MAH and MBH we have:

$$\overline{MA}^2 = \overline{MH}^2 + \overline{AH}^2.$$

and $$\overline{MB}^2 = \overline{MH}^2 + \overline{BH}^2.$$

Whence

$$1°) \quad \overline{MA}^2 + \overline{MB}^2 = 2.\overline{MH}^2 + \overline{AH}^2 + \overline{BH}^2.$$

and using (1), we obtain

$$\overline{MA}^2 + \overline{MB}^2 = 2.\overline{MH}^2 + 2.\overline{OH}^2 + \frac{1}{2}.\overline{AB}^2$$

$$= 2.(\overline{MH}^2 + \overline{OH}^2) + \frac{1}{2}.\overline{AB}^2.$$

Hence:

$$\boxed{\overline{MA}^2 + \overline{MB}^2 = 2.\overline{MO}^2 + \frac{1}{2}.\overline{AB}^2} \tag{3}$$

$$2°) \quad \overline{MA}^2 - \overline{MB}^2 = \overline{AH}^2 - \overline{BH}^2$$

and using (2):

$$\boxed{\overline{MA}^2 - \overline{MB}^2 = 2.\overline{AB}.\overline{OH}.} \tag{4}$$

Now, MO is the median of the triangle MAB and OH is the projection of that median upon the side AB. We can, therefore, state the following two part theorem:

333. Theorem. 1°) The sum of the squares of two sides of a triangle is equal to twice the square of the median relative to the third side together with half the square of the third side.

2°) The difference of the squares of two sides of a triangle is equal to twice the product of the third side and its corresponding median.

– "Question: how do you define your relations?"

– "Messieurs Lebossé and Hémery: we have chosen a unit of length, the relations then becoming numerical, but the equalities do not depend upon the choice of unit."

– "The evidence of Monsieur Georges Foulon, professor of mathematics, who appears before the court to read a page from his work".[1]

METRICAL RELATIONS IN TRIANGLES

91. Consider any triangle ABC with altitudes AA', BB', CC'. The two points A' and C' lie on the circle with diameter AC; by equating the values of the power of the point B with respect to the circle, calculated from the two secants starting at B, we obtain:

$$\overline{BA}.\overline{BC'} = \overline{BC}.\overline{BA'};\qquad(1)$$

the two points A' and B' lie on the circle with diameter AB, and the power of C gives:

$$\overline{CA}.\overline{CB'} = \overline{CB}.\overline{CA'},$$

which can be written:

$$\overline{CA}.\overline{CB'} = \overline{BC}.\overline{A'C}.\qquad(2)$$

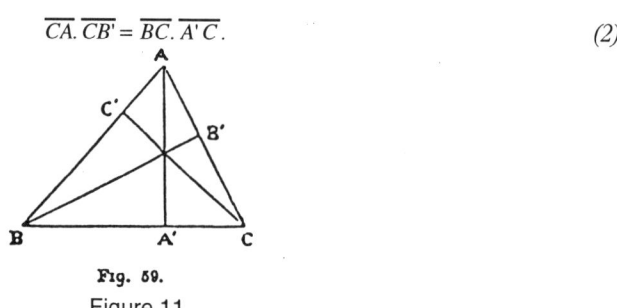

Fig. 59.

Figure 11

(Figure 59 from Foulon)

Adding the two equations (1) and (2) term by term, we obtain:

$$(3)\ \overline{BA}.\overline{BC'} + \overline{CA}.\overline{CB'} = \overline{BC}.(\overline{BA'} + \overline{A'C}) = \overline{BC}^2.$$

since: $$\overline{BA'} + \overline{A'C} = \overline{BC}$$

92. Remark. – If the angle A is a right angle the points B' and C' coincide with the point A and the equations (1) and (3) become:

$$\overline{BA}^2 = \overline{BC}.\overline{BA'}$$

$$\overline{BA}^2 + \overline{CA}^2 = \overline{BC}^2$$

which are the relations for a right angled triangle.

1.   Foulon, *Géométrie*, p. 56-57.

93. Transforming the first part of (3) in terms of vectors with A as origin, we have:

$$\overline{BA} = -\overline{AB}, \quad \overline{BC'} = \overline{AC'} - \overline{AB}, \quad \overline{CA} = -\overline{AC}, \quad \overline{CB'} = \overline{AB'} - \overline{AC}.$$

and the relation becomes:

$$-\overline{AB}.(\overline{AC'} - \overline{AB}) - \overline{AC}.(\overline{AB'} - \overline{AC}) = \overline{BC}^2;$$

collecting terms, we obtain:

$$\overline{AB}^2 + \overline{AC}^2 - \overline{AB}.\overline{AC'} - \overline{AC}.\overline{AB'} = \overline{BC}^2.$$

The four points B, C, B', C' lie on a circle with diameter BC and so:

$$\overline{AB}.\overline{AC'} = \overline{AC}.\overline{AB'};$$

and so the precedent equality can be rewritten in one of the two forms:

$$\overline{BC}^2 = \overline{AB}^2 + \overline{AC}^2 - 2.\overline{AB}.\overline{AC'};$$

$$\overline{BC}^2 = \overline{AB}^2 + \overline{AC}^2 - 2.\overline{AC}.\overline{AB'}.$$

94. Theorem. – In any triangle, the square of one side is equal to the sum of the squares of the other two sides less twice the product of the lengths of two vectors starting from the opposite vertex: one being a side of the triangle and the other the projection of the other side upon it.

With the help of trigonometry, and knowing that the cosine of an obtuse angle is negative, we can now write as the formula:

$$BC^2 = AB^2 + BC^2 - 2\,AB.AC.\cos \overset{\frown}{(BAC)}$$

To a question that is analogous to that posed by the previous witness, M. Foulon has offered an analogous solution.

Exercise 6   Provide a justification for the reasoning of these last-minute witnesses.

---

*
* *

The case having been heard, please make your conclusions

The period surrounding the French Revolution saw the forging of a tool or method which overcame the potential difficulties contained in figures in plane geometry. This had not been realised in the article in the *Encyclopédie* where the demands of rigour are not satisfied. Nor yet in Carnot's work. The difficulty was close to being resolved in the first of Chasles' texts and had been overcome in his second text. In this text the signs +, −, = and 0 cheerfully fulfil different roles, and the ideas of segment and length are often mixed up: it is also a text where mathematical terms are not used consistently and rigorously. But in the text we can see the birth of the idea of a relationship between a line segment and an associated number, although it took many more years before it took on the established form that is taught today at the end of the 20th century. This further development is yet another story!

The court will deliberate

Do not leave the courts of justice until you have examined the exhibits!

– Exhibit A: dossier by Stewart to show the information that was available to Chasles[1]. It contains a figure, and so there can be no doubt about its configuration, contrary to the case with Chasles.

SOME

GENERAL THEOREMS

Of considerable use in the

HIGHER PARTS

OF

MATHEMATICS

By MATTHEW STEWART Minister at Rosneath

MDCCXLVI.

[…]

PROPOSITION II. *Fig. 2* [Figure 12]

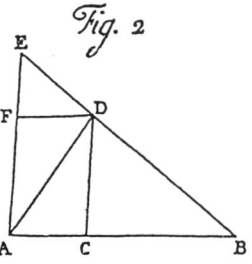

Figure 12

(Figure 2 from Stewart's *Some general Theorems...*)

In the right ligne AB *take any point* C *between the points* A, B *and from the points* A, B, C *let there be drawn right lines to any point* D; *the square of* AD *together with the space to which the square of* BD *has the same ratio that* BC *has to* CA, *will be equal to the rectangle* BAC *together with the space to which the square of* CD *has the same ratio that* BC *has to* BA.

We need to adapt the citation to reflect current usage:

On a straight line *AB*, take a point *C*, between *A* and *B*, and join *A*, *B* and *C* to any point *D*. The square of *AD* plus the square of *BD* multiplied by the ratio of *CA* to *BC* is

---

1.   Stewart, *Some general Theorems...*, title page, proposition II & fig. 2.

equal to the product of AB and AC plus the square of CD multiplied by the ratio of BA to BC.

In other words:

$$AD^2 + BD^2 \cdot \frac{CA}{BC} = AB \cdot AC + CD^2 \cdot \frac{BA}{BC}$$

Today, Stewart's relation is written:

$$DA^2 \cdot \overline{BC} + DB^2 \cdot \overline{CA} + DC^2 \cdot \overline{AB} + \overline{AB} \cdot \overline{BC} \cdot \overline{CA} = 0 \, .$$

We note that C lies between A and B which is a constraint on the figure and, in this case, Chasles' relation is exact. Doubtless this explains why.

Exercise 7   An exercise in algebra. Let A, B and C be three points on a straight line and let D be any other fourth point. Given that s is defined by $s = DA^2 \cdot \overline{BC} + DB^2 \cdot \overline{CA} + DC^2 \cdot \overline{AB} + \overline{BC} \cdot \overline{CA} \cdot \overline{AB}$, show that the sum s is zero

1°) when D lies on the line (ABC),

2°) when D does not lie on the line.

 – **Exhibit B:** a dossier of work by Carnot, placed before the court by the citizen[1].

> § 12. *Whenever it is said of two quantities that they are in direct order or in inverse order; or if it is said of any quantity that it is in a direct sense or in an inverse sense, we mean, at least implicitly, to compare the system to which those quantities belong, to another system – a system taken for comparison or a system taken as a primitive system – without it being necessary to state so explicitly.*
>
> § 13. *I shall use the term correlative to describe systems which can be compared to the same primitive system; that is all those systems that can be considered to be different states of the same variable system which can be transformed into each other by imperceptible degrees.*
>
> § 14. *For two systems to be correlative it is not necessary for there to be a connection between them, that is, they do not really need to be different transformations of the same primitive system; it is sufficient that they should be able to be considered as such, that is that they could be changed from one to the other by means of a mutation that could be imagined to act by imperceptible degrees. Corresponding quantities in the two correlative systems will likewise be called correlative quantities.*
>
> § 15. *If two correlative systems of quantities are such that the same argument, or a sequence of absolutely similar arguments, can be exactly applied to both, I shall say that there exists a direct correlation between the two systems, or that they are directly correlative. But if the arguments should cease to be literally the same, I shall say that there only exists an indirect correlation between the two systems, or the systems are indirectly correlative.*
>
> *[...]*
>
> § 18. *Let us see what happens in general, first when only direct quantities are entered in a formula, and after that what happens when one or more inverse quantities are entered.*
>
> *Let M, N be any two quantities in a primitive system; and let m, n be the corresponding quantities of the second system. Given that, by definition, in so far as the two systems remain directly correlative, m, n will play exactly the same role in the second system as that played by the corresponding M, N in the first system.*

---

1.   Carnot, *De la corrélation des figures de géométrie*, § 12-15, p. 6-7, § 18-20, p. 10-12.

Further, supposing, for example, M > N then if we have similarly m > n, these latter quantities will be in direct order and their difference m − n will be in a direct sense; in other words, if the difference of the two first quantities M and N, is P and the difference of the second two m and n is p, this quantity p will be a direct quantity.

Now, since M > N and m > n, we have

$$P = M - N \qquad and \qquad p = m - n,$$
or $\qquad M = N + P \qquad and \qquad m = n + p.$

Let us substitute the values of M and m in the formulas we have been discussing. It is clear that those formulas, which were alike before the transformation, remain so afterwards, since in the place of M and m, which played the same role [in their respective formulas] we have substituted their values N + P, n + p, which are also of the same form.

Hence, while only direct quantities are entered into the calculations, each formula of the primitive system will remain similar to the corresponding formula of the transformed system because, in practice, changes made to the first, will correspond exactly to those made to the second.

§. 19. Let us now see what happens when we introduce an inverse quantity into the calculation.

Let us suppose that while M > N it always happens that m < n; that is that these two latter quantities are in inverse order with respect to the former. Their difference p will no longer be m − n as before, but n − m; that is it will have become inverse and we shall have in this case, on the one hand

$$P = M - N$$
and on the other $\qquad p = n - m,$
and so $\qquad M = N + P \qquad and \qquad m = n - p.$

If we substitute the respective values of M and m into the corresponding formulas, which have been alike in the two systems up till now, they will cease so to be after this substitution: for, in the one we have put N + P in the place of M, while in the other we put n − p in the place of m. There will be this difference between the new formulas: that the corresponding values P, p will be governed by opposite signs. Hence, in order to change from the formulas of the primitive system to those of the transformed system, it is no longer sufficient, as before, to substitute for the absolute values entered into the first, the corresponding values of the other system; it becomes necessary, in addition, to change the sign of the value of the inverse quantity p.

And since the same reasoning can be applied to all other inverse quantities, it is clear that we can establish this general principle:

§ 20. In order that the formulas of any system of quantities might be immediately applicable to all other systems which are correlative to it, it is necessary, 1° to establish the correlation of absolute values, by substituting for each of the terms of the primitive system the absolute value of the corresponding terms of the other system; 2° to establish the correlation of signs, by changing the signs in the formula, of each of the values whose quantities are found to be in the inverse sense in the second system; and leaving unchanged the signs of those which correspond in the primitive system.

The example which we today refer to as the power of a point with respect to a circle, is very typical of what Carnot has in mind with his idea of correlation[1].

§ 60. For example, when two straight lines $\overline{AD}, \overline{BC}$ (fig. 2) [Figure 13a] intersect at a point K inside a circle, it is known that we have the equation

$$\overline{AK}.\overline{DK} = \overline{BK}.\overline{CK}$$

which is an equation containing two terms.

---

1.   Carnot, *De la corrélation des figures de géométrie*, § 60, p. 35.

Let us consider now that the system changes by small degrees so that the point of intersection, K, moves outside of the circle, in which case the system becomes that shown in fig.3 [Figure 13b]; we shall still have, in the transformed system

$$\overline{AK}.\overline{DK} = \overline{BK}.\overline{CK}\,.$$

However, $\overline{CK}$ *has become inverse; since in the first system we have*

$$\overline{CK} = \overline{CB} - \overline{BK}$$

and in the second, on the contrary, we have:

$$\overline{CK} = \overline{BK} - \overline{CB}\,;$$

and consequently $\overline{CK}$ *must have had a change of sign; but, on the other hand,* $\overline{DK}$ *has also become inverse, since in the first system we have*

$$\overline{DK} = \overline{AD} - \overline{AK}$$

and in the second, on the contrary, we have

$$\overline{DK} = \overline{AK} - \overline{AD}$$

It follows that $\overline{DK}$ *must also undergo a change of sign; and so the final equation must come back to the same as when the intersection point K was inside the circle.*

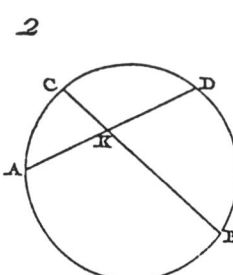

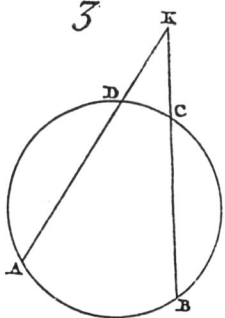

Figure 13a	Figure 13b
(Figure 2 from Carnot's *Corrélation des Figures...*)	(Figure 3 from Carnot's *Corrélation des Figures...*)

– Exhibit C – the power of a point with respect to a circle, according to Lebossé and Hémery[1].

METRICAL RELATIONS IN THE CIRCLE

337. Theorem 1. – If from a point M are drawn two secants MAB and MCD to a circle, we have the relation:

$$\overline{MA}.\overline{MB} = \overline{MC}.\overline{MD}$$

Draw AD and BC (fig. 287 and 288) [Figure 14a and Figure 14b]. In the triangles MAD and MCB, the angles M are either vertically opposite or coincident and the angles at D and B are equal. The triangles are, therefore, similar (1st case) and so:

$$\begin{cases} MAD \\ MCB \end{cases} \qquad \frac{MA}{MC} = \frac{MD}{MB} \qquad \text{so} \qquad MA.MB = MC.MD$$

1.   Lebossé and Hémery, *Géométrie plane*, p.184

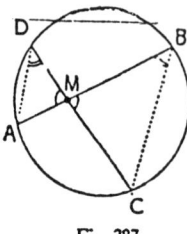

Fig. 287.

Figure 14a

(Figure 287 from *Géométrie plane*)

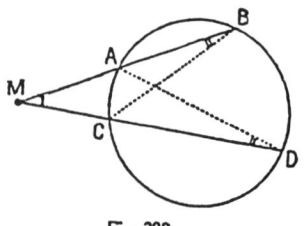

Fig. 288.

Figure 14b

(Figure 288 from *Géométrie plane*)

The two products $\overline{MA}.\overline{MB}$ and $\overline{MC}.\overline{MD}$ have the same absolute value. They are both negative if M lies inside the circle, and both positive if M lies outside the circle. They are, therefore, equal.

The product $\overline{MA}.\overline{MB}$ is called the power of the point M with respect to a circle. It is this which is the subject of the argument put forward by professor Foulon, the second last-minute witness.

– Other exhibits: there are other exhibits which have not been opened. In these, as in the last example, each possible case is considered and it is shown how they can all be treated by a single written formula. We shall cite two examples which can be followed up later:

The common tangents of two circles [5 cases, which the reader should draw, of which one does not have a common tangent].

The plane sections of a cone (see Chapter 9).

&c.

Bibliography

ARGAND, *Essai sur une manière de représenter les quantités imaginaires*, Paris, 1806; 2nd edition ed. Guillaume Jules Houël, Paris 1874. Repr. Blanchard, Paris 1971.

BIOT, *Essai de géométrie analytique*, Paris, 1810.

CARNOT, *De la corrélation des figures de géométrie*, Paris, 1801.

CHASLES, *Aperçu historique sur l'origine et le développement des méthodes en géométrie particulièrement de celles qui se rapportent à la géométrie moderne,* Brussels, 1837; 2nd & 3rd editions: Gauthier-Villars, Paris, 1875, 1889. Repr. Gabay, Paris, 1989.

CHASLES, *Traité de géométrie supérieure*, Gauthier-Villars, Paris, 1880.

DESCARTES, *La Géométrie*, Paris, 1637; English tr., D.E. Smith & M.L. Latham with facsimile of original: Open Court, 1925; Dover, New York, 1954.

DIDEROT & D'ALEMBERT, *Encyclopédie Méthodique Mathématiques*, Ed. Panckoucke, Paris, 1785, reprint ACL, 1987.

FOULON, *Géométrie, Classe de Mathématiques*, Ecole et Collège, Paris, 1911.

LEBOSSÉ & HÉMERY, *Géométrie plane, Classe de seconde des Lycées et Collèges,* Fernand Nathan, Paris, 1947.

STEWART, *Some general Theorems of considerable use in the higher parts of Mathematics*, Edinburgh, 1746.

WESSEL, *Essai sur la représentation analytique de la direction*, (Memoire submitted to Academy of Sciences, Copenhagen, 1797); centenary French translation edition, H.G. Zeuthen, Copenhagen, Paris, 1897.

How did you get on?

Ex. 1   There are 5! = 120 cases altogether but the number of cases to be verified can be reduced by using symmetry.

For the configuration: A, C, D, E, B the "result" is -20.

Ex. 2   By observation of the values of the four products, it can be said that the sum of two of the products is always equal to the sum of the other two.

Ex. 3   The configuration with A, C, D, B equidistant gives:
$$2 + 2 - 6 - 6 = -8$$
As before, the sum of two of the products is equal to the sum of the other two.

Ex. 4   This time, one of the products is equal to the sum of the other two.

It will be seen that here we have a property that can be stated verbally but not as a formula, perhaps showing the superiority of verbal statements to symbolic ones.

Ex. 5   The initial formula can be divided throughout by ED^2 and, since
$$\frac{CD.DB}{ED^2} = \frac{CD}{ED}\frac{DB}{ED} = 1.1 = 1$$
and similarly for the other terms... This may have been Chasles' reasoning but the text does not enlighten us.

Ex. 6   The algebraic measures which are the subject of the equalities that are established are certainly real numbers, positive or negative depending upon the chosen orientation, which is determined by the choice of origin and a point for the abscissa unit on each straight line taken as an axis; the absolute value of the number, is the magnitude which is the ratio of the magnitude which each represents to the magnitude chosen as unit, and the sign of the number is positive or negative according as the segment is in the same or opposite orientation as the unit segment.

Ex. 7   1) Taking D as origin, and a, b, c as abscissae of A, B, C we have:
$$s = a^2.(c-b) + b^2.(a-c) + c^2.(b-a) + (c-b).(a-c).(b-a)$$
which is a null polynomial for all triplets a, b, c.

2) Let O be the orthogonal projection of D upon the line (ABC). It can be verified that:
$$s' = OA^2.\overline{BC} + OB^2.\overline{CA} + OC^2.\overline{AB} + \overline{BC}.\overline{CA}\,\overline{AB} = 0,$$
$$s = s' + OD^2.\overline{BC} + OD^2.\overline{CA} + OD^2.\overline{AB} \quad \text{and}$$
$$= s' + OD^2.(\overline{BC} + \overline{CA} + \overline{AB})$$
$$= s' + 0$$
$$= 0$$

If the reader has been patient enough to work through the last exercise, here is the fully algebracised form of the formula which Chasles was able to derive:
$$EA^2.\overline{BC}.\overline{CD}.\overline{DB} - EB^2.\overline{CD}.\overline{DA}\,\overline{AC} + EC^2.\overline{DA}.\overline{AB}.\overline{BD} - ED^2.\overline{AB}.\overline{BC}.\overline{CA} = 0$$
The verification of this result is an excellent exercise in patience and care in algebraic manipulation for a student.

8

The Brachistochrone Problem

Jean-Luc CHABERT
INSSET St-Quentin

The Problem

Our intention here is to tell the story of the brachistochrone and its remarkable implications. In the contemporary socio-cultural context, the question would essentially be formulated in the following manner: what shape should we make slides in children's playgrounds so that the time of descent should be minimised? The considerable importance of this question is well understood when we consider how children behave, and they want to obtain the best performance, but the question is also important in a more general way, and a great number of scholars have attempted to solve the problem.

Unfortunately the problem appears to be particularly tricky, and it depends upon a number of parameters, including the variable value of the friction between the clothes of the child and the surface of the slide. We shall not attempt to solve that particular problem here, but content ourselves with the theory of the idealised problem, simplifying the situation sufficiently in order to be able to find a solution. In fact we shall replace the child by a perfectly smooth marble, and assume that it rolls down a smooth surface, thus assuming that friction forces are negligible with respect to gravity.

Now, we are simply confronted with the *problem of the brachistochrone* as Jean Bernoulli expressed it in the *Acta Eruditorum* published in Leipzig in June 1696

> *Datis in plano verticali duobus punctis A & B, assignare Mobili M viam AMB, per quam gravitate sua descenden, & moveri incipiens a puncto A, brevissimo tempore perveniat ad alterum punctum B.*[1]

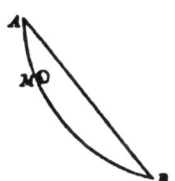

---

1.   Bernouilli (Jean), *Opera Omnia*, T. 1, p. 161.

The expression *brevissimo tempore* is the Latin translation of the term *brachistochrone* from the Greek (brachys = brief, brachisto = quickest, chronos = time, brachistochrone = the shortest time). In a more modern style:

Given two points A and B in a vertical plane, what is the curve traced out by a point subject only to the force of gravity, starting from rest at A, such that it arrives at B in the shortest time?

Common sense suggests that this curve is necessarily situated in the vertical plane containing the points A and B. Common sense also leads us to think that the quickest route is the shortest, and is given by the line segment joining the points A and B. But this is not the case. We know, for example, that a longer journey on a motorway may be faster than going a shorter distance on an ordinary road. Here, in order to try to solve the problem of the brachistochrone, it is necessary to consider all the curves joining points A and B and compare all the corresponding times of travel. Taking everything into account, even under these restrictions, the problem turns out to be a subtle one.

The Brachistochrone Problem, *a priori* a simple game for mathematicians, turns out in the end to be a considerable problem. Indeed, the different approaches tried out in its solution may be considered, in a more or less direct way, as the starting point for new theories. While the true "mathematical" demonstration involves what we now call the Calculus of Variations, a theory for which Euler and then Lagrange established the foundations, the solution which Jean Bernoulli originally produced, obtained with the help of an analogy with the law of refraction on Optics, was empirical. A similar analogy between Optics and Mechanics reappears when Hamilton applied the principle of least action in Mechanics which Maupertuis justified in the first instance, on the basis of the laws of Optics. This correlation finally suggested to de Broglie and Schrödinger the idea of Wave Mechanics as an analogy of Wave Optics.

Galileo and Falling Bodies

In 1638, well before the problem had been explicitly stated, Galileo gave his solution to the brachistochrone problem in the course of the Third Day of his *Discourse on Two New Sciences*[1]. It is here that he studied uniform acceleration – Galileo called it "natural acceleration" – comparing it with uniform motion, and showed that a body falling in space traverses a distance proportional to the square of the time of descent (Theorem II). With regard to bodies moving on inclined planes he deduced:

> *Theorem V. The times of descent along planes of different length, slope and height bear to one another a ratio which is equal to the product of the ratio of the lengths by the square root of the inverse ratio of their heights.*

We interpret this proportionality to be: a body travels a distance L and descends a height H in time t such that:

$$t = kL / \sqrt{H}$$

---

1.   Galileo, *Discourse concerning two new Sciences*, English translation, Henry Crew and Alfonso de Salvio. New York: Dover 1914.

Galileo then proves the following neat result:

Theorem VI. If from the highest or lowest point in a vertical circle there be drawn any inclined planes meeting the circumference, the times of descent along these chords are each equal to the other.

On the horizontal line GH construct a vertical circle. From its lowest point – the point of tangency with the horizontal – draw the diameter FA and from the highest point A, draw inclined planes to B and C, any points whatever on the circumference; then the times of descent along these are equal.

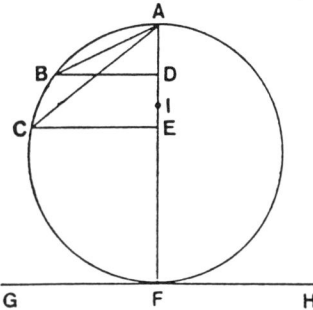

Draw BD and CE perpendicular to the diameter; make AI a mean proportional between the heights of the planes, AE and AD; and since the rectangles FA.AE and FA.AD are respectively equal to the squares of AC and AB, while the rectangle FA.AE is to the rectangle FA.AD as AE is to AD, it follows that the square of AC is to the square of AB as the length of AE is the length of AD. But since the length AE is to AD as the square of AI is to the square of AD, it follows that the squares on the lines AC and AB are to each other as the squares on the lines AI and AD, and hence also the length AC is to the length AB as AI is to AD. But it has been previously demonstrated that the ratio of the time of descent along AC to that along AB is equal to the product of the two ratios AC to AB and AD to AI; but this last ratio is the same as that of AB to AC. Therefore the ratio of the time of descent along AC to that along AB is the product of the two ratios, AC to AB and AB to AC. The ratio of these times to therefore unity. Hence follows our proposition.

Here is a modern version of this passage as given by M. Clavelin:

To prove $T_{AC} = T_{AB}$
Let $AI^2 = AD.AE$. Firstly, $FA.AE/FA.AD = AC^2/AB^2$ and secondly, $FA.AE/FA.AD = AE/AD$. And so $AC^2/AB^2 = AE/AD$. But $AE/AD = AI^2/AD^2$ (multiplying both terms by AD), therefore $AC/AB = AI/AD$. Now, (by Theorem V) $T_{AC}/T_{AB} = AC/AB$. AD/AI; and since $AD/AI = AB/AC$, we finally have $T_{AC}/T_{AB} = AC/AB$. $AB/AC = 1$.

If afterwards, as suggested by Galileo, we pose the problem of determining the inclination of a plane where, starting from rest at a given point A, we reach a given vertical plane in the shortest time, we see immediately that the optimum is given by an angle of 45°.

At the end of the Third Day, Galileo shows that it is also possible to improve on this descent: starting with two points A and C situated on a line inclined at 45 degrees, he proves that the time of descent could be diminished by replacing the segment AC by two consecutive segments AD and DC where D is any point on the arc of the quadrant AC of the circle.

Theorem XXII: If from the lowest point of a vertical circle, a chord is drawn subtending an arc not greater than a quadrant, and if from the two ends of this chord two other chords be drawn to any point on the arc, the time of descent along the two latter chords will be shorter than along the first, and shorter also, by the same amount, than along the lower of these two latter chords.

We omit the geometrical proof and quote only the commentary, which interests us because Galileo here suggests that the curve of shortest descent is the arc of a circle.

> *Scholium: From the preceeding it is possible to infer that the path of quickest descent from one point to another is not the shortest path, namely, a straight line, but the arc of a circle.*

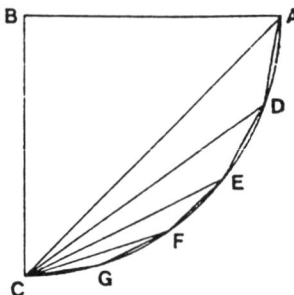

> *In the quadrant BAEC, having the side BC vertical, divide the arc AC into any number of equal parts, AD, DE, EF, FG, GC, and from C draw straight lines to the points A, D, E, F, G; draw also the straight lines AD, DE, EF, FG, GC. Evidently descent along the path ADC is quicker than along AC alone or along DC from rest at D. But a body starting from rest at A, will traverse DC more quickly than the path ADC; while if it starts from rest at A, it will traverse the path DEC in a shorter time than DC alone. Hence descent along the three chords, ADEC, will take less time than along the two chords ADC. Similarly, following descent along ADE, the time required to traverse EFC is less that that needed for EC alone. Therefore descent is more rapid along the four chords ADEFC than along the three ADEC. And finally a body, after descent along ADEF, will traverse the two chords, FGC, more quickly than FC alone. Therefore, along the five chords, ADEFGC, descent will be more rapid than along the four, ADEFC. Consequently the nearer the inscribed polygon approaches a circle the shorter is the time required for descent from A to C.*
>
> *What has been proven for the quadrant holds true also for smaller arcs; the reasoning is the same.*

This result is false, since arguing the case from two to three segments is based on a faulty intuition from arguing from one to two segments. The Brachistochrone Problem is considerably more subtle than the one of the research into optimum inclination of planes, which is a simple problem of the extremum for a function of a single variable.

Exercise 1   Show using calculus that the time of descent is effectively diminished by the introduction of an intermediate point, taking for example the point D in the centre of the quadrant AC.

The demonstration by Jean Bernoulli, in 1697, also derives from an intuitive approach. This approach, an analogy with the law of refraction, leads to the curve solution which one cannot find *a priori*, without an arsenal of sufficiently sophisticated techniques. Let us begin by recalling the first laws of Optics, which are in fact consequences of the principles of optimisation.

Reflection/Refraction

Experience tells us that light travels in straight lines. This phenomenon is stated as a principle: light chooses the shortest path. This formulation led to a real theoretical advance since it allowed Hero of Alexandria in the first century AD to explain the law of reflection, namely, the equality of the angles of incidence and reflection.

Exercise 2   Prove geometrically that the shortest path from A to B which includes touching a mirror is the path taken by a ray of light reflected in the mirror, that is one where the angles made with the mirror are equal.

Hint. To compare different path lengths from A to B, Hero introduced a point B' symmetrical to B with respect to the mirror.

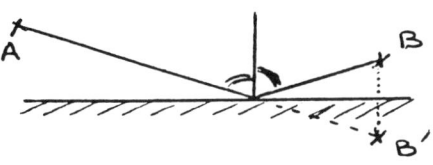

In the case of reflection, the speed remains constant. It is not so for refraction, where the speed of light c/n varies as a function of the index n of the medium traversed. However, the principle stated above could have been stated in the following form as the Fermat's Principle: light chooses the fastest route, which in a homogeneous medium where its speed is constant, is equivalent to the previous principle.

So, to go from A to B, passing from a medium of index n_1 to a medium of index n_2, the trajectory of the light will not be the line segment AB, but a broken line AIB such that the trajectory AIB will have the shortest time of all trajectories from A to B. Using the initial conditions we calculate that the angle of incidence i and the angle of refraction r are related to the respective speeds by the formula:

$$\sin i / v_i = \sin r / v_r,$$

or using the indices n_i and n_r we have the *sine formula*:

$$n_i \sin i = n_r \sin r.$$

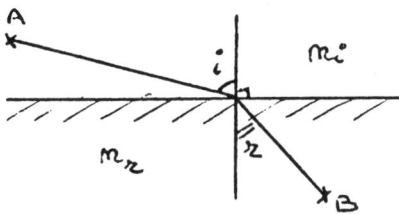

This formula, discovered by the Dutch scientist Snell in 1621, and repeated by Descartes in his *Dioptrics*[1] in 1637, received its correct interpretation with Fermat.

In a letter of the 1st of January 1662 to M De la Chambre, Fermat[2] explains:

> *As I said in my previous letter, M. Descartes has never demonstrated his principle; because not only do the comparisons hardly serve as a foundation for the demonstrations, but he uses them in the opposite sense and supposes that the passage of light is more easy in dense bodies than in rare bodies, which is clearly false. I will not say anything to you about the shortcomings of the demonstration itself ...*

---

1.   An appendix to the *Discourse on Method,* Leyden, 1637.
2.   Fermat, *Œuvres,* t. II, p. 457-463.

*To avoid this embarrassment, and to try to find the true reason for refraction, I indicated to you in my letter that if, in this investigation, we care to use the principle that is so common and so well established, that **nature always tries to take the shortest routes,** we would be more easily able to discover the result.*

Fermat puts his principle to work, and proves the sine formula using his method *'de maximis et minimis'* (see chapter 5).

Exercise 3   Find the law of refraction analytically, by minimising the time t (expressed, for example, as a function of the length $x = HI$).

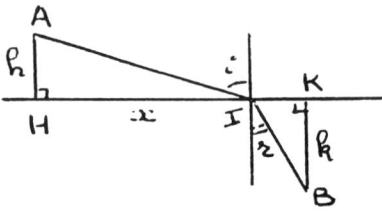

Hint: Put $x = HI$, $h = AH$, $k = BK$ and $a = HK$. We have:

$t = AI.(n_i /c) + IB.(n_r /c)$.

i.e. $t(x) = (n_i/c) \sqrt{h^2 + x^2} + (n_r/c) \sqrt{k^2 + (a - x)^2}$. We can then put $t'(x) = 0$ and note that $\sin i = IH/IA = x/\sqrt{h^2 + x^2}$ and $\sin r = IB/IK = (a - x)/\sqrt{k^2 + (a - x)^2}$

Another example of a non-homogeneous medium where the shortest trajectory is not the quickest occurs in Mechanics, where the effect of gravity is in the vertical direction. And this is the context for Jean Bernoulli's Brachistochrone Problem.

Jean Bernoulli's Demonstration

Jean Bernoulli produced his demonstration in the *Acta Eruditorum* of May 1697[1]. His method typically corresponds to what we now call a discretisation of the problem. He imagines space carved into small lamina, sufficiently fine so that within each one it is possible to imagine that the speed is constant. Within each strip the trajectory becomes the shortest route, and necessarily a segment. The complete trajectory appears as a sequence of segments.

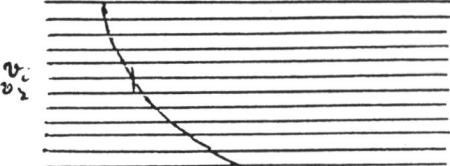

But how can we move from one strip to another? We must always optimise the time of travel. As in the refraction of light, this is done by using Fermat's principle. Thus, if v_i is the speed in a given band and v_r in the band immediately below, the angle i is the

---

1.   Jean Bernoulli, *Opera Omnia*, vol. 1, p. 187-193.

angle made with the vertical by the segment of the trajectory in the first band, and the angle r in the neighbouring band, then they are connected by the rule of sines:

$$\sin i/v_i = \sin r/v_r.$$

If we now imagine that the horizontal strips become progressively thinner, and their number increases indefinitely, the line of segments tends towards a curve. The tangents at each point of this curve approach the sequence of segments. The angle u which the tangent makes with the vertical is then connected to the speed v by the relation:

$$\sin u/v = \text{constant}$$

Here, the speed v of a particle is known; it is the result of the action of gravity and, as we know from Galileo, it is a function of the distance fallen y, according to the formula:

$$v = \sqrt{2gy}$$

And so the rule of sines leads to the equation:

$$\sin u / \sqrt{y} = \text{constant}$$

In particular, for y = 0, the tangent is vertical.

Many readers will recognise here a characteristic equation of a well-known curve of the time, the cycloid or roulette (see Chapter 6). Curiously, this arose, as we know, from a completely different question, since, according to Pascal:

> *It is the path which is made in the air by a nail of a cartwheel when the wheel rolls normally, from the point where it starts to leave the ground, until the continuous movement of the wheel brings it back to the ground after a full turn is completed.*

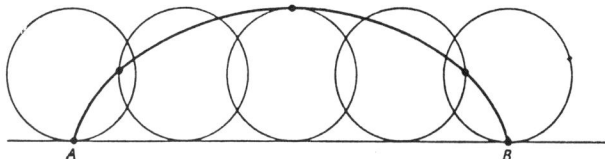

Exercise 4   Verify that the equation sin u / √y = constant = k can be written as an ordinary differential equation, namely:

$$y(1+y'^2) = 1/k^2$$

Hint: y' = cot u and 1/sin² u = 1 + cot² u.

Why is the cycloid the solution to the differential equation $\sin u / \sqrt{y} = constant$?

We can see this with the help of some elementary geometrical ideas. Remember that u is the angle made with the vertical by the tangent to the curve at an ordinate y.

We know from Roberval (see Chapters 5 and 6) that the direction of the tangent can be obtained from kinematics. The cycloid is *"the path made in the air by the nail of a cartwheel when it rolls normally"*, the movement of a point M of the cycloid results from two simultaneous movements: one a rotation about the centre O of the wheel, and a horizontal translation (corresponding to the translation of the centre O), that is a motion in the direction of the tangent MT (perpendicular to OM), and a motion in the horizontal direction MH. But the corresponding infinitesimal displacements have equal lengths (the wheel is rolling without sliding); the tangent to the cycloid is then the bisector of the angle HMT, which passes through the point R diametrically opposite to P (the point of contact of the circle with the ground).

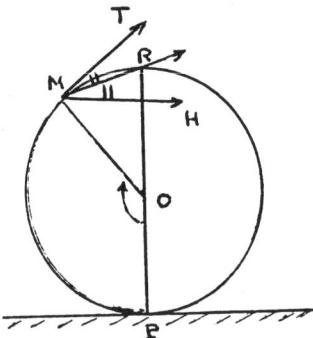 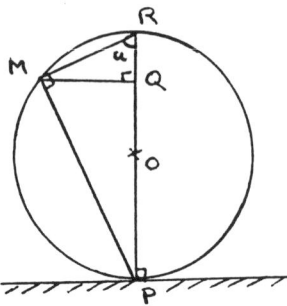

The angle u is therefore the angle MRQ and the ordinate y is the length PQ. Considering the triangles MQP and PMR, we have:

$$\sin u = \cos(\pi/2 - u) = PQ/MP = MP/PR.$$

or,

$$(\sin^2 u)/y = (PQ/MP).(MP/PR).(1/PQ) = 1/PR = 1/2R$$

[Q.E.D]

Exercise 5	Show analytically that the cycloid satisfies the differential equation: $$y(1+y'^2) = 2R$$ *Hint:* Verify, and then use the fact that the cycloid can be represented parametrically by: $$x(t) = R(t - \sin t),\ y(t) = R(1 - \cos t).$$

The Newtonian Construction

We have just seen that the solution to the curve is a cycloid. But how can we construct such a curve, starting from a point A, and arriving exactly at a point B? Newton gave a simple solution in a letter to Montague on the 30th of January 1697:

> **Problem.** *It is required to find the curve ADB in which a weight, by the force of its gravity, shall descend most swiftly from any given point A to any given point B.*

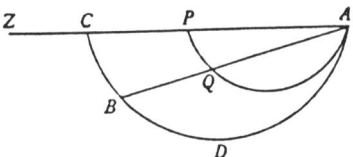

> **Solution.** *From the given point A let there be drawn an unlimited straight line APCZ parallel to the horizontal, and on it let there be described an arbitrary cycloid AQP*

meeting the straight line AB (assumed drawn and produced if necessary) in the point Q, and further a second cycloid ADC whose base and height are to the base and height of the former as AB is to AQ respectively. This last cycloid will pass through the point B, and it will be that curve along which a weight, by the force of its gravity, shall descend most swiftly from the point A to the point B.[1].

In addition to Newton's contribution to the solution of the problem of the brachistochrone, we must also mention Leibniz, and in a lesser role, the Marquis de l'Hospital, and most of all, Jacques Bernoulli, the older brother of Jean:

... my elder brother made up the fourth of these, that the three great nations, Germany, England, France, have given us, each one of their own to unite with myself in such a beautiful search, all finding the same truth.[2]

The method used by Jacques is laborious, but quite general. Also, Jacques, in wanting to show the singular character of Jean's method, extended the problem by posing new questions. Indeed, Jean's method, founded on an analogy, does not work except in a particular case, and cannot be used for more general problems of this type.

Analogies with Galileo

In particular, Jacques Bernoulli put the following question to his brother: given a vertical line, which of all the cycloids having the same starting point and the same horizontal base, is the one which will allow a heavy body passing along it to arrive at the vertical line the soonest? Such a statement reminds us of Galileo's first version, which was about finding the inclined plane through a given point which gave the shortest time to reach a given vertical.

Jean Bernoulli[3] replied and showed that the cycloid in question is the one which meets the given line horizontally. More generally, the cycloid which allows us to achieve the swiftest possible descent to a given oblique line, is the one which meets the line at right angles.

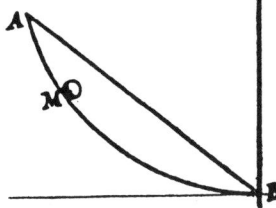
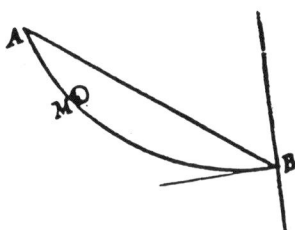

This cycloid which, as we have just said, is a brachistochrone curve, was also known to Huygens from 1659 as the tautochrone curve: bodies which fall in an inverted cycloid arrive at the bottom at the same time, no matter from what height they are released.

1.   Isaac Newton, *Correspondence*, vol. IV (1694-1709), p. 223.
2.   Jean Bernoulli, *Histoire des Ouvrages des Savants,* June 1697, *Opera Omnia*, vol. 1, p. 194-204.
3.   Jean Bernoulli, Letter to M. Varignon, 15 October 1697, *Opera Omnia,* vol. 1, p. 206-213.

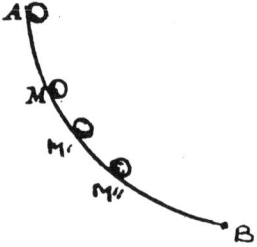

This property was perhaps closer to that observed by Galileo: the equality of the times for the distances on the chords of the same circle. Among the other problems posed by Jacques Bernoulli to Jean are those which are called isoperimetric problems (see Exercise 6 below), which together with the brachistochrone problem are prototypes of optimisation problems which we shall discuss more generally in the next section. The resolution of these problems is then the object – reason or excuse? – for a long dispute between the two brothers; a dispute which developed into a major row, but which gave birth to a new area in mathematics, the Calculus of Variations.

These scientific exchanges between the two brothers were carried out in the form of letters published in the *Journal des Savants*. Here is a sample of Jean's response to some criticisms by Jacques:

> *So there it is, his imagination, stronger and more vivid than those claiming to be sorcerers who believe they have found themselves bodily present at a Sabbath, has seduced him; he is carried along by a torrent of vain conjectures; in a word, he is no longer ready to give reign to reason ...*   (Journal of 8th September 1698)

Exercise 6   a) Show experimentally that the closed curve with a fixed length that encloses the maximum area is a circle. By analogy:

b) Show experimentally that the closed surface with a given area that encloses the largest volume is a sphere. More difficult yet:

c) For a given closed contour, construct the surface depending on this contour which has the minimum area.

Hints. For a) and b) the two questions can also be formulated in the following manner: show that the circle (the sphere) is the closed curve (surface) which of all curves (surfaces) encloses a given area (volume) with the smallest length (area). Think about the fat globules on the surface of soup and soap bubbles in the air; in the two cases there is a minimisation of the surface tension, and there is minimisation of the boundary, namely of the perimeter in the first case and of the area in the second. For c) make the contour with a wire and put it into a soap solution. Pull it out: the surface of the film of soap obtained is the answer to the question. Here, the minimisation of the surface tension guarantees that the area is minimal.

The Calculus of Variations

When we look for the boundary values of a function f of a variable x, i.e. when we look for values of the variable x for which the value f (x) is a maximum or minimum, we look for the points where the graph of f has a horizontal tangent, or we say we look for the values where $f'(x) = 0$ (see above, the exercise on the formula sines). In the case of a function f of two variables x and y, we have to consider the points where the tangent plane is horizontal to the surface which has the equation $z = f(x, y)$. Alternatively we could say we seek the number pairs (x, y) for which the partial derivatives $f_x'(x, y)$ and $f_y'(x, y)$ are zero. Or we can say we are looking for the points where the function f has a stationary value. For example, in the case of the altitude function of the earth`s surface, the horizontal tangent planes correspond to the summits, bowls and saddle points of inflexion. Briefly, in the case of a finite number of variables, the difficulties seem surmountable, and the approach to the problem may be effected with the aid of the differential calculus of Newton and Leibniz. Here, the object which changes is not a number or a point, but a curve, a function, and the corresponding quantity to maximise or minimise is a number depending on this curve or on this function. It is necessary to conceive an extension of the differential calculus. The new theory which was created is called the Calculus of Variations, the variations being those of the function. But, in 1696, this theory had not been formulated and our problem becomes *a priori* somewhat subtle.

A problem in the calculus of variations can be presented generally in the following fashion: we try to find a curve, being the graphical representation of a function *y* of *x*, which minimises or maximises a certain quantity among all the curves constrained by certain conditions (for the Brachistochrone Problem, the curve joining two points A and B). The quantity whose extreme value has to be found (here, the time of the journey) is expressed generally in the form of an integral:

$$I(y) = \int_a^b F(x,y,y')dx$$

where *y* represents the unknown function, *y'* its derivative, *x* the variable and F a particular function.

Exercise 7   Show that, in the case of the Brachistochrone Problem, the function F is:
$$F(x,y,y') = \sqrt{(1 + y'^2)/y}.$$

Hint: $t = \int_0^{x_1} ds/v$ where $v = \sqrt{2gy}$ and $ds = \sqrt{(1 + y'^2)}dx$

Among the typical problems of the calculus of variations, besides the isoperimetric problems considered above, are investigations of the geodesic lines on a surface, i.e. the curves of minimum length joining two points of a surface. Also, the investigation of the shapes of the surfaces of revolution which offer the least resistance to movement, a problem which Newton tackled in 1687 in the *Principia*.

The statement of the Brachistochrone Problem in 1696 could be considered as the definitive origin of the Calculus of Variations, for it is the problem which generated general methods of investigation which were gradually developed in a competitive context. Indeed the reader can take this text to be an offering in the celebration of the tricentenary of the birth of the calculus of variations.

Euler and Lagrange

Jean Bernoulli himself posed the problem of geodesics to Euler. Euler re-worked the ideas of Jacques Bernoulli, simplified them, and finally was the first to formulate the general methods which allowed them to be applied to the principal problems of the calculus of variations. He developed these ideas systematically in 1744 in his *Method for finding plane curves that show some property of Maxima and Minima*[1]. In a way, like Jean Bernoulli, Euler tackles the problem as a problem of limits in an investigation of the ordinary extremum. For this purpose, he divides the question up as follows: he reduces the curve sought to a polygonal line formed by n segments with extremities:

$$A = (x_0, y_0), (x_1, y_1), ..., B = (x_n, y_n).$$

It then becomes a case of finding the values of the ordinates $y_0, y_1, ... y_n$ which give the extremum of the quantity:

$$W(y_0, y_1, ..., y_n) = \sum_{k=0}^{n-1} (x_{k+1} - x_k).F(x_k, y_k, (y_{k+1} - y_k)/(x_{k+1} - x_k)).$$

Making n approach infinity, Euler derived the differential equation:

(*)
$$F'_y - \frac{d}{dx}(F'_{y'}) = 0$$

which satisfies each solution y, and where F'_y and $F'_{y'}$ represent the partial derivatives of F with respect to the variables y and y' respectively. It is only a necessary condition and the method does not establish the existence of a solution. The equation (*), today called the Euler-Lagrange equation, is a second order differential equation in y:

(**)
$$F'_y - F''_{y'x} - y'F''_{y'y} - y''F''_{y'^2} = 0$$

Exercise 8   Using the Euler-Lagrange equation, reconstruct the differential equation that satisfies the brachistochrone curve.

Hint. The calculations are easier if the x and y axes are interchanged. The function F becomes $F(x,y,y') = \sqrt{(1 + y'^2)/x}$ and the Euler –Lagrange equation is written

$$\frac{d}{dx}(F'_{y'}) = 0, \text{ i.e. } F'_y = k: \text{ i.e. } y = k^2 \times (1 + y'^2).$$

Interchanging the axes again, we obtain:

$$y(1 + y'^2) = 1/k^2.$$

In 1760, Lagrange greatly simplified matters by introducing the differential symbol δ, specifically for the Calculus of Variations, corresponding to a variation of the complete function.

---

1.   Euler, *Œuvres*, 1st series, vol. 24, 1952.

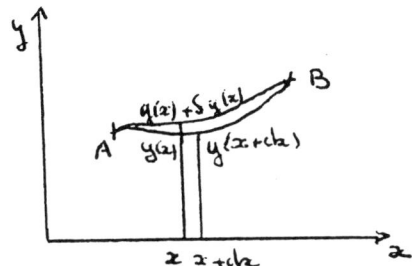

He makes the point of it in the introduction to his *Essay on a new method of determining the maxima and minima of indefinite integral formulas*[1]:

> *For as little as we know the principles of the differential Calculus, we know the method for determining the largest and smallest ordinates of curves; but there are questions of* **maxima** *and* **minima** *at a higher level which, although depending on the same method, are not able to be applied so easily. They are those where it is needed to find the curves themselves, in which a given integral expression becomes a maximum or minimum with respect to all the other curves.*
>
> *The first problem of this type, which the Geometers solved, was that of the* **Brachistochrone**, *or the line of quickest descent, which Jean Bernoulli proposed towards the end of the last century. The solution was only found by considering special cases, and it was not until some time later, and in researching* **Isoperimetral** *curves, that the great Geometer of whom we speak and his illustrious brother Jacques Bernoulli, gave some general rules for solving several other questions of the same nature. But since these rules were not sufficiently general, the famous Euler undertook the task of reducing all investigations of this kind to a general method, in the work entitled:* **Methodus inveniendi lineas curvas maximi, minimive proprietate gaudentes: sive solutio Problematis isoperimetrici lattissimo sensu accepti**; *an original work in which the profound science of the calculus shines through. Even so, while the method is ingenious and rich, one must admit that it is not as simple as one might hope in a work of pure analysis. The author shows himself to be aware of this in article 39 of Chapter II of his book with these words: "Desideratur itaque methodus a resolutione geometrica et lineari libera, qua pateat in tali investigatione maximi minimique, loco Pdp scribi debere -p dP.*
>
> *Now here is a method which only requires a straightforward use of the principles of the differential and integral Calculus; but above all I must give warning that while this method requires that the same quantities vary in two different ways, in order not to mix up these variations, I have introduced into my calculations a new symbol δ. In this way, δZ expresses a difference of Z which is not the same as dZ, but which, however, will be formed by the same rules; such that where we have for any equation dZ = m dx, we can equally have $\delta Z = m\ \delta x$, and likewise for other cases.*

A century later, Mach was able to write in his *Mechanics*[2]:

> *In this way, by analogy, Jean Bernoulli accidentally found a solution to the problem. Jacques Bernoulli developed a geometric method for the solution of analagous problems. In one stroke, Euler generalised the problems and the geometrical method. Lagrange finally freed it completely from the consideration of diagrams, and provided an analytical method.*

1.   Lagrange, *Miscellanea Taurinensia*, t. II, 1760-1761 in *Œuvres*, t. I, p. 333-468.
2.   Mach, *Mechanics, an Historical and Critical Account of its Development*, 1883.

Maupertuis and the Principle of Least Action

We shall make a digression here, the purpose of which will soon become clear. Maupertuis stated his Principle of Least Action in 1744 in a Memoir entitled *The Harmony of the Different Laws of Nature*[1]. He explains and justifies his principle precisely from the law of refraction:

> *In thinking deeply upon this matter, I reflected that light, as it passes from one medium to another, yet not taking the shortest path, which is a straight line, might just as well not take the shortest time. Actually, why should there be a preference here for time over space? Light cannot go at the same time by the shortest path and by the quickest route, so why does it go by one route rather than another? In fact, it does not take either of these, it takes a route that has the greater real advantage:* **the path taken is the one where the quantity of action is the least**.
>
> *Now I must explain what I mean by the* **quantity of action**. *When a body is moved from one place to another, a certain action is needed: this action depends on the speed of the body and the distance travelled; but it depends neither on the speed nor the distance taken separately. The quantity of action is moreover greater when the speed of the body is greater and when the path travelled is greater; it is proportional to the sum of the distances multiplied respectively by the speed travelled over each space. [Since here there is only one body, we disregard the mass.]*
>
> *It is this quantity of action which is the true expenditure of Nature, and which she uses as sparingly as possible in the motion of light. Let there be two different media, separated by a surface represented by the line CD, such that the speed of light in the medium above is m, and the speed in the medium below is n.*
>
> *Let a ray of light, starting from point A, reach a point B: to find the point R where the ray changes course, we look for the point where if the ray bends* **the quantity of action** *is* **the least**: *and I have m.AR + n. RB which must be a* **minimum**.
>
> *Or, having drawn the perpendiculars AC and BD to the common surface,*
> $$m\sqrt{AC^2 + CR^2} + n\sqrt{BD^2 + DR^2} = min.$$
> *Or, since A C and B D are constants,*
> $$\frac{m.CR\ d\ CR}{\sqrt{AC^2 + CR^2}} + \frac{n.DR\ d\ DR}{\sqrt{BD^2 + DR^2}} = 0.$$
> *But CD* being constant, we have d $CR = -$ d DR. We have then,
> $$m.\frac{CR}{AR} - n.\frac{DR}{BR} = 0.\ \&\ \frac{CR}{AR}.\frac{DR}{BR}\ ::n:m$$
>
> *That is to say,* **the sine of the angle of incidence to the sine of the angle of refraction is in inverse proportion to the speed with which the light traverses each medium**.

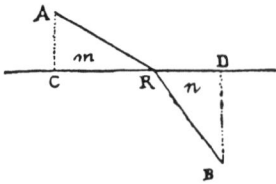

> *All the phenomena of refraction now agree with the central principle, that* **Nature, in the production of its effects, always tends towards the most simple means**. *So this principle follows, that* **when light passes from one medium to another the sine**

---

1.   Maupertuis, "Accord de différentes lois de la nature".

of the angle of refraction to the sine of the angle of incidence is in inverse ratio to the speed with which the light traverses each medium.

And so, for Maupertuis, light is propagated so as to minimise $AR \cdot v_1 + RB \cdot v_2$ and not the quantity $AR/v_1 + RB/v_2$. For these conclusions to agree with the experimental results of the time, and so that his principle would lead to the sine law, Maupertuis has to agree with Descartes that:

> ... *light moves more quickly in denser media, and the whole structure which Fermat built is destroyed.*

It is true that at that time no one knew how to measure the speed of light and no one could find a way of deciding between the different theories. The experimental proof that light travels faster in air than in water was not established until 1850 by Foucault.

In 1746, Maupertuis extended his principle from Optics to Mechanics[1]:

> *When a body is carried from one place to another, the action is greater when the mass is heavier, when the speed is faster, when the distance over which it is carried is longer.... Whenever a change in Nature takes place, the quantity of action necessary for this change is the smallest possible.*

With this general principle, Maupertuis established a kind of union between Philosophy, Physics and Mathematics: Nature works in such a way as to minimise its action; the idea of causality is abandoned in favour of the idea of achieving an aim, characterised by a harmony between the physical world and rational thought.

In Mechanics, Maupertuis' principle of least action operates very effectively. Euler in 1744, then Lagrange in 1760, revised, reformulated and justified it again. It can be stated in this way: When a system changes from an initial configuration P to a final configuration Q, the path taken by the actual movement – or that which is represented in the parametric space – is that which minimises the action necessary to pass from P to Q, in fact that which makes it a stationary value. In mathematical terms, the actual path is the curve joining P and Q in the parametric space for which the action, namely:

$$A = \int_P^Q m \, v \, ds$$

is stationary among all the curves joining P to Q. However, it was Jacobi who, in his Vorlesungen[2], stated the principle in a rigorous manner a century later: it only applies in the case of the conservation of energy and one should only compare the paths where the energy value is a constant.

Optics and Mechanics

The Principle of Least Action, a principle of Mechanics, cannot be strictly applied to the law of propagation of light, contrary to the wishes and beliefs of Maupertuis. Nevertheless, an Optical-Mechanical analogy had previously led Jean Bernoulli to solve the Brachistochrone problem. The movement of a body under gravity, whose speed increases inversely as the square of the height fallen, is compared here with that of the displacement of a ray of light traversing a medium whose refractive index varies in a

1.   Maupertuis, "Les lois du mouvement et du repos déduites d'un principe métaphysique".
2.   Jacobi, *Vorlesungen über Dynamik*.

continuous manner, here again as inversely proportional to the square root of the difference in heights.

Thus Jean Bernoulli wrote[1]:

> ... *For what I am able to show, which is a marvellous thing, is that if a translucent material, starting from a particular light source, continuously changes in density vertically in the same ratio as the speeds acquired by a heavy body which falls from the same particular point, the Curve of swiftest descent will be precisely the same as that of the ray of light, that is both will be the Roulette or the Cycloid.*

He established a very interesting parallel between:

> *the curvature of a wave of light, that is the line which cuts perpendicularly all the rays leaving from a particular point,*

and the curve of equal time,

> *namely that which determines the portions travelled in equal times by all the cycloids described from the same starting point and on the same horizontal base.*

In effect, the curve of equal time

> *is also precisely the same as the wave which is made in the aforementioned translucent material by the radiant point; for both will be perpendicular to these cycloids.*

Jean Bernoulli is able to conclude:

> ... *it follows that these two speculations, taken from two such different areas of mathematics, such as Dioptrics and Mechanics, have a connection between them which is absolutely essential and necessary.*

And, indeed, Hamilton[2] showed in the 1830's how this analogy exists formally at the level of basic principles. He uses it among other things as the starting point of his Dynamics, and extends it to his Optics in non-homogeneous media, since the mechanical principle of least action can in effect be made to correspond with the optical principle of Fermat. Recall the principle of least action: a conservative dynamic system changes from a configuration P to a configuration Q following a path such that the action

$$A = \int_P^Q p \, dl$$

where $p = mv$ represents the moment and dl an element of the path, must be a minimum or a stationary point. As for Fermat's principle, it is formulated in the following way: a ray of light links a point P to a point Q along a path such that the time

$$T = \int_P^Q \frac{n}{c} \, dl$$

where n represents the refractive index of the medium traversed, and c the speed of light in a vacuum, must be a minimum or stationary point.

This is the same type of condition, leading to the same type of equation, all that is needed is to replace the moment $p = mv$ by the index $n = c/v$. Notice, however, that Fermat's condition introduces time as an essential element, whereas the principle of

---

1.   Bernoulli Jean, Sur le Problème des Isopérimètres.
2.   Hamilton "On a General Method in Dynamics" and "Second Essay on a General Method in Dynamics", *Philosophical Transactions of the Royal Society*, 1834, 1835.

least action is not concerned with how movement takes place in time; what we have is a formal analogy.

It is, however, an important analogy since, keeping with the sense of the remarks made by Jean Bernoulli, Malus' theorem in geometrical Optics, according to which, trajectories orthogonal to one remain orthogonal to a whole family of surfaces – the surfaces of a wave, has an exact parallel in Mechanics: if a surface exists – in fact a hyper-surface in a parametric space – to which at each point P_0 the natural trajectories are orthogonal, then the trajectories are orthogonal to a whole family of surfaces (hyper-surfaces), which are the surfaces given by the equation W = constant, where the function W represents the integral of the action from P_0 to P.

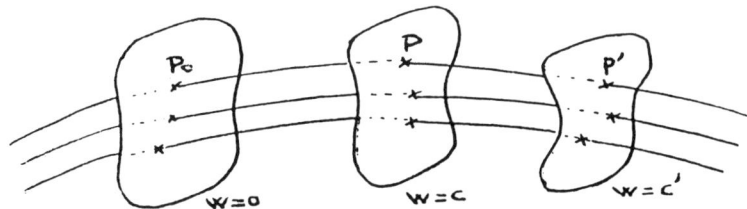

Thus, the surfaces of equal time in Optics are by analogy the surfaces of equal action in Mechanics – for a given energy E. This way of assimilating classical Dynamics to geometrical Optics is obtained by associating an index of refraction with the potential function V:

$$n = \sqrt{1 - V/E}.$$

But of course these surfaces of equal action do not, in the case of Mechanics, correspond to paths of equal time.

Wave Optics and Wave Mechanics

This Optical-Mechanical analogy developed by Hamilton stimulated work by Louis de Broglie and Erwin Schrödinger in the 1920's. However, we should note that this analogy only applies to geometrical Optics when we think in terms of wave surfaces, for the idea of a ray only belongs to geometrical Optics. However, this can be said to be only an approximation to Wave Optics (it does not apply if the objects encountered by light are small in relation to the wavelength); from this we could develop the idea of a Wave Mechanics for which Classical Mechanics would be a macroscopic approximation.

Hence, by introducing the wavelength, inversely proportional to the refractive index n, Fermat's principle can be formulated in terms of the stationary points of the integral:

$$\int_P^Q \frac{1}{\lambda} \, dl$$

In the same way, if according to wave mechanics, one associates with a particle of moment p a wavelength defined by $p = h/\lambda$ where h is Planck's constant, the principle of least action can be formulated in terms of the stationary points of the integral:

$$\int_P^Q \frac{1}{\lambda}\, dl$$

The strict analogy is disturbing (in terms of waves, we talk about the surfaces of a wave, in terms of particles, we talk about orthogonal trajectories). To take this further, here is an extract from an article by Schrödinger[1] which appeared in 1926, in which he appeals to this analogy between Mechanics and Optics.

> *Each point of the surface W_0 is allowed to move in the direction of the positive normal through a distance*
>
> (5) $$ds = \frac{E\,dt}{\sqrt{2(E - V)}}$$
>
> *That is, the surfaces move with a **normal velocity***
>
> (6) $$u = \frac{ds}{dt} = \frac{E}{\sqrt{2(E - V)}}$$
>
> *which, when the constant E is given, is a pure function of position.*
> *Now it is seen that our system of surfaces W = const. can be conceived as a system of wave surfaces of a progressive but stationary wave motion in q-space, where the value of the phase velocity at every point of the space is given by (6). For the normal construction can clearly be replaced by the construction of elementary Huygens waves (with the radius (5)), and then of their envelope. The "index of refraction" is proportional to the inverse of (6), and is dependent on the position but not on the direction. The q-space is thus optically non-homogeneous but is isotropic. The elementary waves are "spheres", though of course, – let me repeat it expressly once more – in the sense of the line-element (3).*
> *The function of action W plays the part of the **phase** of our wave system. The Hamilton-Jacobi equation is the expression of Huygens' principle. If, now, Fermat's principle be formulated thus,*
>
> (7) $$0 = \delta \int_{P_1}^{P_2} \frac{ds}{u} = \delta \int_{P_1}^{P_2} \frac{ds\sqrt{2(E - V)}}{E} = \delta \int_{t_1}^{t_2} \frac{2T}{E}\,dt = \frac{1}{E}\delta \int_{t_1}^{t_2} 2T\,dt,$$
>
> *we are led directly to Hamilton's principle in the form given by Maupertuis (where the time integral is to be taken with the usual **grain of salt**, i.e. $T + V = E = $ constant, even during the variations). The "rays", i.e, the orthogonal trajectories of the wave surfaces, are therefore the **paths of the system** for the value E of the energy.*

A little later, Schrödinger concludes this introduction:

> **We know today, in fact, that our classical mechanics fails for very small dimensions of the path and for very great curvatures.** *Perhaps this failure is in strict analogy with the failure of geometrical optics, i.e. "the optics of infinitely small wave lengths", that becomes evident as soon as the obstacles or the apertures are no longer great compared with the real finite, wave length. Perhaps our classical mechanics is the complete analogy of geometrical optics and as such is wrong and not in agreement with reality; it fails whenever the radii of curvature and the dimensions of the path are no longer great when compared with a certain wave length, to which in q-space a real meaning is attached.*

---

1.   Schrödinger, Quantisierung als Eigenwertproblem II.

Then it becomes a question of searching for an undulatory mechanics, and the most obvious way is the working out of the Hamiltonian analogy on the lines of undulatory optics.

*

* *

It would be right to conclude by revisiting our initial problem of the slides in the playground. We are circumspect, and content ourselves with noticing that in the course of this wander through diverse disciplines, the theme of minimisation or maximisation (briefly the problem of optimisation) is ever present, and should not be underestimated during these unhappy times.

Bibliography

Sources texts

BERNOULLI, *Opera Omnia*, vol. I., p. 161-213, Lausanne and Geneva, 1742.

DESCARTES, *La Dioptrique*, annexe to *A Discourse on Method*, Leyden, 1637. *Œuvres*, Adam & Tannery, vol. VI: republished Paris: Vrin, 1965.

EULER, *Methodus inveniendi lineas curvas maximi minimive proprietate gaudentes sive solutio problematis isoperimetrici*, Lausanne and Geneva, 1744. *Œuvres*, first series, vol. 24, Orel Füssli, Berne, 1952.

FERMAT, *Œuvres*, P. Tannery & C. Henry, vol. II, Paris: Gauthier-Villars, 1894, p. 457-463.

GALILEO, *Discorsi e dimostrazioni matematiche intorno a duo nuove scienze*, Leyde, 1638: *Discourse concerning two new Sciences*, English tr., Henry Crew and Alfonso de Salvio. Dover, New York, 1914.

HAMILTON, "On a General Method in Dynamics", and "Second Essay on a General Method in Dynamics", *Philosophical Transactions of the Royal Society*, 1834, Part II, 247- 308, and 1835, Part I, 95-144. See *Works*, vol. II, Cambridge, 1940.

JACOBI, *Vorlesungen über Dynamik*, posthumous publication, Leipzig, 1866.

LAGRANGE, "Essai d'une nouvelle méthode pour déterminer les maxima et les minima des formules intégrales définies" and "Application de la méthode exposée dans le mémoire précédent à la solution de différents problèmes de dynamique", *Miscellanea Taurinensia*, vol. II, 1760-1761. *Œuvres*, vol. I, Paris, 333-468.

MACH, *Die Mechanik in ihrer Entwickelung*, Prague, 1883: *La Mécanique, exposé historique et critique de son développement*, French translation of the 4th edition, Emile Bertrand, Paris: Hermann, 1904. Reprint: Paris: Gabay, 1987.

MAUPERTUIS, "Accord de différentes lois de la Nature", *Mémoires de l'Académie des Sciences de Paris*, 1744, 417-426. "Les lois du mouvement et du repos déduites d'un principe métaphysique", *Mémoires de l'Académie des Sciences de Berlin*, 1746, 267-294.

NEWTON, *Correspondence*, edited by J. F. Scott, vol. IV (1694-1709), Cambridge: Cambridge University Press, 1967, p. 223.

SCHRÖDINGER, "Quantisierung als Eigenwertproblem", *Annalen der Physik*, 4th series, 1926, vol. 79, 361-376. English translation in *Collected Papers on Wave Mechanics*, New York: Chelsea Publishing Company, 1982.

On the calculus of variations

HILDEBRANDT and TROMBA, *Mathematics and Optimal Forms*, New York, W.H. Freeman & Company, 1985.

9

How may pictures appear to be real?

Didier BESSOT and Jean-Pierre LE GOFF
IREM Caen

We have become so used to seeing images in photographs and on television that we accept them as true representations of the real world. These images, whether fixed or moving, seem to claim that their view of the world is identical to our own view of the world around us. Yet, this is to overlook the fact that cameras have been made to conform to certain conventions about representing the world, conventions that belong to western society. There are many ways of representing the world and this chapter considers something of the history of the conventions that we use.

The question that is at the origin of this story is the following: how can we produce an image on a plane surface that can capture the depth and sense of relief that our binocular vision affords us? We need to be aware of the fact that we come to this topic today from a *point of view* which is the result of several centuries' education about perspective. But before the question came to be seen as concerned with geometry, artists must have come to understand the spatial reality of a surface in a way that presents an illusion to the eye. Transposed into geometrical terms, the question becomes: how can we represent three-dimensional space on a plane surface possessing only two dimensions?

The choices that came to be made and the conventions that arose, were worked out through a long period of rationalisation of vision. The development has its origins in Ancient Greece and Rome, it became 'liberated' during the Renaissance, it used geometrical constructions to represent three dimensional objects on the plane, and ended by being a stimulus to geometry itself, in particular in relation to what we call projective geometry. From a simple question asked about an image which appears to have depth, geometers finally drew lessons about the nature of space itself.

What is light?
Euclid: the beginnings of geometrical optics

European science of the Middle Ages, derived more than its knowledge of geometry and astronomy from Ancient Greece. The Ancients were also interested in optics and, from the earliest times, with the nature of light and human vision. The idea that light travels in a straight line derived from the first observations carried out in the Ancient Worlds of Babylon, Egypt and Greece; representations of the night sky developed from observations of certain alignments, and determining time and seasons came from the use of a shadow stick planted in the ground, the *gnomon*. Doubtless the science of sundials, gnomonics, as well as cartography, contributed to the emergence of ideas about perspective, since certain of their methods consisted in projecting points in space, on to a plane, from a fixed point.

The Greeks developed two concepts of vision: the first was that data about an object was being permanently emitted by its "skin" and penetrated the eye in a reduced form; the other was that the eye had the power to "capture" the forms and colours of an object by means of visual rays which it emitted. What remained of these debates, which had a prolonged echo right up to the Middle Ages and in the Islamic World, was a science of geometric optics, known essentially through Euclid's *Optics*, and which concerned visual rays, emitted or perceived, that were concurrent at a point known as the optic centre. This point was considered to be either on the surface of the eye, at the centre of the pupil, or at the centre of the eye ball. The exact physiology of vision is not important in what follows: the important thing is that the eye is considered to be a point, and perception comes from the use of a cone of linear rays which became known as the *visual pyramid*. A straight line segment appears as a straight line segment, and its position in the plane is dependent on the point of view of the observer and the position of the observed segment; a plane surface also appears as a plane surface. All these geometric conventions led to a theory which conformed to the observed hard data, including the apparent diminution of figures at a distance.

The geometrical knowledge acquired by Greek scientists, up to the *Elements* of Euclid, is the foundation of the development of the idea of perspective, often called natural perspective, and forms an integral part of geometrical optics. This perspective was essentially comparative, and was based upon the angular appearance of magnitudes. An echo of this view was still be found amongst the painters and architects of the Renaissance: thus inscriptions at the top of a column were to be made larger than those at the base if both were to appear to be the same size to the viewer, that is from the angular perspective of the viewer. An engraving by Dürer illustrates the application of this principle (Fig. 1). In order that the successive levels of twists around a column should appear to be equidistant, as seen from a point eye *c*, their actual heights should be determined by an equiangular division of the angle *bac*, to find where the rays of the visual pyramid cut the vertical side of the column.

Exercise 1   Measure the lengths of the segments of the line *ab* produced by the 17 rays of the visual angle and find the numerical law between them.

How can we "see" depth?
Vitruvius: the first laws for its representation

The problem of how to make a "faithful" representation of objects, and especially of objects defined by straight lines and plane surfaces, is the concern of the draughtsman, the architect and the engineer. But it was originally the problem facing painters and other artists, who wished to lay out, on a variety of surfaces, their vision of the world, whether that might be real and observed, or internal and imaginary. The first efforts were prehistoric (like the cave paintings of Lascaux); and Plato's story of the shadows on the walls of the cave have the same origins.

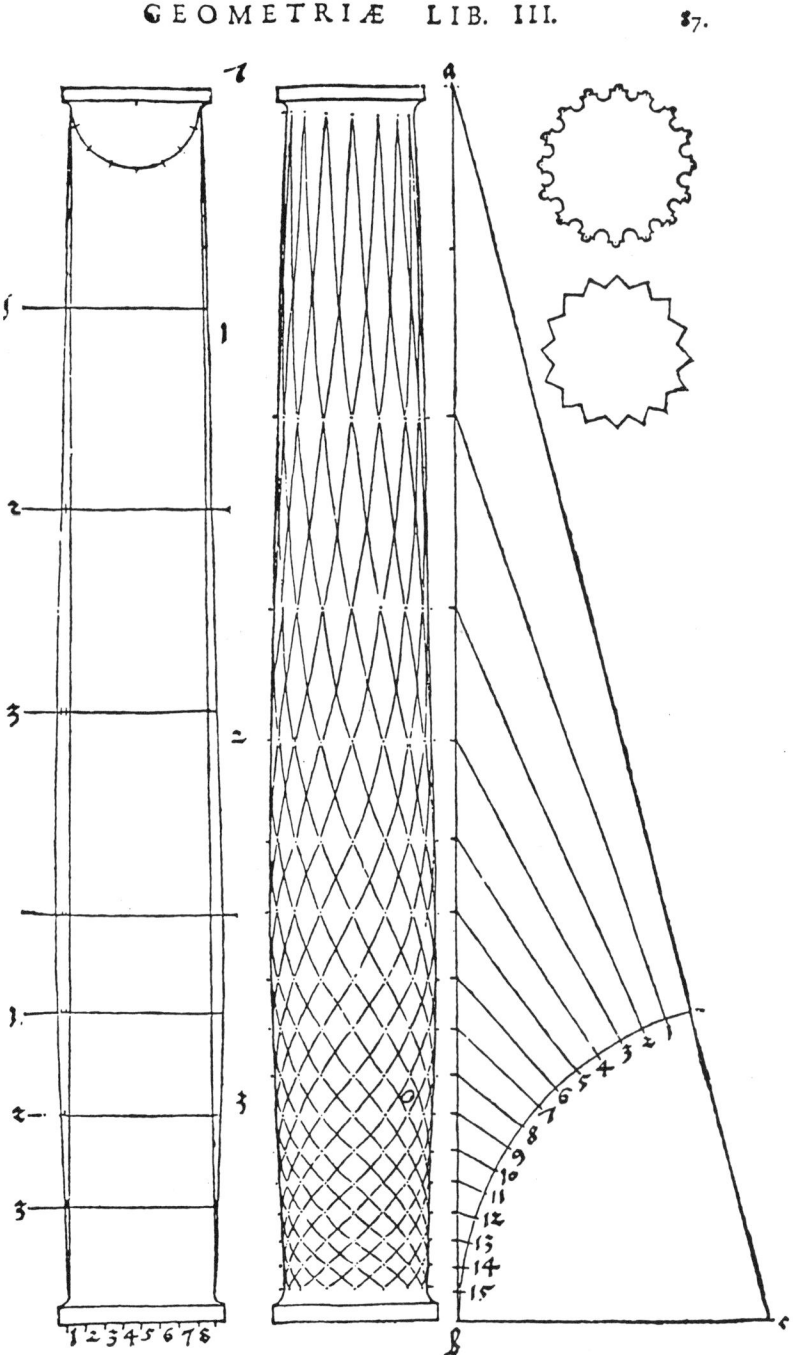

Figure 1

Dürer, *Underweysung der Messung*, Nüremberg, 1525. Plate 7, Book III

Even there, we find creative innovation in the Ancient World: certain frescoes in Pompeii show a desire to give a "realistic" representation to architectural decors. Looking at masks showing comedy and tragedy, we can see this desire for illusion in the works of certain stage designers who wanted to produce flats at the back of the stage to evoke the course of the dramatic action. In the second half of the first century AD, we find Vitruvius referring, in his architectural work *De Architectura Libri X*, to several artists who had produced illusionary decors, such as Agatharcus, Democritus and Anaxagorus.

Study of ancient frescoes does not allow us to confirm with certainty that their creators were using any of the several techniques of linear perspective which were rediscovered several centuries later. However, the use of parallel lines, and the way in which parallel lines perpendicular to the vertical plane appeared as vanishing lines, gave the impression of depth. But if these apparent vanishing lines are extended, we find that there are usually pairs of lines which are symmetrical with respect to a central vertical mirror line at which they converge. Fig. 2 shows an example of this first type of empirical schema which Panovsky calls *fish bone* symmetry.[1] It is probable that such a schema arises from a certain sense of symmetry and that the points of convergence are not in fact construction points: in other words, the vertical symmetry line is an *a posteriori* reconstruction.

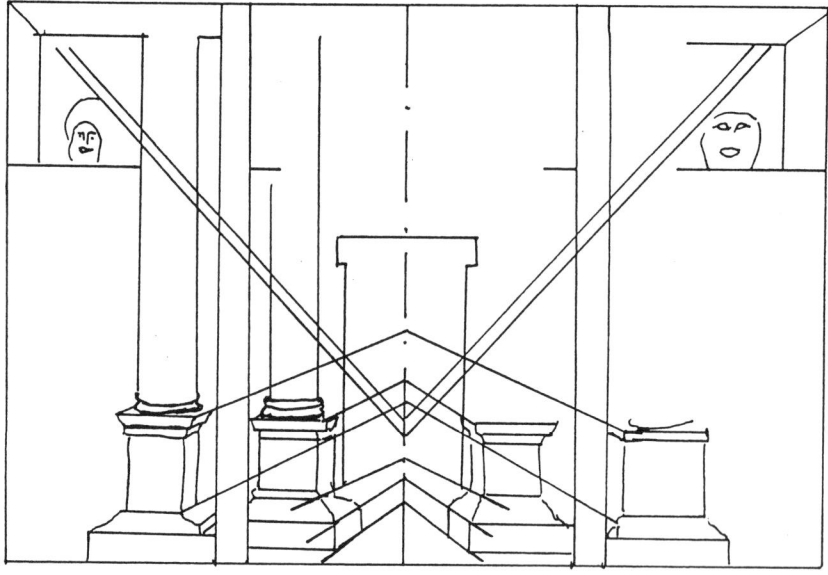

Figure 2
Perspective schema in a Pompeii frescoe (Naples Archeological Museum)

Some frescoes even show representations which have just a single point of convergence, for lines which appear perpendicular to the painted surface (Fig. 3). Today we would call these vanishing lines, but the Ancient World has left no written testimony on the practice of producing such a representation and so we should exercise caution over using such terms as "vanishing lines" or "vanishing points" which come from a retrospective study of these frescoes using ideas which became familiar much later.

---

1.   Panovsky, *La perspective comme forme symbolique*, p. 75.

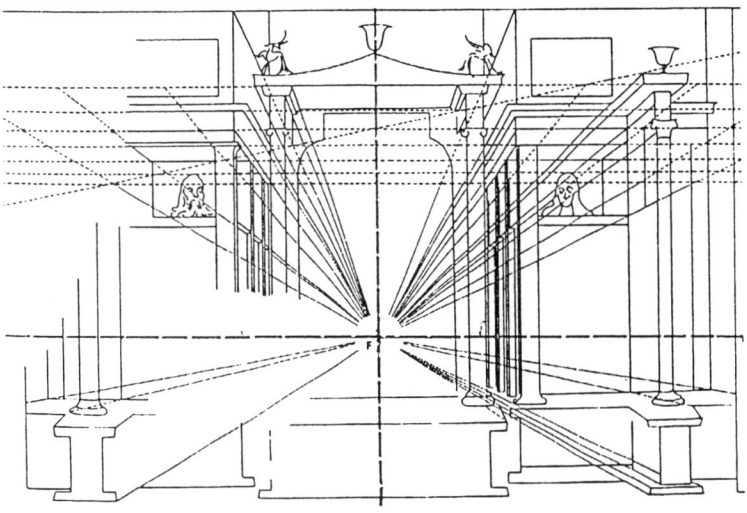

Figure 3
Perspective schema in a frescoe in the Hall of Masks (Roman Palatine)

The Early Middle Ages saw a break with the type of painting typical of the late Greco-Roman period. However, in some Middle Ages drawings in the West we see the same *fish bone* schema, as in the fourteenth century example given by Panovsky[1] (Fig. 4). Although it appears to have the same perspective, it may not have been produced by the same technique: it may be a reminiscence or a rediscovery of earlier drawing methods or, perhaps, a necessary step on the road towards a more sophisticated convention, occasioned by another important rupture with the past: moving from the use of *perspectiva naturalis*, the angular perspective that derives from Greek optics, to a *prospectiva accidentale* or *perspectiva artificialis*, which is a linear perspective and derives from purely geometrical considerations.

Figure 4
Perspective schema in the *Creation of the Stars*, taken from a reredos by Grabow (1379)

---

1. Panovsky, *La perspective comme forme symbolique*, p. 130-131.

The only known evidence about the use of perspective by artists of the Ancient World is to be found in the treatise on architecture by Vitruvius. In Book 1 he describes two projections, which are in fact two orthogonal projections, the one on a horizontal plane called an *ichnography* (the plan or view from above), the other on a vertical plane called an *orthography* (elevation or profile view) (Fig. 5). He completes these two views of a building with a description which he calls a *scenography*, which appears to have been a perspective view, although he gives no indication of its construction. The French architect Claude Perrault, who produced one of a number of editions of Vitruvius in 1673, translates a passage from Vitruvius as follows:

> *The Ichnography is a drawing by ruler and compass of the plan of a building, in a small space, as if it were on the ground; the Orthography represents, also in a small space, the elevation of one of the faces in the same proportions as the work to be constructed; the Scenography has to show, not only one of the faces and the sides, but also the hidden parts, and this by the concurrence of all lines in a central point.* [1]

But this must surely be a retrospective reading of the text, by one who is conversant with the practice of techniques which had by then been used for two centuries, and which he believed to have found their origin in Vitruvius.

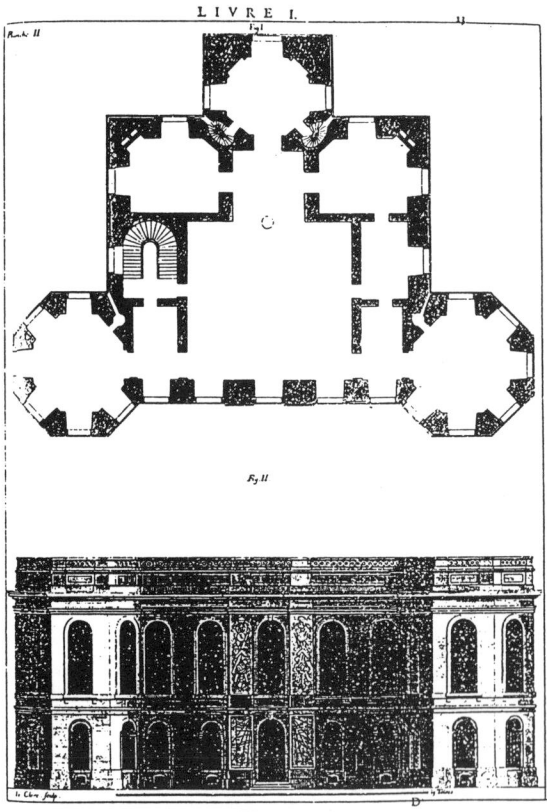

Figure 5
Ichnography and orthography.
Engravings taken from Claude Perrault's 1673 edition of the first century AD
Dix Livres d'Architecture by Vitruvius

---

1.   Vitruvius, *Les dix livres d'Architecture*, Book 1, ch. 2.

Whatever interpretation we may place upon the text by Vitruvius, it remains that the humanist painters of the Italian Renaissance were able to profit from the idea of double orthography. It is certainly at the base of their speculations about perspective. Also, the idea of projecting an object in space on to two orthogonal planes is not found only in the context of linear perspective in fifteenth century drawings: it became the foundation of nineteenth century descriptive geometry and industrial drawing.

Parallel lines appear to be disappearing into the distance with a point in common: can we show this on a drawing?
The emergence of a central point of convergence in the Trecento

Paintings of the Early Middle Ages, both Byzantine and Western, have sacred themes, and are more concerned with hierarchy than with proportion. The height of a person depends more upon his status in the sacred order, or in the social scale, than it does upon his spatial position, relative to the other objects of the drawing. Sculptured figures on Gothic tympanums are a good example. During the thirteenth century there were changes in patronage: in addition to the Church, there were now men of letters and nobles, enriched by trade, whose sensibilities had to be entertained by the artist. The increasing demand for illustrative painting led, as much in Italy as in the countries of Northern Europe, to thinking about a painting or fresco as a sort of scene or story, in which the secular challenges the place of the religious, and which is acted out in a realistic setting. Gilded paintings which were symbolic of the sacred, were replaced by views of the sky and countryside that stretched to the horizon, or by a framed space within which it became necessary to consider how to place lines. This change was not a natural development: in particular, it conflicted with the principles of the "international" style of the later Middle Ages, a style that has since been dubbed "Gothic". These principles influenced the form and structure of a painting, which had to pay more attention to the religious or social status of its subjects than to their relative spatial disposition in the picture. The works of the Limbourg brothers and Jean Van Eyck in Flanders, or those of Giotto and the Lorenzettis in Italy of the Trecento, are illustrations of this period of searching for more realism in art. This led to a systematic method of using a central point where lines orthogonal to the picture converge, as in the Siena (Figure 6) school, or to the use of lateral concurrence points, which derives from angular views, and which is to be found in naturalistic pastoral scenes. Several centuries of visual education may lead us to interpret these points as the early examples of the use of what are today called "vanishing" points. But the same is true here as before: the absence of preparatory drawings or of any theoretical works, if indeed there was a theory, does not permit us to conclude that the practice arose from the use of points of convergence. The convergence that the modern eye believes it detects, may be no other than the result of construction processes that are based on symmetry and similarity.

Whatever may be the case, the convergence of lines orthogonal to the painting towards a single point does not in itself provide an illusion of depth. Painters of the Siena school had, perhaps, understood this, since they often chose to include a square-tiled floor in their paintings and the progressive diminution in the size of the tiles reinforced the illusion of depth (Fig. 6). But what rule should they follow for showing the sizes and relative positions of equal lengths that were at different distances from the observer? This question had to be faced whenever a regular squared tessellation was needed for a floor, and the floor arrangement also determined the way in which lines above the floor had to be drawn. In the particular example of a square tessellation, the question was this: the parallel sides of the squares that are orthogonal to the picture will converge towards a point, but how far apart should we draw the lines for the other parallels so that the diminishing sides appear to be equal in length? The reduction in the length of the sides now becomes a function of their position in the drawing.

Figure 6
Ambrogio Lorenzetti (1290?-1348): **The Annunciation**
A perspective arrangement for the paving slabs.

In an epoch enamoured of proportion, one of the ideas that appealed was to use familiar numerical progressions based on regular intervals, or a given ratio. For this reason many of the painters used an empirical rule of reducing the distance between successive parallel lines by a constant ratio, often two-thirds. Pictures which were drawn using the rule of two-thirds, like Lorenzetti's Annunciation, are instantly recognisable: the diagonals of the square tiles produce a type of spiral (Fig. 7) and this is particularly evident in chequered floor patterns: this is the case in the paved floor of **The Presentation to the Temple** by Giovanni di Paolo (1399-1482) kept at Siena.

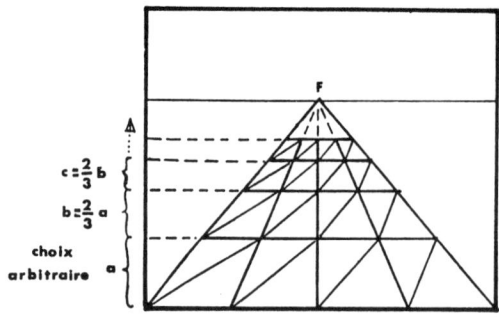

Figure 7
The rule of two-thirds

The vanishing point is placed anywhere within the picture, and the distance (a) between the ground line and the line of the first transversal is chosen arbitrarily. The rule consists in systematically reducing each successive distance between parallels to two-thirds of the previous distance.

Exercise 2	Show that the rule of two-thirds, or any other rule based upon a geometric progression, or even an arithmetic progression, cannot produce diagonal alignments.

Exercise 3	Draw a large number of transversals for different values of (a): what is the limiting position of these horizontal lines in relation to the visual horizon?

These methods do not attempt to reproduce a global spatial reality on the plane, within which we could have a perspective geometry: they are simply geometrical constructions that try to give a faithful representation of what is seen. In order to have a way of representing three dimensions on the plane, we need first of all to have a rationalisation of a portion of space, so that we can see how geometrised elements must be positioned: we must learn to see what seeing is. This rationalisation of how we see three dimensional objects leads to the use of a projection, in fact a double projection, along the lines that were shown by Vitruvius, and rediscovered by the artists and architects of the Italian Quattrocento. The methods led the artists to understand how the size and position of the subjects were to be determined, and according to which geometrical laws.

The position of the artist
Brunelleschi and Alberti: the legitimate construction

Despite certain reservations about representing parallel lines as lines which appeared to converge to a point, (the idea of convergence to a limit had not yet been developed), it soon became apparent that the principal vanishing point was the image of the artist's view–point, and for this reason it would be described by Leon-Battista Alberti of Florence as *the prince of the rays*. It shows, in a sense, the position of the eye in the painting, being its orthogonal projection on to the picture plane (see Fig. 9 below). This point also determines the height of the horizon in the painting, as that of the person looking at the painting. The image is produced by thinking of the painting as a sort of window through which a view is seen, and is formed by the geometrical intersection with it, of the visual pyramid of rays that come from a single, fixed and unmoving eye. Alberti described this as opening *a window on the world*. Since the exact position in the painting of all the objects in space is determined by the intersection of rays that come from the eye, the result depends upon the height of the eye and its distance from the painting. Alberti set out the principle of his new conception in *De Pictura*, a principle that provided a new position for the artists of the Quattrocento. The new position for the artist was not only a new position geometrically, but it also gave him a new position in a sociological sense: the subject is now the vertex of an infinite number of possible visual pyramids, and geometry itself becomes the mechanical working tool of an artist, dedicated to his liberal and noble art. Alberti had published his treatise in Latin in 1435, but such was the interest among artists, that he produced a vernacular translation under the title *Della Pittura* from 1436 onwards, but a printed edition had to wait until 1540. Here is how Alberti defined a picture:

> *And so a painting is no more than an intersection of the visual pyramid, from a given distance, once the centre is positioned and the light rays are determined, represented on a certain surface by the artifice of lines and colour.*[1]

While Alberti was the first to explain the principles of this new method, that would so disturb the whole of the world of visual arts, it was the architect Filippo Brunelleschi who had invented it in the 1420s. This innovation was partly due to the favourable context of commercial expansion, since the Italian cities were increasing their trade around and beyond the borders of the Mediterranean. The Arts flourished thanks to patronage, there was a renaissance of building in towns, which was along grid lines, instead of the hitherto haphazard development, and at the same time new surveying instruments were made for taking sights, and which were based on the use of plane projections, as in map making, surveying and land measurement. On account of the rationalisation of agriculture, the land and the countryside became, so to say, 'squared off'. There was a renaissance of science and technical skills, and the world of tools and machinery, created for civil and military use, demanded a more faithful method of representation, and one that would be sufficiently sophisticated to be able to take account of volume.

In order to construct the appearance of an object in a picture, Brunelleschi imagined the eye of the observer to be placed at a point in front of the picture, and above the ground (which we now call the plan). It is a presentation of perspective (see Fig. 8 and also Fig. 20 below) which is directly inspired by Vitruvius (Fig. 5). A double orthogonal projection allowed Brunelleschi to locate the height of each point image of the visual pyramid intersecting the plane of the picture, and the lateral distance of each point image of the plan, with respect to the principal visual ray, as seen from above.

Brunelleschi left no written account of his method, which was soon to be called the *legitimate construction*, but his principle appeared soon after in the method described by Alberti. The description, as well as the use, of double projection appeared in the first complete treatise on the subject, published around 1470, namely *De Prospectiva pingendi*, by the painter and mathematician Piero della Francesca. This work appeared in Italian and treated the subject in a geometrical way: the perspective should better be described as artificial, and is called linear, conic or central perspective.

In his *De Pittura*, Alberti uses a contracted form of the method called *legitimate* to describe how to draw a view of a square-tiled floor, as seen through the frame of a picture, the frame being parallel to the ground line (Fig. 9 to 11).

---

1.   Alberti, *De Pittura:*

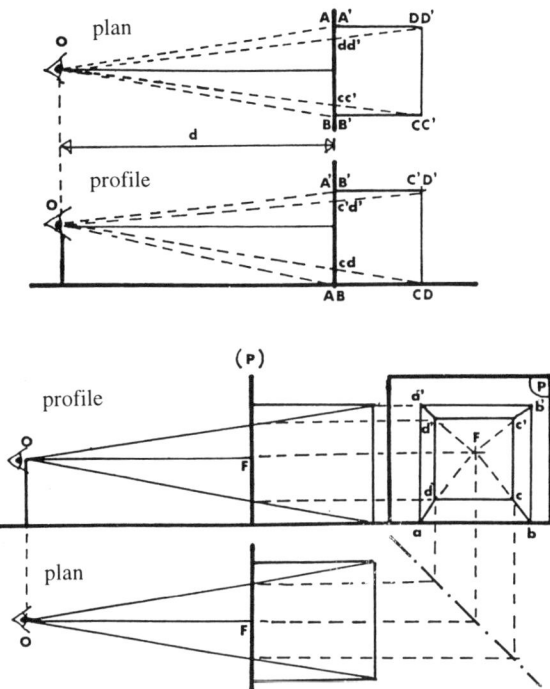

Figure 8

The double projection, a construction method which came to be known
as Brunelleschi's legitimate construction.

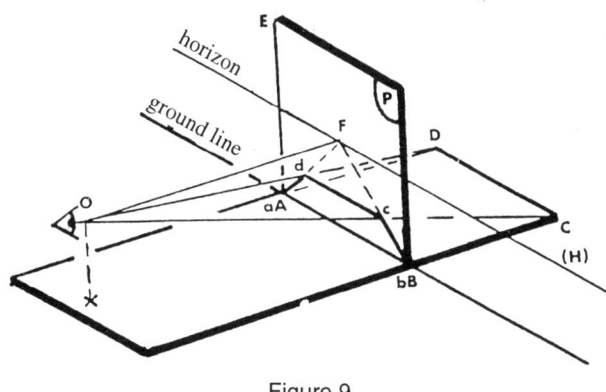

Figure 9

The painter's eye in front of *the window that opens on the world*

In order to make a drawing of a square on the floor, as it would appear to the eye of
an observer at O, in front of a picture (P) (figure 9), Alberti's method is essentially as
follows: in the plane of the picture (Fig. 10), F is the central point (the principal
vanishing point); one of the sides of the picture, here AE, is also taken as a profile view
of the picture; O is the position of the eye in this profile view of the picture which is
found by 'folding it back', with AE as the hinge; AB is the front edge of the square-tiled
floor, coinciding here with the ground line of the picture; H, on AE, shows the height of
the horizontal line through F; AH is then the height of the eye (F or O) and OH is the
distance of the eye from the picture, given by Alberti for positioning O; OB is therefore
considered to be a visual ray, seen in profile, joining O to C, or to any other point on
CD, the back edge of the square-tiled floor, since AB = BC = AD; the intersection of

this ray with the edge AE at R, gives the height in the picture at which the image cd of the line CD must be placed; a guide line through R and parallel to AB cuts the vanishing lines AF and BF at d and c, AF and BF being representations of the lines AD and BC, orthogonal to the picture, and extended indefinitely. Alberti describes his method for a regular tiled floor, but the principal is valid even for a single square: and it was this method which Piero della Francesca followed several years later, the representation of a single square determining the position of every other point in space (see below).

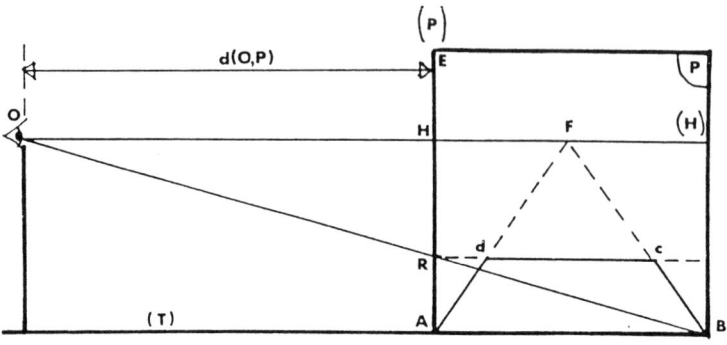

Figure 10
The perspective representation of a square on the ground, according to Alberti.

Alberti's method (figure 11) for the construction of a representation of a regular square-tiled floor follows the same principle, after dividing up AB, here into six equal parts; the visual ray from O and the profile view, determine the heights of the successive tile edges which are parallel to AB. Alberti describes the process in the following way[1]:

> When I wish to make a painting, this is how I go about it: I draw a rectangle, of any size that suits me, which I think of as being an open window, through which I look at whatever it is that I wish to paint; and in this space, I decide how tall I wish the men to be in the painting, and I divide a man's height into three parts, each being proportional to the measure called a 'braccia' [the Florentine braccia was about 58 cm]; because if you measure the height of an ordinary man you will see that he has a height of about 3 braccia; and using parts of the length of one of these braccia, I divide the base line [or ground line], which is the one at the foot of the picture, into as many parts as it will contain: I take this base line to be equal to each of the transverse segments which are parallel to it on the ground in front of me, behind the window, and set back at equal distances behind the base line.
>
> Then, inside the rectangle, and wherever I wish, I mark a point for the end of the central ray, which for this reason I call the central point [or principal vanishing point]. This point will be correctly placed above the base line, if it is no higher than the height of a man which I wish to paint in that position, for in this way, the objects that are painted will appear to be in one and the same plane, to those who are looking at the painting.
>
> Now, once the central point has been determined in the way I have described, I draw straight lines from this point to each of the division points which I have marked on the base line. Once these lines have been drawn, they determine how each successive transversal line will appear, one behind the other, almost up to infinity [...]

---

1.   Alberti, *De Pictura*.

And you should also realise that nothing painted will ever appear identically the same as the artist saw it unless it is viewed from a certain distance. [...] I have discovered this excellent procedure which is as follows: suppose things are as I have already stated, the central point being chosen, and the lines drawn from the central point to the division marks of the base line. Now, for the transversal lines which lie on the ground, this is how I proceed, in such a way that the first one determines the one that follows: I take a piece of paper on which I draw a straight line, which I divide into as many parts as I have divided my base line; then I mark a point above the base line at the same height as the central point; from this point I draw lines to each of the division points of the line I drew. Then I determine the distance which I wish to take as the distance of the eye from the painting, and there I draw a perpendicular, as the mathematicians call it, to cut every line it meets. [...] Thus this perpendicular, at each level where it is cut by the other lines, gives me the position for each transversal line which is at ground level behind the window. And in this way, I am able to determine all the parallel lines of a paved floor, that is squares of side one braccia which form the floor in a painting.

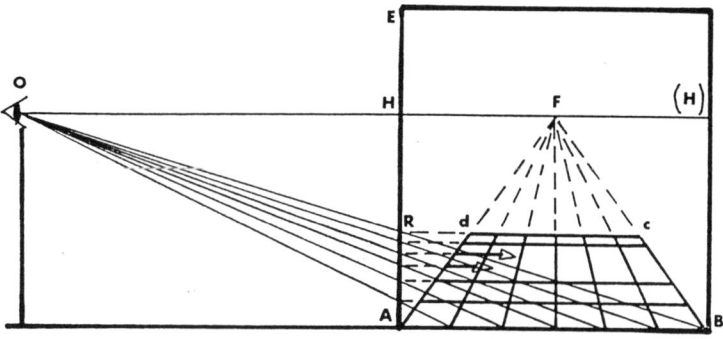

Figure 11
Alberti: setting out a square-tiled floor

Exercise 4   Construct the image of a cube standing on this square as its base.

Alberti justifies the legitimacy of his construction, by the fact that the intersection points of perpendicular lines are aligned in the representation (Fig. 12). This was not, of course, the case when using the empirical rule of two-thirds. In his work, Alberti denounces those paintings which do not possess this property of straight line diagonals, and uses this fact to argue for the superiority and modernity of his method. Given the idea of what constituted vision since the time of Euclid, whose *Optics* stated that the appearance of a straight line should itself be a straight line, perspective needed to be conserved, if not for isolated points whose perspective alignment would be difficult to observe, then at least for those alignments which were clearly visible in reality, like the diagonals of a square-tiled floor. Alberti's method achieves this.

Here is the legitimisation which Alberti gives for his construction (Fig. 12): the diagonals of the squares tiles line up in a straight line in the image, just as they do in reality: *and the fact that the diagonals of several squares drawn in the picture lie on one and the same straight line, is for me the evidence that they have been correctly determined.*[1] However, Alberti does not seem to have realised that the diagonals converge to one point on the horizon (here D_1), since he makes no mention of the fact.

---

1.   Alberti, *De Pictura.*

From this first stage in the theory of perspective, the technique is now available to artists and architects to draw solid shapes. Architects, for example, can now represent all types of structures, by using a square grid reference ground plan, all lengths being shortened in proportion to their distance, which can be determined by the closer spacing of the lines on the ground parallel to the front edge, as we go further up the picture. But history does not stop there: those who practised the drawing arts soon produced a multiplicity of treatises, like so many recipe books, and catalogues of examples to use as models of good practice. Furthermore, the geometers found that there was much to discover about the profound consequences of the way in which artists were now representing the three dimensional world.

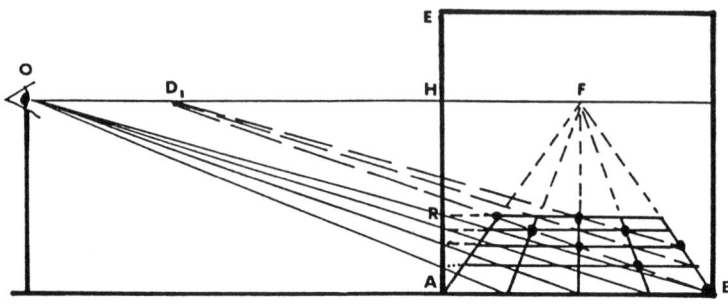

Figure 12
Straight line diagonals in a tiled floor using Alberti's method
A posteriori demonstration of the convergence of the diagonals

Geometry or Art?
Piero della Francesca: establishing the geometrical context

The construction by Alberti, which produced a true representation for a square-lattice floor, and so allowed the true construction of solid objects, by reducing vertical lengths according to their position on the lattice, soon led to other procedures. These appeared soon afterwards in a work by the Italian painter and geometer Piero della Francesca. His *De Prospectiva pingendi*, the first known geometrical treatise on perspective, only appeared in manuscript: it was written in Latin and then translated into Italian and the manuscripts were frequently consulted and partially recopied.

Piero della Francesa proposed, in fact, three methods of construction. The first was internal. Having first constructed the image of a base square by Alberti's method, he appealed, in effect, to the idea of a co-ordinate system for a point within the base square, the abscissa being along the base line and the ordinate along a diagonal (Fig. 13 to 19).

The second method was reserved for cases where the figure was bounded by non rectilineal lines, or where it was not a plane figure. Here Piero della Francesca used a point by point method of producing an image of the original, by the method of double projection, which it is generally thought that he had derived from Brunelleschi's legitimate construction (Fig. 20, see also Fig. 8).

The third method was shown by Piero della Francesca in proposition XXIII of the first book of his treatise as a particular case. This method was not used again in his work but it is an important method in the history of linear perspective, because it introduced what we now call a distance point (Fig. 21).

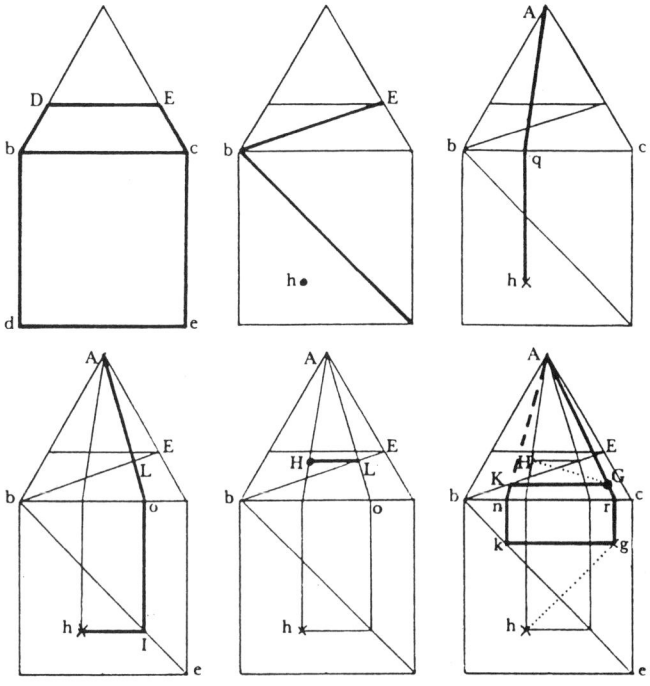

Figures 13 to 18
Piero della Francesca's method for drawing the image
of a triangle within the image of a known square

Here are the steps of the method used by Piero della Francesca to construct the image of a triangle (proposition XVIII, Book I): the real square *bced* with diagonal *be* becomes the trapezium *bcED* with diagonal *bE*. The image of the base square, *bcDE*, is supposed already known, since it can be obtained by Alberti's construction, itself explained and shown to be geometrically valid in proposition XIII of Book I. The figures 13 to 18 show the steps that are needed to construct the image of the vertices of a triangle *fgh*, and figure 19 is Piero della Francesca's own (Figure XVIII of Book I). The construction of *G*, the image of one of the vertices *g* of the triangle, is as follows: the point *g* in the real square is determined by the point *r* on *bc* and the point *k* on the diagonal *be*; the point *k* is itself determined by being on *be* and at *n* on *bc*. The point *K* in the picture, the image of *k*, will be at the intersection of the vanishing line *An* and the diagonal *bE*. Finally, drawing the transversal *KG* parallel to the ground line *bc*, determines the position of *G*, the image of *g* on the vanishing line Ar.

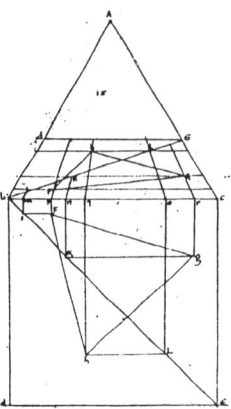

Figure 19

Piero della Francesca: Figure XVIII of Book I.

Construction of the image of a triangle by the method of co-ordinates

Figure 20 shows the construction of an octagon by the method of double projection, which Piero della Francesca sets out in proposition II, fig. XLVI of Book III. The apparent lengths of the sides of the octagon, supposed to be at ground level, are shown in the top diagram; the height in the picture where it is to be drawn, is given by the profile view. The lengths of these shortened sides, and their distance from the bottom of the picture, are found from plan and profile view, and transferred to a third drawing (the scenograph, a perspective view) – by using a horse's hair, says Piero della Francesca. In the example given here, the octagon is simply chosen to teach the method, since the method is not only applicable to rectilineal figures, but can be used to produce perspective drawings of curved figures, be they columns, tores or even human heads.

Exercise 5	Construct the image of a cube standing on its base and viewed from the front, using the method of double projection. Then construct the image of a cube balancing somehow on one of its corners: this problem is proposed by Piero della Francesca in proposition V of Book III !

Piero della Francesca was the first to give a rule for the alternative construction method for making a perspective drawing, using what is called a distance point, although he did not give a justification for this third method. It was only an expedient, which he had perhaps understood in noting that the method of justifying Alberti's construction, led to a point on the horizon (D_1 in Fig. 12). Figure 21, adapted from one by Piero della Francesca (Figure XXIII of Book I), illustrates the method. A is the principal vanishing point, $BCED$ is the known image of a real, full size, rectangle $NOQP$. To set the height of the image of a square of side BC in the rectangle, Piero della Francesca marks a point O on the horizon at a distance from A that is clearly given by the distance of the eye from the painting; he draws OC to cut AB at L; the horizontal line LM is then the apparent rear side of the square.

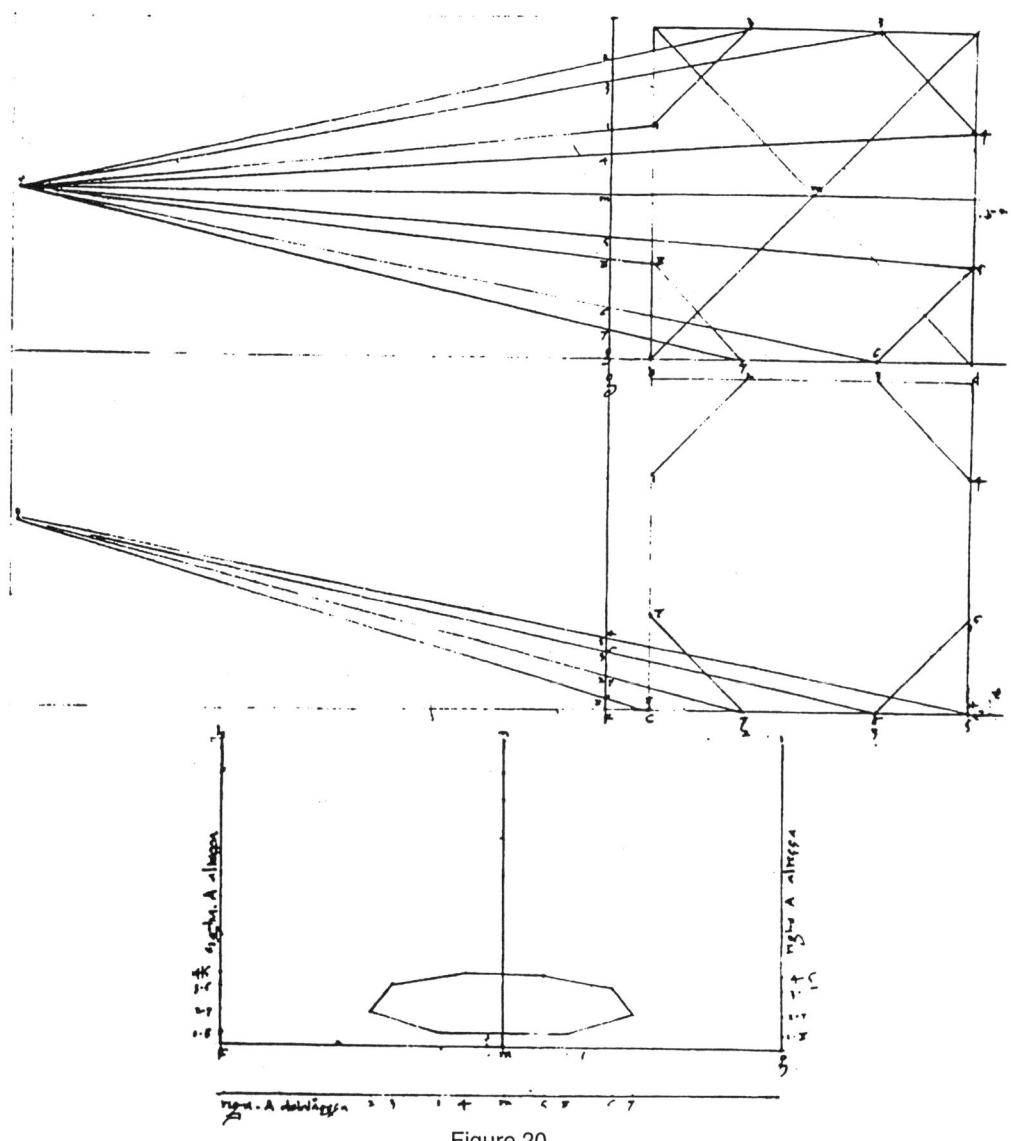

Figure 20
Piero della Francesca: Figure XLVI of Book III
Construction of the image of an octagon by the method of double projection

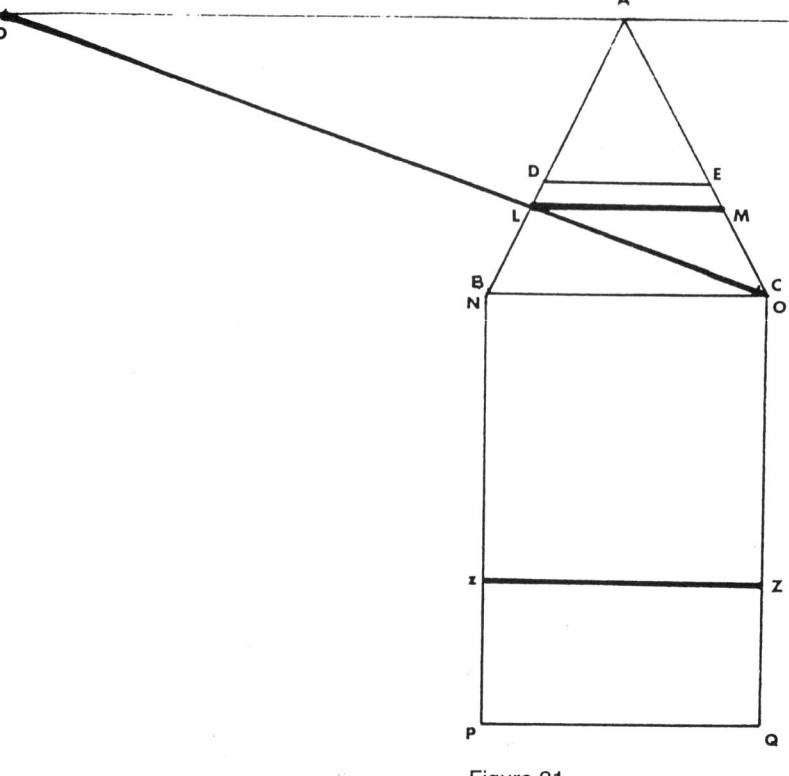

Figure 21
Piero della Francesca: Figure XXIII of Book I (adapted).
The method of using a distance point.

It should, finally, be noted that Piero della Francesca was the first to give a general rule for determining, both geometrically and numerically, the reducing sizes of the appearance of equal lengths, parallel to the ground line, and the same distance apart. Sometime later, Leonardo da Vinci gave a similar rule for determining the apparent lengths for transversals parallel to the ground line, this time for the case where their common distance apart was equal to the distance of the eye of the painter from the picture. In doing so, he came upon the sequence of inverses of integers that is known as the harmonic sequence, and this aspect of the problem was certainly significant. For a painter of the Renaissance, a painting needed to be considered in terms of proportion and harmony, in much the same way as music had been for the Pythagoreans. From the point of view of the history of perspective, this marked the beginning of establishing a central idea, which only became clarified in the 17th century: namely, the scale of perspective reduction.

| **Exercise 6** | Given the state of affairs as set out in figure 22, find the ratios of *cd*, *ef*, *gh*, etc. to *AB* (absolute ratios of reduction, compared to the initial *AB*), and also find the successive ratios *cd* to *AB*, *ef* to *cd*, *gh* to *ef*, etc. (relative ratios of reduction, each length being compared with its predecessor, and given by Piero della Francesca in two cases with figures like this). Also, find the successive ratios for the heights of the lengths seen in profile view, that is: *ce* to *Ac*, *eg* to *ce*, *gj* to *eg*, etc. (the ratios proposed by Leonardo da Vinci using an identical figure). |

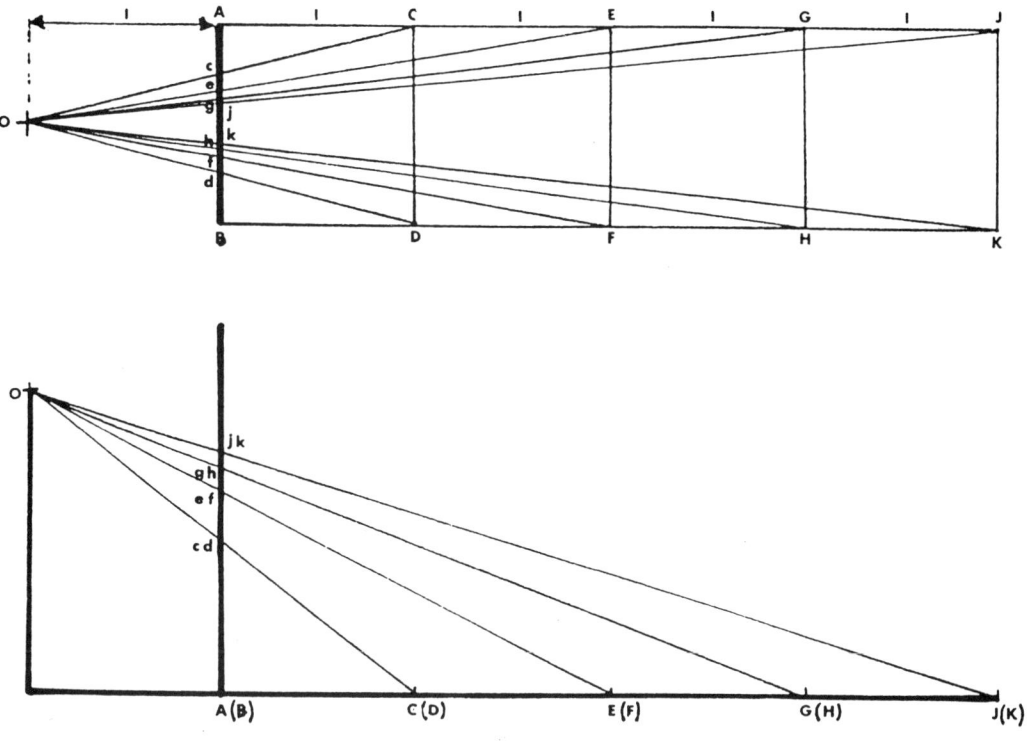

Figure 22
Figure for Exercise 6: after the numerical rule given in Piero della Francesca, Book I

While it may not be possible to claim that the invention of a system for representing space as a mapping of the reals dates from this time, it is nonetheless the case, that the painters of the Italian Renaissance understood, and took advantage of, the idea of double orthogonal projection. Furthermore, they exposed the essential elements in the theory of linear perspective, the first of those that are called artificial perspectives, and which were to have a richer development in the history of architecture and geometry than in painting, where the ideas were the cause of numerous disputes. While Leonardo da Vinci might speak of perspective as the *tiller of the painting*, the idea of perspective was often called into question. To cite two examples: in Italy from the sixteenth century onwards, the mannered style of paintings led to distortions and a certain disequilibrium; in France, in the second half of the seventeenth century, a debate over the primacy of colour, rather than perspective, in a picture, raged at the very heart of the *Académie Royale de Peinture*.

How did the distance point arise?
Viator: the rule of thirds; a time for reinvention

Piero della Francesca's use of a distance point reappeared as part of a systematic construction method in the work of Jean Pélerin, a canon, more commonly known as Viator. His *De Artificiali Perspectiva*, published at Toul in 1505, was the first printed work on perspective, earlier Italian works having appeared only in manuscript form, and it was based on the use of what is called in French, *les tiers points*, the thirds rule. The treatise appeared in bilingual form and was reprinted many times, some in pirate editions like, for example, the *Margarita philosophica* by Gregor Reisch, which appeared in Strasbourg in 1508.

What Viator calls the thirds points method, an idea which he had probably invented independently of Piero della Francesca, doubtless derives from the painting and drawing tradition of Northern Europe, where a painting of the countryside or an architectural drawing, often used an angular view, with two principal vanishing directions (Fig. 24). The foundations of Viator's method are defined as follows[1]:

> *The principal perspective point must be chosen and placed at eye level, this point being called the fixed point, or subject point. Following this, a line [must be] drawn on both sides of the said point; and on this line should be marked two other points, equidistant from the subject, closer for near views, further away for distant views, which points are called thirds points.*

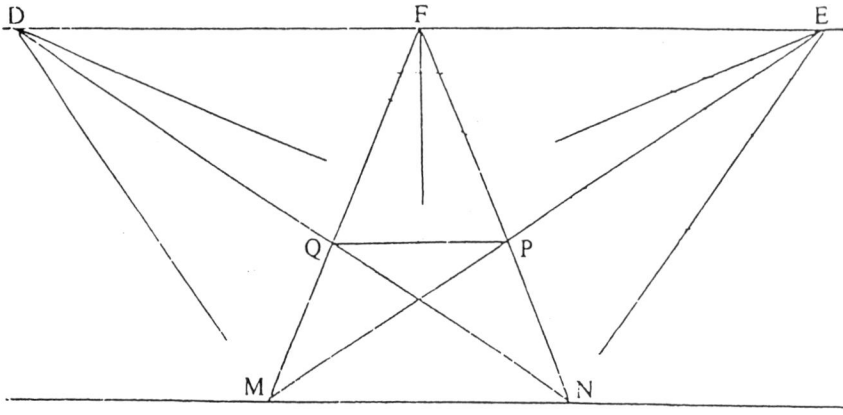

Figure 23
The method of thirds points according to Viator

Figure 23 is an illustration from Viator's work showing the thirds points (the figure in the original does not have letters). After fixing the principal point F and the horizon, Viator places his thirds points, here marked D and E, and then proceeds as follows for setting up the perspective of a square base: the points M and N are joined to F; then M is joined to E and N to D. The lines ME and ND, which appear as diagonals of the square, cut FN and FM at P and Q respectively. PQ is now the far side of the square.

---

1.   Pélerin, called Viator, *De artificiali perspectiva*, fol. 2, v.

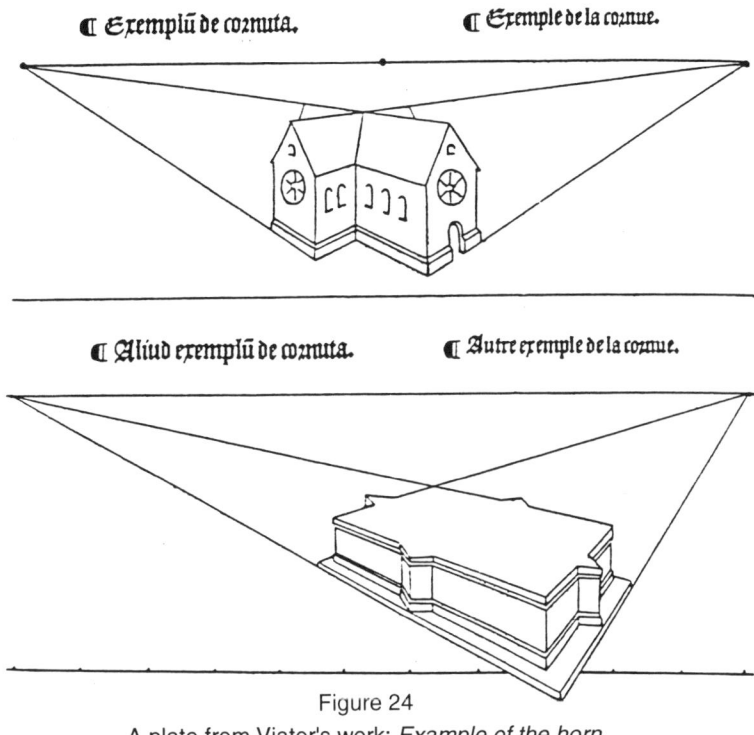

Figure 24

A plate from Viator's work: *Example of the horn*

Viator, just as Alberti, was easily able to extend this construction to deal with a regular square tessellation of a plane base, in such a way as to produce a scale of reduction in length according to distance and elevation. Figure 25 illustrates the process: the diagram is due to Jan Vredeman de Vries, a Flemish artist and theoretician of the 16th century.

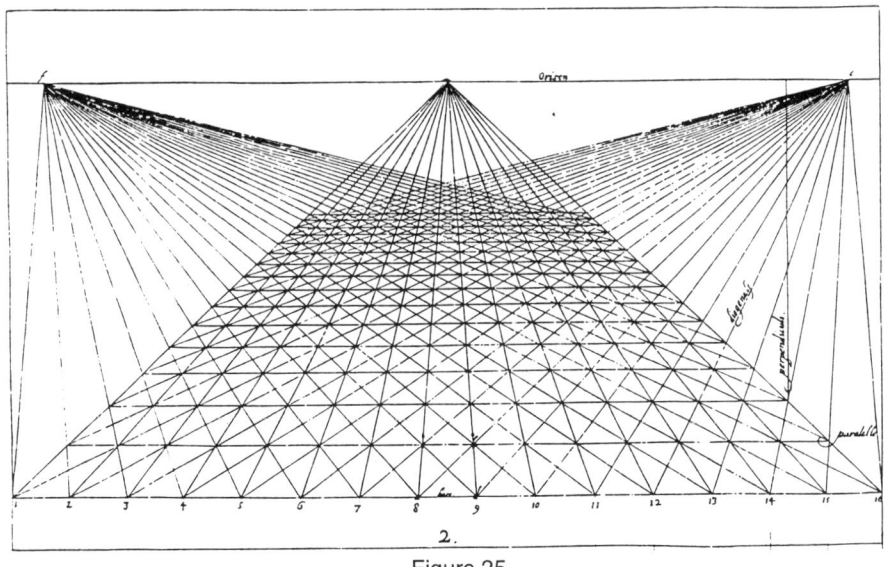

Figure 25

Example showing the construction of a square tessellation according to the method of Viator: a plate from the treatise by Jan Vredeman de Vries (1604)

Viator's thirds points play the same role in perspective construction as our modern distance points (Fig. 27): the lines which converge to these points, being diagonals of a square tessellation, are inclined at 45°, and it is easy to establish that the distance of these points from the principal point, is indeed, the same as the distance of the eye from the painting, ... which you can prove as Exercise 7.

Exercise 7   Prove that the distance of a thirds point is the same as the distance of the eye from the painting.
It is also possible to show this property in space (Fig. 26), namely: if an eye O observes a square on the ground ABCD, the line D_1 B cuts FA at the same point d as the visual ray OD, if and only if OF = FD_1.

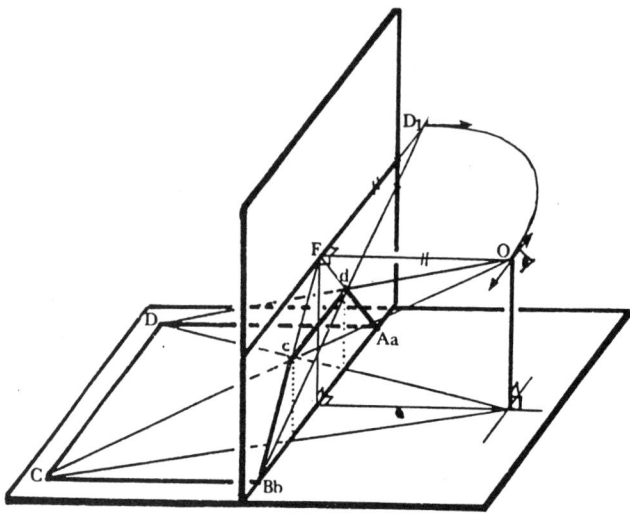

Figure 26
The Viatorian artist in front of his picture, furnished with a thirds point

As well as the principal vanishing point F, fix on the horizon a point D (or D' on the other side of F), such that FD (or FD') is equal to the distance of the eye from the painting. This point D will be found to be the vanishing point for the parallel ground lines, which make an angle of 45° with the principal visual ray (or with the ground line). It is therefore the vanishing point for the diagonals of a square tessellated floor (D_1 in Fig. 12).

A modern reading of Viator's text shows, that his method of establishing the distance of the thirds point from the principal vanishing point, arises from a consideration of the eye-object distance, and not by strict use of the eye-picture distance. The picture is not being thought of any more as a window pane, but we are returning, to a greater or lesser extent, to looking at the objects themselves which are to be painted. The assimilation of the eye-object distance with the eye-picture distance, and with it the reduction in size of objects according to how they appear, as well as a growing interest in the idea of equality, pure and simple, was further developed over the ensuing decades with a fusion of Italian and Northern European traditions.

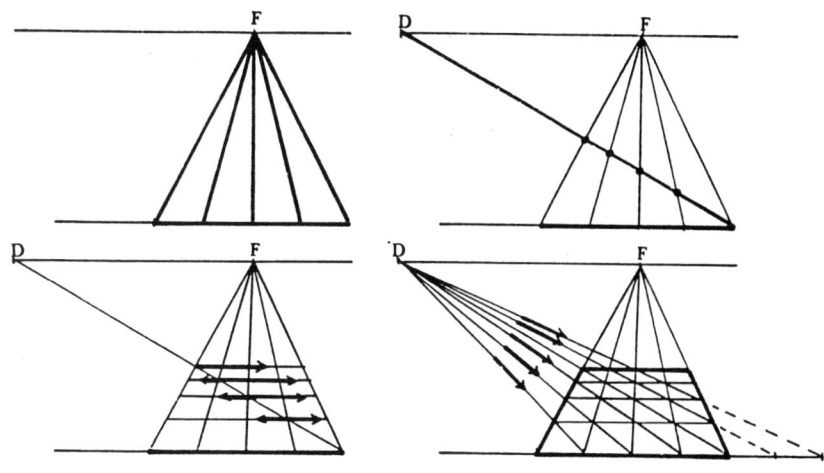

Figure 27
Modern construction using distance points

Painting or Geometry?
The Sixteenth century: diffusion of practice, calling into question, bad habits

A major influence in the diffusion into Germany of Italian methods of perspective, was the appearance of an important practical treatise on geometry by the painter and engraver Albrecht Dürer: *Underweysung der Messung* (*Instruction in measuring*), 1525. This work, intended for practitioners in all branches of the arts, contains in the final section, a perspective method which is inspired by the procedures used in Italy, and particularly those of Piero della Francesca, which Dürer had come to know during his visits to Italy. But Dürer's work also contains two interesting innovations.

The first is an original use of double projection. Alongside the established use of projection to portray architectural works, Dürer also uses projection for strictly geometrical ends. Examples are, the point by point plotting of a conic (Fig. 28), and obtaining the shadow of a cube lit by a point light source (such as a candle), both in plan and central perspective. These are probably the first printed examples of procedures which belong to methods for drawing sections of solids, which were developed later. This science, contained in treatises by Philibert de l'Orme up to those of Amédée-François Frézier, dealing with cutting sections of solids, is known as stereotomy. These methods inspired Gaspard Monge, at the end of the 18th century, in his work on descriptive geometry.

Figure 28 illustrates Dürer's method for constructing an ellipse. Despite the correct geometrical use of double projection, the original engraving shows the ellipse in the shape of an ovoid, for which Dürer proposed the more commonly used name "line of an egg", as if the senses, betrayed by prejudice, are unable to see what reason shows them.

Exercise 8   Construct an ellipse using Dürer's point by point method; have you
laid an egg?
What misconception might have led the painter to imagine the
asymmetry that produced an "egg"?

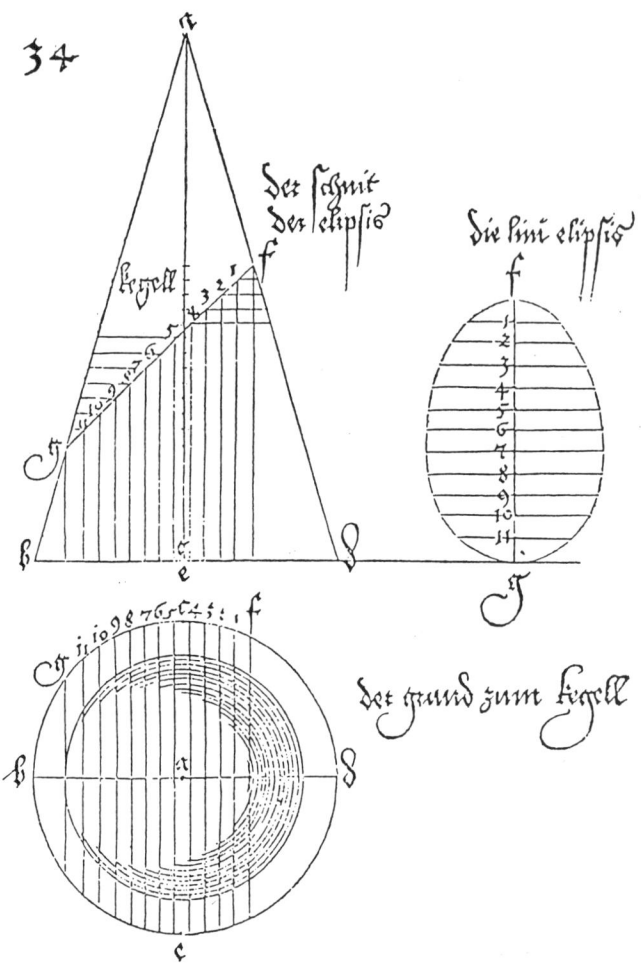

Figure 28
Dürer, *Underweysung der Messung*, Nüremberg, 1525
Book I, Plate 34: Dürer's ovoidal ellipse

The other contribution made by Dürer is directly tied to the idea of linear
perspective. He proposed as many as four models for perspective tables, which would
produce the *perspective view* to an eye in front of *a window open on the world*, using
the pyramid of visual rays (Fig. 29). Just as with the thread of the *intersecting veil*,
which Alberti had imagined in the place of a latticed window, Dürer's tables produced a
working model of a co-ordinate system in which the perspective image could be drawn.

Figure 29
Dürer, *Underweysung der Messung*, Nüremberg, 1538
Last plate of Book IV of the extended second edition

One of the tables that Dürer proposed in the second edition of his work (Fig. 29), uses the principle of point by point plotting of the image, with the aid of a latticed window and squared paper for the drawing.

The sixteenth century was a period of adaptation as much as innovation, during which several alternative processes were developed, some of them erroneous. This was the case with the two rules proposed by the architect Sebastiano Serlio (Fig. 30 and 31). They appear in the second book of his grand treatise on architecture, a book dedicated to perspective drawing, and which was written using notes taken from his master, the architectural painter and theatrical designer Baldassare Peruzzi. It was, however, and paradoxically, considering the way these "errors" were so rapidly taken up by later works, the first work on the subject of perspective to have been printed in the Italian language.

Exercise 9	Explain what invalidates Serlio's two rules, the principles of which are set out in the two diagrams below. In particular, what initial conditions may allow one to be considered as a correct method of projective construction, while the other one cannot be?

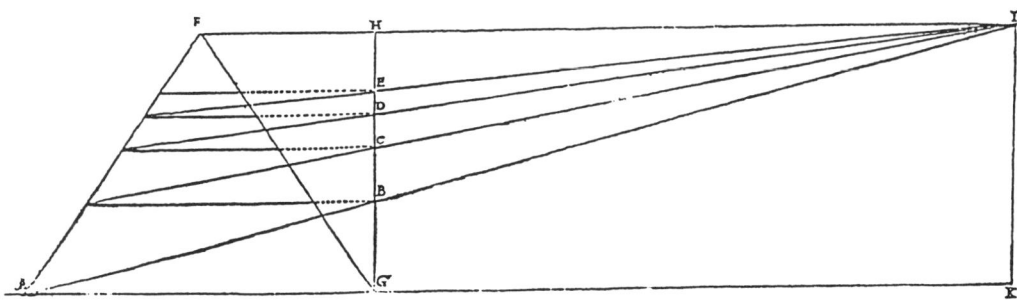

Figure 30
Serlio's first rule

Here is Serlio's first rule. Given F as the principal vanishing point, and AG the side
of a square base which we wish to see in perspective, Serlio extends the ground line AG
and, parallel to it, the horizon line from F, through H, vertically above G. He then fixes
a point I, vertically above K on the ground line, where KG is the distance of the eye
from the painting. He then draws AI to cut GH at B, which determines the level of the
rear side of the square. To find the heights to represent the depths of successive squares,
he uses the same procedure, starting always with the fourth corner of the preceding
square, obtained by the intersection FA with a parallel to AG.

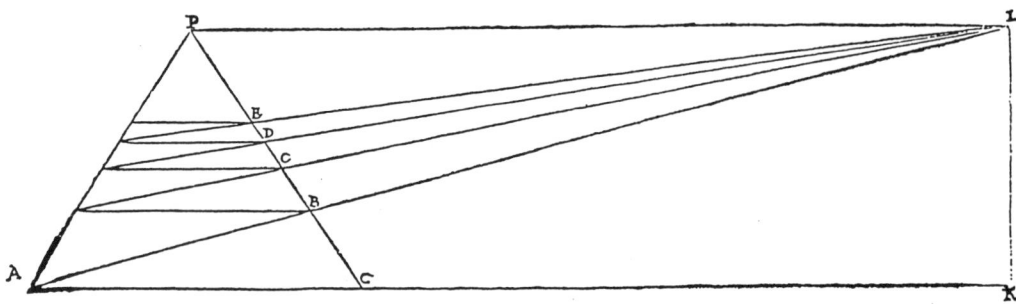

Figure 31
Serlio's second rule

Serlio's second rule is, in essence, as follows. Given the identical initial conditions,
fix I, vertically above K, where KG is the distance of the eye from the picture, and draw
AI to cut FG at B: the line through B, parallel to AG, determines the rear side of the
square. The successive squares are obtained step by step each from its predecessor.

The development of linear perspective is marked in the first half of the sixteenth
century by the multiplication of practical procedures. This proliferation of methods,
often spoilt by inexactitudes, caused people to think about the theoretical basis for
perspective methods, and by the end of the century, there appeared a systematic theory
of perspective, founded on geometry.

Does the same principle apply to different methods?
Vignola-Danti: establishing equivalence

One of the advances in theory in the sixteenth century, is due to Egnatio Danti, in
his commentary on the rules of perspective by the architect Jacobo Barozzi, better
known as Vignola. The work by Vignola-Danti appeared in 1582 at Bologna under the
title of *Le due Regole della prospettiva pratica* ... It consisted of an explanation of
Vignola's two rules, which were in fact, a simplified version of Alberti's construction
and the construction using a distance point. But the work also contained, and this is why
it is interesting, the proof that the two methods are equivalent. The proof appears,
curiously, in a preface to the main work which concerns geometrical optics, and by the
way in which this *third theorem* is stated, it does not appear as a result concerning
perspective but as one in "pure" geometry.

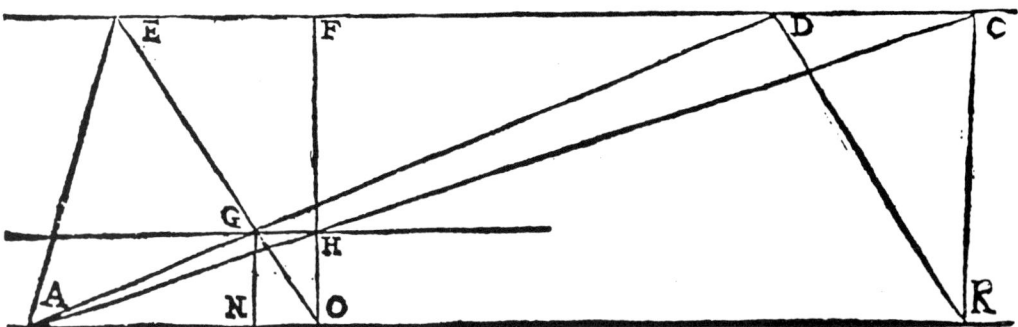

Figure 32
Vignola-Danti: the Figure for *Teorema Terzo*

Vignola-Danti states his equivalence theorem in the following terms (Fig. 32)[1]:
Given two equal and equiangular triangles [EOF & DKC], *arranged in the same way between two parallel lines* [EFDC & OK], *if two other lines* [DA & CA] *are drawn from the two extremities* [D & C] *of the base* [DC] *of one of the triangles and towards the same point* [A] *of the opposite parallel* [OK], *which cuts the two sides* [EO & FO] *of the other triangle, then the line* [GH] *which passes through the two intersections* [G & H] *will be parallel to the base of these two triangles.*

Exercise 10	Prove this theorem. Explain how Alberti's and Viator's constructions are equivalent by identifying each construction in Danti's figure.

Is it possible to have many horizons?
Guidobaldo del Monte and Stevin: from a vanishing point to convergence points

So far, the practice of perspective drawing had used geometrical constructions, but at the end of the sixteenth century, perspective became itself a theory of geometry. The advances in theory were due to the engineers and mathematicians Guidobaldo del Monte and Simon Stevin. The former published his *Perspectivae libri sex* in 1600, and Simon Stevin produced his *De Perspectivis* in 1605, in a Flemish edition, followed by a Latin translation, and then two French editions (1608 and 1634). From being a problem external to painting, perspective became a problem internal to geometry. And so, after a period of uncertainty at the beginning of the sixteenth century, we see the parting of the ways of two different interests: these frequently became referred to as practical perspective, and speculative perspective.

The first of these two authors, with his concern for theory, was a direct follower of the Italian mathematician Federico Commandino, who had published a treatise on perspective as an appendix to his translation of Ptolemy's *Planisphærium*, in order to complete the Ptolemaic theory of stereographic projection (the projection of earthly or heavenly spheres on the plane). The second, was concerned with what had become a

---

1. Vignola, *Le due Regole della prospettiva pratica...* p. 18, French tr. and adapted by J.-P. Le Goff.

subject of general interest at the end of the seventeenth century, the representation of military and civil techniques (architecture and fortifications) in Flanders.

Both stated a new idea, which turned out to be essential for the development of the theory and emergence of projective geometry: it was not only lines orthogonal to the picture, or horizontal lines inclined at 45° to them, which converged to special points on the horizon. It is true that some painters and writers on theory had already used other points of the horizon as vanishing points, but they often committed the error of having non parallel lines converge at the horizon.

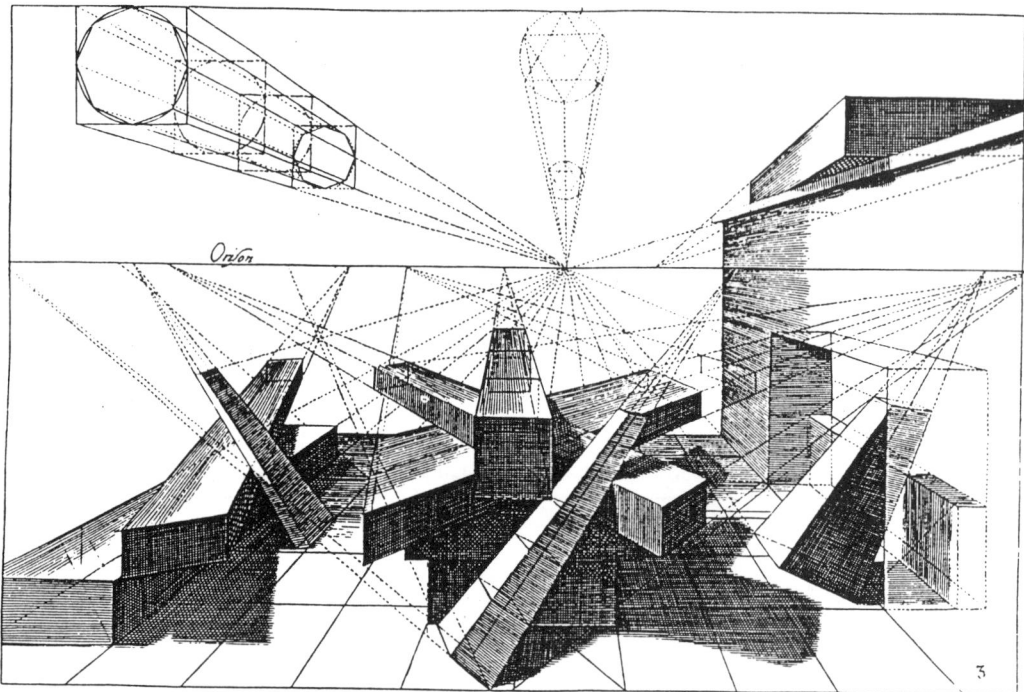

Figure 33
Jan Vredeman de Vries: a plate showing perspective (1560)

Exercise 11   Identify the error (errors) in Fig. 33.

Guidobaldo del Monte devoted the first of his six books on perspective to a step by step proof, from the simplest case to the most general, of the following proposition: every system (we would say set) of parallel straight lines in space, is represented in a picture by a system of lines which are either parallel or concurrent. And the point of convergence of the images of the lines, is where the line in the visual ray, which is parallel to the set of parallels, meets the picture: the images of lines will thus remain parallel, only in the case where the given set of parallel lines in space is itself parallel to the picture. Proofs of this proposition were based, by Guidobaldo del Monte, on the use of properties of similar triangles, that is in the application of what is today improperly referred to as Thales' theorem. But the Italian geometer proposed an alternative justification, using only the properties of incidence in space, an approach that was a novel one among methods of proof. Figures 34 and 35 below, are illustrations that appeared with proposition XXVIII of Book I by Guidobaldo del Monte, and set out to prove the proposition in the particular case, where the set of parallels is at ground level, and not parallel to the ground line of the picture.

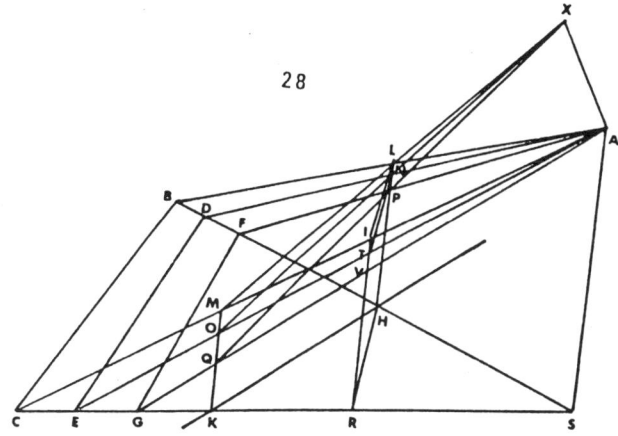

Figure 34
Guidobaldo del Monte: Book I, Figure XXVIII

Guidobaldo's Figure XXVIII, of which this is a copy, serves to illustrate the following statement[1]: *if the eye views parallel lines, no matter how many, situated at ground level, and not parallel to the line of the section* [ground line of the picture], *& if the plane of the section* [the picture] *is erected* [i.e. perpendicularly] *on the base plane, then the images of these lines in the section will be concurrent at one and the same point, at a height above the base plane equal to the height of the eye.*

Let AS be the height of the eye A above the base plane. And let HK be the line of the section; & let the parallel lines situated in the base plane be BC, DE & FG, such that they are not parallel to HK. And also let the section HLMK stand up [perpendicularly] *on the base plane. And also let the apparent lines* [the images of the given lines] *in the section be LM, NO, & PQ. I say that LM, NO, & PQ are concurrent at one and the same point, which, furthermore, is at the same height above the base plane as is the eye A.*

Exercise 12   Prove Guidobaldo del Monte's proposition XXVIII.

Three other aspects of the work of Guidobaldo del Monte and Stevin should be recognised. The Italian theoretician develops, in a geometrical way, the perspective of the theatre which Serlio had set out in his treatise. In the sixteenth century, theatres lost their ephemeral character: instead of occasional *décors* set up in the open air to represent several juxtaposed scenes of actions of different types, we move to a closed setting for a single scene, in which the *décor* is movable or interchangeable, and then to permanently sited buildings, like the Olympic Theatre at Vicenza, conceived by the architects Andrea Palladio and Vincenzo Scamozzi. The town scenes "à l'italienne" of these *décors*, were conceived as so many pictures, no longer being in the plane, but existing in a limited space, by using the idea of foreshortening (figure 36), and giving a coherent image from a particular position in the theatre, usually the "royal box": the perspective avenues of Scamozzi's *décors* for Palladio's theatre become, therefore,

---

1.   Guidobaldo del Monte, *Perspectivæ Libri sex.*

impractical for actors, whose heights to not diminish according to distance as "rapidly" as the buildings in the illustrated street scenes.

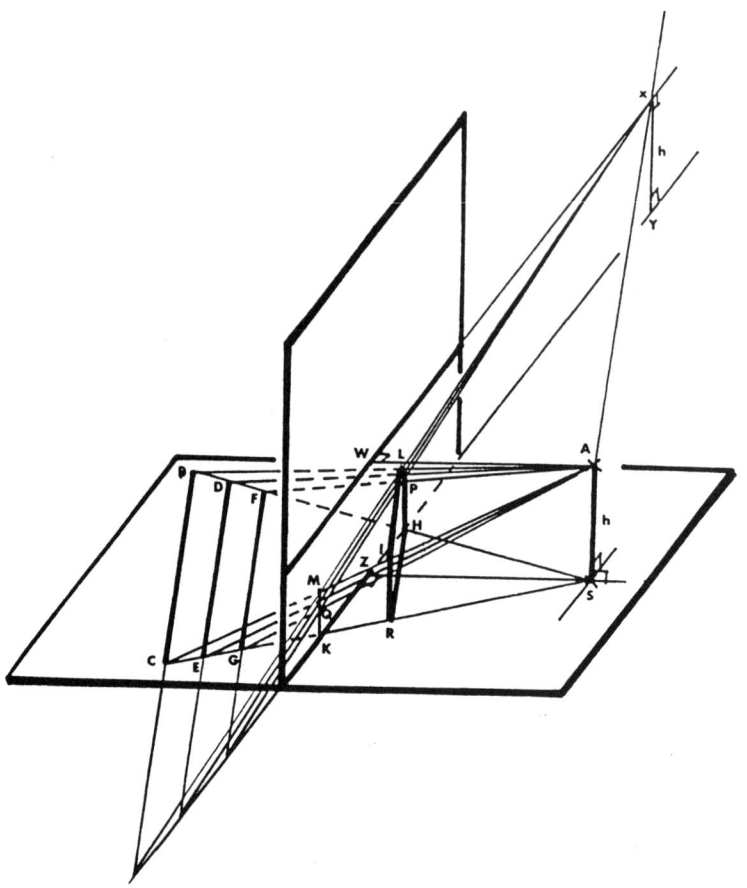

Figure 35
An interpretation of Guidobaldo del Monte's Figure XXVIII

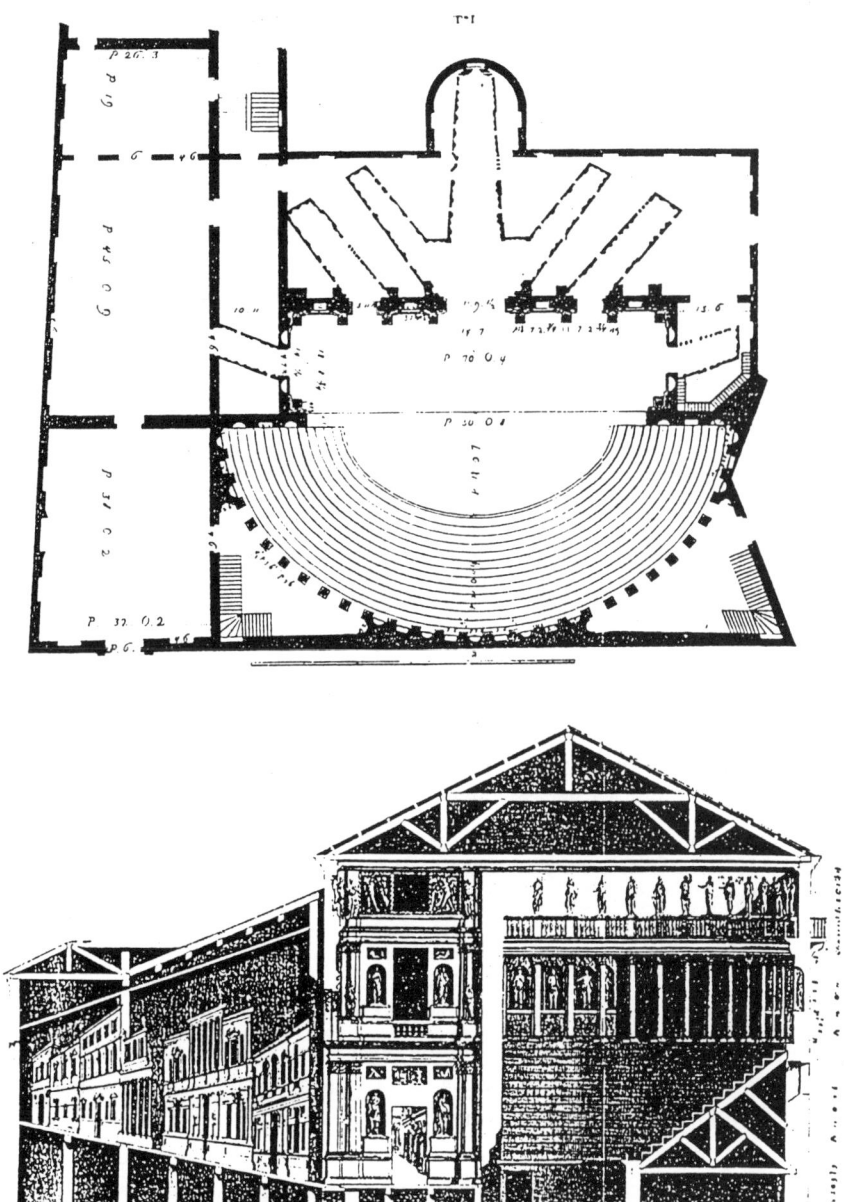

Figure 36
Plan and profile view of the Olympic Theatre at Vincenza

The key idea in this type of illusionist construction developed by Guidobaldo del Monte, is that the image of a point in space, need not be a point in a picture, thought of as the framework of the virtual scene, but may be located at any point of the visual ray connecting the eye to the real point. From this fact, we can have what appears to be a rectangular-shaped box of space, although it has been constructed as the volume of a truncated pyramid. The principal vanishing point is no longer the point F of the picture framed scene, but the point F on the "prince of rays", the summit of the pyramid of the vanishing lines opposed to the visual pyramid (Fig. 37).

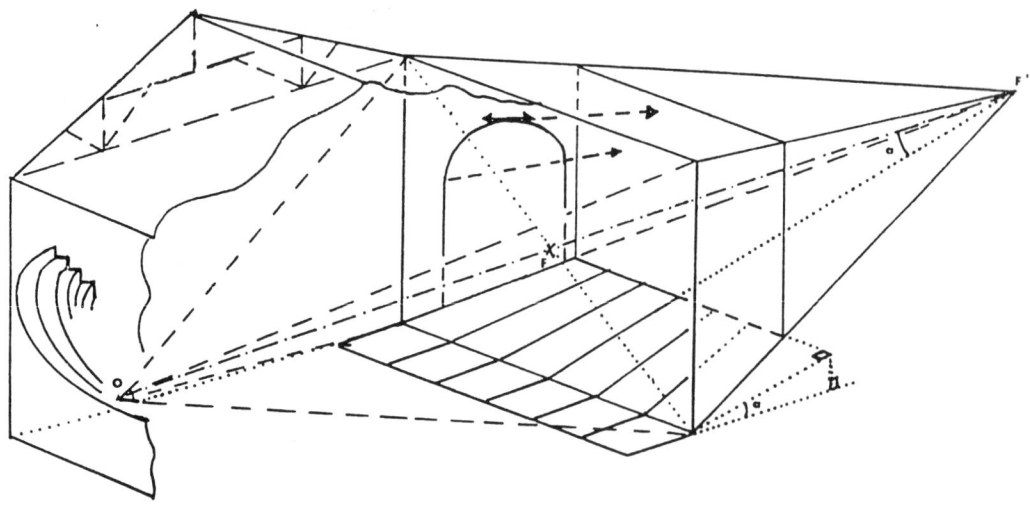

Figure 37
The principle of a scene with foreshortening.

Furthermore, Guidobaldo and Stevin both envisaged the use of tableaux that were inclined to the horizontal base plane, which was related to the need for a theory to deal with the practice of using murals that had to be placed on supports that were not necessarily vertical. These practices, and explanations of the procedures needed to make them work, flourished in the seventeenth century. Finally, Stevin stated, for the first time in geometrical terms, the inverse problem of perspective: given two figures, perspectives one of the other, how are they to be placed so that the perspective can be shown, and where is the position of the eye?

We can see here, then, a "spatialisation" of perspective at the same time as its geometrisation. One of the later consequences of realising that there are a multiplicity of vanishing points, is the idea of a horizon for any plane, which generalises the reference base plane, and on which are the vanishing points of sets of parallel lines on the ground. Being given any plane (P) in space, each set of lines, parallel to each other and parallel to (P), have images which are convergent to a point determined by the direction of the parallels. The set of convergent points, associated with all directions of the plane (P), makes a line which is the horizon of the given plane. This line, parallel to the intersection of the plane with the picture, is the common horizon of all planes parallel to (P).

These aspects of perspectivity were developed during the succeeding centuries, in a context that was more theoretical than practical, and led progressively to the theory of projective geometry, at a time when painters became increasingly less interested in perspective, to the point where they sometimes abandoned it.

Why do we talk of conic perspective?
Linear perspective and the theory of conics: Desargues' synthesis

The history of the theory of perspective from the beginning of the seventeenth century is essentially that of geometry. Girard Desargues, Brook Taylor and Jean-Henri

Lambert are just some of the more important names we may cite. The first, established a true correspondence between perspective scales and scales in the geometral (the ground plane), and he achieved a synthesis between perspective geometry and the theory of conics. The other two built on these developments, to transform the science of perspective into a theory of the projective transformation of figures.

Engineer and architect by training, Desargues became a geometer by reason of his original views, leading to a synthesis of differing practices in geometry (perspective, stereotomy and gnomonics). In 1636 he wrote an *Exemple de l'une des Manières universelles du S. G. D. L. touchant la pratique de la Perspective sans employer aucun tiers point de distance ni d'autre nature, qui soit hors du champs de l'ouvrage.* [Example of one of the universal Methods of S. D. G. L. [i.e. le Sieur Girard Desargues, Lyonnais] touching on the practice of Perspective without the use of any thirds point at a distance, nor any other type of point, lying outside the area of the work]. Although not implied in the title, Desargues was proposing a geometric-numeric method, founded on the use of scales whose division is established by construction, and which followed progressions of the harmonic type (Fig. 38).

Having set out the ground line "ab", the principal vanishing point "g" on the horizon "fe", and "c" the foot of the perpendicular from "g" to "ab", and given the distance D of the eye from the picture (in this case 24 feet), Desargues constructs the images of the transversals in the ground plane, which are parallel to the ground line, and at distances of D, 2D, 3D, etc. from it. To do this, he draws "ga" and "fc", whose intersection determines the height of the first of these transversals, "hd", which cuts "gc" at "t": then he draws "ft", whose intersection with "ga" determines the second transversal, and so on. This construction operates as if the point "f" is a distance point, and produces a scale of separation. The reference line for this scale is no longer the ground line, as with Leonardo da Vinci (see Exercise 6), but the horizon line, since the position of the first transversal, "hd", corresponds to fh = (1/2).fa, that of the second transversal "qn", corresponds to fq = (1/3).fa, and that of the nth, situated at a distance n. D behind the ground line, has an image that is situated below the horizon and with a distance from it of "f" equal to (1/(n + 1)).fa. Here we come back again, but arguing from different premises, to the harmonic sequence of the inverses of integers, which had already been demonstrated by Leonardo da Vinci.

This method allows us to produce constructions, using any point whatever of the horizon, which behaves like a distance point (here the distance is reduced, since "fg" or "ac" represent the 24 feet of the real distance). The procedure relates to the practical method referred to in the title, but its importance is that it demonstrates a real understanding of the theory of perspective. For Desargues, the choice of the distance position for the distance point is not dependent on a prior choice of a reduction ratio, since he suggests placing "f" at the edge of the picture, which implies an *a posteriori* distance reduction ratio equal to "ac/D", a ratio that is not necessarily rational. However, this general idea, leads to the modern method of so called reduced distance points.

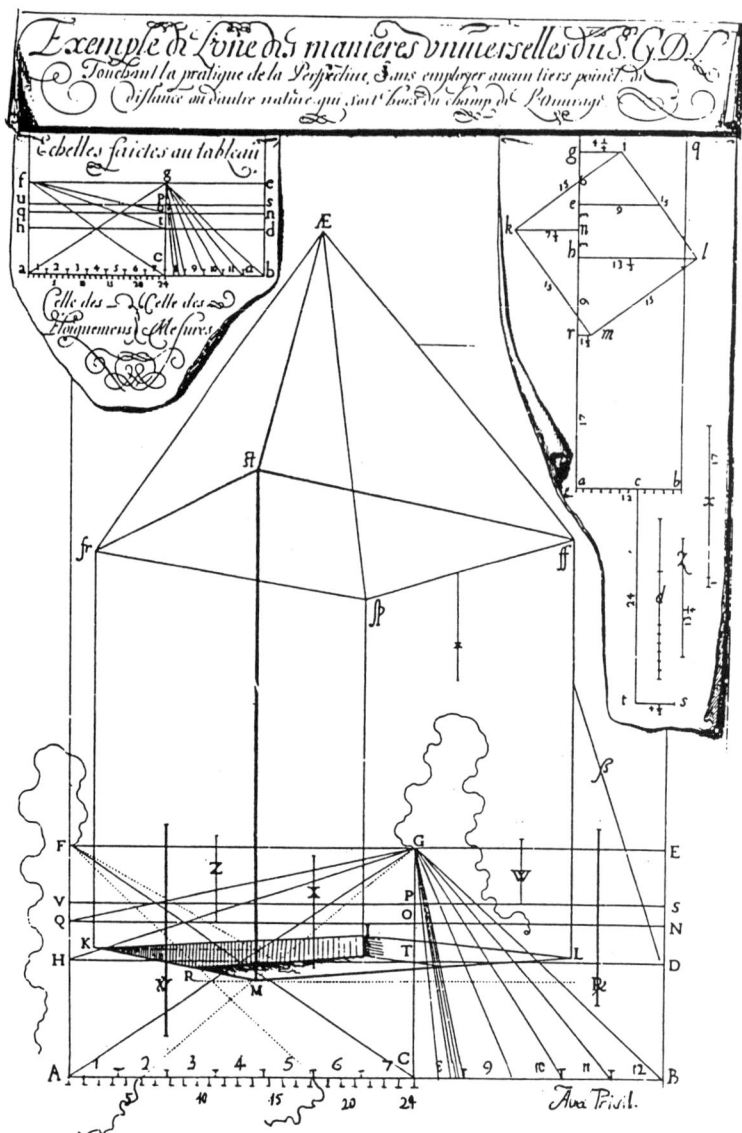

Figure 38

Plate taken from Desargues' *Exemple* of perspective (1636)

Desargues' principle of perspective scales

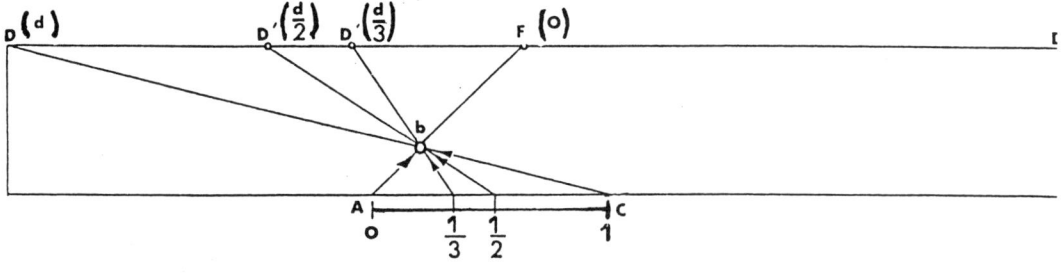

Figure 39
The modern method of reduced distance points.

The way to set up an arbitrary auxiliary point D' on the horizon, instead of the distance point D that the basic method requires, is as follows. If F is the principal vanishing point, and AF a vanishing line, the apparent line Ab of a real line AB seen end on, can be found, either from the intersection of DC with FA, C being one of the points of the ground line where AC = AB, or by the intersection of D'C' with FA, C' being a point on AC such that AC'/AC = FD'/FD.

Moving from space to the plane, becomes for Desargues, a two way process. Stevin proposed the converse problem: the possibility of a perspective relation being set up between two figures (perspective homology). The theorem stated by Desargues concerning triangles in homology, is fundamental for determining the minimal conditions needed to define a perspective. It is found as one of the geometrical propositions that Abraham Bosse includes in his treatise on perspective: *Manière universelle de Mr Desargues pour pratiquer la Perspective par petit-pied, comme le Géométral*. Here it is[1]:

> *Whenever the lines HDa, HEb, cED, [abc], lga, lfb, HlK, DgK, EfK, being in different planes or in the same plane, meet each other in the same points, in any order and at any angle that may be possible; the points c, f, g will lie on the straight line cfg. [...] And conversely, the straight lines abc, HDa, HEb, DEc, HK, DKg, KEf, happening to meet in any way and at any angle, in like points, whether in different planes or the same plane; the straight lines agl, bfl will always converge together at the same point l, lying on this HK. [...] And again, these same straight lines, still being in different planes, if through their points H, D, E, K, there pass other straight lines Hh, Dd, Ee, Kk which converge towards an indefinitely distant point, in other words they are parallel to each other, and they meet one of those planes cbagl, as at the points h, d, e, k, the points h, l, k lie on a straight line, h, d, a lie on a straight line, h, e, b lie on a straight line, k, g, d lie on a straight line, k, f, e lie on a straight line, and c, e, d lie on a straight line. [...]*

1.   Bosse, *Manière universelle de Mr Desargues...* , p. 340 and pl. 154.

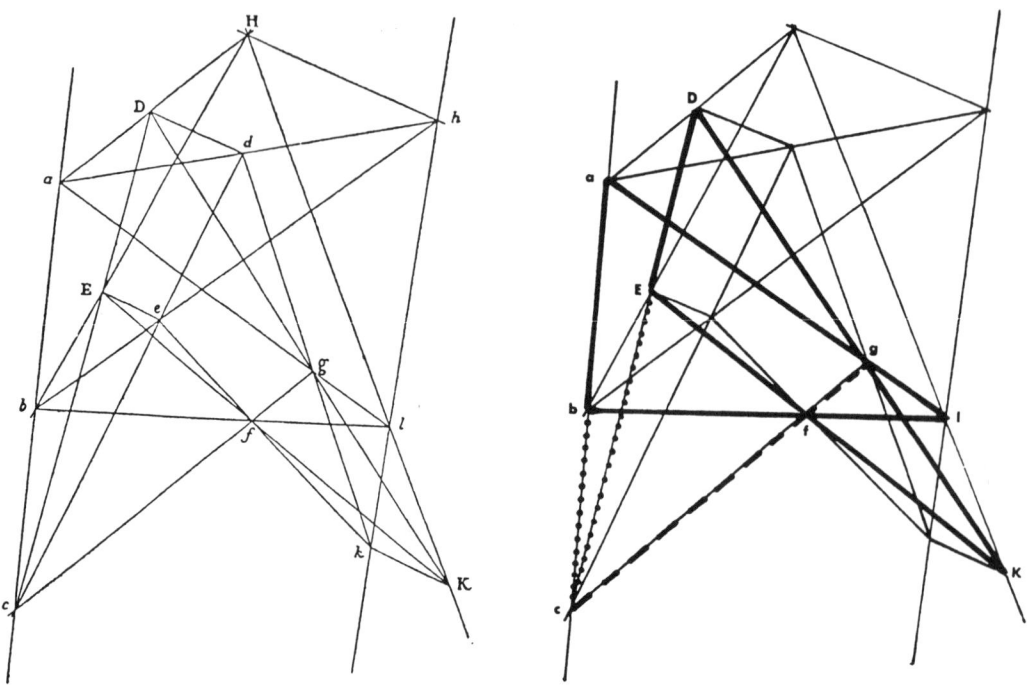

Figure 40
Figure from the treatise by A. Bosse (1647):
the theorem of homologous triangles by Girard Desargues
On the right: a demonstration that the triangles are homologous (in perspective)

In other words: if "abl" and "DEK" are two given triangles, in perspective each to the other, in different planes or in the same plane, that is, if the lines "aD", "bE" and "IK" are concurrent (at "H" from which point the two triangles would be seen as identical), then the intersections of their corresponding sides are collinear, that is, the points "c", common to "ab" and "DE", "f", common to "bl" and "EK", and "g", common to "la" and "KD", are on a straight line "cfg". The converse is also true, which states that if the two vertices "a" and "b" of a given triangle in space "abl" are in perspective with the two vertices "D" and "E" of another triangle "DEK", seen from the position of a point "H", and if the corresponding sides of these two triangles, "ab" and "DE", "bl" and "EK", and "al" and "DK", cut each other at three points "c", "f" and "g" which lie on a straight line, then the third vertex "l" of triangle "abl" is in perspective with "K", seen from this same point "H". The generalisation demonstrates the many configurations of this type which can be seen in the same diagram: in addition to the pair of perspective triangles that Desargues identifies using "D", "E", or "K" as the centre of perspective, there are in addition the triangles "deh" and "DEH", "ehk" and "EHK", "dek" and "DEK", "dhk" and "DHK" in the four configurations shown in figure 41, for which the centre of projection is at infinity and which are cylindrical projections, since the lines "Hh", "Dd", "Ee", "Kk" are mutually parallel.

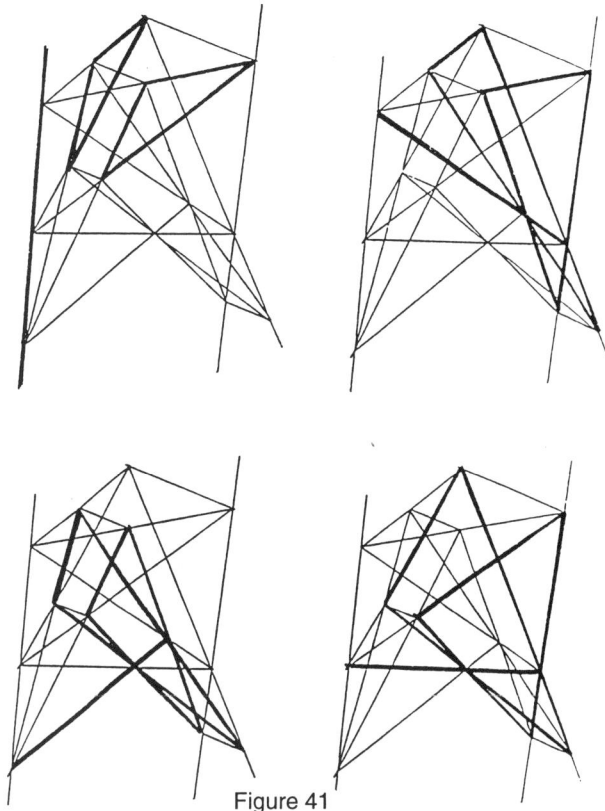

Figure 41
The theorem of homologous triangles by Girard Desargues
The possible choices of pairs of triangles corresponding to cylindrical perspective

The most important theoretical contribution made by Desargues is to be found in his treatise on conics, published in 1639, under the title *Brouillon Project d'une Atteinte aux événements des rencontres du Cone avec un Plan*. [Rough Draft of an Attempt to Deal with the Outcome of a Meeting of a Cone with a Plane.] In it he shows systematically, that the curves called conics, defined since Antiquity as intersections of a right circular cone with a plane, can also be considered as perspectives of the circular base on to the intersecting plane, using the vertex of the cone as the perspective view point. In place of treating each of the sections, circle, ellipse, hyperbola and parabola, case by case, and identifying their properties, Desargues proposes a global approach, based on the perspective transfer of the properties of the circle, and in doing so, forged a tool (involution) of prime importance for the future evolution of geometry. The concept of involution is the equivalent of our modern cross-ratio or anharmonic ratio[1], whose essential property is its invariance under conic projection.

With the way forward already clear, Desargues systematised the contributions of Guidobaldo del Monte and Stevin, in actualising the infinite and in assimilating pencils of concurrent and parallel lines, the latter having a convergence point at infinity, sheaves of planes with a common "axle" and parallel planes, the latter having an "axle" at infinity, and finally cones and cylinders, the latter having a vertex at infinity. In the first half of the seventeenth century, with Desargues, and his immediate disciple Blaise Pascal, geometry became the science of the unlimited, the science of an infinite space made actual in all directions: straight lines and planes, segments of surfaces, all limited

---

1.   See Lehmann and Bkouche, *Initiation à la géométrie*, p. 104 *sq.*

but capable of being indefinitely extended, as they had been with the Greeks, now became ideal objects of a space which is isotropic, continuous and homogeneous, defined later by Kant as *a priori form.*

Is linear perspective the only way of representing three dimensional objects?
From linear perspective to other types of perspective

Parallel to the development of a theory of conic perspective, and doubtless because the invention of this first "realistic" method of representation opened up the field of possible ways of comprehending reality, other conventions for representation appeared, connected with the needs of scientific and technical drawing. Among these were the cross-section, cut-away view, exploded view and the "cavalier" perspective, to name just a few. This latter was often seen, in a more or less approximate way, in numerous books on architecture or in technical treatises like *théâtres de machines*, before being treated theoretically under the names of geometral elevation or military perspective, by two authors of works on linear perspective at the beginning of the seventeenth century. These were the Jesuit Jean Dubreuil and Abraham Bosse, Desargues' engraver and pupil, for whom the geometral and the perspective were two things of the same type, perspective scales being only a sophisticated form of regular geometral scales. And, in fact, we know today that cavalier perspective is a cylindrical projection in an oblique direction, that is a central projection whose centre of perspective is at infinity (Fig. 42).

As for axonometric projection, which is cylindrical projection in a direction orthogonal to the plane of the picture, and is only of interest when the plane of the picture is not parallel to one of the principal planes of the object[1], its use did not become recognised until after the middle of the nineteenth century, often associated with shading of drawings in order to obviate ambiguity. It was not until the beginning of the twentieth century that it became favoured by architects as the main method of representation, following the 1923 exhibition by the *De Stijl* movement, at which Theo van Doesburg and Cornelis van Eesteren exhibited their **Axonométrie de Maison Particulière**. It is perhaps significant that a theoretical treatment of cylindrical projections did not occur until after the invention of central perspective, yet they did show true length representation.

The end of the nineteenth century also saw the invention of descriptive geometry, by Gaspard Monge, a method of representation that systematised double projection (with the elevation folded back on to the plan), and which attempted to deal with such aspects of spatial geometry as the intersections of surfaces, through the methods of pure geometry. But in wishing to compete with analytic geometry, the followers of this new science, conceived as much as a tool for use in engineering as in geometry, had perhaps passed by a quiet revolution taking place in science and technical representation in anglo-saxon countries, namely the practice in industrial design and *glass-box* representation. The first of these used a triple projection of three orthogonal planes, the second a six-fold projection on to the six faces of an imagined cube surrounding the object being represented. These projections are clearly superfluous, but they improve

---

1.   See Locquet and Perrot, *Perspectives cavalières et axonométriques.*

the amount of information that can be taken in at a glance, since they make clear detail inside an object, by showing what can be seen (continuous lines), and what is hidden (dashed lines). In both cases, the draughtsman will choose to use projections that are parallel to the principal planes of the object. These practical forms of descriptive geometry are simplifications; while they are toned down versions of geometric theory, they are better suited to the needs of engineers and craftsmen than Monge's study of descriptive geometry.

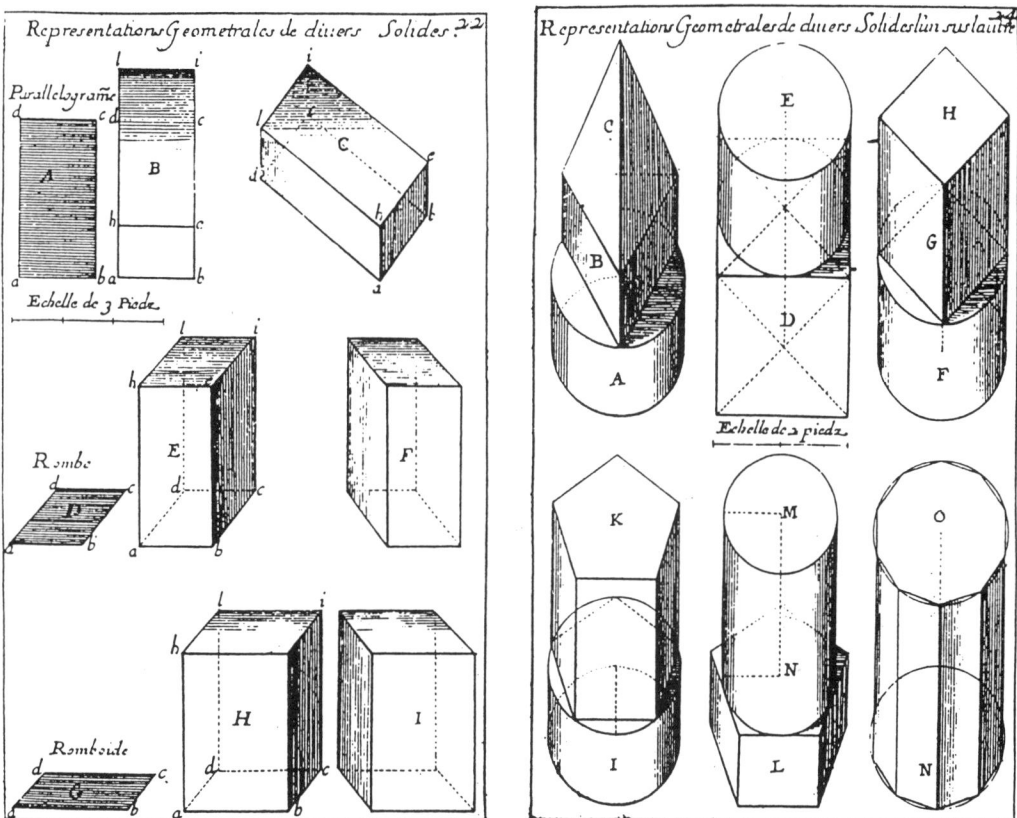

Figure 42
Abraham Bosse: *Traité des Pratiques Géométrales et Perspectives* (1665), Plates 22 and 24

Perspective crosses the frontier of Euclidean geometry

With Brook-Taylor and Lambert, linear perspective breaks free from the use of orthogonal projection and the need to use the geometral, which had so far, always been assumed as essential components of perspective drawing.

Brook-Taylor wrote two books on perspective: *Linear Perspective, or a New Method of Representing justly ...* London, 1715, and *New Principles of Linear Perspective...*, London, 1719. In these works, Brook-Taylor rejects the use of the *prince of rays* in this sense, that he considered central projections on to a picture and above a geometral as forming a certain dihedron, a principle that had its seed in the work of Guidobaldo del Monte, and from which the use of *trompe-l'œil* on decorated panels and ceilings, in the sixteenth and seventeenth centuries, may have been derived. The essential elements of his method, no longer relied on the use of a principal vanishing point or of distance points, but used planes passing through the eye, one parallel to the picture, the other parallel to the geometral. With Brook-Taylor, it becomes possible to say that perspective has freed itself from metric considerations, and that we see the emergence of projective properties recognised as such.

Jean-Henri Lambert considered the question of perspective several times. His *Anlage zur Perspektive* (1752) remained in manuscript form, until very recently. His next work, published in Zurich in 1759, in French and German editions, *La perspective affranchie de l'embarras du plan géométral* [Perspective set free from the encumbrance of the geometral plane], enjoyed a much enlarged second edition, published in 1774. With Lambert, linear perspective became a central projection in the proper sense, to the extent that it removed itself from the geometral plane: all that was required was a point of view and the projection plane. Furthermore, this geometer from Mulhouse was the first to use algebraic equations with perspective transformations. He ended up with a result in terms of a harmonic function, and with a numeric-graphical representation of this result using a hyperbola. The numerical considerations can be found in his 1752 *Anlage zur Perspektive* and they led Lambert to the concept of a perspectograph, which he described, but he is still 'encumbered' by the geometral plane, as this extract shows[1]:

> In general, the distance from S to a given point in the geometral plane, is to the distance from that point to the picture, as is the height OS of the eye, to the elevation of that point in the picture. Or to express it algebraically so that we shall be able to apply the result later; let $bL = x$; $LS = a$; $LH = b$; $OS = c$; then
>
> $$(a + b) : b = c : x \qquad\qquad x = bc / (a + b)$$
>
> and so we have determined the height of each point in the picture.
> Let us now find the point L of the ground line, above which we must place the point h. [...] Let $Hl = d$, $LM = y$, and hence:
>
> $$a : (a + b) = y : d \qquad\qquad y = ad / (a + b)$$

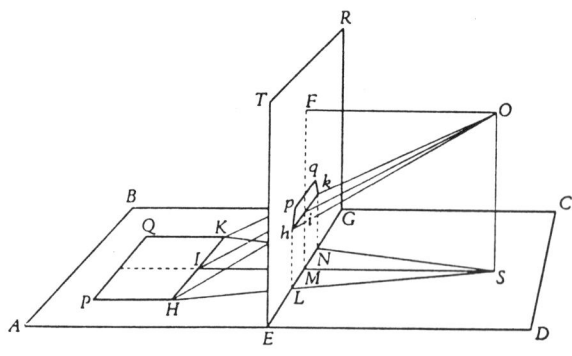

Figure 43
Spatial Figure taken from Lambert's *Essai* (1752)

1.   Lambert, *Essai sur la perspective*, p. 15-16.

Exercise 13   Let O be the principal vanishing point, A the point of view, d the distance OA, and $Oxyz$ a co-ordinate system in space such that Oz is perpendicular to the co-ordinate system Oxy of the plane of projection (Fig. 44). Determine the plane co-ordinates x' and y' of the image M' of a point M (x, y, z), in terms of the co-ordinates of M and the parameter d.

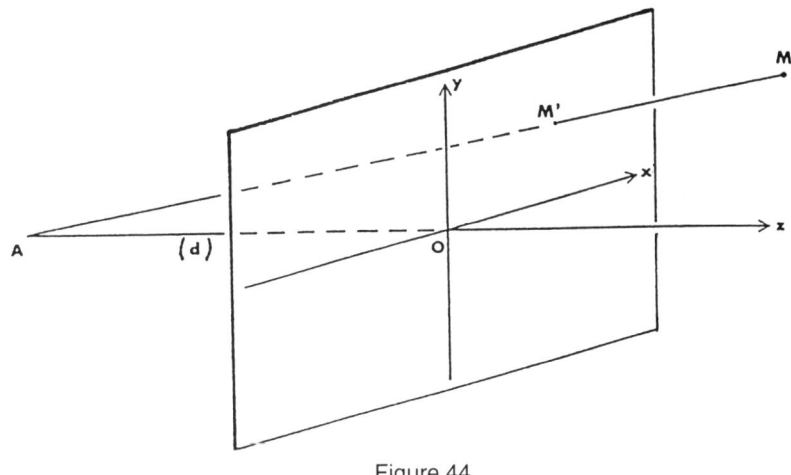

Figure 44
Co-ordinate system in space and the plane of projection for Exercise 13

It is significant that Lambert was one of the first to use perspective geometry as a means of solving problems of plane geometry. These problems in elementary geometry are found in the 1774 *Notes and Additions* to *La Perspective affranchie...* of 1759. Problem III is a noteworthy example[1]:

> *Given a parallelogram ABCD, using only a ruler, construct through a given point P, a line parallel to a given line IE.*

Exercise 14   Solve the construction problem posed by Lambert, given that he proposed a perspective transformation of the real size parallelogram, using IE as the ground line and P as a point on the horizon.

At the end of the period chosen for this history of linear perspective and its derivatives, we see a reversal of the pair geometry/perspective. More generally, throughout each of its episodes, this history is an example of the dialectic between theory and practice, science and art, science and technique, and geometry and perspective. Born out of the practical preoccupations of painters looking for a way of making sense of their visual perception, perspective is also the fruition of a process of rationalisation of vision which had its origins in the theories of optics and geometry developed by the Greeks. With geometry as its starting point, perspective has in its turn enriched geometry. Presented with new configurations, geometers went on to produce new concepts which became the foundations for the higher geometry of Monge, Poncelet and Chasles. Although perspective lost its appeal for painters, perspective

1.   Lambert, *Notes et Additions à La Perspective affranchie*, p. 169, tr. Peiffer, p. 266.

generated projective geometry, the required setting for a transformation from one geometry to a multiplicity of geometries: but what has now become of perspective?

The end point of this history, in the world of mathematics, is projective geometry, which served as a general setting for the redefinition of geometries, based on the idea of groups of transformations, at the end of the nineteenth century. It is here that the non-Euclidean geometries can take their place alongside classical Euclidean geometry, as was asserted by the German mathematician Felix Klein in the famous *Erlangen Programm* of 1872. In the world of the Arts, modern artists distanced themselves from perspective, and painting became again a matter of the surface of the canvas and its frame, and the use of colour and other material. However, the range of artistic practices also generated invention, as in the curved perspective of A. Flocon and A. Barre, or again in the work of Escher, whose work is intimately tied to geometry. Escher's work contains repeated motifs over a Euclidean plane, representations of repeating patterns over a non-Euclidean surface, and representations of impossible figures, based upon paradoxes of perspective. Today, the explosion of new techniques, like video cameras and computer graphics, allows the modern artist to use computer software to present images of three dimensional objects: but, importantly, the mathematical model for these representations, is based on the use of central conic or cylindrical projection.

And so the question that was raised at the beginning of this chapter remains: in confusing vision with representation, the artist only showed impertinence, and in wishing to guide the artist, the geometer has only succeeded in shifting the emphasis.

Bibliography

Sources texts

ALBERTI, *De Pictura*, 1435, tr. J.-L. Schefer, Paris, 1992.

ALBERTI, *De Pittura*, 1436. Translation and adaptation of extracts: J.-P. Le Goff. For extracts of orginal texts with French translation see *Cahiers de la Perspective*, **4**, IREM de Basse-Normandie, Caen.

BAROZZI (called Vignole), *Le due Regole della prospettiva pratica di M. J. B. da V., con i commentari del R. P. M. Egnatio Danti*, Rome, 1611 (1st ed.., Bologne, 1582), third theorem, p. 18. French traduction and adaptation of the extract by J.-P. Le Goff.

BOSSE, *La Maniere universelle de Mr Desargues, pour pratiquer la Perspective par petit-pied, comme le Geometral...*, Paris, 1647. A critical partial repub. in *Œuvres complètes de Girard Desargues*, to appear.

COMMANDINO, *Federici Commandini Urbinatis in Ptolemæi Planisphærium Commentarius. In quo universa Scenographices ratio quambrevissime traditur, ac demonstrationibus confirmatur.*, Venice, 1558. French translation by J.-P. Le Goff, to appear.

DELLA FRANCESCA, *De Prospectiva pingendi*, MS, between 1470 and 1480. First printed edition 1899, C. Winterberg, with German tr. First critical edition, 1942, by G. Nicco-Fasola (G.C. Sansoni, Ed., Florence, repub. 1984). French tr. by M.-F. Clergeau in press. For extracts of the original with tr. by J.-P. Le Goff, see *Cahiers de la Perspective*, **4**, IREM de B.-N., Caen.

DEL MONTE (Burbon), *Perspectivæ Libri sex*, Pesaro, 1600. French tr. by Christian Guipaud, in press. For extracts of the original with tr. by J.-P. Le Goff, see *Cahiers de la Perspective*, **4**, IREM de B.-N., Caen.

DESARGUES, *Exemple de l'une des manieres universelles du S.G.D.L. touchant la pratique de la Perspective, sans employer aucun tiers poinct de distance ou d'autre nature qui soit hors du champ de l'ouvrage*, Paris, 1636. Repub. in *Scholies*, Actes du Séminaire Interdisciplinaire d'Histoire des Sciences du Lycée Malherbe de Caen, **6**, October 1988. Critical new edition in *Œuvres complètes de Girard Desargues*, to appear.

DESARGUES, *Brouillon Project d'une atteinte aux evenemens des rencontres du Cone avec un Plan, par L, S, G, D, L,*, Paris, 1639. Critical new edition by René Taton in *L'Œuvre mathématique de G. Desargues*, Paris, 1951 (repub. 1981, 1988). Critical new edition in *Œuvres complètes de Girard Desargues*, to appear.

DESARGUES, *Trois propositions géométriques*, in *La Maniere universelle de Mr Desargues...* by Bosse, Paris, 1647. Critical new edition in *Œuvres complètes de Girard Desargues*, to appear..

DÜRER, *Underweysung der Messung*, Nüremberg, 1525, repub. with additions, 1538. French tr. by J. Peiffer.

DÜRER, *Géométrie*, Seuil, Paris, 1995.

LAMBERT *La perspective affranchie de l'embarras du plan géométral*, Zürich, 1759, facsimile repub., Ed. Alain Brieux, Paris, 1977.

LAMBERT, *Notes et Additions to La Perspective affranchie (1759)*, Zürich, 1774. French tr. J. Peiffer in *La place de J.-H. Lambert (1728-1777) dans l'histoire de la perspective*, by R. Laurent, Paris, 1987.

MONGE, *Géométrie descriptive, Leçons données aux Ecoles Normales, l'An 3 de la République*, Paris, An VII. Repub. in J. Dhombres, *Leçons de Mathématiques de l'Ecole Normale de l'An III*, Paris, 1992.

PELERIN, (called Viator), *De artificiali perspectiva*, Toul, 1505. Facsimile repub. of 2nd ed. 1509, Nogent-le-Roi, 1978. A critical comparison of all four edition by L. Brion-Guerry, in *Jean Pélerin Viator, sa place dans l'histoire de la perspective*, Paris, 1962.

SERLIO, *Trattato di Architettura*, Book I, Paris, 1545 (simultaneously with Book I), in Italian, with French tr. by Jean Martin. For extracts of the original with French tr. by J.-P. Le Goff, see *Cahiers de la Perspective*, **4**, IREM de B.-N., Caen

VINCI, *Traité de la Peinture*, Fr. edition A. Chastel (ed.), Paris, 1987.

VITRUVIUS, *Les dix Livres d'Architecture*, tr. by Claude Perrault, 1673, repub. Paris, 1986. Tr. into modern French, les Belles Lettres, Paris, 1988.

VREDEMAN DE VRIES, *Perspectiva...*, The Hague and Leiden, 1604-5. Many editions. Facsimile repub., R. Baudouin, Paris.

General works for a first reading

COMAR, *La perspective en jeu; Les dessous de l'image*, Gallimard, coll. "Découvertes" **138**, Paris, 1992.

DELACHET, *La géométrie projective*, P.U.F., coll. "Que sais-je ?" **1103**.

FLOCON, & TATON, René, *La perspective*, P.U.F., coll. "Que sais-je ?" **1050**.

GILBERT, *La perspective en question*, Ciaco Éd., Louvain-la-Neuve, 1987.

Les Cahiers de la perspective, IREM de Basse-Normandie, Université de Caen, 5 numbers have appeared.

General works for further reading

Theory and Practice

BARRE, & FLOCON, *La perspective curviligne*, Flammarion, Paris, 1968.

BONBON, *Perspective scientifique et artistique*, Eyrolles, Paris, 1978.

BONBON *Perspective moderne*, Eyrolles, Paris, 1989.

BONBON, *Perspective inclinée, plongeante, plafonnante, ombres, reflets*, Eyrolles, Paris, 1986.

FRADIN, *Perspective conique*, Éd. Dessain et Tolra, Paris, 1980.

HENNEBICQ, & MOLLE, *La mise en perspective*, Eyrolles, Paris, 1980.

LEHMANN, *Initiation à la géométrie*, P.U.F., Paris, 1988, with an historical appendix by R. Bkouche.

LOCQUET & PERROT, *Perspectives cavalières et axonométriques*, Technique et Documentation Lavoisier, Paris, 1988.

LUDI, *La perspective pas à pas*, Dunod, Paris, 1989.

ROTGANS, *La perspective*, Éd. Dessain et Tolra, Paris, 1988.

History of Art and History of Science

SVETLANA, *L'art de dépeindre*, N.R.F. Gallimard, Paris, 1990.

ARGAN, WITTKOWER, *Perspective et histoire au Quattrocentro*, French tr.: Les éd. de la passion, Paris, 1990.

BAXANDALL, *L'œil du Quattrocento*, N.R.F. Gallimard, Paris, 1985.

BRION-GUERRY, *Jean Pélerin Viator, sa place dans l'histoire de la perspective*, Ed. Les Belles Lettres, coll. *Les Classiques de l'Humanisme*, Paris, 1962.

CHASLES, *Aperçu historique sur l'origine et le développement des méthodes en Géométrie*, Bruxelles, 1837. Repub. of 3rd. (1889) by Lib. Gabay, Paris, 1990.

DAMISCH, *L'origine de la perspective*, Flammarion, Paris, 1987.

DEFORGE, *Le graphisme technique, son histoire et son enseignement*, Ed. Champ Vallon, coll. *Milieux*, Seyssel, 1981.

EDGERTON, *The Renaissance Rediscovery of Linear Perspective*, New York., 1975.

FIELD, & GRAY, *The Geometrical Work of Girard Desargues*, New York, 1987.

FRANCASTEL, *La figure et le lieu*, N.R.F. Gallimard, Paris, 1967.

GOMBRICH, *L'art et l'illusion*, N.R.F. Gallimard, Paris, 1971.

IVINS, *Art and geometry*, Harvard U.P., 1946.

KEMP, *The science of art*, Yale University Press, London, 1990.

KLEIN, *Considérations comparatives sur les recherches géométriques modernes (Programme d'Erlangen)*, 1872. French edition, Gauthier-Villars, Paris, 1974.

KLEIN, *La forme et l'intelligible*, N.R.F. Gallimard, Paris, 1970.

PANOFSKY, *La perspective comme forme symbolique*, French tr., Minuit, Paris, 1975.

POUDRA, *Histoire de la Perspective*, Paris, 1864.

TATON, *L'Œuvre mathématique de Desargues*, P.U.F., Paris, 1951, repub. Vrin, 1981-88.

THUILLIER, *Espace et perspective au Quattrocento*, in *La Recherche* n°160, nov. 1984.

WITTKOWER, see Argan.

WRIGHT, *Perspective in perspective*, Routledge & Kegan Paul, London, 1983.

Bulletin de l'IREM de Lille, **21** and **22**, 1987-88.

Destin de l'art, destins de la science, Actes du colloque ADERHEM de Caen (1986), Caen, 1991. Diffusion: IREM de B.-N., Université de Caen.

In extenso, Recherches à l'Ecole d'Architecture Paris-Villemin, n° 13, Paris, 1990.

How did you get on?

Ex. 1   The actual values for the table depend on the scale used by the printer of this book in reducing Dürer's original drawing. It is easy to set up the table or to check the measured lengths using the following formulas which provide the answer to the second part of the question: the n-th segment cut on ab beginning at a will be made by the angle $nc(n-1)$, a sixteenth of the angle $C = bca$. For example, the length of the segment corresponding to the angle $8c7$ will be proportional to the difference between the two tangents:

$\tan\{[(17-n).C]/16\} - \tan\{[(16-n).C]/16\}$, which for $n = 8$ becomes:

$\tan\{(9.C)/16\} - \tan\{(8.C)/16\}$.

This difference needs to be multiplied by the length of bc to obtain the length of the segment, a length that diminishes with n.

Ex. 2   Take coordinate axes as: x-axis for the ground line, and a perpendicular through F for the y-axis (fig. 45), and let the number of squares be even as in the case of figure 8 (which simplifies the data of the problem without significantly changing it). We then have a series of divisions points on Ox: $-2b, -b, 0, b, 2b$, etc. and a vanishing point F $(0, c)$. The transversals AA', BB', CC', etc. will be given by equations: $y = a$, $y = 5a/3$, $y = 19a/9$, etc. The vanishing line FT has the equation $x/2b + y/c = 1$ so the coordinates of B are $(2b - 10ab/3c, 5a/3)$. The vanishing line FD has the equation $x/b + y/c = 1$ and so E has coordinates $(b - ab/c, a)$ which makes the gradient of OE equal to $ac/(bc - ab)$. But the gradient of EB becomes $(2a/3)/(b - 7ab/3c) = 2ac/(3bc - 7ab)$, which is only equal to $ac/(bc - ab)$ when $c = 5a$. In this particular case, taking two other diagonals, in the second and third rows for example, will show that the particular alignment of O, B and E was fortuitous.

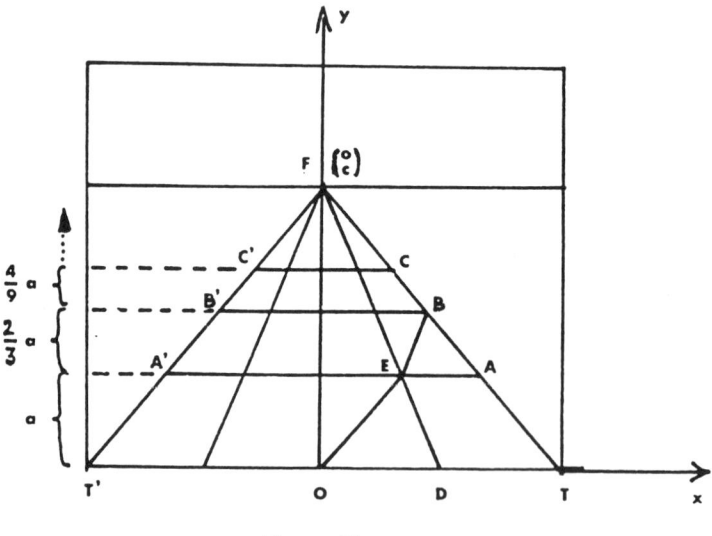

Figure 45

We could also calculate the gradient of OB, which is simpler, giving $5ac/(6bc - 10ab)$, which is not equal to $ac/(bc - ab)$ except in the case of $c = 5a$.

Every choice of reduction ratio q (< 1), giving the equation $y = qa$ for BB', will lead to a similar result for the gradients of OE [$= ac/(bc - ab)$] and EB (or OB), except in the particular case where $c = a(1 + q)/(1 - q)$. In this case, we can show the non-alignment of three other points of the pattern.

Lastly, choosing a reduction in arithmetic progression leads to the same result.

Ex. 3 The sum of the terms a, aq, aq^2, etc. of a geometric progression has a limit $a/(1 - q)$, which fixes a boundary horizon for the transversals, if they are to have the appearance of straight lines seen from above, but in no way provides a resolution of the problem of the diagonals! In the case of figure 8 or 43 ($q = 2/3$), the first interval 'a' which fixes the level of AA' must therefore be chosen to be one third of the eye height, OF = c, if the successive transversals are not the end below or above the line of the horizon.

It does not seem to have been the case in the painting practice of the time that the segment 'a' was always taken to be one third of the eye height, something that would have led to the horizon being the limiting line of the transversals under the two-thirds rule (the infinite sum of a + 2a/3 + 4a/9 + … being 3a). This can be explained partly by the fact that the number of transversals was usually quite limited, and the possible extension beyond the theoretical horizon could be compensated by a 'weak' value for 'a', and partly by the fact that such a conception of a limit was unthought of, and unthinkable, at this time.

Ex. 4 The appearance of the cube is obtained by drawing the reduced lengths on the verticals. The front face will be seen as its 'true' size (almost exactly 1 – 1) like the square EABG, and the back face, hidden if EABG is opaque, or visible if we are looking into an open box, will be the square edcg erected on dc. We only then need to make the joins Ee, Aa, Bb and Gg.

Ex. 5 An answer to the first part of the question is provided by figure 8. For the second cube, we reproduce here the drawing supplied by Piero della Francesca for his proposition 5 (fig. 46):

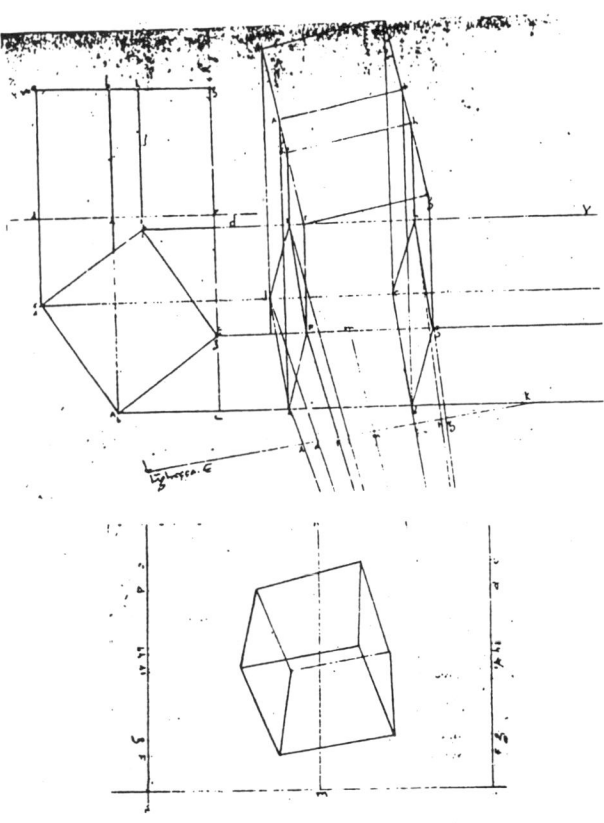

Figure 46
Drawing by Piero della Francesca:
the perspective view of a cube balanced on a vertex

Ex. 6   1°) cd/AB = cd/CD = 1/2; ef/AB = ef/EF = 1/3; gh/AB = gh/GH = 1/4. We
obtain the sequence of inverses of the integers from 2 onwards.
2°) cd/AB = 1/2; ef/cd = (ef/AB).(AB/cd) = (1/3).(2/1) = 2/3; similarly, gh/
ef = (gh/AB).(AB/ef) = (1/4).(3/1) = 3/4. We have the sequence of the ratios of
consecutive integers, the smaller being the numerator.
3°) arguing in the same way: ce/Ac = 1/3, eg/ce = 1/4, gj/eg = 1/5, etc. We
have again the inverses of the integers (the harmonic sequence).

Ex. 7   To understand this, simply consider the situation as viewed from above (that is the
projection on the geometral plane, fig. 47). The diagonals of the square paving are
parallel to the visual ray OD, for which D in the picture is the required distance point
to which all parallels inclined at 45° will converge. We therefore have an isosceles
right angled triangle OFD with OF = FD. For a proof we can proceed like Guidobaldo
del Monte who considered the more general case of convergent points of the images
of sets of parallel lines (see Exercise 12).

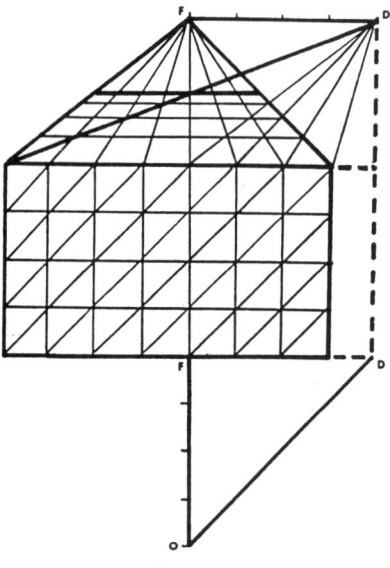

Figure 47
The distance point for lines inclined at 45°

Ex. 8   The profile view chosen by Dürer corresponds to an orthogonal projection onto an
axial plane of the cone, following the direction of the intersection of the sectional
plane with the plane of the base of the cone. The sectional plane is thus seen as the
line gf, which Dürer divides into 12 equal parts by twelve points connected by lines
horizontal to the axis of the cone to one or other of the generators, thence by vertical
lines onto an ichnograph plane. The distances to the axis of the points found on the
generators are the radii of circular sections of the cone taken at the division points of
gf. These distances are used to trace concentric projections of these circles onto the
ichnograph. The two intersections of the vertical lines with the circles to which they
correspond are used to determine segments centred on bd which represent the
'breadth' of the ellipse at each point of gf and so the ellipse can be traced out point
by point.
The ellipse can appear *a priori* more 'pointed' at its upper part because of the greater
proximity of the sectional cut to the vertex of the cone at this point, as if the narrowing
of the cone becomes more extreme towards its vertex. One could also imagine an
opposite argument, by referring to the fact that the angle made by the sectional cut
with the generators which determine the major axis of the ellipse is smaller where the
cut meets the generator further from the vertex.

Ex. 9 The first rule starts by placing a first reduced square at a given distance; this is a strict application of Alberti's short method. But while the latter proposed using a first line of squares set out parallel to the ground line to determine the reduced widths of the succeeding squares, Serlio used the first line to determine the second, and then the second line to determine the third, and so on deeper into the picture. This is where the fault lies and this can be easily seen to be wrong by constructing diagonals to the squares: they will not converge, and if the diagonals of succeeding squares deeper into the picture are drawn they will be seen not to be in line (fig. 48). The construction is therefore not projectively correct. The reason is that the visual rays which mark the distances for the transversals in Alberti's method, which provide in effect the corners of the squares, but only for those at ground level, do not coincide with the lines drawn through the corners of the succeeding lines of squares.

In the second rule, Serlio produces a (con)fusion between Alberti's rule and that of Piero della Francesca using a distance point; he marks the distance levels on the vanishing line PC (fig. 30) and not on an orthogonal to AC at K. But the point I, meant to serve as a distance point, is placed at a distance greater than the picture eye distance since it is vertically above K, and CK is taken to be this distance. However, the construction is projectively correct since the visual rays from I are extensions of the diagonals of the squares they determine. The perspective view obtained is in fact that of an observer situated at a distance IP from the painting.

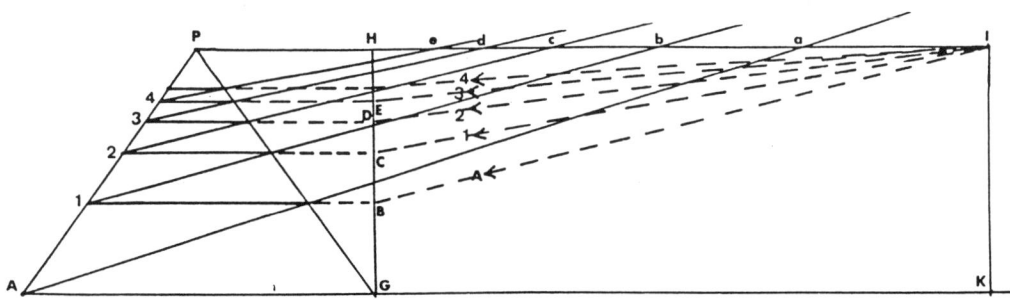

Figure 48
Serlio's first rule leads to apparent diagonals being non-convergent

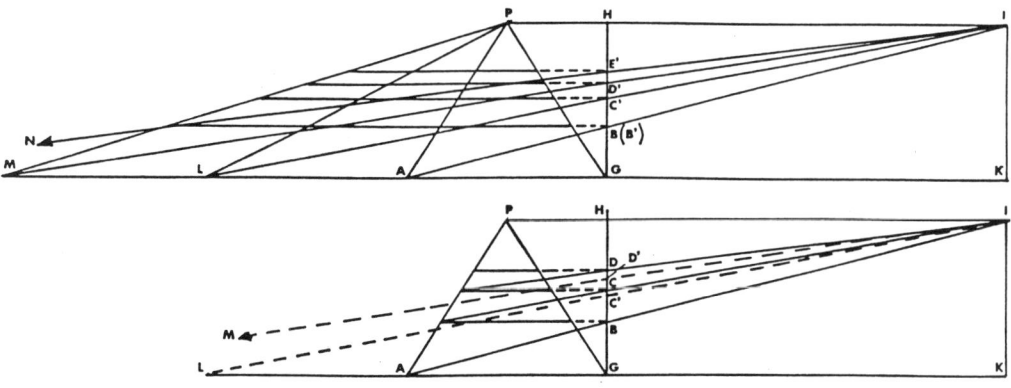

Figure 49
Comparing Alberti's rule with Serlio's

Ex. 10 GH is parallel to AK or EC because AGH and ADC are similar, and so are OGH and OEF. In the configuration GAOED we have GA/GD = GO/GE = AO/ED. In HAOFC, we have HA/HC = HO/HF = AO/FC. Since ED = FC then GA/GD = HA/HC, which establishes GH || DC or GO/GE = HO/HF from which GH || EF.

Alberti's construction introduces a point C such that CF = eye-painting distance: Piero della Francsesca and Viator have the point D such that DE = CF (fig 50).

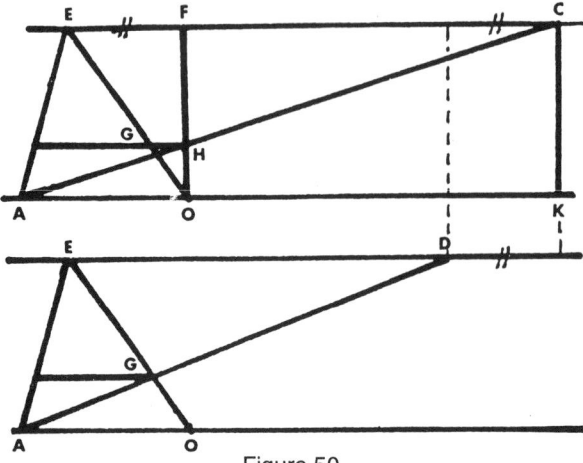

Figure 50

The rules by Alberti and Piero della Francesca as they appear in the Figure by Vignola-Danti

Ex. 11   The three non parallel solids on the ground which appear to be truncated pyramids have edges converging to vanishing points at the horizon. *Either* these edges are parallel in reality (that is the solids are parallelepipeds), in which case their drawn images cannot converge at the horizon, which as its name implies is the position for the vanishing points of horizontal lines, *or* these sides are concurrent in reality which contains, among other objects, three real truncated pyramids (which, by the way, was not part of Vredeman de Vries' conception, since he wanted to illustrate the universal convergence of all sets of parallel lines), and in this case the assumed vertices of all three (truncated) pyramids are to be found at eye level, since they appear on the horizon. But what are we to say about the convergence at the horizon of two sides of the lower face of the centre solid? Does it mean that the base cannot be a parallelogram? And the edges of the base also appear to converge at eye level! We should not confuse coincidence (in geometry) with the accident of coincidence!

Ex. 12   Guidobaldo's proof uses an auxiliary plane LHRI parallel to the given lines BC, DE, FG. Its intersection with the geometral HR is the ground line of the picture in which the image lines LI, NT and PV are parallel. He is then able to establish the property by means of the intersection theorem applied to the different triangles: from A with bases BC and LI, DE and NT; from X with bases MO and LN, OQ and NP, and finally from M, O, Q with bases AX, LI and NT. But the result can be obtained just as easily using incidence properties: the planes ABC, ADE and AFG have a common axis passing through A and parallel to BC; these planes cut the picture along the lines LM, NO and PQ which are parallel to each other if the picture (or its ground line HK) is parallel to BC, and which, in the case of Guidobaldo's proposition XXVIII meet on the common axis of the three planes ABC, ADE and AFG at some point X on that axis, which is parallel to BC and therefore parallel to the geometral containing the three lines BC, DE and FG: this straight line AX is the visual ray from A parallel to the three given lines. The proposition can be generalised to all sets of parallel straight lines in space, its vanishing point being the intersection with the picture of the visual ray belonging to the set of parallels: this point moves to infinity in the case of parallel lines which are parallel to the picture.

Ex. 13   We obtain the transformation formulas:
$$x' = dx/(z + d) \qquad \text{and} \qquad y' = dy/(z + d)$$
which summarise practically six centuries of the history of perspective ...

It is interesting to rework the question using homogeneous coordinates:

Consider a point M in space with the homogeneous coordinates (X, Y, Z, T) with x = X/T, y = Y/T, z = Z/T if T ≠ 0 (that is M is not at infinity), and a point M' in the plane of projection with coordinates (X', Y', T') and x' = X'/T', y' = Y'/T' if T' ≠ 0. We then have the transformation formulas:

$$X'/T' = (d.X)/(Z + d.T) \quad \text{and} \quad Y'/T' = (d.Y)/(Z + d.T)$$

Let this be described by the projection p:

$$M(X, Y, Z, T) \xrightarrow{\ P\ } M'(X', Y', T') = (d.X, d.Y, Z + d.T)$$

whose matrix is given by: $\begin{bmatrix} d & 0 & 0 & 0 \\ 0 & d & 0 & 0 \\ 0 & 0 & 1 & d \end{bmatrix}$

If M is a point at infinity of the space, with coordinates (X, Y, Z, 0), the image point will be M' with homogeneous coordinates (d.X, d.Y, Z) and Cartesian coordinates:

$$x' = d.(X/Z) = d.(x/z) \quad \text{and} \quad y' = d.(Y/Z) = d.(y/z) \quad \text{if} \quad z \neq 0.$$

If z = 0, M is a point at infinity on the projection plane, and M' = M(X, Y, 0, 0).

If z = – d, M is a point of the neutral plane, that is the plane passing through the eye and parallel with the plane of projection, and M with coordinates (X, Y, – d.T, T) has an image M' with coordinates (d.X, d.Y, 0) which is therefore a point at infinity of the projection plane.

Ex. 14   The figure accompanying Lambert's problem III (Fig. 51), shows that he has the idea of finding a horizon through P, parallel to IE, considered as a ground line. CB, DA, CA, BA and CD cut IE at E, F, G, H and I. EP and FP are vanishing lines in the direction CB; GR is an arbitrary vanishing line in the direction of CA, which cuts EP and FP at c and at a. Ha and Ic are therefore vanishing lines in the direction of CD, which meet at Q, necessarily on the same horizon as P. PQ is therefore the sought horizon.

All that is now needed is to make the necessary changes for the particular case when EI is parallel to the sides of the given parallelogram. The problem then becomes one of constructing a parallel to two given parallels through a given point with the use only of a ruler, a problem that approaches that of Problem V in Lambert's *Notes et Additions*. This problem is reproduced here, without solution, for the enjoyment of the reader:[1]

> *AC, BD are lines which meet at a point off the table. Draw with the aid of only a ruler, and without extending the lines, a line passing through a given point E which would cut BD, AC at their point of intersection.*

1.   Lambert, Notes et additions, p. 172-3, tr. Peiffer, p. 268.

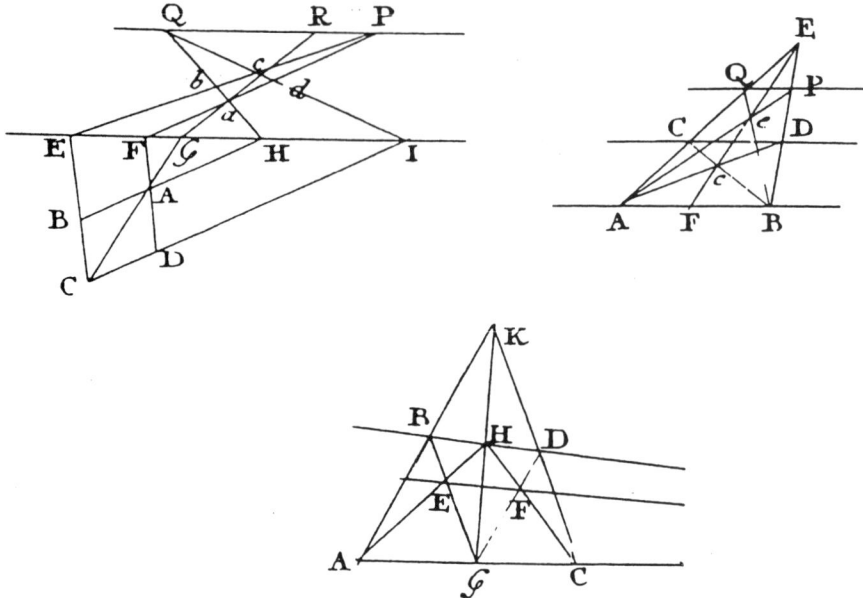

Figure 51

Figures from Plate X, illustrating problems III and V of Lambert's *Notes et Additions*

10

Look at the Stars and become a Geometer!

Monique BELET and André BELET
IREM Toulouse

Shape and size in the universe
The ancient World

Man saw that the celestial vault turned about him and he understood that his ground was fixed. He saw the sun disappear below the horizon, and then reappear. He saw the stars turning above, some making a complete circuit, others disappearing and then reappearing again. He saw how the moon decreased, was gone, and then started to grow again. And man wanted to understand and explain what he saw. Those who walked upon the earth understood that, notwithstanding the hills and valleys, it was essentially flat. Certain stars appeared to pass under the earth and then come back again on the "other side". Perhaps the Earth was a disc floating on an ocean with the sky a hemisphere, or perhaps the Earth was standing on pillars that reached down into the underworld of the dead and some Atlas held it aloft.!

From the sixth century B.C., following the reports of travellers who had seen new and different constellations, the myth that the Earth was flat started to be replaced by the idea of it being a terrestrial sphere, occupying the centre of the universe, considered as a celestial sphere. The idea that the Earth was a sphere was supported by the evidence of two phenomena. The first was that ships appeared to disappear below the horizon as they got further from the shore. The second was that at the beginning and end of eclipses of the moon, the unlit part of the moon had a circular border: might it be the Earth's shadow that eclipsed the moon?

With the Earth as a sphere, the stars can revolve about it in a fixed circular motion which can be described in geometric-mathematical terms. The Earth, changing and perishable, becomes the central point of the universe, and it is from this point on that geometry start gradually to replace mythology as a way of explaining the universe.

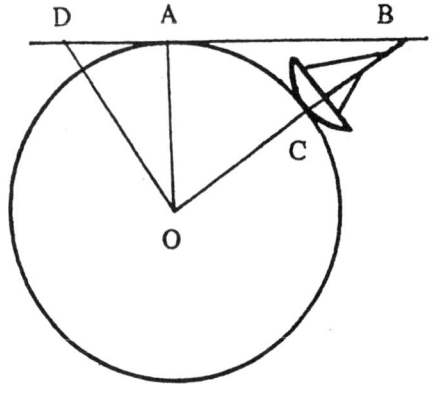

Figure 1

If the Earth is a sphere, how can its radius be found? Eratosthenes, director of the famous library at Alexandria, derived a way of finding the radius of the Earth from the fact that it was a sphere. He knew that at the summer solstice the sun's rays reached the bottom of a deep well at Cyrene (now Assouan), and so the sun was vertically overhead. He also knew that the distance from Cyrene to Alexandria was reckoned by travellers to be 5000 stades, that is about 900 km.

Let A and B be two points on the Earth's surface (Fig. 2). They subtend an angle X_1 at a point S_1 in space, X_2 at a point S_2, etc., and the angle X diminishes as S is taken further from the Earth. If the point S is taken to be at a very great distance, as Eratosthenes supposed in the case of the sun, then the angle X can be considered as negligible and the rays from S to A and B will be parallel.

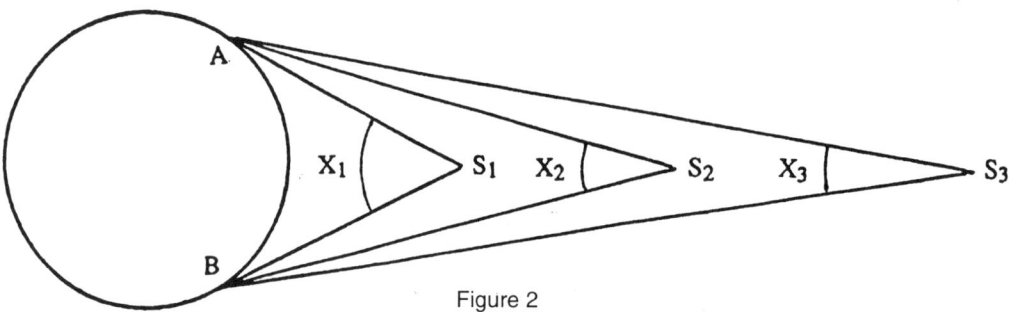

Figure 2

If the angles made by the sun's rays with the vertical turn out to be different at Cyrene and Alexandria, that must be due to the fact that the Earth is round. On the summer solstice, Eratosthenes waited patiently for mid-day with his gnomon (a vertical stick). He measured the length of the gnomon and the length of its shadow on the ground. the ratio of the two gave him a measure of the angle between the vertical and the sun's rays (marked a in Fig. 3).

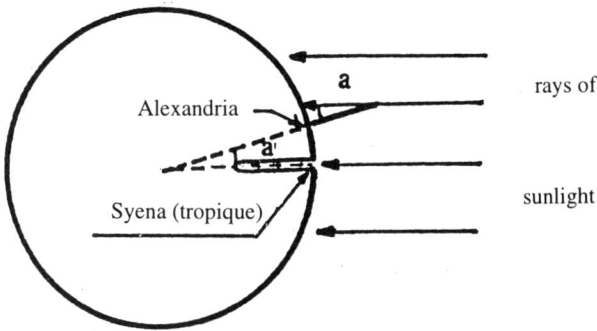

Figure 3

He found that the angle was 1/50 of a full circle, or $7°12'$, which gives the difference in the latitudes of the two places, assumed to lie on the same meridian. This angle of $7°12'$ therefore corresponds to a terrestrial distance of 900 km. A simple rule of proportion will give the circumference of the Earth and hence its radius. The idea that the Earth was a sphere was therefore confirmed by the observations made by Eratosthenes.

Exercise 2	Use the information given here to calculate the circumference of the Earth.

Triangulation: measuring the Earth

Increased travel, following both merchant and military activity, was a stimulus to the art of cartography, and there developed a need for greater precision in land measurement. In the seventeenth century began the methods of triangulation, which replaced simple measurement of distance and angle. Picard used the same principle as Eratosthenes, this time using the two towns of Sourdon (near Amiens) and Malvoisine (south of Paris), known to lie on the same meridian. But his approach to the problem of finding the distance between the towns was quite different from the methods employed in Ancient Greece: the distance, of about 150 km, was found by triangulation. The basic principle was to use the accurate measurement of one side of a triangle on the ground and to measure the angles of the triangle, by taking sightings, and then to calculate the lengths of the other two sides using the Sine Rule: $\frac{a}{\sin A} = \frac{b}{\sin B} = \frac{c}{\sin C}$. If, for example, AB is the known base of a triangle ABC on the ground, the lengths of the sides AC and BC can be found, even if the ground is uneven or other obstacles prevent direct measurement. What is needed is an initial horizontal straight line that can be measured accurately and sightings of the third vertex from each end of the initial line.

In the seventeenth century the length of the base was measured directly with the repeated use of a rod placed end to end along the line. The length of the rod was a *toise*, the standard unit and equal to 1.94 m or about 6 ft. Picard was able to measure the angles accurately using a simple theodolite constructed of a protractor fitted with sight glasses. Altogether some fifty triangles were measured to map the distance from Sourdon to Malvoisine, two triangles having a side in common. Knowing AB, the lengths AC and BC are found (Fig. 4). Using the calculated value of BC and measuring the angles of the triangle BCD, the other sides of the triangle BCD can be found, and the process can be repeated for each consecutive triangle. But this does not give the length along the meridian itself. To do this, Picard also measured the azimuth of each side of the triangle, that is the angle that it makes with South. The preceding process of calculation can be carried out, but this time calculating the length along the meridian. For example, angle BAA_1,which is the azimuth of AB, and angle ABA_1, equal to ABC, can be used with the known side AB to find the length AA_1 of the meridian, and also BA_1, from which CA_1 can be deduced. The whole process can be repeated for triangle CA_2A_1, and so on. Adding up the successive parts of the meridian lengths will give the overall distance required.

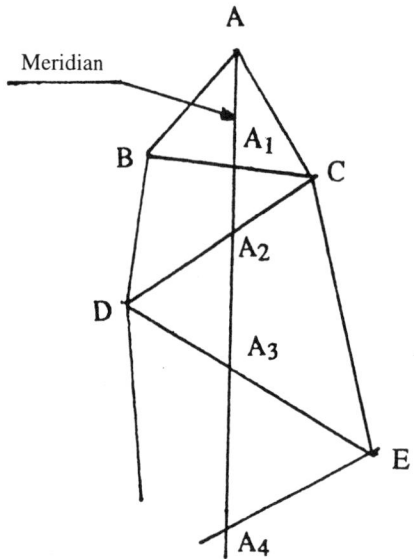

Exercise 3 Find the length of the meridian AA_2 shown in Fig. 4 given the following measurements: AB = 10 km; ABC = 58°; BAC = 72°; BAA_1 = 40°; DBC = 91°; BCD = 42°.

Figure 4

The arc of the meridian that is being measured is, of course, considered to be made up of straight line segments (see Chapter 5). This is an approximation but we should add that, from a theoretical stand point, this was necessary, since the unit of length was rectilinear and had only been used previously to measure straight line distance.

How, in fact, was the azimuth of a side, AB for example, found in practice? The direction AB was found by direct and accurate measurement by sighting B from A. The direction South was found from astronomical sightings: the height of a star, that is its angular height above the horizontal plane, is a maximum in the Northern Hemisphere when it passes above the meridian. All that is required is to follow the path of a star with a telescope, and due South will be found when the star is found to be at its greatest angular height.

Furthermore, it was known how to calculate the latitude of a place by measuring angles of elevation of stars. Using the distance between Sourdon and Malvoisine that he had measured, and the latitudes of the two towns, the abbé Picard was now able to provide a measurement for the length one degree of the meridian which he gave as 57 064 toises. The accuracy, of an order of a few seconds of arc, gave a relative error of a few millionths, and was far better than that obtained from direct land measurement. The accuracy of the result proved invaluable to Newton in confirming his theory of gravitation. We shall see that this accuracy had the effect of making further measures necessary. But, to go forward a century, two astronomers, Delambre and Méchain, repeated the triangulation method to find the distance, this time, from Dunkirk to Barcelona. The work took from 1792 to 1799 and the instruments were by then improved, the seventeenth century theodolite having been replaced by the repeating circle. As its name suggests, the principle of the instrument lay in the repetition of the same measure. An approximation to the required angle measurement is found from these measures. The interest in the instrument lay in the fact that it gave an accuracy that was greater than its gradations. For an instrument graduated in minutes, an accuracy to a tenth of a minute was obtained from ten successive measurements. The expedition to measure the distance from Dunkirk to Barcelona was financed by the Constituent Assembly following the Revolution, with the purpose of establishing an accurate measurement for the new standard metre as a forty thousandth part of the meridian.

The shape of the Earth: a scientific controversy

Between 1683 and 1701, Cassini measured two meridian arcs, the one from Paris to Collioure, going south, and the other northwards, from Paris to Dunkirk. The first gave a measure of 57 097 toises for a degree of the meridian and the second 56 960 toises, from which Cassini deduced that the length of a degree of the meridian shortened as it approached the poles, and that the Earth was flattened at the equator.

At the same time, Newton published his *Principia* with its theory of universal gravitation. According to this theory, if our planet is stable, it is because all its parts are attracted to one another and it will therefore be spherical, if it is immobile, by reasons of symmetry. But it is in motion, revolving about a central axis, and so a centrifugal force, zero at the poles and at its greatest at the equator, will oppose the force of gravitational attraction. We must therefore have an Earth that is flattened at the poles.

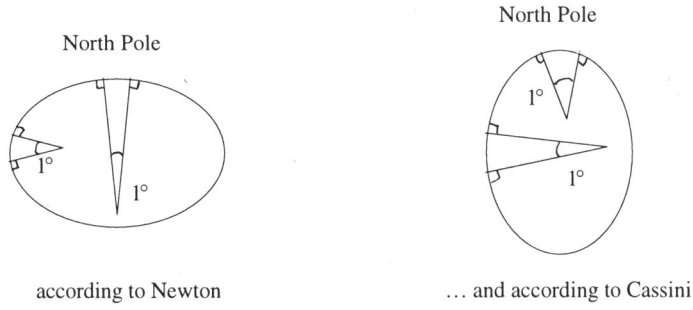

The Earth according to Newton   … and according to Cassini

Figure 5

To settle matters, the French Academy sent an expedition to Peru in 1735 led by Condamine to measure an arc at the equator, and another to Lapland in 1737 led by Maupertius and Clairaut to measure an arc of the meridian near the Pole. It turned out that a degree of the meridian was in fact longer at the equator than at the Pole, thus vindicating Newton. Cassini's measurements, made at latitudes that were not so very far apart, had contained inaccuracies greater than their actual differences. Today we know that the flattening of the Earth is of the order of 1/298, that is that the equatorial radius is greater than the polar radius by about 21.4 km.

How far away is the Moon?

In the third century B.C. Aristarchus of Samos, a contemporary of Eratosthenes, provided an approximation for the distance of the moon from the Earth by using the (angular) distance travelled by the moon during a total eclipse of the moon. The assumption that he made was that the distance travelled was approximately that of the diameter of the Earth.

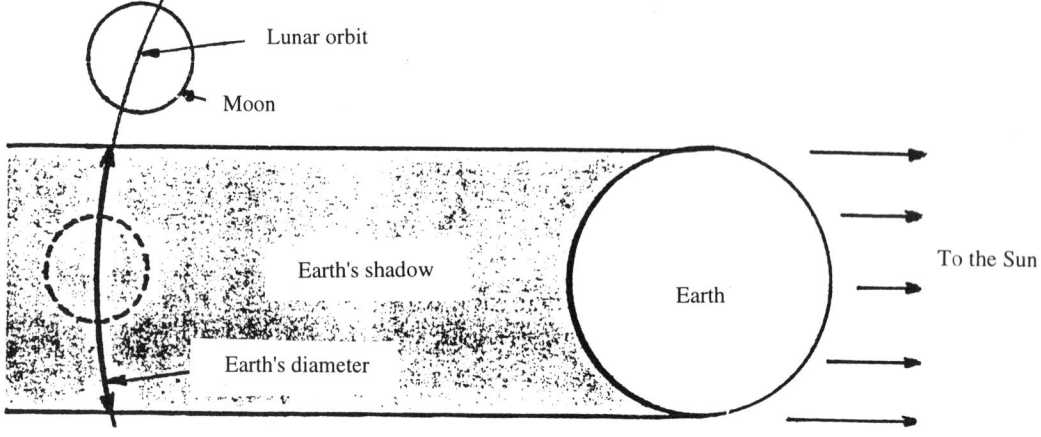

Figure 6

The sun is considered to be so far away that its rays can be taken to be parallel. Since the diameter of the sun subtends an angle of 32' of arc as seen from the Earth, and knowing that the phase of the moon repeats itself after 29.5 days, which can be taken to be the time of its circular orbit, a measurement of the time taken for the moon to pass across the Earth's shadow allows us to calculate the fraction of the moon's orbit that is equal to the diameter of the Earth. From this the radius of the moon's orbit, and so its distance from the Earth, can be found in terms of the terrestrial diameter. Later the distance of the moon was measured by triangulation, using Berlin and Le Cap as two of the vertices of a triangle, the third being the moon itself. This improved on previous measurements and was carried out by Lacaille and Lalande in 1752.

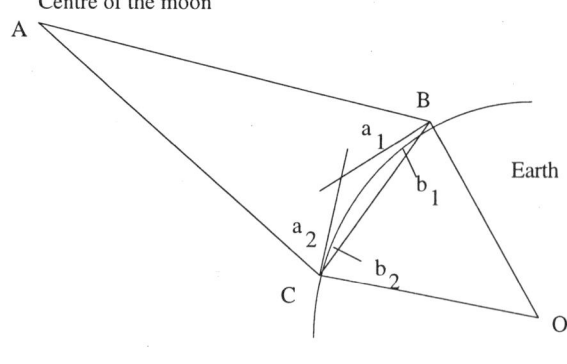

Figure 7

The difference of the latitudes of Berlin and Le Cap being known, the length of BC can be found. The angles a_1 and a_2, the heights of the moon at each place, are measured at the same time when the moon passes over the common meridian of both places. The angles b_1 and b_2 can be deduced from the known latitudes and hence the angles of triangle ABC are known. Knowing angle A and the length BC, the distance AB or AC is then found.

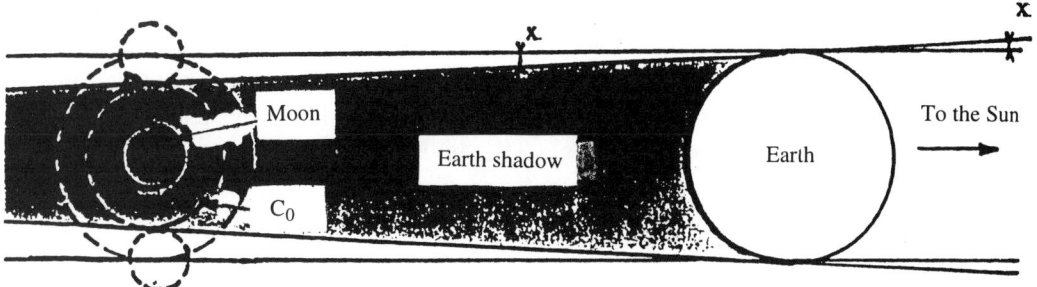

Figure 8

Exercise 4	This is a modern version of Aristarchus's method. It is assumed that the radii of the Sun and the moon as seen from Earth have the same angular distance, marked x in Fig. 8. The radius of the Earth's shadow R_0 in the neighbourhood of the moon is equal to the terrestrial radius R_t minus the lunar radius R_l: so $R_0 = R_t - R_l$. Fig. 9 reproduces a photograph of a lunar eclipse with the remainder of the Earth's shadow drawn in so that it may be measured. The same diagram shows the shadow circle C_0 and the lunar circle C_1 drawn to the same scale. Use the drawing to measure the lunar diameter and the shadow diameter and deduce the Earth's diameter for the drawing. Hence find the actual diameter of the moon.

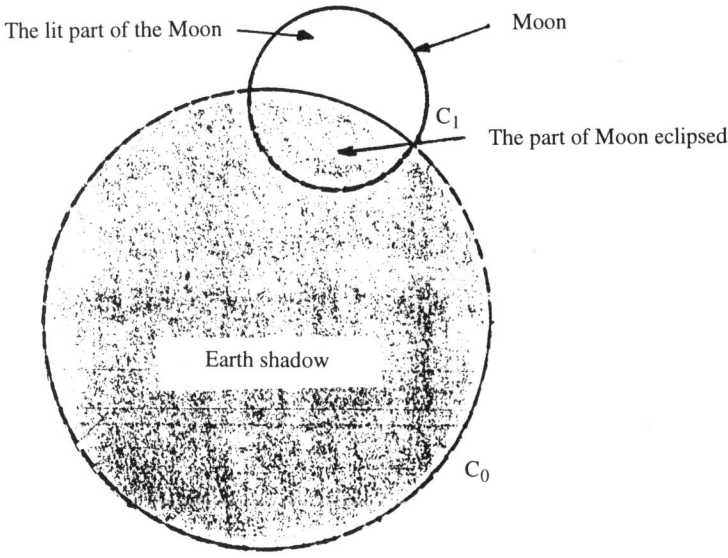

Figure 9

Finding the distance of the Sun in Antiquity

It was Aristarchus again who first gave an estimate for the distance of the Sun from the Earth, albeit a rather inaccurate one. His reasoning was as follows: The time t_1 taken for the moon to change from New Moon, N, to the first quarter, Q, is shorter than the time t_2 for it to go from Q to Full Moon, F (Fig. 10). The first quarter is when the position of the moon is such that exactly half of the moon is illuminated when seen from the northern hemisphere. Let angle TSQ be a; Aristarchus estimated the time interval d, between t_1 and t_2 to be 12 hours (half a day) and we know that $t_1 + t_2 =$ half a lunar cycle = 29.5/2 = 14.75 days.

We have: $\begin{cases} t_2 - t_1 = 0.5 \\ t_2 + t_1 = 14.75 \end{cases}$ from which $\begin{cases} t_1 = 7.125 \\ t_2 = 7.625 \end{cases}$

Assuming that the lunar orbit is circular, the angles and times of rotation are proportional if the speed of the moon is constant; a reasonable assumption for the Greeks, which we also assume here. This gives: 29.5 days corresponds to 360°; 0.5 days corresponds to 6°10′; and since d = 2a, then a = 3°05′. Hence

$$TS = \frac{TQ}{\sin a} = 19 \, TQ.$$

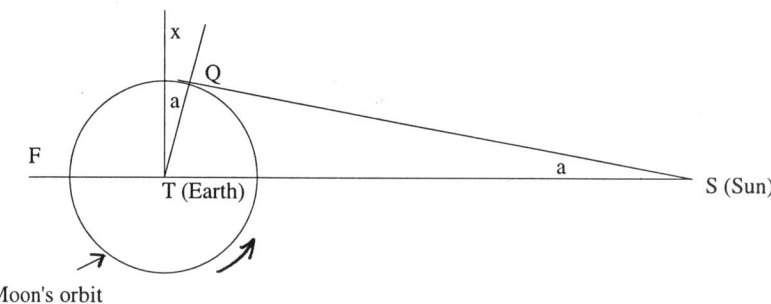

Moon's orbit

Figure 10

Aristarchus thus found that the Sun is 19 times as far away from Earth as is the moon, while the actual value is about 384 times. Why is there such a great error? He had over estimated the value of $d = t_2 - t_1$: in fact d is about 35 minutes, although there is considerable variation.

A further 17 centuries had to pass before a more accurate value for the distance was obtained ... But an answer to that question had to depend on many further discoveries about the relative movements of the stars.

The Almagest or How the Ancient Greeks thought about the movement of the Stars

Ptolemy's work from the second century A.D. collects together and systematises previous knowledge and provides a commentary. It represents an important advance towards experimental science. The current understanding as well as metaphysical considerations favoured the idea of circular motion for celestial bodies. The Earth is the centre of the universe with the Sun and the moon revolving around it. As for the movements of the planets, Ptolemy uses other circles. To simplify the model we shall use two of them: the deferent whose centre is the Earth, and the epicycle whose centre lies on the deferent. For the inner planets (Venus or Mercury), the Sun, the Earth and C, the centre of the epicycle of the planet, always remain in a straight line (Fig. 11).

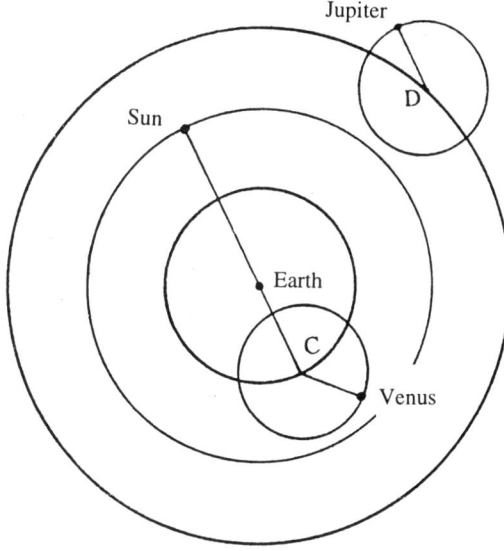

Figure 11

For an outer planet (Mars, Jupiter, Saturn), if D is the centre of its epicycle, we always have the line joining the planet to D parallel to the line joining the Earth to the Sun.

As for the fixed stars, they are arranged on the 'fixed stars sphere' which revolves about an axis through the poles. This model, which agreed quite well with the observations made at the time, prevailed until the beginning of the sixteenth century when Copernicus substituted for it a simpler model.

Sixteenth century discoveries:
The originality of Copernicus

It was, surprisingly, the reading of Ptolemy's Almagest that stimulated Copernicus to find a simpler way of explaining planetary motion. He was struck by the complications necessary to describe Ptolemy's system and it was, perhaps, this that led him to consider a heliocentric model. If the Earth was no longer the centre of the universe, there still needed to be a centre of material attraction. Since the orbits of the planets were uniform circles, they could not all have the same centre. Copernicus thought of the Sun as being in the "middle" of their centres, in order to conform to observations.

The apparent movement of the stars was not due to their own movement but to the rotation of the Earth on its own axis every 24 hours, a truly revolutionary idea at the time. The movements of the planets could now be described by the movement of the Earth in its own orbit and by the use of perspective. In fact, the Copernican system turned out to be almost as complicated as Ptolemy's. He wrote two principal works: the *Commentariolus* in which the theory is sketched out, and his major work, *De revolutionibus orbium caelestium* where his theory is supported by mathematical justification. The work was written in 1532 but not published, despite the entreaties of his friends, until 1543, the year he died. It needs to be said that Copernicus needed to be sure of what he had written. He took the precaution of dedicating his work to Pope Paul III. Himself an ordained canon, Copernicus anticipated the theological objections that would be made to his theory. The prudence of the author is clear in this extract from the preface to *De revolutionibus orbium caelestium* (the manuscript is held at the Krakow University).

> *To the Most Holy Father, Pope Paul III,*
> *I can well imagine, Most Holy Father, that certain men learning that in these books which I have written on the revolutions of the spheres of the world, I attribute certain movements to the Earth, they will immediately call for the condemnation, both of me and of my opinion, yet my opinions do not please me to the extent that I do not take account of the judgement of others. And although I know that the thoughts of the philosopher ought not to be submitted to the judgement of the masses, because his task is to find out the truth in all things, to the extent that God allows to human reasoning, I esteem nonetheless that one should flee from opinions that are entirely contrary to justice and truth ... I came very close to completely suppressing the work which I have completed, for fear of the contempt for the novelty and absurdity of my opinions ... as for what I have accomplished in this study, I submit it to the judgement of Your Holiness, as well as to other learned mathematicians.[1]*

The most radical aspect of the theory, apart from its heliocentricism, was that the planets were no longer seen as individuals, but as part of an entity, the solar system. The Copernican system lets us classify the planets in terms of their distance from the Sun: the inner planets of Venus and Mercury, and the outer planets of Mars, Jupiter and Saturn. And surprise, surprise, this was the very same classification that had been made by the Greeks.

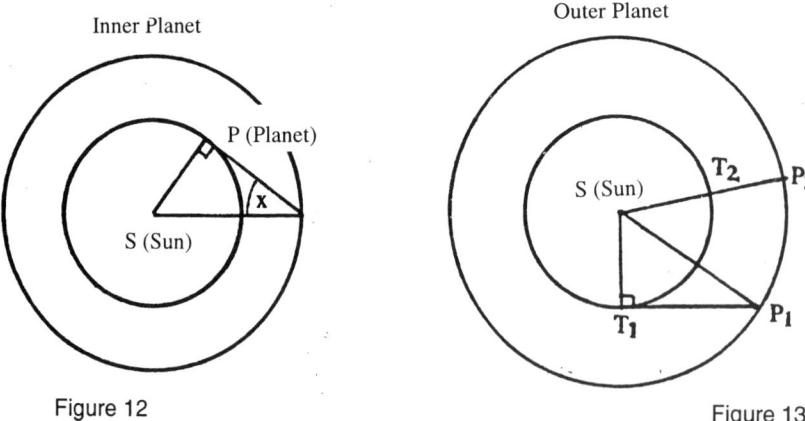

Inner Planet

Outer Planet

Figure 12   Figure 13

The model lets us calculate the ratios of the values of the relative distances of the Earth and the planets from the Sun.

– for the inner planets, measure the angle x, the maximum elongation of the planet with respect to the sun, which gives $\sin x = SP/ST$.

1.   Kuhn, *La révolution copernicienne*, p. 184-185.

– for the outer planets, proceed as follows. Let d be the duration of a synodic period, that is the time elapsed between two successive oppositions ST_2P_2, which is found by observation. Let P_1 be the position of the planet at its western quadrature, that is the position of the Earth T_1 and the planet P_1 is such that the angle $ST_1 P_1 = 90°$ (Fig. 13). Measure the time separating the positions T_1 and T_2. During time t the Earth has made up its angular deficit $ST_1 P_1$ and it gains a complete turn on the planet during time d. We have, in radian measure,

$T_1 SP_1 / 2\pi = t / d$, so $T_1 SP_1 = 2\pi t / d$.

But $ST_1 / SP_1 = \cos T_1 SP_1$ and so $ST_1 / SP_1 = \cos (2\pi t / d)$.

which gives us the ratio SP_1 to ST_1.

Exercise 5	The astronomical unit is the distance of the Earth from the Sun. Find the distance of Mercury from the Sun in astronomical units, given that the maximun angle of elongation of Mercury is 22° 7′.

Tycho Brahe, the observer

While still a youth, Tycho Brahe had observed an eclipse of the Sun. Amazed by what he had seen, he started to study astronomy. He made notes from Ptolemy's Almagest and compared it with his observations. The instruments he used were very simple: first of all, just a pair of compasses with the apex at the eye of the observer and each of the branches directed towards a star. This was replaced later by an instrument that was just as rudimentary: the rod of St. James, a sort of graduated rule on which another piece could slide, which acted as a visor. He also used two armillary spheres and sextants, but no telescopes, which did not come into use in astronomy until after 1690. These simple instruments did not prevent Tycho Brahe from obtaining results that were remarkable for their accuracy (of the order of a minute of arc).

King Frederick II of Denmark let Brahe use the island of Hveen where he had an observatory constructed at the Uranienborg Palace. Brahe spent 21 years there as a tireless and meticulous observer. He established a catalogue of 1000 stars, and made thousands of measurements of the positions of the planets, Mars in particular, which proved invaluable to Kepler. Although he did not entirely accept the heliocentric view of Copernicus, he borrowed some of the ideas: he had the planets turning around the Sun and the whole system revolving around an immobile Earth. Required to quit Uranienborg towards the end of his life, Brahe moved to Prague, where he met Kepler. Despite a number of quarrels between these two men, Brahe decided to leave his entire astronomical heritage to the one man who would be able to draw conclusions from this mass of observations: Johannes Kepler.

Kepler: what are the laws which govern the movements of the stars?

Kepler was both an astronomer and an astrologer who wished to understand the world that God had created. He looked for harmony and beauty, and that is why perhaps he thought of the orbits of each of the planets as inscribed within the five regular polyhedra, all circumscribing each other. This is what he wrote:

> The Earth is the Circle which measures everything else: circumscribe it with the Dodecahedron. The circle around this latter will be Mars: around Mars circumscribe the Tetrahedron. The circumcircle around this latter will be Jupiter: around Jupiter circumscribe the Cube. The circumcircle around this latter will be Saturn: now inscribe the Icosahedron inside the Earth. The circle inscribed within it will be Venus. Inside Venus inscribe the Octahedron. The circle inscribed within it will be Mercury. You will find therein the reason for the number of the planets.[1]

Kepler started out as a theologian, changing to a study of mathematics and astronomy: all his work is imbued with his faith. This was his reasoning: if the world is a divine creation, then it must be perfect, and it must also be comprehensible to the human mind, which is also created by God ... He was in this very close to the Platonists. It was this faith in the intelligibility of the universe that so determined him to find its mathematical laws. The theologico-geometric reasons for circular motion no longer convinced him and Kepler sought physical causes. He believed he had found a law of attraction between two bodies which was proportional to $1/r$, where r is their distance apart. The evolution of his ideas led from Plato to Newton. Kepler's name remains known to us because of the three laws which bear his name; they give a geometrico-temporal explanation of the movements of the planets, so closely observed by Tycho Brahe.

While Kepler had started out as a follower of Copernicus, the observations made by Tycho Brahe which he used were at variance with that theory. In his research he hit upon the idea that the planetary orbits might be ellipses, but raised new problems, since it would contradict the sacrosanct idea of uniform motion. The harmony of the world collapsed, but it was reconstructed with his three laws. What we now call Kepler's first law was discovered, chronologically, after the second. It states: "The orbits of the planets are ellipses, with the Sun at one of the foci." At the beginning of his work, Kepler took these ellipses to be eccentric circles, and this is very close to reality if the eccentricity is small. The second law says that a planet "sweeps out" equal areas in equal time intervals, the radius vector being understood to be the line joining the Sun to the planet. It is interesting to note that Kepler was the first to use the planet-Sun distance as a value for the problem. The third law we now interpret as: if a is the semi-major axis of the orbit and T is the period, then a^3 / T^2 is a constant: it is the same for all the planets. The importance of these laws is that they gave a new foundation for a heliocentric model of the solar system that completely breaks with Copernicanism. Copernicus had restored the Aristotelian idea of uniform circular motion by adapting it to a more suitable model than Ptolemy's, but Kepler introduced an ellipse and non uniform motion. Kepler provided new mathematical laws to interpret the divine order: no more uniform motion, but a law about areas; the orbit was no longer a circle (a divine figure) but an ellipse with the Sun at one of its foci; but a new constant a^3 / T^2 which shows that God treats all the planets in the same way.

Exercise 6	Find the semi-major axis of Jupiter's orbit, given that its orbital period is 11 years 315 days. [Use the astronomical unit = Earth-Sun distance as the unit of length.]

---

1.   Kepler, *Le secret du monde*, p. 26.

Measuring the heavens

Kepler's third law had an important consequence for calculating astronomical distances and, in particular, it allowed for a much more accurate measurement for the distance of the Earth from the Sun. Previously this distance had been found by triangulation, using two points on the Earth's surface as vertices of the triangle and the Sun as the third point, just as Lacaille and Lalande had done for the moon. The procedure was not very accurate because of the extremely small vertex angle at the Sun, the distance of the Sun being so much greater than any measurable distance on the surface of the Earth. Cassini had obtained a value of 22 000 times the Earth' radius for the distance of the Sun using this method.

What was needed was a "passage" of Venus, that is Venus passing across the Sun as viewed from the Earth, and this occurred on 3 June 1769, when measurements were taken at Varda in Sweden (the point A in Fig. 14) and at Tahiti (point B).

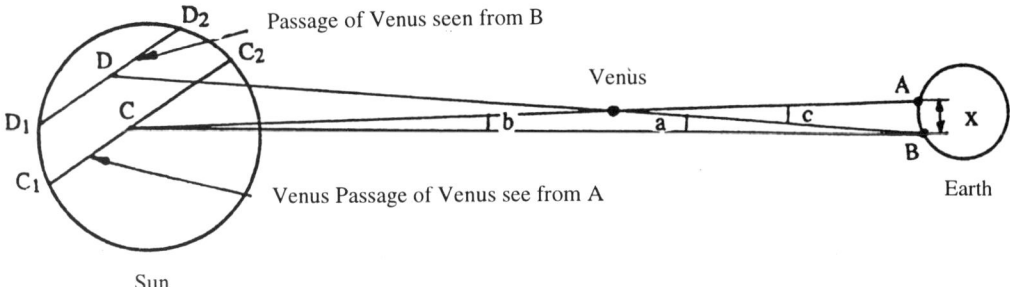

Figure 14

Fig. 15 shows how the passage of Venus would be seen from two different points on the Earth: the observer at A measuring the time taken for Venus to go from C_1 to C_2 and the observer at B measuring the passage D_1 to D_2. It was already known from measurement of other passages that it took 8 hours for a passage of Venus across the diameter of the Sun, which is 32 seconds of arc, and so the angular distances $C_1 C_2$ and $D_1 D_2$ can be found by simple ratio. Also, we have $OD^2 + D_1 D^2 = OD_1^2 = radius^2 = (16')^2$. Hence, OD can be calculated. Similarly, we can find OC and so CD, whose angular distance is marked a in Fig. 14. Since the angles a, b, c are all very small (of the order of minutes of arc) they can be taken to be equal to their tangents, using radian measure.

Let d be the Earth-Sun distance and x the distance AB, taken to be perpendicular to the line BC. Using VS for the Venus-Sun distance and VT for the Venus-Earth distance, we have: $d = x / b$ and $x = VT.c$; so

$b = x / d = VT.c / d$.

$a = CD / d$ and $CD = VS.c$; so $a = VS.c / d$.

This gives $b / a = VT / VS$. We can write $d = x / b$ as $d = x / (ab/a)$; this gives $d = x / (a.VT / VS)$. The distance x can be found from geographical considerations and a is found by observation as described above. Kepler's third law will give the ratio of the semi-major axes of Earth and Venus from a knowledge of the periods of Earth and Venus, and this value turns out to be that the distance VS is 0.73 times the Earth-Sun distance, or 0.73 d, which gives 0.27 d for VT. Hence, VT / VS = 0.27 / 0.73. Using these values in the formula above yields a value for d equal to 149 500 000 km.

A knowledge of the periods of the other planets, and using Kepler's third law, now allows us to calculate the distances of all the planets from the Sun, and so the dimensions of the solar system have been established. But we can do more ...

Exercise 7	Use these values to calculate the distance of the Earth from the Sun. A and B are two points on the Earth 6 500 km apart. The passage of Venus as observed from A is 5 h 56 mn 1 s, and from B is 5 h 44 mn 1 s. The diameter of the Sun is 32′ and assume that d passage across the diameter takes 8 hours.

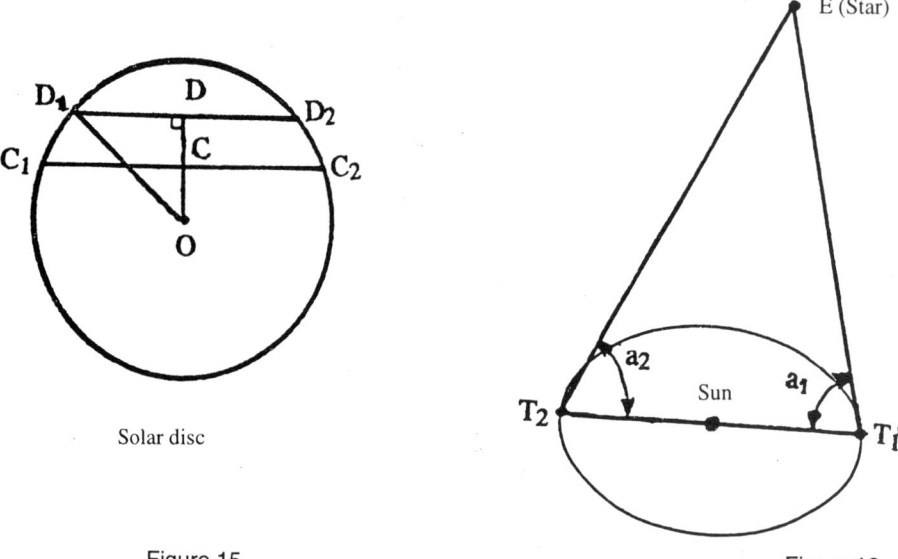

Figure 15   Figure 16

To find the distance to stars, we use triangulation with the diameter of the Earth's orbit for the base of the triangle. The measurements of angles a_1 and a_2 are taken at 6 monthly intervals and the base of the triangle $T_1 T_2$ will be twice the Earth-Sun distance (Fig. 16). The practical difficulty of the method lies in the extremely small size of the vertex angle at the star, always less than 1 second of arc. The first measurement of this type was carried out by Bessel in 1838 on Alpha Centauri, our nearest star. The parallax of this star, that is the difference between the two angles a_1 and a_2, is 0.75″, which is the angle for a marble of 1 cm diameter seen from a distance of 2 750 metres! This star is at a distance of about 4 light years, that is 4.10^{13} km, or about 275 000 times the distance of the Sun. The same procedure can be used to measure the distances of stars up to 40 light years away, which is about the distance of the Pole Star.

Galileo

While Galileo appears to have accepted the heliocentric model of the universe from the start, that did not prevent him from teaching geocentricism at the University of Padua for 17 years. While he was attached to heliocentricism, he lacked any decisive, and therefore communicable, argument in its favour that would allow him to change what he taught. Galileo's strength lay in his wish, just like Kepler, to mathematise the laws of physics. Here is an extract from one of his letters that illustrates this approach. On 16 October 1604 he wrote to Pablo Sarpi:

> *I shall now prove what remains, namely that the distances travelled by a particle under natural movement, are proportional to the squares of the time intervals taken; consequently, the distances travelled in equal time intervals are in the same proportion as the sequence of odd numbers starting from unity.*[1]

In actual fact, what was novel about Galileo was not so much the mathematisation of astronomy, which had already taken place with the Greeks, so much as the addition of arithmetic to the geometry of astronomy. He introduced the idea of the curve as a trajectory, and considered time to be a parameter of its geometry.

In astronomy, although he was not the inventor of the telescope that bears his name – it had been invented in Holland – he was the first to use it for the systematic study of the stars. He certainly improved on the original telescope, which led to research into the shape of lenses and the theory of optics. He directed his telescope towards the moon, and there he discovered that its surface was uneven containing peaks and craters, and not the perfectly smooth surface that it was previously thought to be. Observing Jupiter, he saw four moons which changed places from one day to the next. He was finally persuaded that they revolved about the planet; the idea of stars turning around other stars was completely unthought of in 1610! He looked at Saturn and saw what appeared to be two other stars accompanying the planet ... these were its rings. Two years later, they had disappeared; they were now being looked at side-ways on, and so were invisible from the Earth because of their narrowness. He also noticed that Venus did not always appear to have the same diameter but had phases just like the moon.

In 1612 a polemical dispute erupted with the Jesuit Scheiner. He saw Galileo's observations as scientific work and accepted it as such: the dispute was over the priority that should be given to them. The trial of Galileo by the Church was over a fundamental issue: it concerned the very foundation of truth and reason. The stakes were far from mere obscurantism, but had to take account of the consequences of the court rejecting what appeared to be a complete, definitive, and therefore immutable, description of the world. In 1616, the writings of Copernicus, from which Galileo had drawn his learning, were declared heretical and banned. The prohibition was not lifted until 1822!

Newton: the unifier

Since 1610, scientists had been looking for physical causes which could be expressed in mathematical terms, and Newton belonged to that tradition. The movement of a falling body (on Earth) had been known since Galileo. If it is left to fall from rest, it will travel 4.9 m in the first second, and the distance travelled will be proportional to the square of the time elapsed. It was also known that a body subject to a force, and then left to move freely, will continue to travel with the same constant straight line motion. Also the movement of the moon was known: it had an almost circular orbit around the Earth of 27.3 days, and this period was determined by refernce to the fixed stars. And the young Newton – he was 23 years old – asked himself why it was that the moon did not fall towards the Earth. His reasoning was this: since the moon describes a

---

1.   Dumoulin, *Astronomie en Terminale*, p. 47.

circular orbit, or nearly so, then each second it moves away from a straight line trajectory.

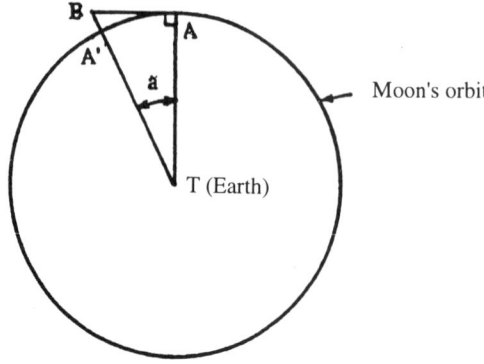

Figure 17

In 27.3 days the moon will describe a complete orbit of 2π radians, and 27.3 days is $27.3 \times 24 \times 3600$ seconds. In 1 second, the moon will turn through an angle a (Fig. 17) where:

$a = 2\pi/(27.3 \times 24 \times 3600) \approx 2.662 \times 10^{-6}$ radians.

Now, TA = TA' = radius of the lunar orbit = 383 000 km.

And TB = TA / cos a

Hence $A'B = TA \left(\dfrac{1}{\cos a} - 1 \right) \approx TA \times 3.54 \times 10^{-12} \approx 1.36 \times 10^{-6}$ km = 1.36 mm

In each second, the moon deviates from a straight line trajectory by 1.36 mm, while in the same period of 1 second a body at the Earth's surface will fall a distance of 4.9 m or 4900 mm. Newton noticed that 4900 / 1.36 is close to 3600, that is that the moon "falls" towards the Earth 3600 less quickly than would a body on Earth. Since 3600 is close to $383\,000^2 / 6384^2$, which is the ratio of the square of the distance of the moon to the square of the Earth's radius, the attraction of a body towards the centre of the Earth appears to be inversely proportional to the square of its distance. Also since all bodies at ground level, irrespective of their size, fall with the same velocity, their attraction must be proportional to their mass.

Newton finally generalised the case for two bodies, which appears in *Principia*, as:

> *That the forces by which the circumjovial planets are continually drawn off from rectilinear motions, and retained in their proper orbits, tend to Jupiter's centre; and are inversely as the squares of the distances of the places of those planets from the centre.*[1]

He also states, as his second Law of Motion:

> *The change of motion is proportional to the motive force impressed; and is made in the direction of the right line in which that force is impressed.*[2]

Today we know this as $F = m.\gamma$. The force of attraction is therefore $F = G.m.m'/d^2$, where G is a constant and m, m' are the masses of the two bodies whose centres of gravity are at a distance d apart. This result did not only challenge the laws of physics but also the very foundations of the subject itself. Up till then there had been the laws

1.   Newton, *Principia*, Book III, Prop. I, v. II, p. 406.
2.   Newton, *Principia*, Book I, Law II, v. I, p. 13.

for physics on Earth, where the distance travelled by a body in flight was proportional to the square of elapsed time, and laws for the heavens, where the planets travelled along elliptical orbits. Newton's bold claim was that all these cases were examples of the same universal gravitation. The moon then becomes simply a heavy body under free fall.

In order to justify his claim, he supposed that the attraction the sun exerted on the planets was a function of "his" universal gravitation, and he looked for the possible trajectories that would result from such a hypothesis. The calculations depended upon his previous work on fluxions, what we now call differential calculus, and resulted in showing that the orbits of the planets would have to be ellipses, with the Sun at a focus, just as Kepler had predicted.

Now everything fell into place: phenomena which had been observed and considered unconnected could now be explained by the theory of universal gravitation. This was the case with the flattening of the Earth at the poles, which was verified by the Peru and Lapland expeditions of 1737. The tides could now be explained in terms of the joint attraction of the Sun and moon on the oceanic masses, the precise nature of the tides being largely determined by the shape of the coast and the consequence of resonance. It was also possible to explain the precession of the equinoxes, the slow westward motion of the equinoctial points along the ecliptic, caused by the greater attraction of the Sun and the moon on the excess of matter at the equator, such that the times at which the Sun crosses the equator come at shorter intervals than they would otherwise do. Another consequence was Halley's prediction of the return of the comet that is now called after him. Halley, a friend and contemporary of Newton, noticed that the written accounts of the comets of 1531 and 1606 appeared to conform to the observations of the trajectory of the comet of 1682. It seemed to him that it must be the same object, following an elliptical orbit, with considerable eccentricity, and he predicted that it would return in 1758 ... which it did. It was not until after that date that universal gravitation started to become accepted in France. It must be said that translations of *Principia* had been delayed, because of the influence of Descartian ideas, and also because the idea of an instantaneous force acting at a distance presented problems. The first translation of *Principia* into French was by the Marquise du Châtelet with Clairaut reading the manuscript. We should add that it was the Frenchman Leverrier, in the nineteenth century, who used the law of universal gravitation to correctly predict the existence of Neptune.

In his work, Newton stated properties that were not to be challenged until Einstein in 1905. He wrote:

> *Absolute, true, and mathematical time, of itself, and from its own nature, flows equably without relation to anything external." and "Absolute space, in its own nature, without relation to anything external, remains always similar and immovable.*[1]

He also stated the foundations for reasoning in modern physics:

> *Therefore to the same natural effects we must, as far as possible, assign the same causes.*[2]

While Newton's work resolved many problems, it also raised many questions. In particular, the value of G, the universal constant of gravitation, remained an unknown quantity.

---

1. Newton, *Principia,* Book I, Definitions, Scholium I, II, v. I, p. 6.
2. Newton, *Principia,* Book III, Rule II, v. II, p. 398.

Finally, proving the hypotheses

Observations led to hypotheses to explain what was seen. As these were challenged, new hypotheses were accepted to describe and explain the natural world. However scientific proof is another matter, and if we have hardly cited here any evidence to support the theories, there is a good reason. The proofs did not arrive until long after the theories had been constructed!

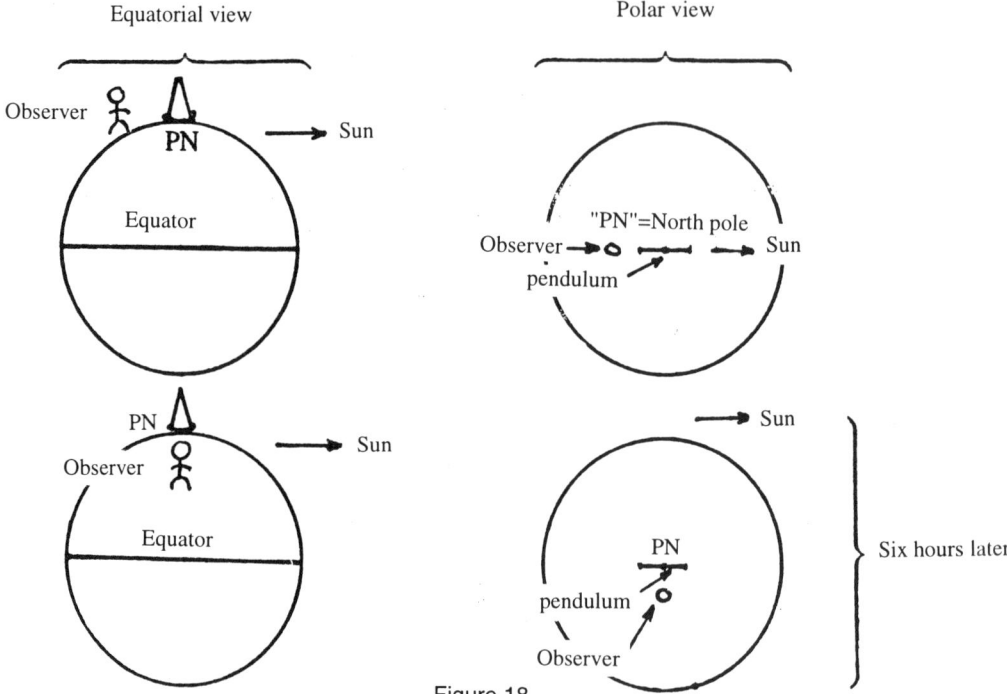

Figure 18

The rotation of the Earth

The rotation of the Earth, an idea that had been advanced by Copernicus, was not proved experimentally until 1851, by Foucault with his 70 m long oscillating pendulum hung in the Pantheon in Paris.

The idea for this experiment is quite simple. If a pendulum swings freely at the North Pole, it will always do so in the same direction; as an example we could let it swing in the plane containing the Sun. But the Earth turns upon itself, and this means that the line from the Earth to the Sun appears to move, from the point of view of an observer on Earth, and that is why we see the Sun rise in the morning and set in the evening. This will be the same for the plane in which the pendulum swings, which will make a complete rotation, with respect to the ground, in 24 hours. It is this apparent displacement that proves the rotation of the Earth (fig. 18). In fact, in Paris, the pendulum took 31 hours 50 minutes to make a complete rotation; the problem is a little more complicated because the pendulum was not hanging at the North Pole.

Movement of the Earth around the Sun

Another hypothesis of Copernicus, that the Earth travelled around the Sun, was not definitively proved until 1728 by Bradley, who discovered the phenomenon of aberration. This can be simply explained in terms of vector addition. When rain is falling on a car's windscreen it appears to do so obliquely, and the more so as speed is increased, so that the apparent movement of the drops of rain is affected by the movement of the observer.

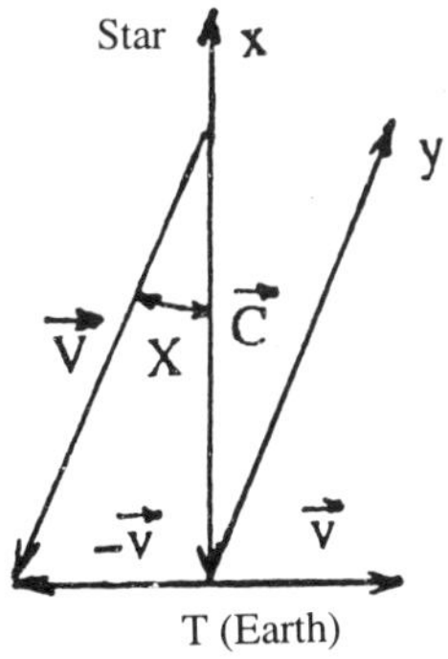

Figure 19

The same argument holds for light reaching the Earth from a star. If c is the speed of light coming from a star and v the velocity of the Earth along its orbit, the light from the star will appear to have a velocity $\vec{V} = \vec{c} - \vec{v}$ (Fig. 19) to an observer on Earth. The light will appear to be coming from the direction Ty instead of Tx, with an apparent change in the angle of direction equal to X where $\tan X = v/c$. The velocity of the Earth around the Sun changes direction, with a complete rotation of 360° in a year, and there will be a similar rotation of Ty about Tx in the same period. This means that a star will appear to describe a small circle, as seen from the Earth, if it lies in a direction that is perpendicular to the plane of the Earth's orbit about the Sun (the ecliptic). If the star is not in that direction, it will appear to describe an ellipse, that reduces to a line if it lies on the ecliptic itself. In all cases, the major axis of the ellipse has an angular length of 40.92 seconds of arc. How is this figure arrived at? The Earth travels in a circle, or nearly so, of radius 149 500 000 km.

This gives it an average speed of $\dfrac{149\,500\,000 \times 2\pi}{365.25 \times 24 \times 3600} = 29.76$ km/s. Hence, $\tan X = 29.76/c$ and so $X = 20.46''$ (taking c = 300 000). Therefore the major axis will be $40.92''$.

The evidence for universal attraction and calculating G

In 1798 Cavendish was able to demonstrate that two metal spheres of several kg mass attracted each other, and he was able to measure the very small attractive force that Newton had predicted. The value of G was found and is now known to be 6.67×10^{-11} in the MKS system. That means that two masses, each of 1 kg, at a distance apart of 1 m, will attract each other with a force of 6.67×10^{-11} newtons. It is therefore possible to "weigh" the Earth.

Exercise 8	Given the radius of Earth as r = 6370 km, G = 6.67×10^{-11} and g, the acceleration due to gravity at the Earth's surface, equal to 9.81 m/s^2, calculate the mass M of the Earth.
	Note: g is the value of the force of attraction of the mass of Earth on a unit mass of 1 kg at ground level. We then have g = G.M.1/r^2 since all the mass of the Earth can be considered as concentrated at its centre.

The Twentieth century and revolutionary new ideas

The word revolutionary is not too strong to describe the major impact of Einstein's thought upon our conception of the universe. It was not simply the movement of the stars, but the very geometry of space itself that was challenged. Euclid was no longer sufficient to explain all our observations. Worrying experiments showed that geometry alone was no longer sufficient. For example, it had been established that it was impossible to detect the movement of the Earth relative to the "ether", that all pervading substance which was thought to be a medium for the transmission of light and radiation. The ether was simply an invention of the human mind, a very useful one to provide a geometry of the world, but one that could not explain everything. Einstein replaced it with his theory of relativity, which did not require the presence of an ether, and which was rather more than a geometrical description of the universe.

Einstein and general relativity

It would not be possible here to give even a short résumé of the theory of relativity. We shall restrict ourselves to the applications of the results of that theory to the geometry of the universe. For Einstein, general relativity causes the universe to be curved, positively. We shall see how that is possible. According to his theory, the presence of an object causes rays of light passing by it to be curved, like the orbit of a planet. Yet, and unexpectedly, light continues to travel along the shortest path.

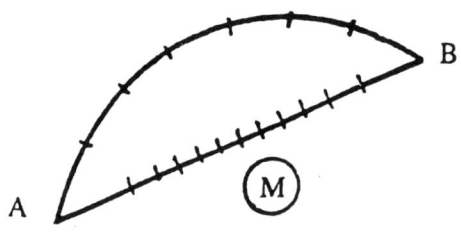

Figure 20

The theory of relativity tells us that objects appear shorter, when they are in movement, or when they are close to large masses. What is meant by the distance between two points A and B, as it is normally understood, is the number of units of length, metres say, that can be interposed between A and B. If a large mass M is near A and B, and if we place the metre lengths between A and B, these will shrink as they become closer to M, according to the theory of general relativity, so that more units will be needed (Fig. 20). On the other hand, if a path is chosen that goes round M, the units of length will not shrink so much, and so fewer units are needed to join A to B along the curved path. Obviously we would not want to choose a path that deviates so much from M that more units of length are needed to complete it. For each mass M, and for each pair of points A and B, there exists a path for which the number of units of length is a minimum. This is called, by definition, a geodetic line, or a geodesic. It is said that space curves positively in the presence of large masses. Suppose now that there are three points A, B and C and a large mass M. From what has been said before, there will be a geodesic from A to B (Fig. 21), the same from B to C, and another from C to A, and the "triangle" ABC has an angle sum greater than π radians: it is clear that its angle sum is strictly greater than that of the Euclidean triangle shown by the dotted line in the diagram.

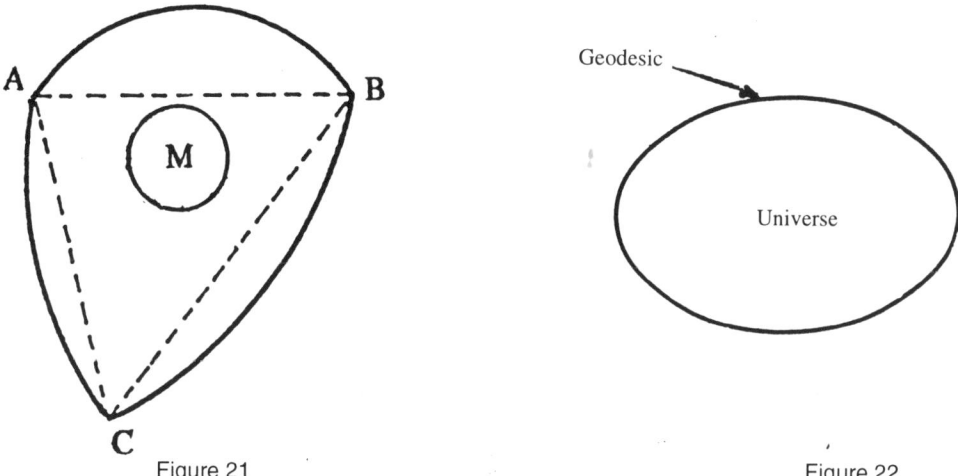

Figure 21   Figure 22

The whole universe can now be thought of as being a single large mass. A geodesic will encompass it in the same way that a great circle goes round a sphere (Fig. 22). Therefore the universe of general relativity is bounded but without limits, like a sphere. A ray of light emanating from one point, will return to its point of departure ... after many billions of years, in the same way as a terrestrial traveller journeying along a meridian. The universe of general relativity is no longer Newton's space of pure geometry, but a space composed of singularities called masses.

Hubble and the expanding Universe

The American astronomer Hubble, who had been a lawyer before moving to Cambridge to study physics and mathematics, made an amazing discovery in 1929. All the galaxies appeared to be moving away from us, and their speed of flight became greater as they moved further away from the Earth. Hubble made his discovery using spectography, which measures the properties of light. He found that light coming from distant galaxies had frequencies that were lower than those from the nearer galaxies, and so he concluded that they were moving away. The analogy is with the pitch of the whistle of a train, which has a lower tone when it travels away from us than it did on approach. From a measurement of the variation in frequencies, it became possible to estimate the speed of flight.

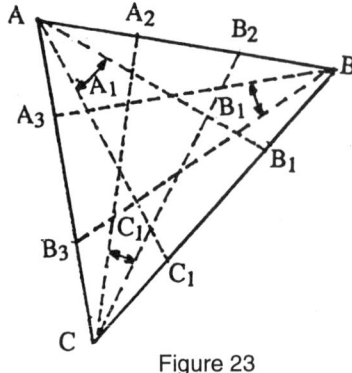

Figure 23

The universe can be thought of like a cake swelling up in a hot oven: two raisins in the mixture will move apart faster as they get further away from each other. The more that time passes, the larger will become the universe. Conversely, if we go back in time the universe will shrink and the galaxies will come closer together. It is possible to conceive of a certain moment in time when the diameter of the universe was zero: the moment of the original "big bang" which started the expansion of the universe, the consequences of which we witness today in the way in which the galaxies are flying apart. Hubble made another observation: the speed of light is finite, and although very large in terrestrial terms, it is relatively slow on a universal scale – the further away an object is, the further back in time is the image we see. What we see today as galactic masses at a distance of 500 000 000 light years is how they were 500 000 000 years ago, presumably nearer than they are now!

At equal distances, the angle of observation of a part of the universe is therefore smaller than the angle by which it would be seen today. Also, the further away an object is, the younger it is, because of the time taken for light from it to reach us. The further back in time the smaller is a galaxy, and we see it relatively smaller than it is today, and so the angle of observation is diminished. It is said that the universe is curved negatively. By this effect alone, the angle sum of a triangle will be less than π. In fig. 23, the observer at A sees the segment BC as it used to be; then it was only the segment $B_1 C_1$, and so now subtends the angle A_1 at A. Similarly for the observers B and C where the opposite segments subtend angles B_1 and C_1 respectively. It is clear that $A_1 + B_1 + C_1 < \pi$.

Geometries as models of the Universe

The effect of general relativity, as well as the Hubble effect, produced a change in the way in which the universe is thought about, and above all in our thinking about the geometry of the universe: Euclidean geometry is no longer suitable. In the nineteenth century Riemann constructed a mathematical model for a non-Euclidean geometry, and this provides a suitable model for the space of general relativity. A different model of non-Euclidean geometry was constructed by Lobachevsky, contemporaneously with Riemann, and this geometry fits Hubble's exploding universe., although invented long before.

The foundations of Euclidean Geometry

Like all geometries, and all mathematical theories, Euclidean geometry rests on a number of "postulates", or agreed assumptions, which are then used to prove other results. An example is that in Euclidean geometry it is accepted that one, and only one, straight line can be drawn to a given straight line, through a point not on that line. How are postulates of a theory to be chosen? Today we say that the postulates must not be self contradictory. Euclid chose his postulates in accordance with his observation of the physical world around him, what we today call Euclidean space. In Euclidean geometry, if two lines (xy) and (x'y') are parallel, then every line (zz') cutting these two lines will produce equal internal alternate angles, shown here as xAz and z'By', and yAC and BCA, in Fig. 24. This means that the sum of the internal angles of the triangle ABC, is equal to the value of the angle xAy, a straight angle (or π radians), and this is true for any plane triangle, and is the basis of Euclidean geometry.

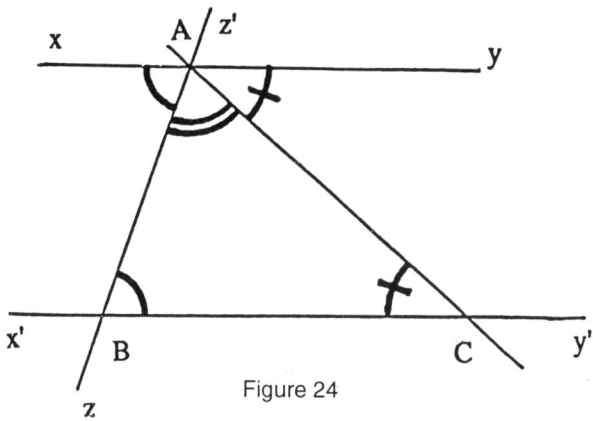

Figure 24

Another result in this geometry is that through any two points A and B, there is only one straight line, and that this line is the shortest distance between the points. Such a line is called a geodesic, and in Euclidean geometry, geodesics are straight lines.

Riemann geometry, or positive curvature

The postulates in this geometry are:

– Through any point not situated on a "line" it is not possible to draw any "line" parallel to that "line". The word "line" is in fact a geodesic, and the word "parallel" means that there is no point in common.

– The sum of the angles of a triangle is always strictly greater than a straight angle. Here again, we can think of a "triangle" as three points joined by geodesics.

There is a real and very simple example of Riemann geometry, and that is the surface of the sphere. If instead of being on a plane, we are on the surface of a sphere, then geodesics are great circles, whose radius is equal to the radius of the sphere itself. Suppose A and B are two points on the surface of the Earth and on the same latitude. The arc of the circle of latitude that joins them has a radius r, but the arc of the great circle passing through A and B has radius R, equal to the radius of the Earth (Fig. 25). The parallel of latitude is longer than the great circle route, since R > r, and the latter is a geodesic (Fig. 26). It can also be seen that through any point not on a given great circle, it is not possible to draw any great circle that does not intersect it.

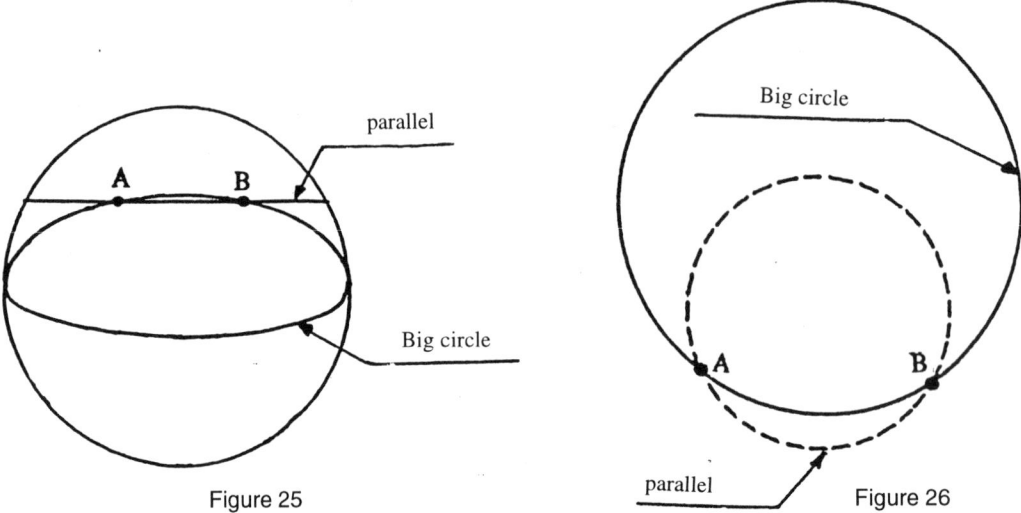

Figure 25   Figure 26

Let A be a point on the Earth on longitude 0° and B and C points on longitudes 90° and – 90° respectively on the equator (which is a geodesic). The great circle passing through A, whatever its latitude, will cut the equator at B and C (fig. 27). Every spherical triangle has an angle sum strictly greater than π radians. We take the angle between two geodesics to be the angle between their tangents at the point of contact. In the case of a meridian, it makes a right angle with the equator.

The spherical triangle PBC (fig. 28) has two right angles and we can see that its angle sum is certainly greater than π radians. Let $A + B + C = \pi + e$, where e is called the spherical excess. We can see that on any given sphere, the larger the area of a triangle, the greater will be the spherical excess, and that the two are proportional. Further, if the sphere has a radius of 1 unit, and the angle is measured in radians, then the area of the triangle is equal to the excess. More generally, for any sphere of radius R, we have $A + B + C = \pi + ((\text{area of } ABC)/R^2)$ for any spherical triangle ABC. The proof is as follows:

Let A', B', C' be points diametrically opposite A, B, C respectively

area ABC + area A'CB = $2AR^2$ (this is a lune with spherical angle A); similarly,

area ABC + area CAB' = $2BR^2$, and

area ABC + area C'AB = $2CR^2.$

Now, area C'AB = area CA'B' by symmetry about the centre, and area ABC + area A'CB + area CAB' + area CA'B' represents the visible part of the sphere, with area $2\pi R^2$. By adding the three equations above, and simplifying, we get:

2 area ABC + $2\pi R^2$ = $2R^2$ (A + B +C), which implies

$A + B + C = \pi + (\text{area } ABC)/R^2$.

The sphere is an example of a two-dimensional space with positive curvature. To construct it we need a fixed centre point in three dimensions, and we take all points of our space at a fixed distance R from that centre. To construct a three-dimensional space along similar lines, we need to have a fixed centre in four-dimensional space for a "hyper-sphere", to determine the properties of the three-dimensional space. We can imagine it, but it is impossible to visualise!

The geometry of space with positive curvature has the effect of dilating angles. Consider an object seen as AB on a sphere, as viewed from a pole P (fig. 30).

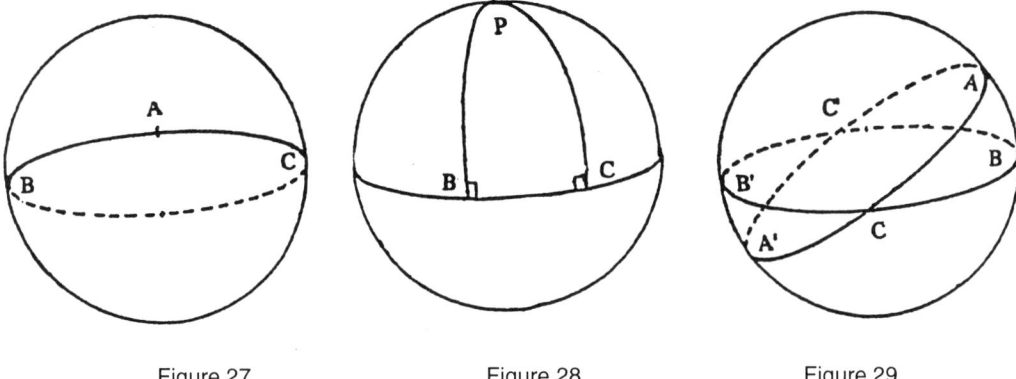

| Figure 27 | Figure 28 | Figure 29 |

It is seen with a width of angle a, and so is a nearer object A'B'. However, if straight lines PA, PB are drawn from P, the Euclidean angle APB will be b, which is less than a. In other words, angles in space with positive curvature are wider than they would be in Euclidean space. If light travels along the surface of the sphere, then points at the antipodes appear as close together as if they were close to us.

Lobachevsky Geometry or negative curvature

The postulates of this geometry are:

– Through any point not on a geodesic it is possible to draw an infinity of geodesics which are their "parallels".

– The sum of the angles of any triangle is always strictly less than π radians.

The words parallel, triangle and size of an angle have the same meaning here as in Riemann geometry. An example of such a geometry is the surface of a hyperboloid of revolution, seen in the massive cooling towers of power stations.

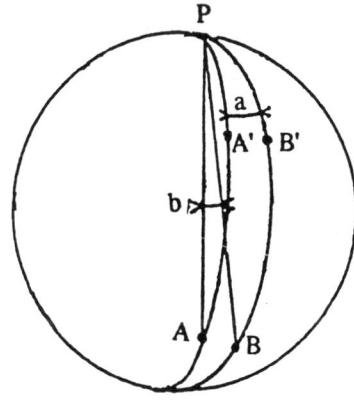

Figure 30

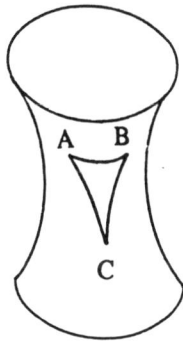

Figure 31

Here again, it is difficult to imagine a three-dimensional space having negative curvature. We notice that such a space is unbounded, which is not the case for the sphere or any other space having positive curvature. In similar situations of observation, a negative curvature diminishes the angle through which an object is viewed: in fig. 32, the object AB seen from the point O, subtends an angle a using the geodesics of Euclidean geometry; but it is seen through the smaller angle b using light rays passing along geodesics of the hyperboloid. A final remark should be made: the word "curvature" in geometry does not have the same meaning as in normal use. For example, if a triangle is drawn on a piece of paper, it will be flat. The paper can then be rolled up to form part of a cylinder and the triangle will then be on a curved surface, at least in the popular sense of the term. But the triangle will not be curved in geometry. The proof: the sum of the angles of the triangle remain equal to π radians (fig. 33). It is assumed that the straight lines of the triangle, which become helical, are geodesics of the cylinder and angles do not change when the paper is rolled up, unless the surface is stretched.

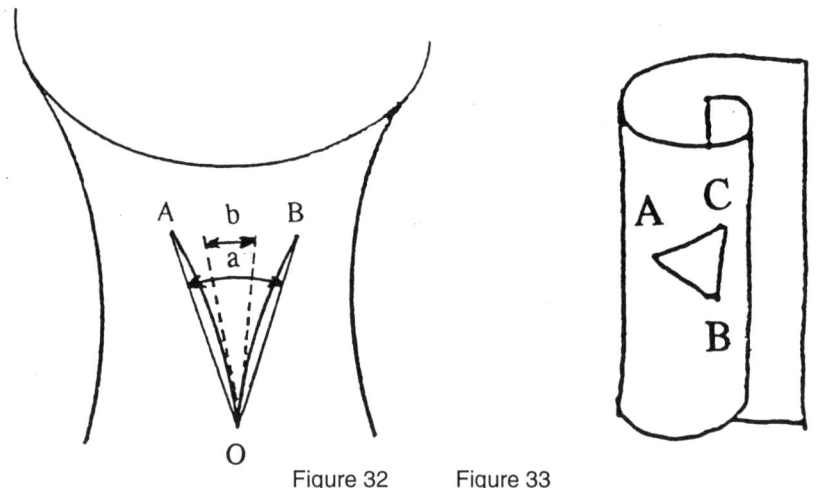

Figure 32   Figure 33

Conclusion

What is the true nature of the universe? Does it have positive curvature, negative curvature, or is it a Euclidean space?

The intensity of the effect of relativity depends upon the mean density of matter in the universe; the greater the mass, the greater the curvature. The Hubble effect, which leads to negative curvature, is greatest when velocities of expansion are very large. What effect does the one have on the other?

If the universe is curved positively, its density will be such that over many billions of years the gravitational effect will impinge on the galactic expansion, and the universe will collapse upon itself. Under Newton's law of attraction, all matter will come together and we shall have a "big-bang" in reverse, an implosion instead of an explosion. If on the other hand, the Hubble effect is the dominant effect over relativity, that is the mean density of the universe is less, then the universe will continue to expand indefinitely and become more spread out and colder.

In the last century, Lobachevsky searched in vain for examples of triangles formed by distant stars where the angle sum differed from π radians. At the present state of knowledge, it is not known which of the two effects impinges on the other, and it is not even known if the one has any effect on the other! If the two effects exactly cancel each other out, then we have a Euclidean space for the universe, and it looks as if this might be the case, in which case we return to the two centuries old geometrical description for the universe. Physical reality is simply in accord with the geometrical model that is chosen, and one model is only better than another model if it agrees with observations, those observations depending upon the degree of accuracy of their measurement.

It can be easily understood why the Euclidean model has prevailed during more than 2000 years. If non-Euclidean models are to be preferred, they have to depend for their acceptance upon extremely accurate measurements. We shall end by quoting Einstein on this subject:

> *How is it that mathematics, the product of human thought and independent of all experience, should conform in such an excellent way to observed reality? Is human reason itself, and without having recourse to experiment, able to discover the properties of real objects by its own activity?*
>
> *To this question, in my opinion, the answer must be the following: as far as the laws of mathematics refer to reality, they are not certain; and as far they are certain, they do not refer to reality.*
>
> *... But on the other hand it is certain that mathematics in general, and geometry in particular, owe their existence to our need to know about how real objects behave. The term geometry, which means land measure, already proves it. For measuring the land deals with propositions about the relative positions of certain objects in nature, that is parts of the land, rope, surveying poles, etc...*
>
> *... Solid bodies behave, as far as the possibilities of their relative positions, as bodies in three-dimensional Euclidean geometry; the propositions of which therefore contain statements about the behaviour of bodies which can be considered rigid.*
>
> *... Geometry is therefore completely and manifestly a science derived from experience; we can even consider it to be a branch of the much older science of physics. Its statements rest essentially on induction from experience, and not solely upon logical deduction. We shall call such complete geometry, practical geometry, and in what follows we shall distinguish it from pure axiomatic geometry. The question of knowing whether the real geometry of the world is Euclidean or not has a precise meaning, and if the answer can only be found through experiment. All measurement of length in physics is practical geometry in this sense; and it is the same for the measurement of geodesic length in astronomy, if one adds the experimental proposition that light travels in a straight line – in a straight line in the sense of practical geometry[1].*

---

1.   Einstein, *L'ether et la théorie de la relativité...*, p. 14-16

Bibliography

Sources texts

ACKER, *Initiation à l'astronomie,* 3rd ed., Masson, Paris, 1982.

DUMOULIN, *L'Astronomie en Terminale*, IREM de Limoges, 1986.

EINSTEIN, *L'éther et la relativité. La géométrie et l'expérience*, tr. Solovine, 1921. Repr. Gauthier-Villars, Paris, 1921.

KEPLER, *Le secret du Monde*, tr. Segonds, Les Belles Lettres, Paris, 1984.

KUHN, *La Révolution Copernicienne*, to celebrate the 500th anniversary of the birth of Copernicus, Fayard, Paris, 1973.

LEBLANC, Lomont, *L'Astronomie au quotidien*, Atlas, Paris, 1987.

NEWTON, *Principia*, Motte-Cajori; Motte's translation, 1729; revised by Cajori, University of California Press, Berkeley, California, 1934, 1962.

WALUSINSKI, *Ciel, passé, présent*, Etudes Vivantes, Paris, Montreal, 1981.

General works for further reading

COUDERC, *Histoire de l'Astronomie classique,* 1st edition 1945, 7th edition 1982, PUF, Paris.

FLAMMARION, *Astronomie populaire*, Flammarion, Paris 1955; republished 1975.

GALILÉE, *Le messager des étoiles*, tr. Hallyn, Seuil, Paris, 1992.

GALILÉE, *Dialogue sur les deux grands systèmes du Monde*, tr. Fréheux, Seuil, Paris, 1992.

GEYMONAT, *Galilée*, tr. Rosset and Martin, Seuil, Paris, 1992.

HAYLI, *Histoire de l'Univers*, Hachette, Paris, 1980.

REEVES, *Patience dans l'azur*, Seuil, Paris, 1981.

SHEA, *La révolution galiléenne*, tr. de Gandt, Seuil, Paris, 1992.

SIMON, *Kepler, astronome astrologue*, Galimard, Paris, 1979.

VERDET, *Une histoire de l'astronomie*, Seuil, Paris, 1990.

How did you get on?

Ex. 1   11.287 km. For 100m above the sea: 46.98 km. Island: 185.48 km.

Ex. 2   45 000 km.

Ex. 3   13.345 km.

Ex. 4   Compare with the true value...

Ex. 5   0.38 of an astronomical unit.

Ex. 6   5.2.

Ex. 7   141 687 000 km. Compare this with the true value...

Ex. 8 5.96×10^{24} kg.

11

Proving the Fifth Postulate: True or False?

Jean-Luc CHABERT
INSSET St-Quentin

*... if God really exists, and if he really has created the
world, then he has made it according to Euclid...*

Dostoevsky, *The Brothers Karamazov*

Given a point not on a given straight line, there is one and only one line through the
point and parallel to that line; this is the form by which the Fifth Postulate is usually
known. This statement has always been considered to be true since the time of Euclid,
not least because it states a fact about our own space. Our own experience verifies it to
be a natural law of space whenever we draw lines. So much so that the problem of the
Fifth Postulate is not about its truth but concerns its proof.

Although Euclid puts it a little differently, what he states is essentially the same:

Let the following be postulated: [...]
*5. That, if a straight line falling on two straight lines make the interior angles on the
same side less than two right angles, the two straight lines, if produced indefinitely, meet
on that side on which are the angles less than the two right angles.*[1]

This seems to be a theorem of geometry capable of proof, and not to prove it would
offend against the canons of mathematical practice. It is little wonder then, that a search
for a proof of the Fifth Postulate has occupied the energies of so many mathematicians
over the centuries. But these efforts have been constantly checked: there always seems
to be some little detail missing, a detail that is as important as the fifth postulate itself.
D'Alembert's view was that the inability to prove the Fifth Postulate constituted *"the
real scandal of geometry."*

Our purpose here will be to acquaint the reader with the elements of this scandal.
We shall start by showing the position that the postulate occupied in Euclid's work and
ultimately its place in Western Geometry. Then we shall examine attempts that have
been made to remove the problem, either simply by attempted proofs, or more modestly

---

1.   Heath, *The Thirteen Books of Euclid's Elements*, vol. 1, p. 155.

by replacing the postulate with a simpler axiom, or again by modifying the definition of parallel lines. We shall also see how every attempt at chasing away the problem results in it returning again at full gallop. And the problem became resolved, neither by changing the definition of the Fifth Postulate, nor by proving it, but in a third way. The non-existence of a proof was established and also, paradoxically, the impossibility of not having the Fifth Postulate.

Euclid's Postulates

We know very little about Euclid the man, not even whether he really existed. However his *Elements,* consisting of thirteen books, occupied a major place, not only among his own works, but also in mathematics at all times. Most likely written around 300 B.C., the text is the oldest we know in which a body of geometrical knowledge has been set out in such an axiomatic-deductive manner.

The First Book of *The Elements* starts with a list of definitions of which the last is:

> 23. *Parallel straight lines are straight lines which being in the same plane and being produced indefinitely in both direction, do not meet one another in either direction.*[1]

This definition says nothing about the existence of parallel lines since it defines a negative attribute. It is possible to test whether two lines are concurrent but it does not say that there may be lines which are not.

The definitions are followed by "demands" or "postulates":

> *Let the following be postulated:*
> 1. *To draw a straight line from any point to any point.*
> 2. *To produce a finite straight line continuously in a straight line.*
> 3. *To describe a circle with any centre and diameter.*
> 4. *That all right angles are equal to one another.*
> 5. *That, if a straight line falling on two straight lines make the interior angles on the same side less than two right angles, the two straight lines, if produced indefinitely, meet on that side on which are the angles less than the two right angles.*
> 6. *That two straight lines do not enclose a space.*[2]

The first three postulates have an instrumental character. They express the theoretical possibility of carrying out certain elementary constructions that are usually referred to as "ruler and compass" constructions: to draw a line, to extend it and to draw a circle of given centre and radius, and in this respect they are of a fundamental character. But what is the next postulate for? Is there really anything here to postulate, or indeed to prove? In fact, as Hadamard says:

> *To prove that right angles are equal to one another, or rather as a preliminary to the proof, one would be obliged to forget that such an equality must necessarily exist.*

As for the last postulate, this is often given in the form: 'through any two points there is at most one straight line' whereas, considering its construction, the first means: 'through any two points there is at least one straight line.' But the Fifth Postulate, or the Parallel Postulate, or Euclid's Postulate, or even sometimes the 11th Postulate,

---

1.   *Ibid.*, p. 154.
2.   *Ibid.*, p. 154-155 [Postulate 6 is not included in the Heath edition but is in Peyrard's French translation.]

depending on the editions[1], occupies a place apart, although this may not be immediately evident from its statement. Before examining various attempts at proof, we shall first of all look at the role that this postulate played in Euclid's own work.

The place of the Fifth Postulate in Euclid

We have already raised doubts as to the existence of Euclid and we might also doubt the existence of the Fifth Postulate itself. In fact we do not possess any original text of The Elements, having to rely instead on Greek, Latin or Arabic manuscripts which came to light in Europe from the 12th century. These present a number of variants, so that we cannot be certain what belongs to the original Euclid, and what belongs to his various commentators, translators or copyists, and the many editions reflect this lack of certainty. The quotations given here will be from Peyrard[2, 3] but there are variations. For example, it is often recognised that the 6th postulate was not among the original postulates but was, unnecessarily, interpolated to improve on the proof of I 4 and does not figure in Heiberg's edition.[4] Also Tannery argues that only the first three postulates about constructibility belong to Euclid.

Whether or not Euclid had placed it at the head of his work, the Fifth Postulate is present throughout the 48 propositions of Book I. It is there implicitly, constantly in Euclid's thoughts, as he avoids its use as long as possible in the chain of deductions, which he is able to do right up as far as propositions 27 and 28.

> *Proposition 27: If a straight line falling on two straight lines make the alternate angles equal to one another, the straight lines will be parallel to one another.*

Remember the possible cases of alternate angles:

internal alternate angles   external alternate angles

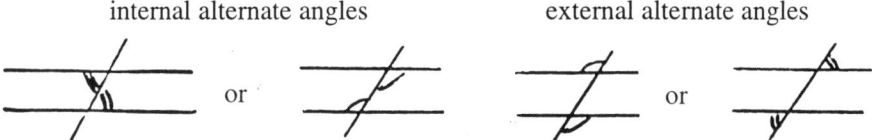

> *Proposition 28: If a straight line falling on two straight lines make the exterior angle equal to the interior and opposite angle on the same side, or the interior angles on the same side equal to two right angles, the straight lines will be parallel to one another.*

Here are the various possibilities:

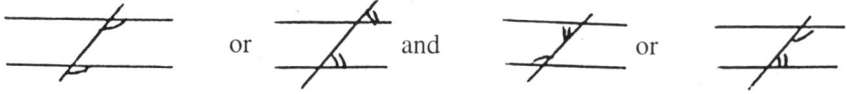

---

1.   It appears as Axiom 12 in Todhunter's edition of 1862, still in use in English schools well into this century [tr.].
2.   Peyrard, F. *Les Œuvres d'Euclide*, Paris,1819. Republished, Blanchard, Paris, 1966.
3.   Quotations in this English translation will be from the Heath 1925 edition which is a translation of Heiberg's text with later corrections. [tr.]
4.   English readers are referred to Heath, vol. 1, p. 249.

The proof of these two propositions, which at once establishes the existence of parallel lines, does not require the Postulate but rests instead on Proposition 16:

> *In any triangle, if one of the sides be produced, the exterior angle is greater than either of the interior and opposite angles.*

But from here on, with hardly an exception, the next twenty propositions require the use of the Postulate or of another which derives from it. For example, the converse of 27 and 28:

> *Proposition 29:* **A straight line falling on parallel straight lines makes the alternate angles equal to one another, the exterior angle equal to the interior and opposite angle, and the interior angles on the same side equal to two right angles.**

follows from the Postulate using reductio ad absurdum. This is also the case in proposition 30:

> **Straight lines parallel to the same straight line are also parallel to one another.**

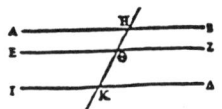 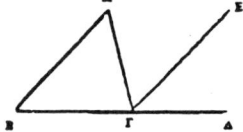 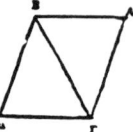

and in propositions 32 and 33:

> *32:* **In any triangle, if one of the sides be produced, the exterior angle is equal to the two interior and opposite angles, and the three interior angles of the triangle are equal to two right angles.**

> *33:* **The straight line joining equal and parallel straight lines (at the extremities which are) in the same directions (respectively) are themselves also equal and parallel.**

This last expresses the fact that **parallel lines are equidistant**. Note also that proposition 30 shows that **through a given point there passes at most one line parallel to a given line**, and this became the formulation of Playfair's axiom in the 18th century.

Therefore, taking for granted the existence of parallel lines, this allows us to say that **through a given point one and only one straight line can be drawn parallel to a given straight line**. And it is this last formulation that is often used for the Fifth Postulate. And indeed it is equivalent, in the sense that it can be used instead of the Postulate in proofs, accepting the other postulates and of course those other postulates that are implicit in the axiomatic system, as for example the fact that a circle passing through the centre of another circle intersects it in at least one point. (Prop. 1) The statements of propositions 30, 32 and 33 are also equivalent to the Fifth Postulate. But this sort of question, namely the equivalence of axiomatics, is rather recent and had no place at that time, Geometry being then principally a science of objects in space, and the only requirement for a postulate was that it was suitable for that Geometry.

Having explained what was the Fifth Postulate and its origin, we shall now pass on to look at a number of attempts that have been made to prove it. These efforts were too numerous for us to be able to review all of them. We shall not explore the point of view of Greek commentators like Posidonius and Geminus of the first century B.C., nor that of Ptolomy of the second century A.D., nor even the proof of which 6th century Proclus was so proud. But we shall want to retain the idea of parallels that Aganis developed.

The question of a definition for parallel lines

For Aganis (6th century), two parallel lines were coplanar and equidistant: that is two lines in the same plane for which the shortest distance joining a point of one line to the other line was the same irrespective of the choice of point. It is clear that two such lines can never intersect and it is certainly the sense in which Euclid uses the idea of parallel lines, that is coplanar and non intersecting. On the other hand, if Euclid has proved that two parallel lines are equidistant in proposition 33, it is by use of the Fifth Postulate. The concept of parallels in Aganis' sense is therefore a stronger property which is, at least in appearance, better adapted to the theory because it avoids the need for recourse to the Postulate.

But the question of the existence of parallels remains. In fact if Euclid has proved the existence of parallels in proposition 31 without the use of the Postulate, Aganis could only proceed by making an appeal to the principle of existence, that is to a postulate such as: **the line of points equidistant from a straight line and situated on the same side of that line is a straight line**. And this is a statement that is itself equivalent to the postulate. And Aganis let himself fall into the trap in declaring that: **through a point not on a given straight line there can always be drawn a straight line parallel to the first**.

[The attentive reader will have become aware that all statements equivalent to the Fifth Postulate as we meet them are printed in bold.]

Al-Ḥayyām's proof
thanks to a philosophical principle

The first Arabic translations of *The Elements* date from the 9th century. Among those whose commentaries have contributed to the theory of parallels were al-Gauharī (9th century), Tābit ibn Qurra (10th century), Ibn al-Hay tam (10th – 11th century), al-Ḥayyām (11th – 12th century) and Nasīr ad-Dīn at-Tūsī (13th century). For his part, al-Ḥayyām noticed that the first 28 of Euclid's propositions were independent of the Postulate; he therefore proposed to prove it by inserting after these a further eight new propositions. The proof according to al-Ḥayyām which is given below is taken from the *Opuscule sur l'explication des postulats problématiques du Livre d'Euclide*, translated into French by K. Jaouiche.[1] Its premises are remarkably close to the fundamental statements formulated by Saccheri in the 18th century. After the first two of al-Hayyam's new propositions, we come to the third which says, essentially: **if a symmetrical quadrilateral possesses two right angles, then the other two are also right angles**.

It will be seen in the course of the proof that there are three possibilities to consider: the two other angles are right angles, they are less than a right angle, or they are greater than a right angle. This examination of different possible cases was taken up later by Saccheri and Lambert. Incidentally, prior to the proof proper, the author raises a certain number of philosophical considerations and it is by use of an observation of this type, presented as "*fundamental impossibility*" or "*philosophical principle*" that he completes

---

1.   Jaouiche, *La théorie des parallèles en pays d'Islam*, 1986, p. 185-199.

his proof: **two straight lines, cutting another straight line making two right angles, can not converge (or diverge) on both sides simultaneously**. Put otherwise, situations of this type are not judged to be acceptable:

> *"And we have to admit twenty eight propositions of the Book of The Elements because they do not need this premise. But the one that needs it is the twenty-ninth, where we shall identify the principles of parallel lines. It is allowable to him that wishes it, to consider the first proposition of this chapter as the twenty-ninth of Book I in order that it may be integrated within the body of the Book of The Elements, so please it God. It is time to begin the true and causal proof of this proposition with the help of God: may He grant us success since He guides and satisfies him who leans on Him.*
> ***Proposition 1.*** *Let AB be a straight line and draw AG perpendicular to AB and BD perpendicular to AB and equal to the straight line AG. They are parallel as Euclid shows in proposition 28. Draw GD.*
>
> *I say that the angle AGD is equal to the angle BDG …*

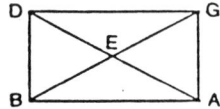   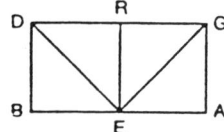

> ***Proposition 2.*** *Redraw the figure ABDG, divide AB in two halves at E and draw ER perpendicular to AB.*
> *I say that GR is as RD and ER is perpendicular to DG …*
> ***Proposition 3.*** *Redraw the figure ABGD. I say that the angles AGD, BDG are right angles."*

Exercise / Proof	Find the assertions which are not validly founded in the proof given below of al-Hayyām's Proposition 3. [The reader may wish to leave this at a first reading.]

Al-Hayyām begins with an extended construction:

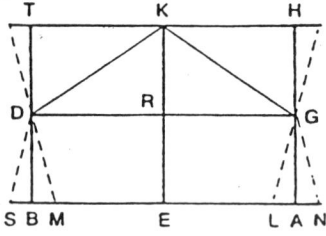

> *Let us divide AB in two halves at E, let a perpendicular ER stand up and produce it further in a straight line. Make RK equal to RE and construct HKT perpendicular to EK. Produce AG and BD which will cut HKT at H and T because AG and EK are parallel. Now the distance between any two parallels never varies. Therefore AG will extend to infinity parallel to EK, and HK will extend to infinity parallel to RG. They will necessarily meet.*
> *Join GK and DK. The straight line GR is as RD and RK is common and perpendicular. The two bases GK and KD are therefore equal and the angles RGK and KDR are equal. The remaining angle HGK is as KDT. And the two angles GKR and DKR are equal. The*

two remaining angles GKH and DKT are therefore equal. And the straight line GK is as
DK. Therefore GH is as DT and HK is as KT.
If the two angles AGD and BDG are right angles then the proposition is true.

Al-Ḥayyām is therefore thinking along the lines of a proof by contradiction. He starts by dealing with the case where they could be acute angles:

If they are not right angles then each of the angles must be either less than a right angle
or greater than a right angle. Suppose first of all that each is less than a right angle. The
plane HD can be superposed on the plane GB. RK can therefore be superposed on RE
and HT on AB. Hence the line HT is as the line NS, for the angle HGR is greater than the
angle AGR. The line HT is therefore greater than AB. In the same way if we produce the
two lines to infinity in this manner, each of the lines joining them will be greater than
the other and so on. The two lines AG and BD will therefore move apart from each other.
In the same way, if we produce AG and BD in a straight line on the other side, they will
necessarily spread out from each other, which can be shown by a similar proof and
seeing the similarity of the cases of both sides which follows from superposition. So we
have two straight lines cutting a line making two right angles, and whose distance apart
increases on both sides of that line, which is impossible first when one considers the
rectilineal character of a straight line and when one realises what is meant by the
distance between two lines. And it is this that the philosopher has concerned himself with.

Al-Ḥayyām deals in a similar way with the obtuse angle case and is then able to conclude his proof.

Comment. Besides the use of "fundamental impossibility" explicitly stated, we also see that the proof is dependent on the intersection points H and K, clearly visible in the figure, but whose existence has not been established, other than by appeal to the idea that the distance between parallels never varies. It would, in fact, be sufficient to have **the distance between two parallel lines is bounded**, an assertion that is equivalent to the Postulate.

Al-Ḥayyām went on to introduce an idea of parallels in a further sense: two lines perpendicular to a third line. He showed that they were parallel and had no further difficulty in deducing the Postulate. More than the accuracy of its argument, the interest in al-Ḥayyām's text lies in its approach to the question. It throws up the connection between the Postulate and the sum of the angles of a quadrilateral and a proof based upon an earlier contradiction which results in a hypothesis according to which this sum would be greater or less than four right angles.

At-Tūsī's proof
(the Archimedes-Eudoxus axiom and Pasch's axiom)

In the 13th century Nasīr ad-Dīn at-Tūsī proposed a proof, the essence of which was contained in Proposition 6, that **a perpendicular and an oblique line to a given line will intersect:** it is in fact a particular case of the Postulate which can be deduced with little difficulty. The proof is interesting because of its use of two axioms, the one explicitly stated being substantially the Archimedes-Eudoxus axiom: two unequal lengths being given, one can always find a multiple of the lesser which surpasses the greater. The other, now known as Pasch's axiom, is a clearly stated property though not explicitly called on: any straight line entering a triangle must leave it.

The proposition and proof are taken from the *Opuscule qui délivre des doutes concernant les droites parallèles*.[1]

> **Proposition 6.** *If two straight lines of unlimited length cut each other than at right angles and if a perpendicular is raised on one of them, then that perpendicular, if it is extended, will cut the other line on one of its sides, namely on the side of the acute angle contained between that perpendicular and the straight line cut by that perpendicular.*

Exercise / Proof	The proof given by at-Ṭūsī can be taken as an exercise but may be omitted at a first reading.
	We note the use of earlier propositions: proposition 5, that two lines perpendicular to a third, that is a particular pair of parallel lines, make equal internal alternate angles with any other line, and proposition 4, that the opposite angles of a rectangle are equal. Their proof depends upon proposition 3, in fact the same proposition 3 as al-Ḥayyām's which at-Ṭūsī holds, follows from the same philosophical principle.

Let AB and GD be two straight lines which intersect at a point E not at right angles. And let HR be the perpendicular raised on the straight line GD. I say that if I extend this perpendicular it will cut the line AB on one of its sides.
The proof is:
Let us suppose that of the two unequal angles AEG and GEB, which taken together are equal to two right angles by virtue of proposition 13 of The Elements, *the angle AEG is acute.*

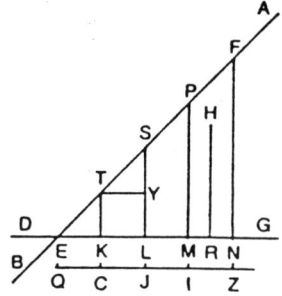

Let T be any point whatever on the line AE. Lower the perpendicular TK onto the line GD as was shown by proposition 12 of The Elements. *The point K must necessarily be found, either between E and R, or at the point R, or beyond that point on the same side as G.*
If the point K is found between E and R, let us take a length equal to EK and let QC be that length. Extend the line on the side C. Take successive lengths along the line equal to QC, as was shown in proposition 3 of The Elements, *until the collection of multiples of QC surpasses the length ER and let the length obtained be QZ. Suppose that these multiples are the parts QC, CJ, JI, IZ, each of these being equal to the length EK.*
Take now on the line AE successive lengths equal to the length TE, the number of them being equal to the number of those parts; these are ET, TS, SP, PF. Now from the points S, P, F lower perpendiculars SL, PM, FN to the line GD, as was shown by proposition 12 of The Elements.
Draw from T the perpendicular TY to the line SL. In the triangles ETK, TYS the angles ETK, ESY, the one internal, the other external, are equal by virtue of the preceding proposition, the two perpendiculars TK, SL raised on the line LK being cut by SE. Furthermore, the angles EKT, TYS are right angles and the sides ET and TS are equal as was shown in proposition 26 of The Elements *and the side YT is equal to the side EK. But the quadrilateral YTLK is a rectangle since the angles L. K. Y are right angles by hypothesis and the angle T is equally a right angle as is shown in the preceding*

1.   *Ibid.*, p. 206-226.

proposition. The opposite sides YT, LK are therefore equal as was shown in the fourth of these propositions. The lines EK, KL are therefore equal.

From the foregoing equality and other analogous ones, at-Tūsī deduces:

The point N is found necessarily outside the interval contained between E and R on the side of G, and the perpendicular HR is therefore found in the interior of the triangle FNE. Therefore, if we extend the perpendicular RH, which is parallel to the perpendicular FN, until it comes out of the triangle FNE, it must necessarily cut the side AB.

If the point K does not lie between E and R, then, as at-Tūsī says, the result is even more evident. He can therefore conclude:

One can therefore see that the intersection of the perpendicular and the oblique line takes place on the side of the acute angle, that is the angle AER. As for the proposition used in this theorem by which it is possible to find multiples of the lengths of the shorter of two lines laid end to end which will surpass that of the greater, this has been used here and we have said that it is self evident. The author of The Elements *used it in the first proposition of Book X in such a way that it could be used for all types of magnitudes, without it having been the object of a postulate in any part of his work.*

Wallis's Proof
Movement and similarity

At the Renaissance the first European commentators on geometry, like le Commandin, Clavius, Cataldi, Borelli and Vitale, brought nothing really new to Greek and Arabic results. Only Wallis produced a really original contribution. He started with a new postulate, which he considered to be more natural, that

given a figure, another figure is possible which is similar to the given one and of any size whatever.

The proof by Wallis given here is taken from Heath.[1]

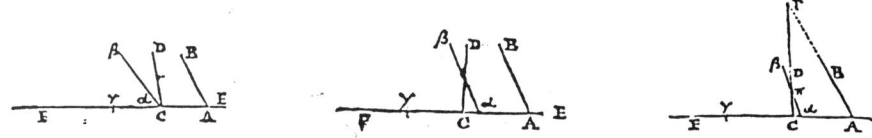

He first proved (1) that, if a finite straight line is placed on an infinite straight line, and is then moved in its own direction as far as we please, it will always lie on the same infinite straight line, (2) that, if an angle be moved so that one leg always slides along an infinite straight line, the angle will remain the same, or equal, (3) that, if two straight lines, cut by a third, make the interior angles on the same side less than two right angles, each of the exterior angles is greater than the opposite interior angle.
(4) If AB, CD make, with AC, the interior angles less than two right angles, suppose AC (with AB rigidly attached to it) to move along AF to the position αγ, such that α coincides with C. If AB then takes the position αβ, αβ lies entirely outside CD (proved by means of (3) above).
(5) With the same hypotheses, the straight line αβ or AB, during its motion, and before it reaches C, must cut the straight line CD
(6) Here is enunciated the postulate stated above.

1.   Heath, *Thirteen Books of Euclid's Elements*, vol. I, p. 210-211.

(7) Postulate 5 is now proved thus.

Let AB, CD be the straight lines which make, with the infinite straight line ACF meeting them, the interior angles BAC, DCA together less than two right angles.

Suppose AC (with AB rigidly attached to it) to move along ACF until AB takes the position of αβ cutting CD in π.

Then αCπ being a triangle, we can, by the above postulate, suppose a triangle drawn on the base CA similar to the triangle αCπ. Let it be ACP.

[Wallis here interposes a defence of the hypothetical construction.]

Thus CP and AP meet at P; and, as by the definition of similar figures the angles of the triangles PCA, πCα are respectively equal, the angle PCA being equal to the angle πCα and the angle PAC to the angle παC or BAC, it follows that CP, AP lie on CD, AB produced respectively.

Hence AB, CD meet on the side on which are the angles less than two right angles.

In this proof two points merit consideration: first the use of similarity which appears original and was Wallis's starting point and second, the use of an argument based on geometrical movement, which was to be taken up and used in the 17th century (see chapter 5, 6).

Lazare Carnot appears to have been convinced by this principle of similarity since he wrote in his *Géométrie de position* in 1803:

> *The theory of parallels relates to a fundamental concept, which gradually appears to me to be of the same order of clarity as that of perfect equality or that of superposition, namely similarity. It seems to be that one can regard it as a principle of fundamental evidence that, what exists full size, like a ball, a house, a drawing, can be made small and conversely.*

In a better argued manner, starting from the laws of physics, Pierre-Simon de Laplace also arrived at the principle of similarity in his *Exposition du Système du Monde* (1796). In Book V, Chap. V (On the discovery of universal weight), he writes

> *proportionality is a postulatum even more natural than Euclid.*

for it is found in the laws of gravitational attraction as well as in those due to electric and magnetic attraction; furthermore

> *the simplicity of the laws of nature allow us only to observe and understand ratios.*

In fact, as Saccheri observed, it was sufficient in Wallis's proof to suppose that:
Being given a triangle ABC and a line DE, there exists a triangle DEF such that ABC and DEF are equiangular one with the other.

As for the idea of movement in geometry, it can be said that as a general rule it does not appear explicitly in Euclid. Where it is implied it is only in order to show the coincidence of figures in order to prove their equality. It is therefore a very limited idea of movement in which the intermediary stages are not in consideration, as they are with Wallis, but just the initial and final positions.

In this regard, we may recall Ibn al-Haytam who, in his *Livre expliquant les postulats d'Euclide dans les Eléments,*[1] "proves" the existence of equidistant lines – that is parallel lines in the strong sense – by making a line AB, perpendicular to a straight line D, move so that the point A describes D and the point B describes D', equidistant from D.

---

1.   Jaouiche, *La théorie des parallèles en pays d'Islam*, p. 161-175.

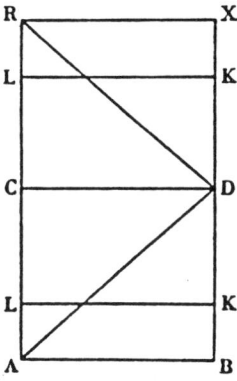

But Saccheri showed how it is possible to avoid movement simply by supposing that **there exist two unequal triangles with equal angles** (and just two such are sufficient).

Saccheri's proof
The hypotheses of the right angle, the acute angle and the obtuse angle

Saccheri's work, Euclides ab omni naevo vindicatus, (*Euclid cleansed of every flaw*), appeared in 1733. Just like al-Hayyām, Saccheri took as known and established the first 28 propositions and then introduced the quadrilateral ABDC in which the angles A and B are right angles and the sides AC and BD are equal. This quadrilateral formed the base of all his work. After showing that CD = AB, CD < AB or CD > AB according as the equal angles C and D are right angles, obtuse or acute, he considers each of these three hypotheses.

Definitions. Since the straight line joining the ends of the two equal perpendiculars to the same straight line (which we shall call the base) make equal angles with these perpendiculars, therefore there are three hypotheses to be distinguished according to the species of these angles. And I shall call the first of these the hypothesis of the right angle, the second the hypothesis of the obtuse angle and the third the hypothesis of the acute angle, respectively.

Proposition V. If the hypothesis of the right angle is proved true in a single case, it is true in every other case.

Exercice

Proof Let the join CD make right angles with any two perpendiculars AC, BD standing upon any straight line AB. Then CD will be equal to this AB (Prop. III). Assume in AC, and BD produced two sects CR, DX equal to these AC, BD; and join RX. We may easily show that the join RX will be equal to this AB, and that the angles at it are right angles. And the first indeed by superposition of the quadrilateral ABDC upon the quadrilateral CDXR, applied to the common base CD.

Also we may proceed more elegantly thus. Join AD, RD. It follows (Eu I. 4) in the triangles ACD, RCD, the bases AD, RD will be equal and likewise the angles CDA, CDR, and certainly ADB, RDX because equal remainders from a right angle. Whereby in turn (Eu

I. 4) in the triangles ADB, RDX, the base AB will be equal to the base RX. Therefore (Prop. IV) the angles at the join RX will be right, and so we abide in the same hypothesis of the right angle.

Since now the length of the perpendiculars can be similarly increased infinitely, under the same base AB, the hypothesis of the right angle always subsisting, it only remains to be proved that the same hypothesis will always abide in any case of diminution of those perpendiculars; which indeed is thud evinced.

Assume in AR, and BX any two equal perpendiculars AL, BK, and join LK. If the angles at the join LK are not right, nevertheless (Prop. I) they will be equal to each other. Therefore they will be toward one part, as suppose toward AB obtuse, and toward RX acute, since certainly the angles here at each of those points are (Eu I. 13) equal to two rights.

But it also holds that the perpendiculars LR, KX, those standing upon RX, will be mutually equal. Therefore (Prop. III) LK will be greater indeed than the opposite RX, and less than the opposite AB.

But this is absurd; because AB, and RX have been shown equal. Therefore the hypothesis of the right angle is not changed by any diminution of the perpendiculars, whilst abides the one posited base AB.

But neither is the hypothesis of right angle changed for any diminution, or greater amplitude of the base; since manifestly may be considered as base any perpendicular BK, or BX, and therefore may be considered in turn as perpendiculars that AB, and the equal opposite sect KL, or XR.

Therefore is established that if even in a single case the hypothesis of right angle be true, always in every case it alone is true. Quod erat demonstrandum.

Saccheri then deals with the other two hypotheses:

Proposition VI. *If even in a single case the hypothesis of obtuse angle is true, always in every case it alone is true.*

Proposition VII. *If even in a single case the hypothesis of acute angle is true, always in every case it alone is true.*

Saccheri also shows that in the case of the hypothesis of the right angle (obtuse/acute), **the sum of the angles of any quadrilateral is equal to** (greater than/less than) **four right angles; the sum of the angles of any triangle is equal to** (greater than/less than) **two right angles; an angle inscribed in a semi-circle is a right angle** (obtuse/acute).

The hypothesis of the right angle being equivalent to the Postulate, it was a question of rejecting the other two. To do this he sought a way of showing that both would lead to a contradiction. He achieved this easily enough for the obtuse angle by using Euclid I. 16, since this presupposes the possibility of indefinite extension of straight lines according to the second postulate. (Regarding this see Legendre's proposition XIX below.)

Seeking a way of rejecting the acute hypothesis, Saccheri set up a chain of valid deductions. He showed in particular that the relative positions of two lines separate into three categories: (1) lines that cut, (2) lines than have one perpendicular in common (and, as is shown in the drawing, they diverge on each side of the perpendicular, contrary to al-Ḥayyām's assumption) and (3) lines that are asymptotic.

But a little later Saccheri declares:

> *the hypothesis of the acute angle is false because it is repugnant to t*he nature of the straight line.

In effect, according to him, two straight asymptotic lines would have at their common point at infinity, a common perpendicular, which could not be.

We see here the genesis of statements that would be formulated later by Lobachevsky when he developed his imaginary geometry. We also recognise Lambert, another precursor of non-Euclidean geometry who, in the hypothesis of the acute angle, established a relation of proportionality between the area of a triangle and its deficit, that is the difference between the sum of its angles and 180°.

A proof by Legendre
The excess and deficit of a triangle

Legendre proposed several proofs of the Fifth Postulate, each one more ingenious than the last. Moreover Lobachevsky studied them and they were perhaps what finally convinced him of the reality of non-Euclidean geometry. Here is one of Legendre's proofs taken from his *Eléments de Géométrie* (3rd edition).

The first proposition excludes the hypothesis of the obtuse angle used by Saccheri since, as clearly appears in the construction, he uses the infinite character of the straight line. As for the second proposition, it excludes the hypothesis of the acute angle; all very much to the point… There remains therefore only the hypothesis of the right angle.

PROPOSITION XIX

LEMMA

The sum of the three angles of a triangle cannot be greater than two right angles

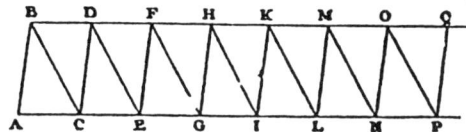

If it is possible, let ABC be a triangle in which the sum of the three angles is greater than two right angles
On AC produced take CE = AC; make the angle ECD = CAB, the side CD = AB; join DE and BD. The triangle CDE will be equal to the triangle BAC since they have an equal angle contained between two sides equal to each other; hence we have the angle CED = ACB, the angle CDE = ABC and the third side DE equal to the third side BC.
Since the line ACE is a straight line, the sum of the angles ACB, BCD, DCE is equal to two right angles; now it was assumed that the sum of the angles of the triangle ABC is greater than two right angles; therefore we have CAB + ABC + ACB > ACB + BCD + ECD; subtracting the common ACB from both parts and using CAB = ECD, there remains ABC > BCD; and because the sides AB, BC of the triangle ABC are equal to the sides CD, CB of the triangle BCD, it follows that the third side AC is greater than the third side BD.
Imagine now that the line AC is indefinitely produced such that we have a sequence of equal and similarly placed triangles ABC, CDE, EFG, GHI, etc.; if the neighbouring vertices are joined by straight lines BD, DF, FH, HK, etc., we shall produce a sequence of intermediary triangles BCD, DEF, FGH, etc. which will all be equal to each other since they have a corresponding equal angle contained between corresponding equal sides. Thus we have BD = DF = FH = HK, etc.

This being done, since we have AC > BD, let the difference AC − BD = D; it is clear that 2D will be the difference between the straight line ACE, equal to 2AC, and the line, straight or broken, BDF itself equal to 2BD; whence we have AE − BF = 2D. In the same way we shall have

AG − BH = 3D, AI − BK = 4D, and so on. Now, no matter how small this difference D may be, it is evident that repeated a sufficient number of times it will become greater than any given length. We can therefore imagine the sequence of triangles prolonged sufficiently far so that AP- BQ > 2AB; and we shall then have AP > BQ + 2AB. Now, on the contrary, the straight line AP is shorter than the angled line ABQP which joins the same extremities A and P, so that we always have AP < AB + BQ + QP, or AP < BQ + 2AB. Hence the hypothesis with which we started is absurd; hence the sum of the three angles of a triangle can not be greater than two right angles.

PROPOSITION XX

THEOREM

In every triangle, the sum of the three angles is equal to two right angles.

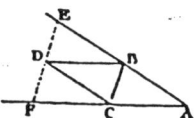

Having just proved that the sum of the three angles of a triangle cannot be greater than two right angles, it remains to prove that this sum cannot be less than two right angles. Let ABC be the proposed triangle, and let the sum of the angles, if it is possible, be equal to 2P − Z, where P stands for a right angle and Z for some quantity by which it is supposed that the sum of the angles is less than two right angles.

Let A be the smallest of the angles of the triangle ABC; on the opposite side of BC make the angle BCD = ABC, the angle CBD = ACB; the triangles BCD, ABC are equal, having a common equal side BC adjacent to two corresponding equal angles. Through point D draw some straight line EF which meets the two extended arms of the angle A, in E and F.[1] Since the sum of the angles of the two triangles ABC, BCD is 2P − Z, and that of each of the triangles EBD, DCF cannot exceed 2P, it follows that the sum of the angles of the four triangles ABC, BCD, EBD, DCF, cannot exceed 4P − 2Z + 4P, or 8P − 2Z. If we take away from that sum the sum of the angles at B, C, D which is 6P, since the sum of the angles formed at each of the points B, C, D is 2P, what remains will be equal to the sum of the angles of the triangle AEF. Hence the sum of the angles of the triangle AEF will not exceed 8P − 2Z − 6P, or 2P − 2Z.

Hence, while we have to add Z to the sum of the angles of the triangle ABC in order to make two right angles, we have to add at least 2Z to the sum of the angles of the triangle AEF in order to make two right angles.

Using triangle DEF we can similarly construct a third triangle, such that we would have to add at least 4Z to the sum of its three angles in order that the sum should total to two right angles; and by use of the third triangle we could construct a fourth in the same way, such that its sum would need at least 8Z added to it to make two right angles, and so on.

Now, however small Z may be in relation to the right angle P, the sequence Z, 2Z, 4Z, 8Z, etc., whose terms increase by a ratio of 2, will lead to a term equal to or greater than 2P. We would arrive therefore at a triangle such that one would need to add a quantity equal to or greater than 2P to the sum of its angles in order that the total sum should come to 2P. The consequence of this is visibly absurd; hence the hypothesis with which we started cannot be substained; that is to say that it is impossible for the sum of the angles of the triangle ABC to be less than two right angles: it cannot be greater than two right angles by virtue of the preceding proposition; hence it is equal to two right angles.

[1] It is assumed that A is the smallest of the angles of the triangle ABC, so that it follows that it is less than, or not greater than, a third of a right angle, to make the possibility

that a straight line drawn through D will meet both the two sides AB, AC produced, more sensible

Comment: An attentive reading of this last proof will identify the sole assertion that is insufficiently argued: **Through any point situated inside an angle a straight line can always be drawn which meets both sides of the angle.**

The non-proof by Lobachevsky
Non-Euclidean Geometry

Convinced since the 1820s of the possibility of a geometry in which the Fifth Postulate was not verified, Lobachevsky brought his ideas together in his *Etudes géométriques sur la Théorie des Parallèles.* He in fact started from what Saccheri had touched upon: being given a straight line BC, the straight lines passing through a given point A can be divided into three categories: secants which cut BC, non-secants which do not cut BC, and parallels to BC, that is the two non-secant boundaries, those which form a frontier between the two preceding categories.

16 – All straight lines which in a plane go out from a point can, with reference to a given straight line in the same plane, be divided into two classes – into cutting and non-cutting. The boundary lines of the one and the other class of those lines will be called parallel to the given line.

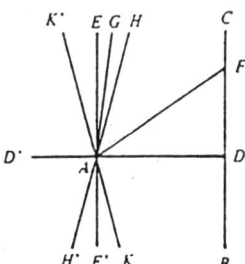

From the point A let fall upon the line BC the perpendicular AD, to which again draw the perpendicular AE. In the right angle EAD either will all straight lines which go out from the point A meet the line DC, as for example AF, or some of them, like the perpendicular AE, will not meet the line DC. In the uncertainty whether the perpendicular AE is the only line which does not meet DC, we will assume it may be possible that there are still other lines, for example AG, which do not cut DC, how far soever they may be prolonged. In passing over from the cutting lines, as AF, to the non-cutting lines, as AG, we must come upon a line AH, parallel to DC, a boundary line, upon one side of which all lines AG are as such as do not meet the line DC, while upon the other side every straight line AF cuts the line DC.
The angle HAD between the parallel HA and the perpendicular AD is called the parallel angle (angle of parallelism), which we will here designate by Π(p) for AD = p.[1]

The straight line ΛK, symmetrical to AH with respect to AD is also a non-cutting boundary and so parallel to BC. Thus through the point A pass two lines parallel to BC, each related to a direction, except in the Euclidean geometry case where Π(p) = π/2. This concept of parallel lines corresponds to Saccheri's asymptotic lines: as for the non-cutting lines, they correspond to lines which have a common perpendicular.

---

1.   English tr. by G.B. Halsted in Fauvel J. and Gray J., *History of Mathematics: A Reader*, London, 1987 p. 524-525 (© All rights reserved).

When considering, however, the case $\Pi(p) < \pi/2$, Lobachevsky obtained the following proposition whose proof recalls both Legendre's introduction of the deficit of a triangle and the successive perpendiculars of at-Tūsī together with Pasch's axiom.

23 – Given any angle α a distance p can always be found such that $\Pi(p) = \alpha$

Exercise / Proof The following proof may be used as an exercise for the reader.

Let two straight lines AB and AC form an acute angle α at their intersection A. Take on AB any point whatever B'; from this point lower a perpendicular B'A' on to AC; make A'A" = AA'; raise at A" a perpendicular A"B" and continue thus until we arrive at a perpendicular CD which does not reach AB. This must necessarily happen since if the sum of the three angles of the triangle AA'B' is equal to $\pi - \alpha$, that sum in the triangle AB'A' will be equal to $\pi - 2\alpha$; in the triangle AA"B" it will be less than $\pi - 2\alpha$, and so on, until eventually it becomes negative, in which case it becomes impossible to form a triangle.

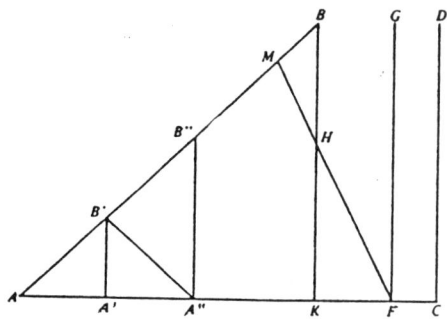

The perpendicular CD could be that line that forms the boundary between the perpendiculars nearer point A which cut AB, and those perpendiculars further from A which do not cut AB. In all cases there must exist such a boundary perpendicular, FG say, where one passes from cutting perpendiculars to non-cutting perpendiculars. Take now through the point F the perpendicular FH making an acute angle HFG with FG and situated on the same side of FG as the point A. From some point H on the straight line FH drop a perpendicular HK onto the line AC whose extension will meet the line AB at some point B, and thus form a triangle AKB into which the extension of FH will protrude; this extension will therefore meet the hypotenuse AB at some point M. The angle GFH being arbitrary and as small as is wished, FG will then be a parallel to AB and we have AF = p. It can be easily see that, as p diminishes, the angle α increases, and that it approaches $\pi/2$ as p tends to zero. On the other hand, as p increases, the angle α diminishes, and it approaches closer and closer to 0 as p tends to infinity.[1]

Lobachevsky went on to develop a geometry which is called hyperbolic or non-Euclidean. He proved new relationships between the angles and the side of triangles and other major formulae. These results were also obtained independently by others, notably Bolyai and Gauss.

---

1.   Lobachevsky, *La théorie des parallèles*, p. 20-21

Some formulae in non-Euclidean geometry

The angle of parallelism $\Pi(x)$ satisfies:

$$\tan\left(\Pi(x)/2\right) = e^{-x/k} \text{ (k > 0 being an arbitrary constant)}$$

Euclidean geometry corresponds to $\Pi(x) = \pi/2$, that is $k = \infty$. Two parallel lines are asymptotes and the distance between them diminishes proportionately with $e^{-x/k}$ as x increases.

In a triangle with angles A, B, C and sides a, b, c the following hold:

$$\sin A \tan\Pi(a) = \sin B \tan \Pi(b)$$

$$\cos A \cos \Pi(b) \cos \Pi(c) + \sin \Pi(b) \sin \Pi(c) / \sin \Pi(a) = 1$$

$$\cot A \sin C \sin \Pi(b) + \cos C = \cos \Pi(b) / \cos \Pi(a)$$

$$\cos A + \cos B \cos C = \sin B \sin C / \sin \Pi(a).$$

As k tends towards ∞ these formulae give a first approximation to the classical formulae.

From the point of view of our work on the Fifth Postulate, we can say that at this stage of development, in the first half of the 19th century, there was a growing conviction of two things: that it is not possible to prove the Postulate without admitting another equivalent postulate, and that it might be possible to construct geometries in which the postulate does not hold. [Through a point not on a line there can be two parallels to that line, and an infinite number of non intersecting lines.] That this might be so remained for the time being a matter of personal conviction rather than certainty.

The Lobachevsky test

Obstacles that he came across in his attempts at proof, led Lobachevsky to think that what he wanted to prove did not lie in the given problem and that he needed to have recourse to experimentation. This is what he proposed to test the value of the constant k in the formula $\tan\left(\Pi(x)/2\right) = e^{-x/k}$. Consider a triangle ABC whose side BC = a is equal to the diameter of the Earth's orbit and whose vertex A is a fixed star in direction perpendicular to BC. Let 2p be the maximum parallax of the star A.

Then:

$\Pi(a) > BCA = \pi/2 - 2p$;

let $\tan\left(\Pi(a)/2\right) > \tan[\pi/4 - p]$;

that is $e^{-a/k} > (1 - \tan p)/(1 + \tan p)$;

whence: $a/k < \tan 2p$.

Given the parallax of Sirius as $1'' 24'''$,

Lobachevsky deduced that: $a/k < 0.000\ 006$.

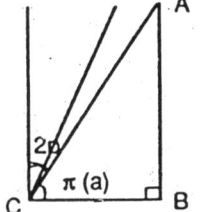

Hence, we do not know the value of the constant k but we can deduce that it is very large relative to the diameter of the Earth's orbit. If Euclidean geometry is true, k is infinitely large and there must exist stars whose parallax is infinitely small.

The contrary case, where there does not exist any parallel to a given straight line through a given point, corresponds to the case of spherical geometry. In geometry on the surface of a sphere, great circles play the part of straight lines but they always intersect and the sum of the angles of triangles formed by arcs of great circles is always strictly greater than two right angles. But this is an example where we may be said to have strayed from our study which has concerned unlimited lines, since these lines are all of finite length.

Proof from Klein and Beltrami
The power of a model

Inspired by Cayley's work in which he devised a new metric for the plane based on a conic, Felix Klein[1] showed, in 1871, how to set up a hyperbolic, or non-Euclidean, geometrical model for the Euclidean plane or Euclidean space.

In order to restrict ourselves to the plane, and also to make the reading of the text easier, we have adapted Klein's text by substituting plane for space, curve for surface, and line for plane, wherever they occur. Otherwise the text is identical to the original (except of course that it was in German).

> *Let there be given any arbitrary curve of the second degree as a fundamental curve. The curve defines a plane and two given points of the plane will determine two points of the curve where the straight line joining them intersects the curve. The two given points, together with the other two points of the curve, have a certain harmonic cross-ratio and the logarithm of this cross-ratio, multiplied by a certain constant c, will be said to be the distance of the two given points. In the same way, given two lines, we can draw two tangents to the fundamental curve from their point of intersection. These tangents, together with the given lines, have a certain harmonic cross-ratio. The logarithm of the cross-ratio multiplied by a certain constant c' is what we shall call the angle of the two lines.*
>
> *Following these definitions, the points of the fundamental curve are at an infinite distance from all other points: the fundamental curve is thus where points infinitely distant are situated …*

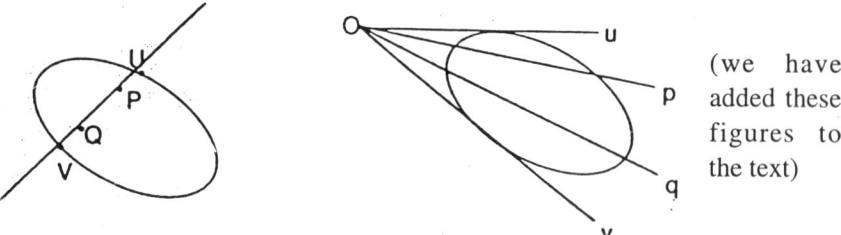

(we have added these figures to the text)

> *We obtain a Geometry corresponding to Hyperbolic Geometry, by taking a real fundamental curve and considering points in its interior. This restriction to points inside the curve comes about naturally: for supposing that one finds oneself inside the curve and that one can only change one's position in the plane by means of two-dimensional linear transformations which represent movements in the plane according to the determined metric, then one would never be able to leave the interior of the second degree curve, which is situated at infinity (in the determined metric). Beyond the fundamental curve, there would exist then another portion of the plane about which*

1.   Über die sogenannte Nicht-Euklidische Geometrie, *Mathematische Annalen*, **4**, 1871, p. 573-625, French translation of extracts in *Bulletin des Sciences Mathématiques*, **2**, 1871, p. 341-351.

one would know nothing, and which would not be noticed, since any two straight lines would never cut it, if one did not assume such a portion of the plane... If we now restrict ourselves to those constructions that do not leave the inside of the curve, by using the corresponding determined metric, those constructions will absolutely be subject to the laws of Hyperbolic Geometry, established in general for constructions in the plane. Every straight line, for example, has two real points at infinity; for every straight line passing through the interior of the curve will cut that curve in two real points. Through a point one can draw two parallels to a line, namely the two lines that join that point to the two intersection points of the given line and the fundamental curve. A triangle whose vertices are at infinity, that is to say whose vertices lie on the fundamental curve, is one whose angle sum is zero. For any two lines meeting on the fundamental curve (any two parallels) contain a zero angle, etc. Finally, the constant c, by which the harmonic cross-ratio must be multiplied to give the distance between two points, represents the characteristic constant mentioned above, which occurs in Hyperbolic Geometry.

To put it briefly, for a particular circle (or ellipse) fixed in a plane, points inside the circle represent points of a hyperbolic plane to which can be ascribed the following measure: the distance between two points P and Q is equal to c |ln (P, Q, U, V)| where U and V are the points of intersection of the straight line PQ with the circle and where (P, Q, U, V) is the cross-ratio $\frac{PU}{PV} : \frac{QU}{QV}$. The lines of the hyperbolic plane are thus chords, and two lines are parallel if and only if the corresponding chords have a common extremity. The isometries of this plane are given by projective transformations of the ordinary plane (preserving alignment) which leave the circle invariant. [A similar model was developed by Beltrami from 1868.]

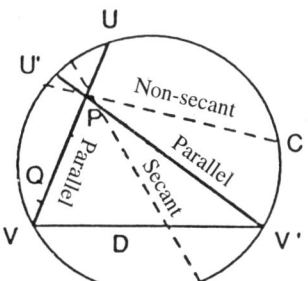

This model of the hyperbolic plane is remarkable in that its description uses only the Euclidean plane and its associated properties. Hence, if Euclidean geometry is free from contradiction, in particular if the Fifth Postulate may be imposed as an axiom, then non-Euclidean geometry is itself free of contradiction, and in particular, the Fifth Postulate can be denied. We have here, then, a valid proof that Euclid's axiom cannot be proved starting only with the other axioms, whether explicit or implicit.

Better still, using the contrary hypothesis of the negation of the Fifth Postulate, Lobachevsky demonstrated the existence of surfaces in hyperbolic space which were finally confirmed to be valid models of Euclidean planes, namely horospheres on which horocycles play the part of straight lines. [We could say that a horocycle is a curved line whose normals are all parallel to each other.] And so the non-contradiction of the axioms of Euclidean geometry is, in its turn, a consequence of the non-contradiction of the axioms of non-Euclidean geometry.

Poincaré's model

Somewhat later, Poincaré proposed a different model for the hyperbolic plane. The "points" of this model are those of an open half-plane bounded by a straight line D and the "lines" of the model are half-lines perpendicular to D together with semi-circles whose centres lie on D. The "distance" between two points is given by the expression k |ln (P, Q, U, V)| where U and V are the points of intersection that the "line" through P, Q makes with the line D.

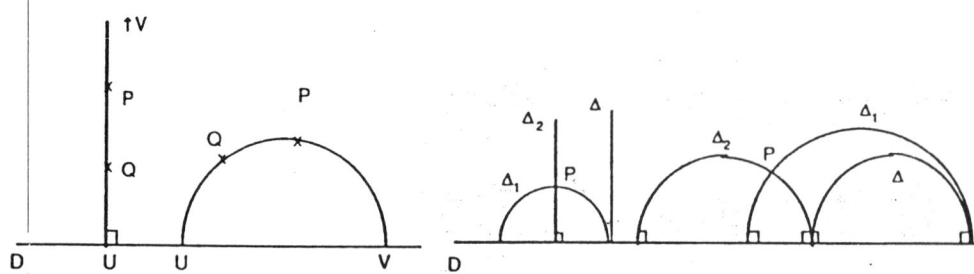

Thus, through a point P not on a "line" Δ there are two "lines" parallel to Δ: the semi-circles Δ_1 and Δ_2 with centres on D which pass through P and are tangents to Δ. (Points on D itself do appear in the model.) "Isometries" are inversions with respect to a circle with centre on D and compositions of these inversions.

* *
*

It is not then possible either to prove or disprove the Fifth Postulate, to the extent that one at least of the axiomatic systems is non-contradictory. In his *Foundations of Geometry* in 1899, Hilbert clearly and precisely identified the different axioms of Euclidean geometry, particularly those that had been used implicitly. The non-contradiction of these can be easily dealt with thanks to the use of Cartesian methods, and depend essentially upon the non-contradiction only of the axioms of the reals.

As for the non-contradiction of the axioms of the reals, we can only hope and pray.

Bibliography

Sources texts

GRAY, *Ideas of space euclidean, non-euclidean, and relativistic*, Clarendon Press, Oxford, 1979.

GREENBERG, *Euclidean and non euclidean geometries, Development and history*, Freeman, San Francisco, 1980.

HEATH, *The Thirteen Books of Euclid's Elements*, 3 vol., Cambridge, 1908: 2nd ed. Cambridge 1925: Dover New York, 1956.

HEIBERG, *Euclidis Elementa*, Teubner, Leipzig, 1883. Fr. tr. Vitrac, Presses Universitaires de France, Paris, 1990.

JAOUICHE, *La théorie des parallèles en pays d'Islam*, Contribution à la préhistoire des géométries non-euclidiennes, Vrin, Paris, 1986.

KLEIN, Ueber die sogenannte Nicht-Euklidische Geometrie, *Mathematische Annalen*, **4**, 1871, pp. 573-625. Fr. tr. of extracts in *Bulletin des Sciences Mathématiques*, **2**, 1871, pp. 341-351.

LEGENDRE, *Eléments de Géométrie*, 3rd ed., Paris, 1800. Repr. in "Réflexions sur les différentes manières de démontrer la théorie des parallèles", *Mémoires de l'Académie Royale des Sciences*, vol. XII, 1833, 367-410.

LOBATCHEVSKY, *Etudes géométriques sur la théorie des parallèles*, Berlin, 1840. Fr. tr. J. Houël, 1866. Repr. *La théorie des parallèles*, Monom, Coubron, 1980.

PEYRARD, *Les Oeuvres d'Euclide* traduites littéralement d'après un manuscrit grec très ancien, resté inconnu jusqu'à nos jours, Paris, 1819. Repr. Blanchard, Paris, 1966.

ROSENFELD, *A history of non-euclidean geometry, Evolution of the concept of geometric space*, Springer, New York, 1988.

SACCHERI (le Père), *Euclides ab omni naevo vindicatus*, Milan, 1773. English tr., Halsted, Girolamo Saccheri's Euclides Vindicatus, Chicago, Open Court, 1920. Repr. Amer. Math. Soc., 1986.

WALLIS, De Postulato Quinto et Definitione Quinta, *Opera Mathematica*, t. II, Oxford, 1693 (pp. 665-678). Fr. trad. avec commentaires J.-L. Chabert & J. Neuberg, *Wallis, le Cinquième Postulat et la Similitude*, IREM de Picardie, 1986.

General works for further reading

BONOLA, *La geometrica non euclidea*, Bologne, 1906. Engl. Trad. 1912, reprint Dover, New-York, 1955.

CHABERT, *La Préhistoire des Géométries non euclidiennes*, IREM de Picardie, 1986, 129 p.

CHABERT, « Les Géométries non euclidiennes », *Repères*, n° 1, 1990, 69-91.

PETIT, *Le Géométricon*, Belin, Paris, 1982.

PONT, *L'Aventure des parallèles*, Lang, Berne, 1982.

EFIMOV, *Géométrie supérieure*, 1978. Fr. trad. Mir, Moscou, 1981.

GONSETH, *Les fondements des mathématiques : de la Géométrie d'Euclide à la Relativité générale et à l'Intuitionisme*, 1926. Reprint, Blanchard, Paris, 1974.

12

A Desperate Search

Jean-Pierre FRIEDELMEYER
IREM Strasbourg

Examples of calculations for examining nature and for understanding all that exists, each mystery ... each secret.

This is the appearance of the Egyptian text known as the Rhind Papyrus. It is one of the oldest mathematical texts, being about 4000 years old. We know its age because the writer, the scribe A'h–mosè (or Ahmes) tells us *"this book was copied in the year 33, in the fourth month of the inundation season, under the majesty of the king of Upper and Lower Egypt, 'A-user-Rê', endowed with life, in the likeness of writings of old made in the time of the king of Upper and Lower Egypt, 'Ne-ma'et-Rê'"*[1].

The Rhind Papyrus contains 87 problems with their solutions. For example, problem 31 states:

A quantity, its $\frac{2}{3}$, its $\frac{1}{2}$ and its $\frac{1}{7}$ added together become 33.

and we recognise here what we would call the equation: $x + \frac{2}{3}x + \frac{1}{2}x + \frac{1}{7}x = 33$. The solution is easily found as $x = \dfrac{33}{1 + \frac{2}{3} + \frac{1}{2} + \frac{1}{7}} = 33 \div \frac{97}{42} = \frac{1386}{97}$.

But the solution offered in the papyrus is more laborious and much longer (the equivalent of a whole page of calculations). The answer is written in the form of an integer followed by a string of signs which indicate the inverses of integers:

$$14 \; \frac{1}{4} \; \frac{1}{97} \; \frac{1}{56} \; \frac{1}{679} \; \frac{1}{776} \; \frac{1}{194} \; \frac{1}{388}$$

1. Chace, *The Rhind Mathematical Papyrus*, p. 27.

and it can be verified that this answer (adding up all the fractions), is indeed the same as the value of the unknown found earlier.

This solution is longer and more laborious because the only calculation algorithms available to the Egyptians were multiplication and division by 2. And the answer was written in this form because they only had symbols for integer fractions with numerator 1, with the exception of $\frac{2}{3}$ for which they had a special symbol.

This example is characteristic of problems that can be expressed in the form of an equation, problems that have been posed since the very beginnings of mathematics, and raises the questions:

– what algorithms exist to find the solution(s), accurately and if possible automatically?

– in what form, and with what symbols, can the solutions be written?

The answers to the first question depend strictly on those that are given to the second, and *vice versa*. Up until the nineteenth century, the history of algebra was essentially the same as the history of the answers to these two questions. The search for an efficient algorithm went along with a desire for a general method that could be applied to all equations of the same type. Contrary to the approach adopted by the scribe Ahmes, who offered numerous similar examples so that the method could be learnt like a circus trick, mathematicians gradually came to develop general methods of solution:

– First through geometry, since all sorts of relations, that were independent of any particular numerical values, could be established in geometrical figures. It is likely that those who worked on such figures were led to write their calculations, first using abbreviations, and then by using mathematical symbols which became the start of an algebra of symbols and letters.

– Increased practice of manipulation of signs and symbols led to the development of new algorithms, which gave rise to methods and solution procedures that were so effective that it was thought at one time that their power was unlimited. Although this view was illusory, it nonetheless produced valuable results.

– The mathematicians of the Italian Renaissance found formulas for solving equations of the third and fourth degree using radicals (that is, writing solutions in terms of $\sqrt{}$, $\sqrt[3]{}$, $\sqrt[4]{}$, etc.) and it must have seemed that nothing would stop their successors from solving equations of higher degree. Algebra, as a tool, appeared to have reached its maximum efficiency and it would only be a short time before a method for solving quintics would be uncovered; it was simply a matter of wisdom and ingenuity.

– But the quintic resisted attempts to solve it. Furthermore it turned out that it was impossible to solve it by the use of radicals. The effort undertaken to arrive at this impasse was not however in vain. On the contrary, it resulted in a revolution in the conception of the nature of mathematical activity: a change from mathematics being about numbers and quantities, to mathematics being about relations and structures.

And so we have four chapters in this rich and eventful history:

1. The methods of false position.
2. The birth of algebra
3. Methods using radicals (or surds).
4. The emergence of the concept of a group.

Finding truth through falsehood: the methods of false position

How can we find the solution to an equation, if we have to use our own thought processes, but without the help of symbolism? How, for example, would we tackle problem 31 from the Rhind papyrus, or the following problem, posed by Francès Pellos, a gentleman of Nice at the end of the fifteenth century?

> The tax collected on grain sold at Nice comes to 2000 florins, for each bushel that is to be use for milling or for consumption, the buyer must pay his due; ordinary citizens must pay a quarter of a groat [12 groats = 1 florin] per bushel and bakers must pay one groat and a quarter. If you are told that the division is: one third is bought by bakers and two-thirds by ordinary citizens, I ask you how many bushels must pass through Nice in order that the tax should come to 2000 florins?[1]

Pellos solves the problem as follows: if the number of bushels were 12, then the bakers would have 4 and would pay 5 groats, and the citizens would have 8 and pay would 2 groats. This makes a total of 7 groats which is $\frac{7}{12}$ of a florin for the tax-collector. A simple rule of three can then be used to find the number of bushels needed to produce 2000 florins:

12 divided by $\frac{7}{12}$ multiplied by 2000 which gives 41 142 bushels and $\frac{6}{7}$.

Pellos has used what is called a method of false position. In this case we have an example of simple false position, but it is also possible to use double false position.

Simple False Position

This is used whenever we have a problem that can be translated into algebra as an equation of the type $ax = b$. But the lack of symbolism makes it difficult or impossible to calculate the value of the coefficient a and so prevents us from finding the value of the unknown x by the simple division of b by a. Because of proportionality, the use of a false position x_1 avoids the need for calculating the value of a, since the statement provides a way of calculating $b_1 = ax_1$. And so the value of x can be found uniquely from the given values and the false position:

$$x = \frac{x_1}{b_1} \cdot b \text{ since } \frac{x}{b} = \frac{x_1}{b_1} \text{ (see fig. 1)}$$

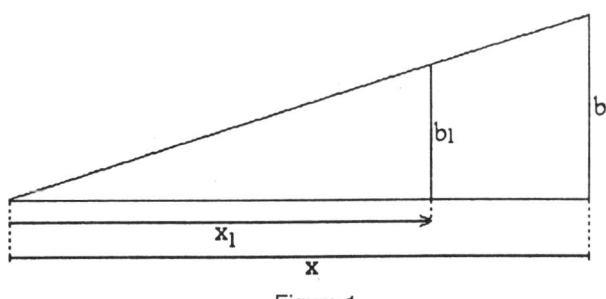

Figure 1

1. English adaptation from Francès Pellos, *Compendion de l'abaco*, p. 182, cited by Speisser, *Equations du premier degré*, 1982, p. 11 (© All rights reserved).

Exercise 1   a) Use the method of simple false position to solve problem 29 of the
Rhind Papyrus which states: *"A quantity and its $\frac{2}{3}$ are added*
together, and $\frac{1}{3}$ of the sum is added; then $\frac{1}{3}$ of this sum is taken and
the result is 10. What is the quantity?".

This would now be written as:
$(x + \frac{2}{3}x) + \frac{1}{3}(x + \frac{2}{3}x) - \frac{2}{3}[(x + \frac{2}{3}x) + \frac{1}{3}(x + \frac{2}{3}x)] = 10$
b) Use the method of simple false position to solve this problem from
the Pratica d'Arithmetica by Francesco Ghaligai (1552): *"Find a*
number such that if you take away $\frac{2}{3}$ of it, what remains is $\frac{3}{4}$".

Double False Position

There are many problems, however, that do not admit of solution by the method of
simple false position, since their statement does not translate immediately into an
equation of the type $ax = b$, but into an equation of the type $ax + b = c$. In this case there
are two coefficients, a and b, which are not explicit in the formulation of the problem.
An example of this type of problem from the fifteenth century is given by Luca Pacioli[1]:

> *Divide 44 ducats between three people in such a way that the second should have double*
> *the first's plus 4 and the third should have as much as the other two together plus 6.*

Here, using x for the first person's share, we have to solve the equation:
$x + (2x + 4) + (3x + 10) = 44$. The lack of algebraic processes makes it difficult to
render this as an equation of the type $ax = b$. Using a first false position of $x_1 = 8$
produces a sum of 62, which is too great by an excess $e_1 = 18$ over the 44 ducats
available for sharing. Using a second false position $x_2 = 6$ produces a sum of 50, which
is too great by an excess of $e_2 = 6$. In which case, says Pacioli, to find the exact value of
x, multiply the first excess by the second hypothesis, and subtract from it the product of
the second excess and the first hypothesis, then divide the result by the difference of the
excesses. As we would write it:

$$x = \frac{e_1 x_2 - e_2 x_1}{e_1 - e_2} = 5$$

For Luca Pacioli, the whole calculation can set out according to the schema:

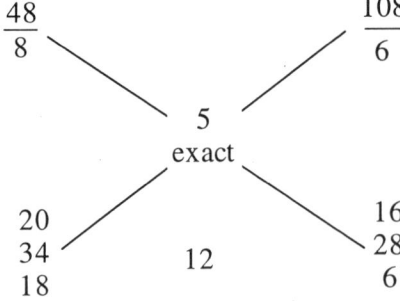

$x = 5$ is the correct answer, but why?

---

1.   Luca Pacioli, *Summa* fol 98 verso to 111 verso, 1494, quoted by M. Cantor, *Vorlesungen über*
Geschichte der Mathematik, vol. II, p. 318.

The proposed equation is of the form $ax + b = c$ with a and b unknowns. A false position x_1 will produce $ax_1 + b = c_1$ which differs from c by an error e_1, that is $e_1 = c_1 - c = a(x_1 - x)$: a second false position x_2 gives, in same way, an error e_2 equal to $c_2 - c = a(x_2 - x)$. From this we get:

$$\frac{e_1 x_2 - e_2 x_1}{e_1 - e_2} = \frac{x_2 a(x_1 - x) - x_1 a(x_2 - x)}{e_1 - e_2} = \frac{x(ax_1 - ax_2)}{e_1 - e_2} = x$$

Of course, it must be understood, that in the absence of an algebraic symbolism, mathematicians at this time were unable to offer a justification for their methods in this form. Instead they had recourse to geometric arguments, but were obliged to consider three cases according to whether the differences e_1, e_2 were excesses or deficits. Calculations with negatives were not possible, and in any case they were without meaning in this context.

We present here Fibonacci's justification for the method, taken from his *Liber Abbaci* (1202), chapter 13. The case concerns two positive excesses, as in the example above. In Figure 2 let the length ab be the exact value of x to be found. In its place take a (known) value $x_1 = $ length af. (Of course, Fibonacci does not introduce a value x (nor x_1, x_2) which we have used in order to provide an algebraic interpretation to help us understand his geometric argument.) This leads to an incorrect result which is too great by an excess, the difference being $e_1 = gi = bf$.

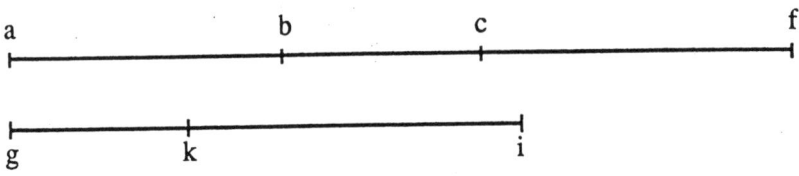

Figure 2

Another value ac $= x_2$ produces a different excess $e_2 = gk = bc$.

So we have: ki $= e_1 - e_2$, cf $= x_1 - x_2$, bc $= x_2 - x$.

We also have $\dfrac{ik}{kg} = \dfrac{cf}{cb}$ which becomes: $\dfrac{e_1 - e_2}{e_2} = \dfrac{x_1 - x_2}{x_2 - x}$.

Rules for calculating proportions, already known since they were developed by Euclid in *Elements* Book V, lead to the conclusion:

$$x = x_2 - \frac{e_2(x_1 - x_2)}{e_1 - e_2} \text{ or } x = \frac{e_1 x_2 - e_2 x_1}{e_1 - e_2}.$$

Exercise 2	Use the method of double false position to solve the following problem, posed by Ben Ezra in the eleventh century:
	"Suppose that a man went into an orchard and picked some apples. But the orchard has three gates and each is protected by a watchman. The man then shares the apples with the first watchman and gives him two extra. Then he shares his pickings with the second watchman and gives him two extra. Finally he shares what is left with the third watchman, also giving him an extra two, and leaves the orchard with just one apple. How many apples had he picked?"[1]

1.   Spiesser, *Equations du premier degré*, p. 13.

A complete and detailed analysis of the methods of false position, consisting of n first degree equations in n unknowns, shows both the extraordinary ingenuity of the mathematical calculations and at the same time the narrow limits within which the system operates. The method cannot be adapted to solving problems which lead to equations of higher degrees. The reader will not be surprised to learn that the main tool for solving equations of the second degree lies in a geometrical translation of the problem to a situation that represents products and squares, by use of the areas of a rectangle and squares. In Chapter 13 there is a discussion of how Euclid employed a geometrical treatment of equations of the type $x(a - x) = K$. Geometrical methods of this type remained the main method of proof for as long as the absence of symbols prevented the use of the sort of algebraic methods of solution that we know today. Algebraic methods were more favoured by Arabic mathematicians and first of all by Diophantus of Alexandria.

The Birth of Algebra
Diophantus: Using an unknown quantity

The history of what comprised algebra up to the beginning of the nineteenth century only really began with Diophantus, a mathematician from Alexandria of the third century A.D. The contribution that Diophantus made to the existing mathematics (Babylonian and Egyptian as well as Greek) was an innovation that can be compared in importance and significance to the invention of zero.

The use of a sign (the "*zero*") to indicate the absence of anything was a considerable conceptual leap whose traces can be found in the name "*zero*" itself, which the Arabs called "*sifr*" to mean "*empty*". In the same way, the use of a word and a corresponding symbol to stand for something which is unknown, allowed calculations to be carried out using this word and its symbol just as if it was a known value. This was a step towards abstraction and a determinate symbolism. Up to then the formulation of a problem, the equation, was entirely static: its resolution involved a quasi-ritualistic process of a number of routine calculations. This is clearly seen in the case of the method of false position: the nature of the problem itself is never considered, but just calculations with arbitrarily posed values, blindly working through the statement of the problem word for word, and arriving at a solution by means of an algorithm whose justification lies **external** to the problem itself and to the numbers it contains. This is also true for the solutions of problems of the second degree which are to be found throughout the mathematics literature of all the early civilisations (Babylon, Egypt, but also India and China) and which are carried out by a succession of calculations, the logic of which is never made explicit.

In order to appreciate better the contribution made by Diophantus, let us look at the same problem, solved first by the Babylonians, and then by Diophantus: "*Find the dimensions of a rectangle of area 96 and half of the perimeter is 20*". This problem is in what Neugebauer[1] calls the *"normal form"* of a problem of the second degree for the Babylonians, which were passed on through hundreds of such problems "written" on more than 300 clay tablets during an immense period of time from 1800 B.C. to 300 B.C. (The same problem treated by Euclid can be found in Chapter 13.)

The Babylonian method: this consists of a series of operations to be carried out on the given numbers according to instructions laid down by the scribe:

1. Divide the semi-perimeter by 2: $20/2 = 10$

2. Square this result: $10^2 = 100$

3. Subtract the given area: $100 - 96 = 4$

4. Extract the square root: $\sqrt{4} = 2$

5. The length is $10 + 2 = 12$; the width is $10 - 2 = 8$

Hence the solution method for the Babylonians, insofar as we are able to judge, consists of a simple **chain of numerical operations**. Even the idea of calculating on an unknown is entirely absent. The *Arithmetica* of Diophantus on the other hand introduces an unknown which Diophantus calls *"arithmos"* together with a sign ς for its representation. He also introduced **symbols** for square (Δ^y) and cube (K^y), for known numbers units in M and a sign for subtraction: $\bigwedge$.

The Diophantine method: The problem itself is stated in general terms:

> *To find two numbers whose sum and product are equal to given numbers.*

But the resolution of the problem starts from the same numerical example.

Diophantus' Approach

> *Let us suppose that the sum of the two numbers is 20 units and that their product is 96. Let the excess of the numbers be 2 Arithmos. From this, since the sum of the numbers is 20 units, if we divide it into two equal parts, each of the parts will be half the sum or 10 units. So, if we add to one part and take away from the other half the excess of the numbers, that is 1 Arithmos, we shall have again the sum of the two numbers being 20 units and their excess 2 Arithmos. In consequence, let us suppose that the greater number is 1 Arithmos more than the 10 units which is the half of the sum of the numbers; then the smaller number will be 10 units less 1 Arithmos and it follows that the sum of the numbers is 20 units and the excess is 2 Arithmos.*
>
> *The product of the numbers must be equal to 96 units. Now their product is 100 units – 1 square Arithmos, which we equate to 96 units, and the Arithmos becomes 2 units. In consequence, the greater number will be 12 units and the lesser will be 8 units, and these numbers satisfy the proposition.[2]*

This solution by Diophantus uses operations with an unknown quantity, but it is not one of the unknowns of the problem, but an auxiliary unknown by which all the unknowns of the problem can be found once it has been determined. The appearance of the auxiliary unknown, combined with the use of symbols, was a real conceptual change and marks the birth of algebra. The scope of this change remained somewhat limited because of the need to reduce all questions to a single unknown, the Arithmos, but it was nevertheless extremely effective. To confirm this, one has only to look through the 189 problems dealt with by Diophantus in the six books of his *Arithmetica* and we offer here as an example problem 17 of Book V.[3] (In 1968 a twelfth century Arab manuscript

---

1.   Neugebauer, *The exact sciences in Antiquity*, p. 149.
2.   Ver Eecke, *Diophante d'Alexandrie*, p. 36-38.
3.   *Ibid.*, p. 214.

found in a library at Meshed (Iran) proved to be a translation from the Greek of four further books of the *Aríthmetica* of Diophantus.)

Exercise 3   Add the same number to each of two given numbers such that each of them makes a square. You could take 2 and 3 to be the two numbers, like Diophantus, and suppose that the number added to both is equal to 1 Arithmos. The problem translates into the system of unknowns x, α, β:

$$x + 2 = \alpha^2 \text{ and } x + 3 = \beta^2$$

Classifying equations: Al-Kwārismī

It was the Arabs who first gave us the word *Algebra* which derives from a book written in 830 by the astronomer Mohammed ibn Musa al-Kwārismī with the title *Hisab al-gàbr w'al-muqàbalah*. The word *"al-gàbr"* means here *"restoring"* or *"uniting what has been broken"*. This word was used to describe the practice of bone setting. The word passed into Spanish where *Algebrista* describes a bonesetter (see Don Quixote, Book II, Chap. 15: *"Llegàron à un pueblo, donde fuè ventura hallàr à un Algebrista con quièn se curõ el Sanson desgraciado"*: *"Coming to a village, the two companions had the great fortune to find an Algebraist who was able to treat the unfortunate Sanson")*. We have not yet arrived at using symbols for equations, nor the use of letters for numbers, known or unknown. Here, restoring, just as with *"comparison"* (muqàbalah), referred to the *"re-equilibration"* of the two sides of an equation, as is done with a balance. Consider, for example, the equation which we would write as

$$2x^2 - 13x + 8 = x^2 + 3.$$

In a context where numbers were necessarily strictly positive, this equality can only be thought of concretely in the following form: there is no equality between $2x^2 + 8$ and $x^2 + 3$, the first expression being too great by $13x$.

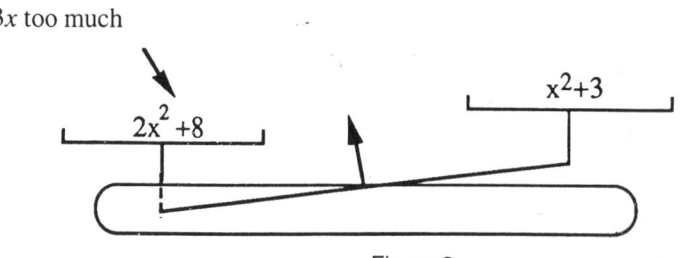

Figure 3

We *"restore"* the equation by adding $13x$ which, by *Al-gàbr* becomes:

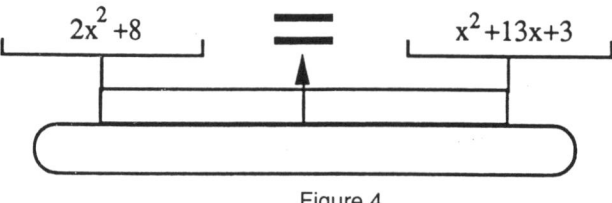

Figure 4

and then by *muqàbalah*

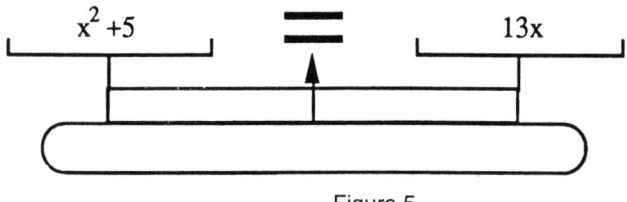

Figure 5

In this way, equations of degree two or less were reduced by Al-Kwārismī to six standard types, given here with their equivalent in modern symbolism:

Squares equal to roots	$ax^2 = bx$
Squares equal to a number	$ax^2 = c$
Roots equal to a number	$bx = c$
Squares and roots equal to a number	$ax^2 + bx = c$
Squares and a number equal to roots	$ax^2 + c = bx$
Roots and a number equal to squares	$bx + c = ax^2$

The geometric representations that were used to demonstrate the methods and justify them did not allow for the use of negative numbers or zero. Today all six cases would be reduced to the single form $ax^2 + bx + c = 0$.

Let us consider the fourth type. Al-Kwārismī first shows the method of solution by use of an example, and then goes on to give a geometrical proof of the method. (In all cases, the equation is amended so that the coefficient a becomes 1.)

Al-Kwārismī takes the example $x^2 + 10x = 39$ and explains the method of solution as follows:

> *Take half the value of the roots, here 5 – you multiply it by itself, this gives 25; add on the 39, this makes 64; you take the root which is 8, from which you subtract the half of the roots 5, this leaves 3 and this is the root of the square that you need.*[1]

The explanation is of a particular case, but the method is general. If the equation is $x^2 + px = q$, the solution x (there is only one, the positive case) is given by calculating the

value of the discriminant $\left(\frac{p}{2}\right)^2 + q$ and then finding $x = \sqrt{\left(\frac{p}{2}\right)^2 + q} - \frac{p}{2}$.

---

1.   Al-Kwārismī cited by Tropfke; *Geschichte der Elementar Mathematik*, p. 419.

The justification of the method is given by using the following geometrical illustration:

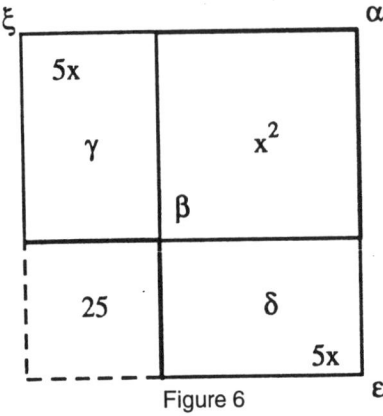

Figure 6

Let $\alpha\beta$ be a square equal to x^2 to which are added two rectangles γ and δ to the sides, corresponding to the roots ($px = 10x$) which make a gnomon which produces a square $\xi\varepsilon$, which only has the small square 25 missing from a corner. But the large square $\xi\varepsilon$ is equal to

$$x^2 + px + \left(\frac{p}{2}\right)^2 = \left(x + \frac{p}{2}\right)^2 = \left(\frac{p}{2}\right)^2 + q$$

and so $x = \sqrt{\left(\frac{p}{2}\right)^2 + q} - \left(\frac{p}{2}\right)$

Although the method was effective for equations of the second degree, it could not easily be adapted to equations of higher degree, there being no comparable geometric representation.

Omar Khayyam certainly tried to treat cubic equations in a similar way, but he was only able to propose a geometric solution by the use of the intersection of conics. There was increasing use of calculating techniques throughout the Middle Ages which relied on the use of signs and symbols and this led the Italians of the Renaissance to make a decisive step in this direction. We shall not present here that long history of signs and notation which many contributed to. We remember the names of Widman for the signs + and – (1486), Rudolff for the radical sign $\sqrt{}$ (1525) and Recorde for the equals sign = (1556). The decisive contribution was made by Viète with the introduction of a literal representation for numbers at the end of the sixteenth century.

Calculations with letters: Viète and Descartes

With Viète we come to a decisive moment in the history of algebra where letters are used to distinguish known from unknown quantities. Viète used the letters A or other vowels for unknown quantities, and B, C, G or other consonants for known quantities. For example, the equation that we would write as $x^3 + ax = b$, would be presented by Viète as: *"Si A cubus + D plano in A, æquetur, Z solido".*[1] We note that Viète remained a prisoner of geometric representation, which demanded respect for the principle of homogeneity, by which we can only add together magnitudes of the same dimension. Here A must be multiplied by a coefficient D *"plano"* (of dimension 2) and the sum is given as Z *"solido"* (of dimension 3). Descartes improved on Viète in two respects in his *Géométrie* of 1637. Firstly in rejecting the need to adhere to homogeneity:

> *It is not however the same thing when unity is determined, because unity can always be understood, even when there are too many or too few dimensions; thus, if we are*

1.   Viète, *De æquationem recognitione et emendatione tractatus duo*, p. 97.

required to extract the cube root of aabb – b we must consider the quantity aabb to be divided once by unity, and the quantity b multiplied twice by unity.[1]

His second contribution was to use the letters at the beginning of the alphabet for known quantities and the letters at the end of the alphabet for unknowns, a practice that still survives today.

Used as we are today to using letters in calculations, we find it difficult to appreciate the major change that took place in that development and even in the concept of mathematics itself. The use of literal symbols was not just a simple matter of presentation, like the use of a more efficient type face. The use of a letter set the mathematician free from his attachment to the concrete. The *arithmos* used by Diophantus and Al-Kwarismi's *res* [thing] were only used for either whole numbers or geometric magnitudes. But the letters a, b, x, y are independent of the concrete objects that they are intended to represent and gradually start to take on a life of their own. They gave rise to operations that had been previously invalid because they did not have a concrete representation. And while it remained impossible to think of *"a number which is increased by three to give one"*, or of *"a number whose square added to two makes one"*, equations like $x + a = b$ could be manipulated without scruple, and results like $x = b - a$, or obtaining $x = \sqrt{a}$ from $x^2 = a$, became a part of the practice of manipulation of symbols, and thereby opened a door on a radical modification of the concept and meaning of mathematics itself. The sense of mathematics was no longer to be found through a connection with an exterior concrete reality, but in an internal consistency of the rules and calculations. From this came a host of new numbers created by the mathematics itself: negatives, imaginaries, quaternions, octonions, etc. (see Chapter 13). The problem of the solution of equations has now been provided with remarkable new tools which, apparently no equation will be able to resist.

Using radicals to solve equations: cubics and quartics

To solve equations of higher degrees, the temptation at first was to attempt to produce a more general version of the solution for the quadratic. This is the approach in an anonymous fourteenth century manuscript[2] which gives several formulas and examples for solving cubics and higher degree equations. For example, the author gives

the solution to $ax^3 + bx^2 + cx = k$ as as $x = \sqrt[3]{\left(\dfrac{c^3}{b}\right) + \dfrac{k}{a}} - \dfrac{c}{b}$.

and the author supports this with the example $x^3 + 60x^2 + 1200x = 4000$ which has the exact solution $x = \sqrt[3]{12000} - 20$. In fact the formula is correct for all cubics for which the coefficients satisfy the condition $b^2 = 3ac$. In the same manuscript there are similar formulas for quartics and even for equations of the fifth degree. We must marvel at the author's ingenuity, if not for his rigour!

Exercise 4	Establish the condition between the coefficients that makes the solution a correct one.

1.   Descartes, *Géométrie,* p. 299. (Dover edition, p. 6-7).
2.   Cantor, *Vorlesungen über Geschichte der Mathematik*, vol. 2, p. 161.

Who should have the credit for being the first to produce an algorithm giving the exact solution to a cubic equation? The answer is a tricky one. According to Guillaume Libri[1] the credit should go to a certain Scipione del Ferro, a native of Bologna, where he taught mathematics between 1496 and 1525. But he died before publishing his discovery although he had confided it to his pupil Antonio Maria Fior who used the information in proposing problems to, among others, Tartaglia, in the popular mathematical contests that were in vogue.

The rest of the story is provided by the two claimants to the discovery, Tartaglia and Cardan.[2]

Eight days before the time limit for solutions imposed by Fiore, Tartaglia found the solution on 12 February 1535. At Cardan's insistence, Tartaglia revealed his solution to him in cryptic verse, as follows:

> *When the cube with the things*
> *is equal to a certain number* $x^3 + px = q$

> *Find two others which differ by this last* $U - V = q$

> *Then take this to be always so that*
> *their product is equal to* $U.V = \left(\dfrac{p}{3}\right)^3$
> *the cube of a third of the things*

> *Finally, the general result*
> *of their cube roots, subtracted* $x = \sqrt[3]{U} - \sqrt[3]{V}$
> *will yield the principal thing*[3]

For example, to solve the equation $x^3 + 6x = 2$, we put $U - V = 2$ and $U.V = 8$. The value of U (and $-V$) is found from the quadratic equation $t^2 - 2t - 8 = 0$ which gives $U = 4$ and $V = 2$ and so $x = \sqrt[3]{4} - \sqrt[3]{2}$.

Although it is not possible to be precise about the origin of this formula, there is some evidence in written records to guide us. In a manuscript, probably written by Regiomontanus in 1456, we find the following problem solved under the title: *"Cubus census et res equantur numero"*[4] which was: Given an initial capital of 100 Gulden, a certain man gains an interest of x at the end of the first year, and for the next two years the gains are proportionally the same; at the end of the third year he has altogether 265 Gulden. How much did he gain in the first year?

Curiously, Regiomontanus did not solve the problem in the usual way with interest problems by considering: $100\left(1 + \dfrac{x}{100}\right)^3 = 265$.

Instead, he undertook the year by year calculation that led to an equation that we would write as:

$$100 + 3x + \frac{3x^2}{100} + \frac{x^3}{10000} = 265 \text{ or } x^3 + 300x^2 + 30000x = 1650000.$$

1.   Libri, *Histoire des Sciences mathématiques en Italie*, tome 3, p. 148.
2.   Tartaglia, *Quesiti et inventioni diverse*.
      Cardan, *Ars Magna*.
3.   Tartaglia, *Quesiti et inventioni diverse*, p. 266.
4.   Tropfke, *Geschichte der Elementar Mathematik*, p. 443.

which corresponds to an equation of the type described in the title (a cube plus things equals a number):

$$x^3 + a_2x^2 + a_1x = a_0 \qquad (1)$$

Regiomontanus compares this with the equation:

$$(x+c)^3 = x^3 + 3cx^2 + 3c^2x + c^3 = a_0 + c^3 \qquad (2)$$

which he describes geometrically by the decomposition of a cube, and which is identical to (1) if

$$c = \frac{a_2}{3} = \sqrt{\frac{a_1}{3}} \text{ that is, if } a_2{}^2 = 3a_1$$

and x is then obtained by use of the formula:

$$x = \sqrt[3]{a_0 + \left(\frac{a_2}{3}\right)^3} - \frac{a_2}{3} = \sqrt[3]{2650000} - 100.$$

We can now have a better understanding of the formula given in the anonymous manuscript cited earlier and also the probable origin of Tartaglia's rule given above. In effect, the equation (2) can be written as

$$(x+c)^3 = x^3 + 3cx(x+c) + c^3 = a_0 + c^3$$

and comparing it with (1) we get two equations in two unknowns x and c:

$$(x+c)^3 = a_0 + c^3 \text{ and } 3c(x+c) = a_1$$

By putting $x + c = v$, we get: $v^3 - c^3 = a_0$ and $vc = \left(\frac{a_1}{3}\right)$ or $v^3c^3 = \left(\frac{a_1}{3}\right)^3$

Exercise 5 Solve the following equations, using the Cardan-Tartaglia method: $x^3 + 9x = 100$; $x^3 + 4x = 1$; $x^3 + 3x = 14$

Following further requests from Cardan, Tartaglia gave him more explanations by letter, pledging him not to divulge the secret. But Cardan broke his promise when he published the method in his *Ars Magna* in 1545. Even though Cardan credited Scipione del Ferro as well as Tartaglia as having discovered the method before him, Tartaglia was annoyed because he himself has wished to publish the method in one of his own works. But in the meantime Cardan had, in 1542, been able to examine the posthumous writings of de Ferro, held by his son-in-law, and to note the complete agreement of de Ferro's solution with that of Tartaglia; he therefore no longer felt obliged to consider that the solution was the property of Tartaglia.[1]

In the same *Ars Magna* (Chap. 39) Cardan gives the solution to the quartic equation, found by his pupil Ludovico Ferrari, who died young in 1565. Ferrari's method consists of eliminating the cube term in $x^4 + ax^3 + bx^2 + d = cx$ by use of the substitution $x = t - (a / 4)$ to produce an equation of the form $t^4 + pt^2 + q = rt$.

Start by putting the t^4 term on its own and then changing each side of the equation into a square by introdusing of a parameter x:

$$t^4 + 2t^2x + x^2 = (2x - p)t^2 + rt + x^2 - q$$

The left hand side is a complete square by construction and the right hand side will be one if its discriminant is zero, that is if

$$r^2 - 4(2x - p)(x^2 - q) = 0$$

---

1.   Cantor, *Vorlesungen über Geschichte der Mathematik*, tome 2, p. 480 *sqq.*

which determines the value of x by the solution of a cubic.

Example: Consider the equation $t^4 - 24t - 2 = 0$. We need to find x so that the equation becomes

$$(t^2 + x)^2 = 2t^2x + 24t + x^2 + 2$$

where the right hand side should be a complete square. This will be so if $\Delta = 144 - 2x(x^2 + 2) = 0$ which gives $x^3 + 2x - 72 = 0$ and this is true for $x = 4$.

Hence $(t^2 + 4)^2 = 8t^2 + 24t + 18 = 2(2t + 3)^2$

and the four roots are:

$$t_1 = -\sqrt{2} + i\sqrt{2 + 3\sqrt{2}} \qquad\qquad t_3 = \sqrt{2} + \sqrt{3\sqrt{2} - 2}$$

$$t_2 = -\sqrt{2} - i\sqrt{2 + 3\sqrt{2}} \qquad\qquad t_4 = \sqrt{2} - \sqrt{3\sqrt{2} - 2}$$

Exercise 6   Use Ferrari's method to solve the equations:
$$x^4 = 9x^2 + 24x + 16; \quad x^4 = 7x^2 + 24x + 15; \quad x^4 + x = 1$$

Using radicals to solve equations: quintics and higher degree equations

With the highly acclaimed discovery of solutions for cubics and quartics, the Italian mathematicians had opened up an apparently fertile line of research, namely the search for general methods for the solution of polynomials of whatever degree. Many mathematicians engaged on the project, improving existing methods and attempting to find solutions for higher degree polynomials: Descartes, Hudde, Tschirnhaus, Bézout, Euler, each producing a new method for cubics or quartics. But all failed in the case of the quintic. Euler felt convinced that a general method did exist, having shown that there existed a whole class of quintic equations having solutions of the form:

$$A\sqrt[5]{v} + B\sqrt[5]{v^2} + C\sqrt[5]{v^3} + D\sqrt[5]{v^4}.$$

For example, the equation $x^5 = 2625x + 61500$ has a root equal to:

$$\sqrt[5]{75(5 + 4\sqrt{10})} + \sqrt[5]{225(35 + 11\sqrt{10})} + \sqrt[5]{225(35 - 11\sqrt{10})} + \sqrt[5]{75(5 - 4\sqrt{10})}$$

Two and a half centuries passed in such speculations; an end had to come. In 1770 and 1771 Lagrange[1] attacked the problem from a different angle. Instead of looking directly for a method of solution of equations of the fifth degree, he made an inventory of all the existing methods for solving equations of a lower degree, and sought to understand what were the mathematical properties of these equations and what were the underlying ideas in the methods that allowed their solution. In this way he immediately made several fundamental discoveries. We shall present them here, first of all by means of an example: $t^4 - 24t - 2 = 0$, which we shall refer to as (1). The equation has four roots, two real and two complex:

$$a = \sqrt{2} + \sqrt{3\sqrt{2} - 2} \qquad\qquad c = -\sqrt{2} + i\sqrt{2 + 3\sqrt{2}}$$

$$b = \sqrt{2} - \sqrt{3\sqrt{2} - 2} \qquad\qquad d = -\sqrt{2} - i\sqrt{2 + 3\sqrt{2}}$$

---

1.   Lagrange, *Réflexions sur la résolution algébrique des équations*.

(This example, as well as the other numerical examples, were constructed by the present author in order to make it easier to understand.)

Why is equation (1) solvable with radicals, and in general any fourth degree polynomial?

Because, says Lagrange, there exists a function on the roots a, b, c, d which can only take three distinct values from the set of all $4! = 24$ possible permutations of these four roots.

In order to understand the reasoning advanced by Lagrange, we first of all recall some classical results on polynomial equations and their roots.

a) If $P(x)$ is a polynomial of degree n with real coefficients, then there exist n roots, real or complex, distinct or not.

We shall use: $P(x) = a_n x^n + a_{n-1} x^{n-1} + \ldots + a_1 x + a_0$ and $x_1, x_2, \ldots x_n$ for the roots.

b) There are relations between the coefficients and the roots which are called elementary symmetric functions of the roots:

$$\sigma_1 = \sum_{i=1}^{n} x_i = -\frac{a_{n-1}}{a_n}; \ \sigma_2 = \sum_{i<j} x_i x_j = +\frac{a_{n-2}}{a_n}; \ \sigma_3 = \sum_{i<j<k} x_i x_j x_k = -\frac{a_{n-3}}{a_n}; \ \text{etc}\ldots$$

up to $\sigma_n = (-1)^n x_1 x_2 \ldots x_n$.

c) It can be shown that all other symmetric functions on the n roots can be written as rational expressions using only the σ_i. For example:

$$x_1^3 + x_2^3 + \ldots + x_n^3 = \sigma_1^3 - 3\sigma_1\sigma_2 + 3\sigma_3.$$

d) Non symmetric functions of the roots have expressions (Lagrange talks of values) which change with permutations of the roots. A priori, there are n! permutations of the roots and so n! distinct values for a function of n roots. For example, in the case of a cubic, the function $\phi(x_1, x_2\ x_3) = x_1 + 2x_2 + 3x_3$ will take the six following values, using permutations of the roots:

Id:	$x_1 + 2x_2 + 3x_3$	(2, 3)	$x_1 + 2x_3 + 3x_2$
(1, 2)	$x_2 + 2x_1 + 3x_3$	(1, 2, 3)	$x_2 + 2x_3 + 3x_1$
(1, 3)	$x_3 + 2x_2 + 3x_1$	(1, 3, 2)	$x_3 + 2x_1 + 3x_2$

[We use the usual notation for permutations where $(1, 2, 3, \ldots, n)$ represents a cyclic permutation where 1 is replaced by 2, 2 is replaced by 3, etc, and n is replaced by 1.]

e) If there is a function of the roots which takes fewer than n! distinct values, say m distinct values where $m < n!$, then these m values can be found from an equation of degree m, the coefficients of which can be found by means of the elementary symmetric functions.

For illustration consider the equation $t^4 - 24t - 2 = 0$ and the function on the four roots (for the moment supposed unknown): $\phi(a, b, c, d) = ab + cd$. It turns out that ϕ is invariant for the following eight permutations:

$S_0 = \text{Id}; S_1 = (3,4); S_2 = (1,2); S_3 = (1,2)(3,4); S_4 = (1,3)(2,4)$

$S_5 = (1,3,2,4); S_6 = (1,4,2,3); S_7 = (1,4)(2,3)$

For example, S_6 applied to ϕ produces $dc + ab$, that is the same expression as $ab + cd$, as can be shown in this diagram:

$$\begin{pmatrix} 1\ 2\ 3\ 4 \\ a\ b\ c\ d \\ 4\ 3\ 1\ 2 \\ d\ c\ a\ b \end{pmatrix}$$

From this it follows that of the 24 total permutations, there are only three *"truly"* distinct values, which are:

$\phi_0 = ab + cd$; $\phi_1 = ac + bd$; $\phi_2 = ad + bc$

These can be calculated by using the results:

$a + b + c + d = \sigma_1 = 0$

$ab + bc + cd + ad + ac + bd = \sigma_2 = 0$

$abc + abd + acd + bcd = \sigma_3 = 24$

$abcd = \sigma_4 = -2$

We have:

$\phi_0 + \phi_1 + \phi_2 = \sigma_2 = 0$.

$\phi_0\,\phi_1 + \phi_1\phi_2 + \phi_2\,\phi_0 = (ab + cd)(ac + bd) + (ac + bd)(ad + bc) + (ad + bc)(ab + cd)$

$$= \sigma_1\sigma_3 - 4\sigma_4 = 8.$$

$\phi_0\,\phi_1\,\phi_2 = (ab + cd)(ac + bd)(ad + bc) = \sigma_1{}^2\sigma_4 - \sigma_3{}^2 = 24^2 = 576.$

And so ϕ_0, ϕ_1, ϕ_2 are the solutions of the cubic: $x^3 + 8x - 576 = 0$, whose roots are

$x_1 = 8 \qquad x_2 = -4 + 2i\sqrt{14} \qquad x_3 = -4 - 2i\sqrt{14}$

Since we have no other information about ϕ_0, ϕ_1, ϕ_2 we choose $\phi_0 = 8 = ab + cd$.

The complementary relations are:

$abcd = -2$; $a + b + c + d = 0$; $(a + b)cd + (c + d)ab = 24$

and we can conclude:

1) ab and cd are roots of $z^2 - 8z - 2 = 0$, because $ab + cd = 8$ and $abcd = -2$: hence $ab = 4 - 3\sqrt{2}$ and $cd = 4 + 3\sqrt{2}$.

2) since $c + d = -(a + b)$ we deduce that $a + b = \dfrac{24}{cd - ab} = \dfrac{24}{6\sqrt{2}} = 2\sqrt{2}$.

This implies that a, b are roots of the quadratic: $u^2 + 2\sqrt{2}u + 4 - 3\sqrt{2} = 0$. The discriminant simplifies to $18 - \sqrt{2}$ and the roots are:

$$a = \sqrt{2} + \sqrt{3\sqrt{2} - 2} \text{ and } b = \sqrt{2} - \sqrt{3\sqrt{2} - 2}.$$

Similarly, starting with $cd = 4 + 3\sqrt{2}$ and $c + d = -2\sqrt{2}$ we can find

$$c = -\sqrt{2} + i\sqrt{2 + 3\sqrt{2}} \text{ and } d = -\sqrt{2} - i\sqrt{2 + 3\sqrt{2}}.$$

In a general way, Lagrange demonstrates the following properties:

• If $x_1, x_2, \ldots, x_n$ are the n roots of an equation of degree n, and $\phi\,(x_1, x_2, \ldots, x_n)$ is a rational function of these n roots, then the number of *"distinct values"* that ϕ can take when considering all possible permutations of these n roots, is a factor of n!

• If cubic and quartic equations are solvable with radicals it is because there exists a function of the roots which has *"fewer"* distinct values.

For a cubic, $x^3 + px^2 + qx + r = 0$, the function $(x_1 + jx_2 + j^2x_3)^3$ takes only two distinct values: $T_1 = (x_1 + jx_2 + j^2x_3)^3$ and $T_2 = (x_1 + j^2x_2 + jx_3)^3$ where $j = e^{\frac{2\pi i}{3}}$. It follows that $T_1 + T_2$ and T_1T_2 are symmetric functions of the roots which can be computed entirely from the coefficients of the equations. The values of T_1 and T_2 therefore allow us to find the three roots of the cubic by solving:

$$\begin{cases} x_1 + x_2 + x_3 = -p \\ x_1 + jx_2 + j^2x_3 = \sqrt[3]{T_1} \\ x_1 + j^2x_2 + jx_3 = \sqrt[3]{T_2} \end{cases}$$

For quartic equations, we have seen through the use of an example, the existence of a function $\phi(a, b, c, d) = ab + cd$ which only takes three distinct values, showing how the solution is possible by use of auxiliary equations constructed from symmetric functions of the ϕ_i and which Lagrange called resolvent equations.

It follows that if a solution to a quintic, or higher degree equation, is to be found, it becomes necessary to find a function of the roots that takes less than 5 values (not all being symmetric), which will lead to an equation of degree less than 5, which can then be solved.

Lagrange's memoir had an immense influence upon his successors. By its complete treatment of the question, although he was not able to solve the problem, he indicated the way in which progress could be made, he identified the tasks to be done, and he presented the tools for the job. More specifically, he set the research task of finding a function of five variables that took only three or four distinct values when it acted on the 5!, that is 120, possible permutations.

Above all, Lagrange introduced the idea that to solve an algebraic equation was to be concerned with those permutations that conserved or modified a given function of the roots. This was the first intuition into the fundamental role that is played by the group of an equation, and its subgroups, in its solution: an idea that Galois was to take up sixty years later.

In the same year that the second part of Lagrange's memoir appeared, 1771, there appeared another memoir by Vandermonde[1] which explored the same question but from a different perspective. It is not possible to give a detailed analysis of this memoir which touched on several of Lagarnge's results, although it did not have the breadth nor the precision of Lagrange's work. It did, however, point to a result that had not been clearly and explicitly identified by Lagrange: all radicals appearing in the final formulas to be solved can be expressed as rational functions of the roots. For example, the discriminant of the quadratic $ax^2 + bx + c = 0$ is in fact equal to $a^2 (x_1 - x_2)^2$ where x_1, x_2 are the roots of the equation. Also, Vandermonde proved his ability to penetrate the subject by giving an explicit solution to the 10th degree equation

$$\frac{x^{11} - 1}{x - 1} = x^{10} + x^9 + \dots + x^2 + x + 1 = 0.$$

Following the work done by Vandermonde, and most of all that of Lagrange, we can identify two currents of research. In one, mathematicians pursued the path opened up by Lagrange, and tried to find the key to the general solution of all algebraic equations of

---

1.   Vandermonde, *Mémoire sur la résolution des équations.*

whatever degree in the study of sets of permutations. This was the path followed by Ruffini, Cauchy and Abel. Ruffini studied the properties of functions of five variables and showed that there was no rational function that could take only 8, 4 or 3 values when the five variables are permuted in all 120 possible ways. Cauchy generalised this result by proving that

> *The number of different values of a non symmetric function of n variables cannot be less than the greatest prime number p contained in n, unless it is equal to 2.*

All these results, together with his own research, enabled Abel to prove the impossibility of solving an equation of degree 5 or higher by the use of radicals.

The other direction of research was to try to understand how and why certain classes of equations (the cyclotomic equations, whose roots are related in a certain way) could be solved by means of radicals. This was the research undertaken by Gauss and Abel.

It is remarkable that Galois was able to utilise the results of both research areas to achieve a marvellous synthesis.

The solution of the cyclotonic equation by Gauss

In a letter to his friend and colleague Gerling, Gauss wrote as follows:

> *By thinking very deeply about the relation existing between the set of roots (of the equation $\dfrac{x^p - 1}{x - 1} = 0$) for the arithmetical reasons, I succeeded in grasping very clearly the nature of this relation, while on a holiday in Brunswick, the morning of that day (before getting up) – it was the 30 March 1796 – in such a way that I was able immediately to apply it to the particular application to a polygon of 17 sides and to carry out the numerical calculations.*

This remarkable discovery decided the young Gauss (it was just before his 19th birthday) on a career in mathematics, and he started to keep a scientific journal on that day.

The equation in question, $\dfrac{x^p - 1}{x - 1} = 0$, is still called the **cyclotonic equation** by reason that its roots $e^{i\frac{2k\pi}{p}}$ can be plotted in the complex plane as the vertices of a regular p-sided polygon inscribed in a unit circle. We shall look at the main ideas put forward by Gauss, starting with the example of the 17 sided polygon to which he referred in his letter.

If we let r be the root $e^{i\frac{2\pi}{17}}$, the other roots are given successively by the powers r^m. But instead of taking these in the order of the exponents, Gauss ordered them according to the powers of 3, taken modulo 17, shown in the following table:

k	0	1	2	3	4	5	6	7	8	9	10	11	12	13	14	15	16
m = 3^k	1	3	9	10	13	5	15	11	16	14	8	7	4	12	2	6	1

Since, for example, $3^6 = 17 \times 42 + 15$, the value k = 6 is mapped to m = 15, and we can see that all the integers from 1 to 16 appear as a value for m but in a different order. The number 3 is said to be a primitive root modulo 17, because 16 is the least value for which $3^k = 1$ (mod 17). (By contrast, 2 is not a primitive root modulo 17, since its powers produce the sequence: 1, 2, 4, 8, 16, 15, 13, 9, 1, which then repeats when we reach

$$2^8 = 256 = 17 \times 15 + 1)$$

We can use this ordering to produce functions of the roots which remain invariant under certain permutations and so allow us to construct intermediate resolvent equations (in Lagrange's sense of the term), by grouping the roots into sums containing 8, 4, 2, 1 roots respectively, corresponding to the factors of 16. Using $\omega_h = r^h$ for the root, we have firstly the two cycles, or periods, of eight terms:

$$p = \omega_1 + \omega_9 + \omega_{13} + \omega_{15} + \omega_{16} + \omega_8 + \omega_4 + \omega_2$$

$$p' = \omega_3 + \omega_{10} + \omega_5 + \omega_{11} + \omega_{14} + \omega_7 + \omega_{12} + \omega_6$$

Note that p, as well as p', contains at the same time both ω_a and its inverse $\omega_a^{-1} = \omega_{17-a}$ Regrouping we have:

$$p = (\omega_1 + \omega_{16}) + (\omega_2 + \omega_{15}) + (\omega_4 + \omega_{13}) + (\omega_8 + \omega_9)$$

$$p' = (\omega_3 + \omega_{14}) + (\omega_5 + \omega_{12}) + (\omega_6 + \omega_{11}) + (\omega_7 + \omega_{10})$$

Now, p and p' can be found since $p + p' = -1$ (the sum of all the roots except 1) and $p.p' = -4$ using the relation $\omega_a . \omega_b = \omega_c$ where $c = a + b \pmod{17}$. This means that p, p' are roots of the quadratic equation $x^2 + x - 4 = 0$.

Regrouping the roots in p, and also in p', again we have:

$$\begin{cases} p_1 = \omega_1 + \omega_{13} + \omega_{16} + \omega_4 \\ q_1 = \omega_9 + \omega_{15} + \omega_8 + \omega_2 \end{cases}$$

$$\begin{cases} p'_1 = \omega_3 + \omega_5 + \omega_{14} + \omega_{12} \\ q'_1 = \omega_{10} + \omega_{11} + \omega_7 + \omega_6 \end{cases}$$

We can again calculate $p_1 + q_1 = p$ and $p_1 q_1 = p + p' = -1$; $p'_1 + q'_1 = p'$ and $p'_1 q'_1 - 1$. And so:

p_1 and q_1 are roots of $x^2 - px - 1 = 0$

p'_1 and q'_1 are roots of $x^2 - p'x - 1 = 0$

Since $p_1 = 2\cos\frac{2\pi}{17} + 2\cos\frac{8\pi}{17}$ (because $\omega_a + \omega_{17-a} = 2\cos\frac{2a\pi}{17}$) we know that p_1 is the positive root of $x^2 - px - 1 = 0$ and so q_1 is the negative root.

In the same way, we see that $q'_1 = 2\cos\frac{12\pi}{17} + 2\cos\frac{14\pi}{17}$ is negative.

Regrouping for the last time:

$p_2 = \omega_1 + \omega_{16}; q_2 = \omega_{13} + \omega_4$, we have: $p_2 + q_2 = p_1$ and $p_2 q_2 = p'_1$

which allows us to calculate, among others, $\omega_1 + \omega_{16} = 2\cos\frac{2\pi}{17}$, and so by replacing the different values of p, p_1, etc. we get:

$$p = \frac{\sqrt{17} - 1}{2} \; ; p' = -\left(\frac{\sqrt{17} + 1}{2}\right)$$

$$p_1 = \frac{p + \sqrt{p^2 + 4}}{2} = \frac{1}{4}(\sqrt{17} - 1 + \sqrt{34 - 2\sqrt{17}}); q_1 = \frac{1}{4}(\sqrt{17} - 1 - \sqrt{34 - 2\sqrt{17}})$$

$$p'_1 = \frac{p' + \sqrt{p'^2 + 4}}{2} = \frac{1}{4}(-\sqrt{17} - 1 + \sqrt{34 + 2\sqrt{17}}); \; q'_1 = \frac{1}{4}(-\sqrt{17} - 1 - \sqrt{34 + 2 + \sqrt{17}})$$

and finally

$$\frac{1}{2} p_2 = \cos \frac{2\pi}{17} = -\frac{1}{16} + \frac{1}{16}\sqrt{17} + \frac{1}{16}\sqrt{34 - 2\sqrt{17}} + \frac{1}{8}\sqrt{17 + 3\sqrt{17} - \sqrt{34 - 2\sqrt{17}} - 2\sqrt{34 + 2\sqrt{17}}}$$

which allows us to calculate all ω_a.

The method presented here for n = 17 can be generalised to any n provided that the resolvent equations are of degree n_i where n_i is one of the prime factors in the decomposition of (n − 1). A remarkable aspect of this example is that $\cos \frac{2\pi}{17}$ is obtainable entirely from a sequence of quadratic equations (here n − 1 = 16 = 2.2.2.2) and so the length of $\cos \frac{2\pi}{17}$ can be constructed by a straight edge and a compass, and this enabled the young Gauss to establish that a polygon of 17 sides is constructible by "ruler and compass" methods. More generally, his research enabled him to affirm:

> For it to be possible to divide a circle geometrically into N parts, N has to be 2, or a power of 2, or a prime number of the form $2^m + 1$ or even the product of a power of 2 and one or more different prime numbers of that form.

But, for whatever integer n, the cyclotonic equation is always solvable in terms of radicals, since each root is a power of one and the same root $e^{i\frac{2\pi}{n}}$.

Exercise 7   Find the equations for solving the equation $\frac{x^7 - 1}{x - 1} = 0$. Note that 3 is a primitive root modulo 7.

The emergence of the idea of the group

The brilliant mathematician Galois was able to synthesize all these results and to understand and explain the unique and underlying reason for them: the existence of the set of permutations with those properties that we call the group of the equation. Galois' text is very dense and general, which makes it difficult to read. We shall try to grasp the main ideas by looking at the general equation of the quartic:

1) $x^4 + px^3 + qx^2 + rx + s = 0$,

and the particular equation which has been solved above:

2) $x^4 - 24x - 2 = 0$.

Let a, b, c, d be used to refer to the roots in each case. Galois starts by pointing out that to ask the question whether an equation is solvable makes no sense unless the field in which the solution is to be found is stated. Thus equation 2) has no solutions at all if I restrict myself to the rationals, which however are sufficient for writing the equation (in this case the coefficients are all integers). But from the simple reading of the equation I am unable to tell if I need rationals, real or complex numbers to solve it. The only *"numeric"* information available which I have about the roots is given by the equalities:

$a + b + c + d = - p$

$ab + ac + ad + bc + bd + cd = q$

$$abc + acd + bcd + abd = -r$$

$$abcd = s$$

in the case of equation 1), and the same relations for equation 2), taking $p = q = 0$, $r = -24$, $s = -2$.

This lack of information illustrates the fact that we are unable to distinguish between the 24 possible permutations of a, b, c, d, which all produce the same values for the coefficients p, q, r, s, nor can we distinguish between the 24 following squares. (Imagine that a, b, c, d represent different colours for the corners of the squares.) First we have the eight isometries of the square:

d b	c b	d a	c a	b d	a d	b c	a c
a c	a d	b c	b d	c a	c b	d a	d b

now interchange just b and c

d c	b c	d a	b a	c d	a d	c b	a b
a b	a d	c b	c d	b a	b c	d a	d c

and then interchange just c and d

c d	b d	c a	b a	d c	a c	d b	a b
a b	a c	d b	d c	b a	b d	c a	c d

Let G be the set of these 24 permutations and G_1 the eight permutations given in the first row. For a quartic equation we have seen that the function of the roots $\phi(a, b, c, d) = ab + cd$ only takes three distinct values from the above 24 permutations. The invariance of $ab + cd$ can be illustrated in all the squares of the first row by the fact that (a, b) and (c, d) lie on the diagonals of the square. The two other rows of squares correspond to the values $ac + bd$ and $ad + bc$ respectively. These three values of ϕ can thus be found from an auxiliary cubic equation (see p. 321-324). The choice of one of these values provides the supplementary information necessary that restricts us to one of the rows. At this point, there exists one of two possibilities:

– either the solution of the auxiliary equation introduces radicals which change the domain of rationality needed to express the value of $ab + cd$, as is the general case, equation 1). The group for that equation is formed from the set of all 24 permutations. The adjunction of the irrational value $ab + cd$ allows me to restrict myself to G_1 which is called a sub-group of G.

– or the solution of the auxiliary equation produces a rational value for $ab + cd$, as in equation 2) where it takes the value 8. In that case G_1 is produced immediately, and the group of the equation is G_1 (or a sub group of G_1 if there is more *"rational information"*). In both cases we only have to choose between the eight permutations of the first row of squares, those that fix the value of $ab + cd$. In what follows, we shall just consider the example equation 2).

Let us consider the function of the roots $\theta(a, b, c, d) = ab$. This function is invariant for the first four permutations of G_1 and changes to cd for the following four. We have seen (p. 000) that it can be calculated if we include $\sqrt{2}$. The choice of $ab = 4 - 3\sqrt{2}$ now restricts us to a new subgroup G_2 formed from the four permutations which fix the value of ab (a and b lying always on the same diagonal.)

d b	c b	d a	c a
a c	a d	b c	b d

Finally bringing in a final irrational $\sqrt{3\sqrt{2}-2}$ fixes the values of a and b of the group G_3:

d b	c b
a c	a d

and the choice of sign in front of $i\sqrt{3\sqrt{2}-2}$ fixes the values of c and d, arriving at the complete determination of the four roots, that is to the choice of a unique square:

d b
a c

The situation demonstrated by these examples is, in fact, completely general and was the basis for two fundamental theorems stated by Galois:[1]

> *Proposition 1: Let there be a given equation with m roots a, b, c, ... There will always be permutations of the letters a, b, c, ... which have the following properties:*
> *1. that every function of the roots which remains invariable under substitution in that group can be known rationally*
> *2. reciprocally, every function of the roots that can be determined rationally will be invariable under these substitutions.*

> *Proposition 2: If to a given equation is added the root r of an irreducible auxiliary equation there are two possibilities:*
> *1. Either the group of the equation does not change, or it can be partitioned into p groups each belonging respectively to the group of the proposed equation when each of the roots of the auxiliary equation is added.*
> *2. These groups possess the remarkable property that it is possible to move from one to the other by using the same substitution of letters in all the permutations of the first.* (For example, interchanging b and c, or interchanging c and d, in the three rows of squares shown in the example above.)

If we return now to the example of the quartic equation, the reader will notice that the group G_1 is identical to the set of eight transformations of the plane that preserve the square abcd:

id: Identity

d b
a c

S_1:   Reflection at diagonal [ab]

c b
a d

S_2:   Reflection at diagonal [cd]

d a
b c

S_3:   Rotation of π about the centre of the square O

c a
b d

---

1.   Galois, *Mémoire sur les conditions de résolubilité...*, p. 421 and 423.

S$_4$:   Reflection at the vertical, bisector of [ac]

b	d
c	a

S$_5$:   Rotation of $(+ \pi / 2)$ about O

b	c
d	a

S$_6$:   Rotation of $(- \pi / 2)$ about O

a	d
c	b

S$_4$:   Reflection at the horizontal, bisector of [bc]

a	c
d	b

These eight transformations can be combined as shown in the following table defining what is called the group of the square:

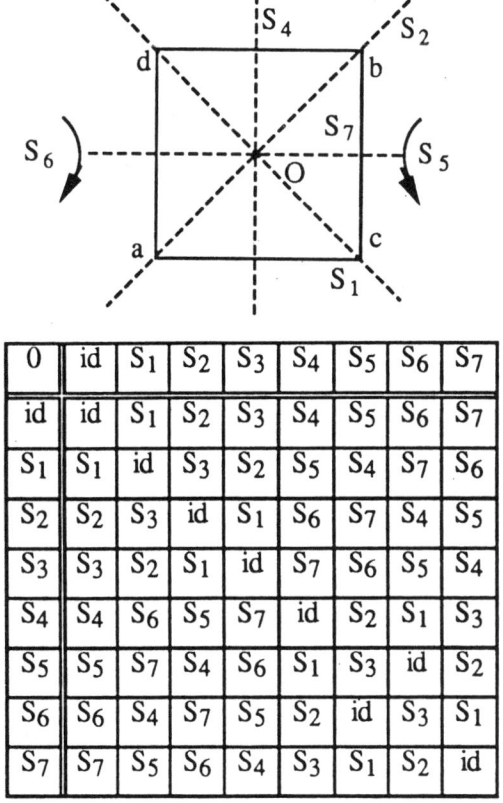

0	id	S$_1$	S$_2$	S$_3$	S$_4$	S$_5$	S$_6$	S$_7$
id	id	S$_1$	S$_2$	S$_3$	S$_4$	S$_5$	S$_6$	S$_7$
S$_1$	S$_1$	id	S$_3$	S$_2$	S$_5$	S$_4$	S$_7$	S$_6$
S$_2$	S$_2$	S$_3$	id	S$_1$	S$_6$	S$_7$	S$_4$	S$_5$
S$_3$	S$_3$	S$_2$	S$_1$	id	S$_7$	S$_6$	S$_5$	S$_4$
S$_4$	S$_4$	S$_6$	S$_5$	S$_7$	id	S$_2$	S$_1$	S$_3$
S$_5$	S$_5$	S$_7$	S$_4$	S$_6$	S$_1$	S$_3$	id	S$_2$
S$_6$	S$_6$	S$_4$	S$_7$	S$_5$	S$_2$	id	S$_3$	S$_1$
S$_7$	S$_7$	S$_5$	S$_6$	S$_4$	S$_3$	S$_1$	S$_2$	id

Figure 7

Exercise 8   Construct the group of the equation $\dfrac{x^7 - 1}{x - 1} = 0$.

Conclusion

We can now understand better why it is that the cyclotonic equation $\frac{x^n - 1}{x - 1} = 0$ is always solvable in terms of radicals. Its roots correspond to a regular polygon which possesses a great many symmetries. On the other hand, any given equation may be one whose roots do not possess sufficient symmetries, and so makes it impossible to solve algebraically. The theory of groups invented by Galois is a sort of second view for detecting hidden symmetries, and not only those in equations. In less than a century, the theory of groups became what Guillen called *"a veritable extra sense, a means by which scientists are able to become aware of all the beautiful symmetries that nature conceals, and which had escaped them when they only had their five senses available to them: The theory of groups allows them to uncover symmetries of objects through the simple observation of the equations that describe them."*[1] As a consequence, that theory forms the basis of most of the models for physics, chemistry, and of course mathematics, so that in a certain way the dream of the scribe Ahmose, quoted at the beginning of this chapter, has been realised: *"... calculations for examining nature and for understanding all that exists, each mystery ... each secret."*

---

1.   Guillen, *Des ponts vers l'infini*, p. 92.

Bibliography

Sources texts

CANTOR, *Vorlesungen über Geschichte der Mathematik*, 2 vol., Johnson Reprint Corporation, New York, B.G. Teubner Verlag Gesellschaft, Stuttgart, 1965.

CARDAN, *Ars Magna sive de regulis algebraisis*, Nuremberg, 1545.

CAUCHY, Mémoire sur le nombre des valeurs qu'une fonction peut acquérir, lorsqu'on y permute de toutes les manières possibles les quantités qu'elle renferme, *Journal de l'Ecole Polytechnique*, 17e cahier 10, 1815, 1-28.

COUCHOUD, *Recherches sur les connaissances mathématiques de l'Egypte pharaonique*, Thèse; Editions du Léopard d'Or, Paris, 1993.

GALOIS, Oeuvres mathématiques d'Evariste Galois, in *"Journal de mathématiques pures et appliquées"*, **11**, 1846; See also E. Galois, *Ecrits et mémoires mathématiques*, Edited by R. Bourgne and J.P. Azra, Gauthiers Villars, Paris, 1962.

GAUSS, *Recherches arithmétiques*, appeared in 1801 under the title *Disquisitiones Arithmeticæ* (Fleischer – Lipsiæ 1801); Fr. tr. A. C. M. Poullet-Delisle, Courcier, Paris, 1807; republished Blanchard, Paris, 1979.

GUILLEMOT & SPIESSER, "Du rouleau de cuir au problème 31 du papyrus Rhind", *Actes du colloques Inter-Irem*, Montpellier, 1985, p. 179-192.

GUILLEN, *Des ponts vers l'infini*, Bibliothèque Albin-Michel Sciences, Paris, 1992; translation of *Bridges to Infinity,* Jeremy Tarcher, 1983.

LAGRANGE, Réflexions sur la résolution algébrique des équations, *Mémoires de l'Académie royale des Sciences et Belles-Lettres de Berlin*, 1770, pp. 134-215, 1771, pp. 138-253.

LIBRI, *Histoire des Sciences mathématiques en Italie*, 4 vols. Paris, 1828-1841.

NEUGEBAUER, *The Exact Sciences in Antiquity*, Dover, New York, 1957.

RADFORD, *Diophante et l'algèbre présymbolique*, L'ouvert. N° 68, Sept 92, Revue de l'IREM de Strasbourg.

SPIESSER, *Equations du Premier degré*, IREM de Toulouse, 1982.

TARTAGLIA, *Quesiti et inventione diverse*, Brisciano, Venice, 1542.

TROPFKE, *Geschichte der Elementar-Mathematik Band I*, 4th ed. (ed. K. Vogel, K. Reich, H. Gericke), De Gruyter, Berlin and New York, 1980.

VANDERMONDE, Mémoire sur la résolution des équations, *Mémoires de l'Académie Royale des Sciences de Paris* (1771), 1774, pp. 365-416.

VER EECKE, *Diophante d'Alexandrie. Les six livres Arithmétiques et le Livre des Nombres Polygones*, Desclée de Brouwer, Liège. Repr. Blanchard, Paris, 1959.

VIETE, *De aequationum recognitione et emendatione tractatus duo*, Ed. A. Anderson, Paris, 1615.

Reference texts

CARREGA, *Théorie des corps – La règle et le compas*, Herman, Paris, 1981.

CHACE, *The Rhind Mathematical Papyrus*, Oberlin, Ohio: Mathematical Association of America, 1927-1929, abridged reprint in 1 vol., *Classics in Mathematics Education* 8, Reston, Virginia: The National Council of Teachers of Mathematics, 1979, 1986.

MUTAFIAN, *Equations algébriques et théorie de Galois,* Vuibert, Paris, 1980.

SERRET, *Cours d'algèbre supérieure*, 2 vol., Paris, 1854.

VERRIEST, *Evariste Galois et la théorie des équations algébriques*, Gauthiers-Villars, Paris, 1951.

WUSSING, *Die Genesis des abstrakten Gruppenbegriffs*, VEB, Deutscher Verlag der Wissenschaft, Berlin, 1969; English tr., MIT Press, 1984.

How did you get on?

Ex. 1   a) Let the false position be $x_0 = 27$, because of the fractions.
The sum becomes $b_1 = 20$ and so $x = 27 \times \dfrac{10}{20} = \dfrac{27}{2} = 13\dfrac{1}{2}$

b) Use false position $x_0 = 6$:

$$x_0 - \frac{2}{3}x_0 = 2; \quad x = 6 \times \frac{\frac{3}{4}}{2} = \frac{9}{4} = 2\frac{1}{4}$$

Ex. 2   (Method proposed by Ben Azra)

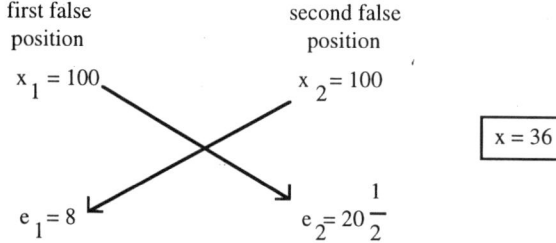

first false
position

second false
position

$x_1 = 100$

$x_2 = 100$

$\boxed{x = 36}$

$e_1 = 8$

$e_2 = 20\dfrac{1}{2}$

Ex. 3 $\beta^2 - \alpha^2 = 1$. Diophantus puts $\beta + \alpha = m$ so $\beta - \alpha = \dfrac{1}{m}$. Hence $\beta = (1/2)(m + 1/m)$ and $\alpha = (1/2)(m - 1/m)$ from which we get $x = \alpha^2 - 2 = \beta^2 - 3 = (m^4 - 10m^2 + 1)/(4m^2)$. Diophantus takes $m = 4$ and finds $x = 97/64$.

Ex. 4   Let $h = \dfrac{c}{b}$; $\Delta = \left(\dfrac{c}{b}\right)^3 + \dfrac{k}{a}$; $x = \sqrt[3]{\Delta} - h$, then:

$$x^3 \quad = \Delta - 3h\sqrt[3]{\Delta}(\sqrt[3]{\Delta} - h) - h^3$$
$$= \Delta - 3xh\sqrt[3]{\Delta} - h^3$$

hence x will be a solution of $ax^3 + bx^2 + cx = k$ on condition that
$-3ah\sqrt[3]{\Delta} = -b\sqrt[3]{\Delta} + bh - c$.
That is, if the coefficients are rational, and $\sqrt[3]{\Delta}$ is irrational:
$3ah = + b$ *and* $bh = c$
this gives: $bh = + 3ah^2 = c$ and so: $b^2 = 3ac$.

Ex. 5   a) $x^3 + 9x = 100$ has one real root $x = \sqrt[3]{50 + \sqrt{2527}} + \sqrt[3]{50 - \sqrt{2527}}$
but $2527 = (19\sqrt{7})^2$ and $50 \pm 19\sqrt{7} = (2 \pm \sqrt{7})^3$ and so $x = 4$.

b) $x = \sqrt[3]{t_1} + \sqrt[3]{t_2}$ with $t_1 = \dfrac{1}{18}(9 + \sqrt{849})$ and $t_2 = \dfrac{1}{18}(9 - \sqrt{849})$; c) $x = 2$

c) $x = 2$.

Ex. 6   a) the equation decomposes into $x^2 - 3x - 4 = 0$ and $x^2 + 3x + 4 = 0$
b) the cubic auxiliary equation is $2y^3 + 7y^2 + 30y - 39 = 0$ which has a solution $y = 1$.

this yields the solutions: $x = \dfrac{3 \pm \sqrt{21}}{2}$ or $x = \dfrac{-3 \pm i\sqrt{21}}{2}$

c) the auxiliary equation is $y^3 + 4y - 1 = 0$ already solved in Exercise 5. One of the solutions will be: $x = \dfrac{1}{2}\sqrt{\sqrt{4y^2 + 16} - y} - \sqrt{y}$ where y is the solution found in Exercise 5b.

Ex. 7   We have the table:

k	0	1	2	3	4	5	6
3^k (mod 7)	1	3	2	6	4	5	1

Let $\omega_k = r^k$ with $r = e^{2i\frac{\pi}{7}}$ and let: $p = \omega_1 + \omega_2 + \omega_4$, $p' = \omega_3 + \omega_6 + \omega_5$

Then $p + p' = -1$ and $p.p' = 2$, so p, p' are the roots of $x^2 + x + 2 = 0$ which gives:

$$p = \frac{-1 + i\sqrt{7}}{2} \; ; p' = \frac{-1 - i\sqrt{7}}{2}$$

and since $\omega_1\omega_2 + \omega_2\omega_4 + \omega_4\omega_1 = p'$ and $\omega_1\omega_2\omega_4 = 1$, then ω_1 , ω_2 , ω_4 are the roots of

$x^3 - px^2 + p'x - 1 = 0$ and ω_3 , ω_5 , ω_6 are the roots of $x^3 - p'x^2 + px - 1 = 0$.

Ex. 8   The group of the equation $\dfrac{x^7 - 1}{x - 1} = 0$ is cyclic. It is generated by the permutation $\pi = (1,3,2,6,4,5)$ obtained by using the primitive root 3 (see Exercise 7). Thus, $\pi^2 = (1,2,4)(3,6,5)$ which leaves p and p' invariant.

13

Are Imaginary Numbers Real?

J.-P. FRIEDELMEYER
K. VOLKERT
IREM Strasbourg

In *2001: A Space Odyssey* Frank Poole had to leave his space craft Explorer I to carry out repairs to a faulty component AE-35. To do so he had to pass through an air lock which allowed him to leave an environment which closely matched that on earth, for the weightless and airless conditions of outer space. The conditions for life on each side of the air lock are totally incompatible, so that it would be impossible for him to pass from one side to the other without preparation time, or else he would face being sucked out violently into interstellar space. The air lock itself is in contact with the two worlds, it has an opening into both of them which is only for passing through and it is made so that it has no effect on either environment.

It is a little like that with the history of mathematics. Without suitable preparation it is not possible to pass from the mathematical world of today to that of the mathematics of the past, even more so in the case of the distant past. For a voyage into the past there is no less need for deconditioning if we are to avoid incomprehension of a mental universe so very different from our own, where the evidence that we gather can only serve as a window on the actual difficulties that our ancestors encountered in trying to solve a mathematical problem. Applying the rigour and cold precision of our definitions and theorems to the thoughts and texts which were often intuitive, hesitant, and even empirical, we cannot comprehend the difficulties they had to disentangle ideas which seem to us so clear and simple. And furthermore, we cannot comprehend the richness of a problem, and the brilliant imagination of the mathematicians who solved what is for us no longer a problem!

Take for example the solution of algebraic equations. According to the chosen number field (just the integers, or the rationals, or the reals, or complex numbers, etc.) the number of roots of, for example, the polynomial $ax^2 + bx + c$ will vary from zero to two. For a polynomial of degree n with real coefficients it is not all evident, *a priori,* that it has always at least one complex root, which can then be written as a+ib, with a and b real. While the solutions of equations of the 3rd and 4th degree by the Italians Cardano, Bombelli and others, had led mathematicians to become used to working with new *"magnitudes"* of the form a+b$\sqrt{-1}$, there was nothing to prevent them from thinking that the solution of equations of degree five or more (not yet then solved!) might not require the use of other *"magnitudes"* more or less *"impossible"* or *"imaginary"* but not necessarily of the form a+b√-1 (The notation i for √-1 was not introduced until 1777 by Euler and did not come into regular use until taken up by Gauss. For that reason we shall write √-1 in all cases for work preceding Gauss.)

Apart from the historical interest, what gives a particularly significant and interesting character to this question, from the epistemological point of view, is that for equations of greater than the fourth degree, their roots not being, in general, expressible by surds, there arises a real problem of *existence*. Mathematicians had been drawn by their own practice to ask questions about the nature of objects, about which they knew neither their existence, nor what might be their *"possible form"* they were *"vera umbrae umbra"* (*"veritable shades of a shade"*) as Gauss was to put it later.[1]

What remains today of those questions and problematics? There remains a theorem, certainly an important one (often referred to as the *fundamental theorem of algebra*), which is mainly due to d'Alembert and states:

> In the field *C* of complex numbers, every non constant polynomial with real coefficients has at least one (complex) root

a property that can be summarised in modern terms by the expression:

> the field *C* is algebraically closed.

Here we see how questions gave way to a precise theorem, itself rigorously proved in a branch of mathematics (Algebra) and that can only be cause for congratulation: it is the normal outcome for mathematical research. Simply, from a historical point of view, such a statement obscures the centuries of research which it necessitated and renders a property, that had for so long stimulated the ardour and imagination of mathematicians, banal and without mystery. Furthermore, it runs the risk of making the researches of the mathematicians incomprehensible, since their inquiries took place in an altogether different culture.

In order better to understand the problem, and to avoid misrepresenting the sense that mathematicians before the 19th century gave to the word *"Imaginary"*, we shall:

1. study an example of *"Imaginaries"* in their own sense and without any particular reference to our own normal use of the term. This will allow us to pass through the 'deconditioning air lock' of which we spoke earlier, by presenting situations where conditions for applying the theorem, and hence its result, had not by then been realised.

2. take a voyage in time which retraces the emergence of an intuition arising out of practice. Through solving more and more elaborate equations it is possible to go just a little further to consider equations not yet known to have solutions.

3. study one of the proofs proposed for the theorem. This study will enable us to show the way in which the history of this problem is related to several other histories, such as the history of:

   • the construction of our present concept of number

   • the solution of equations

   • the relation (bifurcation) between Algebra and Analysis.

---

1.   Gauss, *Werke* 3, p. 14.

This will also allow us to show the relative nature of the concept of rigour, by considering the multiplicity and diversity of proofs which will show how rigour is dependent upon the context in which it is found.

Galois: "Where all sorts of imaginaries can be imagined"

With his visionary genius, the young Galois, anticipating today's mathematicians even in 1830, proposed a study of a numerical field that was totally *"imaginary"*. This would not bear any relation to an intuitive Analysis based on numerical magnitudes, but did relate to what was universal practice from the beginning of the 19th century. He wrote a short text *"On the Theory of Numbers"* which appeared in 1830. We shall present the essential ideas by looking at a simplified example based on the set of integers modulo 3 which we shall call **E**. For **E** the addition and multiplication tables are as follows:

+	0	1	2
0	0	1	2
1	1	2	0
2	2	0	1

×	0	1	2
0	0	0	0
1	0	1	2
2	0	2	1

The multiplication table immediately shows that there is no element of **E** whose square is 2: in other words the equation: $x \in$ **E** and $x^2 = 2$ has no solution. Galois says:

> *The roots of this congruence must be thought of as types of imaginary symbols, since integers cannot satisfy them, symbols whose use in calculations is often as useful as that of the imaginary $\sqrt{-1}$ in ordinary analysis.*[1]

Let us now add to **E** an imaginary number which Galois denotes by i and which satisfies $i^2 \equiv 2$ [modulo 3]. Extend the operations on **E** and we obtain a finite field $\mathbf{F_9}$ of nine elements of the form $a + ib$. The reader can easily construct the addition table. We give below the multiplication table:

×	0	1	2	i	2i	1 + i	2 + i	1 + 2i	2 + 2i
0	**0**	0	0	0	0	0	0	0	0
1	0	**1**	2	i	2i	1 + i	2 + i	1 + 2i	2 + 2i
2	0	2	**1**	2i	i	2 + 2i	1 + 2i	2 + i	1 + i
i	0	i	2i	**2**	1	2 + i	2 + 2i	1 + i	1 + 2i
2i	0	2i	i	1	**2**	1 + 2i	1 + i	2 + 2i	2 + i
1 + i	0	1 + i	2 + 2i	2 + i	1 + 2i	**2i**	1	2	i
2 + i	0	2 + i	1 + 2i	2 + 2i	1 + i	1	**i**	2i	2
1 + 2i	0	1 + 2i	2 + i	1 + i	2 + 2i	2	2i	**i**	1
2 + 2i	0	2 + 2i	1 + i	1 + 2i	2 + i	i	2	1	**2i**

---

1.   Galois, *Ecrits et Mémoires mathématiques*, p. 114.

As can immediately be seen from the diagonal of the table only 0, 1, 2, i and 2i have a square root in $\mathbf{F_9}$. For example, $1 + i$ does not have a square root so there is no solution here for $x \in \mathbf{F_9}$ and $x^2 = 1 + i$.

Consider now the general equation (1): $x \in \mathbf{A}$ and $ax^2 + bx + c = 0$ where $\mathbf{A}$ is some given field, the coefficients a, b, c belonging to $\mathbf{A}$ with a non zero. Different situations can arise:

1. We have seen that for $\mathbf{A}$ equal to $\mathbf{E}$ the equation $x \in \mathbf{E}$ and $x^2 = 2$ has no solution. Since -1 is congruent to 2 [mod 3], this is equivalent to: $x \in \mathbf{E}$ and $x^2 + 1 = 0$ has no solution.

2. If $\mathbf{A}$ is equal to $\mathbf{F_9}$ but the coefficients belong to $\mathbf{E}$ it can be shown that equation (1) always has a solution.

Exercise 1	Prove that the equation $ax^2 + bx + c = 0$, the coefficients a, b, c belonging to $\mathbf{E}$ with a non zero, has a solution for all $x \in \mathbf{F_9}$.

3. If $\mathbf{A}$ is equal to $\mathbf{F_9}$ and the coefficients a, b, c also belong to $\mathbf{F_9}$ then equation (1) does not always have a solution, as we have shown for $x \in \mathbf{F_9}$ and $x^2 = 1 + i$. It therefore becomes necessary to create a new set, containing $\mathbf{F_9}$, constructed with a new imaginary I where $I^2 = 1 + i$, etc ...

A new, even more paradoxical, situation arises with the skew-field $\mathbf{H}$ of quaternions invented by Hamilton. $\mathbf{H}$ consists of elements of the form: $q = a + ib + jc + kd$ where a, b, c, d are reals and i, j, k are *"imaginaries"* whose multiplication properties are as given in the following table:

$\times$	i	j	k
i	-1	k	-j
j	-k	-1	i
k	j	-i	-1

We can see that the equation $x \in H$ and $x^2 + 1 = 0$ in this case has *three* distinct solutions i, j, and k even though it is an equation of the second degree!

Euclid: "Where we find a problem that is sometimes impossible"

Greek geometry was, above all, a geometry of ruler and compass so it is not surprising that a great part of Euclid's *Elements* concerns equations of the second degree, albeit treated in a geometric manner. The general theory of equations of the second degree can be found in Book VI, propositions 28 and 29, but the way they are stated could lead a modern reader astray. It is worth making some preliminary remarks.

Consider an equation of the form $x(a - x) = K$ where a and K are given positive real numbers. Geometrically this is equivalent to constructing a rectangle of area K with sides x and $a - x$ on a line AB of length a. In Book I, 44 and 45, Euclid shows how to construct a parallelogram equal to a given rectilineal figure (fig. 1 and 2).

Figure 1   Figure 2

Additional conditions can placed on the parallelogram: it can be smaller by a parallelogram LBCM (fig. 3) or greater by a parallelogram BLMC (fig. 4) and should be similar to a given parallelogram.

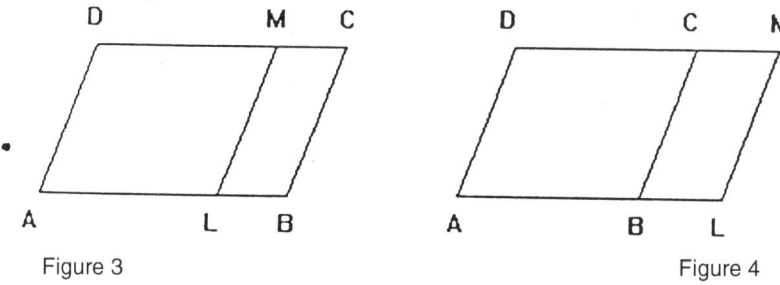

Figure 3   Figure 4

This is stated in Book VI, proposition 28:

> *To a given straight line to apply a parallelogram equal to a given rectilineal figure and deficient by a parallelogrammic figure similar to a given one: thus the given rectilineal figure must not be greater than the parallelogram described on the half of the straight line and similar to the defect*[1].

In order to simplify the problem and to make it easier by using modern algebraic notation, we shall consider this theorem using rectangles and squares. The problem can then be stated as: Given a line AB of length a, to construct a rectangle of area K and a square side by side on AB. (See fig.5) In other words, if we let the side of the square be x, we have $x(a - x) = K$ or $x^2 - ax + K = 0$. If we put $a - x = y$ this becomes the problem of finding a solution to $\begin{cases} x+y =a \\ xy =K \end{cases}$

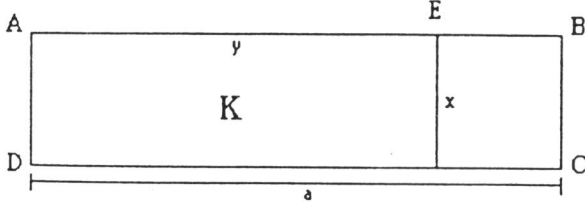

Figure 5

1.   Heath, *The Thirteen Books of Euclid's Elements*, vol. 2, p. 260.

Here is the way that Euclid solved the problem:

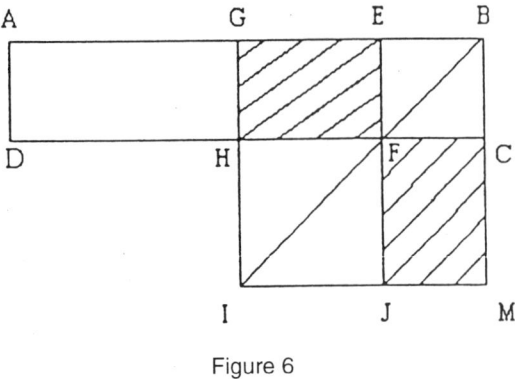

Figure 6

Analysis (fig. 6)

Consider that the construction has been done. On half of AB construct a square BGIM. Since BEFC is a square then so is FJIH whose area is $\left(\frac{a}{2}-x\right)^2$ .Hence the figure BGHFJMB, called a *gnomon* by the Greeks, is equivalent to the area K because the rectangles GEFH and FCMJ have the same area. Algebraically, this equivalent to

$$\left(\frac{a}{2}\right)^2 - \left(\frac{a}{2}-x\right)^2 = -x^2 + ax = K.$$

Construction (fig. 7)

Given the straight line AB, let G be its centre and construct GI perpendicular to AB. On GI construct the square HIJF equal to $\left(\frac{a}{2}\right)^2 - K$, which is a given quantity (equivalent to what today we call the discriminant). Completing the figure we get, side by side, the rectangle AEFD of area K and the desired square EFBC.

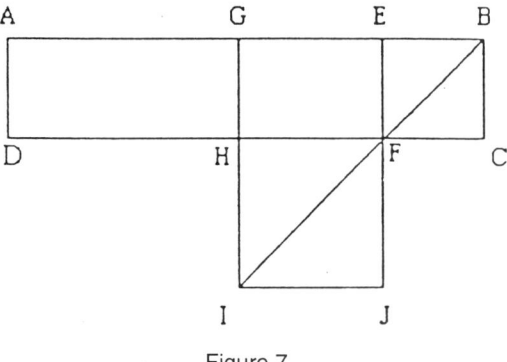

Figure 7

Discussion

The construction is only possible if the square constructed on half AB is greater than K. That is why, in the statement of the proposition, Euclid says: *"the given rectilineal figure must not be greater than the parallelogram described on the half of the straight line."*[1]

---

1.   Heath, *The Thirteen Books of Euclid's Elements*, vol. 2, p. 260.

The immense advantage of this geometric treatment over the numerical treatment given, for example, by the Babylonians is that it is entirely *general*. The Babylonians could only illustrate their method of solution by giving specific numerical examples. On the other hand, the geometric solution does have drawbacks: numbers treated geometrically condition our perception of number through the idea of *magnitude*. In practice only those magnitudes were considered that can be drawn with ruler and compasses. From this arose a mass of difficulties that were not really resolved until the 19th century: the status of negative numbers, the laborious liberation of analysis from its geometrical setting and above all the near impossibility of the numerical solution of equations other than those of the second degree or of those deriving from them. And we can verify this, fifteen centuries after Euclid, from Fibonacci.

Exercise 2	Using ruler and compasses, construct a rectangle and square side by side on a line AB = 10 cm such that the rectangle has an area of 20 cm^2.

Fibonacci: "Where we realise that not everything is contained in Euclid"

Euclid's approach, and that of Greek mathematicians generally, determined the method of constructing, solving and proving that was to be followed for centuries. The Arab mathematicians such as Al-Khwarizmi and Omar Khayyam only dealt with equations whose solutions could be constructed and proved according the canon of Euclidean geometry. In Book X of the *Elements* Euclid made an inventory of those magnitudes that could be constructed by ruler and compasses from some arbitrarily chosen length **u** which we shall take as a unit and Euclid called a *rational* line. He obtained fifteen sorts of lines:

• *rational* lines which are commensurable with unity or whose squares are commensurable with squares constructed on the unit (the Greek term *rhetos* means: 'line expressible'). We refer to these as a and $\sqrt{a}$

  • *medial* lines, corresponding to $\sqrt{\sqrt{a}} = \sqrt[4]{a}$

  • *binomial* and *apotome* lines: $\sqrt{a} + \sqrt{b}$ and $\sqrt{a} - \sqrt{b}$

  • the *bimedials* $\sqrt[4]{a} + \sqrt[4]{b}$

  • the *fourth bimedials* and *apotomes* a + $\sqrt{b}$ and a – $\sqrt{b}$

  • the *majors* $\sqrt{a + \sqrt{b}}$ and the *minors* $\sqrt{a - \sqrt{b}}$

  • the *fifth binomials* $\sqrt{a} + b$ and *apotomes* $\sqrt{a} - b$

  • and, finally lines that can be expressed by

$$\sqrt{\sqrt{a} + b}; \ \sqrt{\sqrt{a} - b}; \ \sqrt{\sqrt{a} + \sqrt{b}}; \ \sqrt{\sqrt{a} + \sqrt{b}}$$

In the 13th century Jean de Palerme, philosopher to Frederick II, put the following question to Fibonacci:*"find a number such that its cube together with two times its square and ten times itself equals 20"* which we would write today as: find X such that $X^3 + 2X^2 + 10X = 20$. Fibonacci's reply was:

> *On thinking about this point I thought that the solution should come from what is contained in Euclid Book X and because of this I became interested in digging around in Book X itself [...]. This Book X deals with fifteen varieties of straight lines. Of these fifteen lines, two are called calculable or rational. The thirteen others are called illogical or irrational.*
>
> *And while I was reflecting passionately on these fifteen numbers and on the differences between them, I found that none of them could correspond with a single one of those ten roots of which he had spoken above which, with two squares and the cube, could make twenty, as is demonstrated geometrically in what follows.*[1]

Exercise 3   Prove that the equation $X^3 + 2X^2 + 10X = 20$ does not have a positive rational solution nor solutions of the form $\sqrt{n}$ where is rational but not a square.

And so Fibonacci proved geometrically that the root of this equation was not one of the magnitudes that had been studied up to the present by geometers and found himself, like the Pythagoreans before him, in the uncomfortable position of considering a magnitude not previously known and contemplating its nature. He did not doubt the existence of this magnitude and went as far as to give an extremely precise value for it although, unfortunately, he did not leave behind any indication of the methods that he had used. He ends as follows:

> *And because this question cannot be solved with one of the numbers already known, I have tried to come to the solution by use of approximation. And I found that one tenth of the ten named roots [...] by approximation is 1 and 22 first minutes and 7 second minutes and 42 third minutes and 33 fourth minutes and 4 fifth minutes and 40 sixth minutes.*[2]

The result is given in sexagessimals and we can verify the extreme accuracy of the approximation which appears all the more remarkable to us because he did not have used any of the methods which were developed later.

Exercise 4   Solve the equation $X^3 + 2X^2 + 10X = 20$ and verify Fibonacci's solution. To solve we can use Cardan's method after suppressing the X^2 term by substituting $y = X - \dfrac{2}{3}$ or use a numerical method of approximation.

Cardan, Girard, Descartes and many others: "where imaginary numbers take a small step towards reality"

The approximate solution given in the above equation is expressed, as was usual at that time, in sexagessimal form. However Fibonacci was one of the principal mathematicians to encourage the use of the decimal form and zero in the West. This greatly facilitated the writing of numbers and the setting up of algorithmic calculations and made it possible to write equations in symbolic form. The Italian mathematicians of the 16th century were responsible for a decisive step forward in the numerical solutions of equations.

---

1.   Fibonacci, *Sritti di Leonardo Pisano*, p. 228; tr. from Latin by Jean Boyé.
2.   *Ibid.*

Cardan, in his *Ars Magna* published in Nuremberg in 1545, posed the following question:

> If someone says to you, divide 10 into two parts, one of which multiplied into the other shall produce 30 or 40, it is evident that this case or question is impossible. Nevertheless, we shall solve it in this fashion. Let us divide 10 into equal parts and 5 will be its half. Multiplied by itself, this yields 25. From 25 subtract the product itself, that is 40, which [...] leaves a remainder -15. The root of this added and then subtracted from 5 gives the parts which multiplied together will produce 40.
>
> These, therefore, are $5+\sqrt{-15}$ and $5-\sqrt{-15}$.[1]

Today we would express this as needing to find two numbers X and Y such that
$$\begin{cases} X+Y=10 \\ XY=40 \end{cases}$$
or to find X given $(10 - X)X = 40$.

The problem is the same form as that proposed by Euclid in Book VI, but to construct its solution we would need to construct a square of area $(10/2)^2 - 40 = -15$. (fig. 8).

As long as we remain attached to Euclidean magnitudes, we can go no further.

We would need to subtract a square of area 40 from the square GBMI whose area is 25. Cardan uses a symbolic calculation to *"imagine"* a solution.

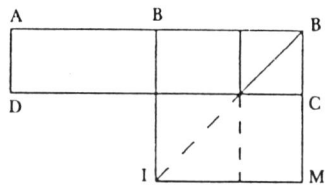

Figure 8

PROOF
That the true significance of this rule may be made clear, let the line AB which is called 10, be the line which is to be divided into two parts whose rectangle is to be 40. Now since 40 is the quadruple of 10, we wish four times the whole of AB. Therefore, make AD the square on AC, the half of AB. From AD subtract four times AB. If there is a remainder, its root should be added to and subtracted from AC thus showing the parts (into which AB was to be divided). Even when such a residue is minus, you will nevertheless imagine $\sqrt{-15}$ to be the difference between AD and the quadruple of AB which you should add and subtract from AC to find what was sought.
That is $5 + \sqrt{25 - 40}$ and $5 - \sqrt{25 - 40}$ or $5 + \sqrt{-15}$ and $5 - \sqrt{-15}$.

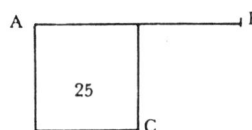

Multiplying $5 + \sqrt{-15}$ by $5 - \sqrt{-15}$ the imaginary parts being lost, gives 40. Therefore the product is 40. However, the nature of AD is not the same as that of 40 or AB because a surface is far from a number or a line. This, however, is closest to this quantity which is truly imaginary [sophistica] since operations may not be performed with it as with a pure negative number, nor as in other numbers.[2]

---

1. Cardan, *Ars Magna De Arithmetica*, ff.66r in D.E.Smith, *A Source Book in Mathematics*, McGraw Hill 1929: New York, Dover, 1959, p. 201-202 (© All rights reserved).
2. *Ibid.*

It may be interesting to note that from a young age Cardan was the subject of hallucinations. He wrote:

> *This happened from when I was four years old until I was seven: and every day from the second hour of the day until the fourth, or if I got up or woke up later, I thought I could see pictures by the end of the bed, shaped like little copper rings, showing trees, animals, men, cities, soldiers arrayed for battle, instruments of war, and of battle and other things, which went up and down one after the other. [...] As for me, although I was so young, I knew very well that it was a miraculous vision, so I assured them I could not see anything, fearing that if I told them of it the vision would leave me or some ill would befall me for having revealed such a secret.* [1]

It certainly needed a fertile mind to be able to calculate with numbers that were truly *sophistica*. These calculations did not lead to contradictions and allowed mathematicians to solve hitherto unresolved problems even if they were not entirely sure of the status of their solutions. In particular, their inability to find solutions to equations of degree five led them to believe that solutions might exist in an ideal number world.

Cardan was able to observe that equations of degree three could have three roots and that fourth degree equations could have four. A century later Albert Girard (from Lorraine) was the first to assert that while certain algebraic equations with rational coefficients had as many roots as the degree of the equation, others had fewer and it would be of some advantage to use the term *"impossible solutions"* for those that were missing, so that the total of the roots and the impossible solutions would always be equal to the degree of the equation.

> *All algebraic equations have as many solutions as the denomination of the highest term [...] It could be asked what point is there in solutions that are impossible? My reply is threefold: the sureness of the general rule, and that there can be no more solutions and for its utility.* [2]

This utility shows itself especially in the relations between the coefficients of the equation and the roots. Girard uses the term *"faction"* for the symmetric relations between roots. So for the equation $X^4 = 4X - 3$ the solutions are $a = b = 1$ (double), $c = -1 + \sqrt{-2}$ and $d = -1 - \sqrt{-2}$ and the "factions" are $a + b + c + d = 0$, $ab + ac + ad + bc + bd + cd = 0$, $abc + abd + acd + bcd = 4$, and $abcd = 3$. And the fact that the product of the two last roots is 3 gives a sort of *existence* or a certain reality to these impossible solutions in relation to the coefficients.

A little later, Descartes wrote in his *Géométrie* in 1637:

> *Neither the true nor the false roots are always real; sometimes they are only imaginary; that is, while we can always conceive of as many roots for each equation as I have already assigned, yet there is not always a definite quantity corresponding to each root so conceived of.* [3]

The term *"imaginary"* is introduced here for the first time. It comes into mathematical vocabulary with a meaning that becomes specialised only after the belief became general that the only imaginaries that we meet in the solution of equations will be of the form $a + b\sqrt{-1}$, with a, b real. Furthermore, as long as this conviction remained unsupported by proof, mathematicians vascillated between the terms *"imaginary"* and

---

1.   In Dedron and Itard *Mathematics and Mathematicians I*, p. 202.
2.   Girard, *Invention nouvelle en l'Algèbre*, ff 1 r.
3.   Descartes, *La Géométrie*, p. 380.

"impossible". Euler, for example, in his 1770 *"Algebra"* demonstrates the impossibility of complex numbers:

> *Since all possible numbers that can be imagined are either greater than or less than or equal to zero, it is evident that the roots of negative numbers cannot be counted among possible numbers. So we are obliged to say that there are impossible numbers. Hence we have had to come to terms with such numbers, that are impossible by their very nature and which, by habit, we call imaginary because they exist only in the imagination.*[1]

And even later, in Klügel's Mathematical Dictionary (1836) we read:

> *We use the term impossible or imaginary magnitude for all expressions for which it is impossible to find a real value, for example Arccos x, Arcsin x for x > 1*[2]

And the term 'imaginary magnitudes' becomes used as the label for a large bag into which everything was thrown that was not know sufficiently well. So for the mathematicians of the 18th century the problem came to this: is it possible to reduce all impossible or imaginary roots to the form $a + b\sqrt{-1}$? This is the sense of the fundamental theorem of algebra in the 18th century. The existence of the roots was not in question; one wished only to prove that the roots were or could be transformed into $a + b\sqrt{-1}$. Hence it is possible to distinguish three equivalent forms of the fundamental theorem:

1. An equation of the nth degree has n roots (real or complex).

2. Each root of any equation is either real or can be expressed in the form $a + b\sqrt{-1}$.

3. All algebraic equations or all polynomials can be factorised

    – either with real linear factors $X - a$ or real quadratic factors $X^2 + pX + q$,

    – or with complex linear factors.

This last form of the theorem became more prevalent during the 18th century with the development of the integral calculus. In order to integrate a rational function it is necessary to split it into partial fractions whose denominators are either of the form $(X - a)^n$ or of the form $(X^2 + pX + q)^m$ with n,m natural numbers and a, p, q real. In other words it became imperative to be able to factorise the denominator according to form 3.

In a memoire which appeared in 1702, Leibniz conjectured that this factorisation might not always be possible. For the sake of argument he considered the factorisation of

$$X^4 + a^4 = (X^2 - a^2\sqrt{-1})(X^2 + a^2\sqrt{-1})$$
$$= (X + a\sqrt{\sqrt{-1}})(X - a\sqrt{\sqrt{-1}})(X + a\sqrt{-\sqrt{-1}})(X - a\sqrt{-\sqrt{-1}})$$

and this last form never produces a real factor no matter which pair are taken together. Leibniz did not seem to be aware that $\sqrt{\sqrt{-1}}$ can indeed be written in the form $a + b\sqrt{-1}$.

In fact:

$$\sqrt{\sqrt{-1}} = \frac{1}{2}\sqrt{2}(1 + \sqrt{-1}) \text{ et } \sqrt{-\sqrt{-1}} = \frac{1}{2}\sqrt{2}(1 - \sqrt{-1})$$

$$X^4 + a^4 = (X^2 + a\sqrt{2}X + a^2)(X^2 - a\sqrt{2}X + a^2)$$

1.   Euler, *Algèbre, Opera Omnia*, Series Prima I, p. 55.
2.   Klügel, *Mathematisches Wörterbuch* V, p. 555.

In a letter to Nicholas Bernoulli dated 1-11-1742, Euler formulated a theorem for the factorisation of polynomials with real coefficients in exactly the form that Leibniz had taken to be false. Contradicting Bernoulli who had proposed as counter example $X^4 - 4X^3 + 2X^2 + 4X + 4$, Euler showed that it could be factorised as the product of two polynomials with real coefficients, namely

$$\left[X^2 - (2+a)X + 1 + \sqrt{7} + a\right]\left[X^2 - (2-a)X + 1 + \sqrt{7} - a\right] \text{ where } a = \sqrt{4 + 2\sqrt{7}}$$

Exercise 5	Factorise $X^4 - 4X^3 + 2X^2 + 4X + 4$ as the product of two second degree polynomials with real coefficients, and so establish Euler's result.

A few days later, on the 15-12-1742, Euler wrote to Goldbach to say that he had reworked the formulation of his theorem and added that he had not, at the present, proved it completely but only "almost" proved it like certain of Fermat's theorems. In this letter he also mentions that for a real polynomial the imaginary roots can be regrouped in pairs so that when multiplied together they produce real second degree factors. Goldbach was sceptical and offered a new counter example: $X^4 + 72X - 20$, which Euler promptly factorised.

To summarise, mathematicians, wishing to carry out formal calculations with $\sqrt{-1}$, were forced to:

1. became familiar with complex numbers to the point where they were willing to accept them as having a place among integers, rationals and reals;

2. in their solutions of equations never encountered expressions other than those that could be composed of reals together with $\sqrt{-1}$ so that they became convinced that all imaginaries could be expressed in this form.

3. in working on important theoretical problems, such as the integration of rational functions, needed to be able to answer the important question: can we always factorise the denominator into real factors of first or second degree?

The times were ripe for an attack on a proof of an affirmative answer to this last question; an answer that was fundamental to the parallel progress of analysis and algebra.

The Proof

A hundred or more proofs of the fundamental theorem of algebra can be counted since the first tentative one offered by De Gua in 1741, which relied on the false hypothesis that every algebraic equation could be solved by means of surds, up to the constructive proofs of which one of the latest is that proposed by Kneser in 1981. The theorem has not ceased to fuel the imagination of mathematicians, which points to its conceptual richness as much as the difficulty of providing a proof to satisfy every one. Compare this entry in the *Encyclopédie des Sciences Mathématiques,* appearing at the beginning of this century:

> *These proofs differ from one another essentially according to the doctrine that one holds about the foundations of Arithmetic. From that point of view such a proof is either rigourous or not. [...] It depends upon whether one assumes the **analytic** notion of continuity, and also whether one considers that all proofs of the fundamental theorem, to be legitimate, ought to provide a means of calculating the roots of f(z) from its*

> *coefficients to such an accuracy as one wishes, the existence of these results deriving from the calculation process itself: thus even the purpose of the proof of the fundamental theorem can change.*[1]

The principal stumbling block lies, in effect, in a question that transcends the topic of algebra: it is the very topological structure of the set of reals themselves. Despite a prodigious development of algebra in the 18th century that infiltrated all branches of mathematics, including geometry, ideas about real number and magnitude remained unclarified. For a long time the ultimate foundation was Euclidean space which alone gave *sense* to these ideas. This conception found its most complete expression in Kant. In *"The Critique of Pure Reason"*, in 1781, we read:

> *All conceptions, and with them all principles, and whatever is accepted a priori, therefore derive from empirical intuition, that is to say from what experience tells us is possible. Without that they have no necessary value but they are only a simple creation on the the imagination and thought with their respective representations. Taking only mathematical concepts, for example, everything is imagined by pure intuition; space has three dimensions, only one straight line can be drawn between two points, etc.... Whereas all these principles and the representations of objects with which this science concerns itself are, a priori, produced in the mind, they have, however, absolutely no meaning if we can not always demonstrate their meaning with phenomena (empirical objects). Also it is vital to be able to give sensual meaning to an abstract concept, that is to show intuitively an object to which it corresponds, because without that the concept would not have, as we say, any sense, that is to say, any meaning. Mathematics fills this condition by the construction of a figure which is a phenomenon present to the senses (even though produced a priori). The concept of quantity in this same science finds its support and sense in number, and these in the digits and counters of the calculating table, or in the marks and points set out before our eyes.*[2]

The idea of continuity can be seen only to have a meaning related to the idea of space (or time). The most characteristic example of this is in the intermediate value theorem whose proof depended for a long time on a simple geometrical visualisation. Thus Cauchy in his 1821 *Cours d'Analyse* stated and proved the theorem as follows:

> **Theorem IV** *If the function $f(x)$ is continuous with regard to the variable x between the limits $x = x_0$ and $x = X$, and if b stands for a value intermediate between $f(x_0)$ and $f(X)$, the equation $f(x) = b$ will always be satisfied by one or more real values of x which lie between x_0 and X.*
>
> **Proof** *In order to establish this proposition, it is sufficient to notice that the curve which has $y = f(x)$ for its equation will meet, one or more times, the line that has $y = b$ as its equation in the interval between the ordinates that correspond to the abscissae x_0 and X; now it is evident that that is what will happen in the assumed hypothesis.*[3]

The need for a *purely analytic* proof was in general not appreciated. It was however formulated by Bolzano in 1817:

> *There is absolutely no objection to this geometric theorem, neither against its justice nor its evidence. But it is also manifest that it contains an intolerable fault against good practice, a fault which consists in wishing to deduce truths that are purely mathematical or general (that is to say truths in arithmetic, in algebra or in analysis) from considerations that belong only to an applied (or special) part, namely geometry.*[4]

---

1.   *Encyclopédie des Sciences mathématiques*, vol. I_2, p. 190.
2.   Kant, *Critique of Pure Reason*, p. 218.
3.   Cauchy, *Cours d'analyse algébrique*, p. 50.
4.   Bolzano, *Rein analytischer Beweiss...*, p. 137.

This need was not appreciated simply because analysis had not yet been separated from geometry, being seen only as a most efficient tool of investigation. The fundamental theorem of algebra, because it was stated in purely algebraic terms, contributed to the separation of analysis from algebra in that it required the attention to be fixed on the *topological* properties of the set of reals or on complex numbers from the point of view of *continuous functions*. That is why the history of the proofs of this theorem largely coincides with the history of that separation, the earlier proofs staying attached to geometric views and becoming more and more pure as awareness developed of their numerous implicit assumptions. That awareness only became acquired with the theoretical construction of the reals by Dedekind and Cantor in the second half of the 19th century.

A whole book would be required to study the proofs and their evolution. Many of them would be difficult to understand without a word by word commentary in order to express the proof in today's terminology. We can become aware of this by reading the following extract, which is d'Alembert's proof of 1746:

> *Proposition I. Let TM be any curve whose coordinates are given by TP = z, PM = y and where y = 0 or ∝ when z = 0. If z is taken positive or negative but infinitely small, the value of y in terms of z can always be expressed by a real quantity when z is positive; and when z is negative by a real quantity or a quantity $p+q\sqrt{-1}$, where p and q are both real..* [1]

The reader may well understand that it is better to explore a single significant proof rather than carry out a survey of several to the detriment of precision and clarity. We have chosen what has come to be called Gauss's first proof dated 1799 (he produced four altogether), titled: "*A new proof of the theorem: All integer algebraic functions of one variable can be decomposed into real factors of the first or second degree.*" This proof was presented by Gauss for his doctoral thesis at the University of Helmstedt.

Gauss: or "how a young mathematician gave lessons to the most established and famous mathematicians of his time and developed a new conception of mathematical rigour"

Gauss concluded his proof with the phrase: "*I discovered the elements of this proof at the beginning of October 1797.*" Gauss was born on 30th April 1777 at Braunschweig and at the age of twenty had already been active in mathematics as is shown by his *Disquisitiones Arithmeticae* published in 1801, section VII of which concerns the construction of regular polygons including the case of the 17-sided polygon.

How could a young mathematician impose himself on a mathematical community which did not know him, criticising his elders and showing lacunae and weak points in their proofs? He would have to be sure of his ground – and he was! Gauss never published anything of which he was not absolutely certain, preferring to leave to others the glory of new discoveries which he might know but was unwilling to publish if he judged them not sufficiently clear or well developed.

---

1.   d'Alembert, *Recherche sur le calcul intégral*, p. 183.

The criticisms Gauss made on previous ideas on the fundamental theorem were generally well founded but it would take too long to study them here in detail. His principle objection concerned the entirely new idea of *existence*.

> *Certain authors,* he wrote,[1] *take it as a sort of Axiom that every equation in fact admits of roots which, in default of other possibilities, are* **impossible**. *What they mean exactly by possible or impossible magnitudes is not sufficiently clearly explained. If the expression* **"possible magnitudes"** *is to mean the same thing as "real", and "impossible" the same as "imaginary", then this cannot in any sense be taken as an axiom, but requires a proof. However, these expressions do not seem to be taken in that sense; the axiom must rather be taken in the following sense:* **"While we may not be yet sure that there are necessarily m magnitudes, real or imaginary, which satisfy a given equation of degree m, we wish, nevertheless, to admit it is so in the following: for if it should happen that one is unable to find sufficient magnitudes, real or complex, the matter should remain open if we say that the others are impossible."** *If it is recommended to use such an expression in place of saying simply that there are not enough roots, then I can say nothing against it; but if it is used as if to say that they exist in fact, and if one says that, for example, the sum of the roots of the equation $x^m + A\,x^{m-1} + B\,x^{m-2} + \ldots = 0$ is $= -A$, even though among them there are some which are "impossible" (which in the end means that some are missing) then I cannot approve. For impossible roots accepted in that sense are like roots, and so the theorem can not in any way be accepted without proof; furthermore, in this matter, one might wonder if there could be any equations which have impossible roots.*

From our point of view, all proofs before Gauss did not pose the question of *existence* of the roots of an equation but their *form,* whether or not they were of the type $a + b\sqrt{-1}$. Girard's hypothesis is accepted implicitly as an axiom, without ever being justified. It was even thought for a long time that there was a veritable hierarchy of imaginary numbers among which complex numbers of the form $a + b\sqrt{-1}$ were the simplest. In the 18th century analytical methods developed which were not related to solutions, and theoretical expressions were used which had not been explicitly evaluated, and it was then that the following question was seriously posed:

Do all imaginary magnitudes have the form $a + b\sqrt{-1}$

The answer to this question would not take us very far in itself. Recall how in our earlier example of the field $\mathbf{F}_9$ one could keep adding *"imaginary"* numbers, which were more and more complicated, and yet there always remained new equations with coefficients expressed in terms of these imaginaries which would not have solutions. Here, it is not the same thing. Once the imaginary number $\sqrt{-1}$ has been introduced then nothing new can be generated: all equations with real or complex coefficients will have only roots which are complex: that is what is meant by saying that the field $\mathbf{C}$ is algebraically closed.

The principal novelty in Gauss's proof[2] lies in that he does not seek to calculate a root, nor to specify its form, but to *prove its existence.*

Let there be given an equation with real coefficients:

$$X = x^m + Ax^{m-1} + Bx^{m-2} + \ldots + Lx + M = 0$$

---

1. Gauss, *doctoral thesis* §3, tr. Friedelmeyer.
2. The proof is given, in English, in Struik, D. J.*"A Source Book in Mathematics 1200-1800"*, p. 115.

We know that:

(a) if a is a real root of X, then X can be divided by (x − a),

(b) if a + ib is a complex root of X, then X can be divided by the real factor
$(x^2 − 2ax + a^2 + b^2)$.

Gauss wished to prove that X can be completely decomposed into a product of factors of the type (a) or (b). He wanted to establish his proof without the use of complex numbers and he started by establishing these two *Lemmas*:

> *Lemma 1. If m is an arbitrary positive integer, then the function*
> $sin\ \phi.\ x^m − sin\ m\phi.\ r^{m\ −\ 1}x + sin\ (m − 1)\phi.\ r^m$
> *is divisible by $x^2 − 2\ cos\ \phi.\ rx + r^2$.*

> *Lemma 2. If the quantity r and the angle ϕ are so determined that the equations*
> *(1) $r^m\ cos\ m\phi + Ar^{m\ −\ 1}cos\ (m − 1)\phi + B\ r^{m\ −\ 2}cos\ (m − 2)\phi + ...$*
> *$+ Kr^2\ cos\ 2\phi + Lr\ cos\ \phi + M = 0,$*
>
> *(2) $r^m\ sin\ m\phi + Ar^{m\ −\ 1}sin\ (m − 1)\phi + B\ r^{m\ −\ 2}sin\ (m − 2)\phi + ...$*
> *$+ Kr^2\ sin\ 2\phi + Lr\ sin\ \phi = 0,$*
> *exist, then the function*
> *$x^m + Ax^{m\ −\ 1} + Bx^{m\ −\ 2} + ... + Kx^2 + Lx + M = X$*
> *will be divisible by the qudratic factor $x^2 − 2\ rx\ cos\phi + r^2$, unless r sin $\phi = 0$.*
> *If r sin $\phi = 0$, then the same function is divisible by $x − r\ cos\ \phi$.*

Exercise 6   Prove these two lemmas.

Gauss can now proceed:

> *The previous theorem is usually given with the aid of imaginaries, see Euler, Introductio in Analysin Infinitorum, I, p. 110; I found it worth while to show that it can be demonstrated in the same easy way without their use. Hence it is clear that, in order to prove our theorem, we only have to show: If some function X of the form*
> *$x^m + Ax^{m\ −\ 1} + Bx^{m\ −\ 2} + ... + Kx + Lx + M$*
>
> *is given, then r and ϕ can be determined in such a way that the equations (1) and (2) are valid.*

Using polar coordinates, Gauss set up a plane whose points were given by r and ϕ with reference to an axis GCG', ϕ being measured clockwise. Then he considered surfaces defined by the equations;

$(r, \phi) \to T(r, \phi) = r^m\ sin\ m\phi + Ar^{m\ −\ 1}sin\ (m − 1)\phi + ... + L\ r\ sin\ \phi$

$(r, \phi) \to U(r, \phi) = r^m\ cos\ m\phi + Ar^{m\ −\ 1}cos\ (m − 1)\ \phi + ... + L\ r\ cos\ \phi + M$

The intersections of these surfaces with the plane (r, ϕ) are given by

$T(r, \phi) = 0$ which gives a curve C_T, and

$U(r, \phi) = 0$ which gives a curve C_U.

To prove the theorem, it is necessary to show that the curves C_T and C_U have at least one common point.

1. C_T has at least one point. This comes from the intermediate value theorem, since r can be chosen so that $r^m\ sin\ m\phi$ is greater than the sum of all the other terms of T. (r^m can be made greater in absolute value than any other power of r.) But $r^m\ sin\ m\phi$ can take both positive and negative values (it depends only upon ϕ).

The fixed plane must therefore be intersected by T and so C_T has at least one point. The same argument applies to C_U.

2. C_T and C_U are each composed of branches separated by m asymptotes corresponding to solutions of sin mϕ = 0 and cos mϕ = 0 respectively, that is for $\phi = \dfrac{k\pi}{2m}$. In fact, for r tends to infinity,

$$\frac{U}{r^m} = \cos m\varphi + \left(\frac{A}{r}\right)\sin(m-1)\varphi + \left(\frac{B}{r^2}\right)\cos(m-2)\varphi + \ldots \sim \cos m\varphi$$

$$\text{and } \frac{T}{r^m} = \sin m\varphi + \left(\frac{A}{r}\right)\sin(m-1)\varphi + \left(\frac{B}{r^2}\right)\sin(m-2)\varphi + \ldots \sim \sin m\varphi$$

Since cos mϕ and sin mϕ can be expressed as a rational function of cos ϕ and sin ϕ it follows that the equations C_T and C_U can be written as polynomials in x and y. These will be algebraic curves of order m.

Let us illustrate this by means of an example. Consider the equation $X^5 - 4X - 2 = 0$, which is an equation that can not be solved in terms of surds.[1] Using the above notation, we get

$U(r,\phi) = r^5 \cos 5\phi - 4r\cos\phi - 2 = 0$

$T(r,\phi) = r^5 \sin 5\phi - 4r\sin\phi = 0$

Using the results $\cos 5\phi = 16\cos^5\phi - 20\cos^3\phi + 5\cos\phi,$

$\sin 5\phi = 16\sin^5\phi - 20\sin^3\phi + 5\sin\phi,$

we get the following Cartesian equations:

C_U: $x^5 - 10x^3y^2 + 5xy^4 - 4x - 2 = 0$ (fig. 9)

C_T: $y(y^4 - 10y^2x^2 + 5x^4 - 4) = 0$ (fig. 10)

Exercise 7   Sketch the curves with equations

C_U: $x^5 - 10x^3y^2 + 5xy^4 - 4x - 2 = 0$
C_T: $y(y^4 - 10y^2x^2 + 5x^4 - 4) = 0$
(Both equations can be solved for y.)

---

1.   Mutafian, *Équations algébriques et théorie de Galois*, p. 242.

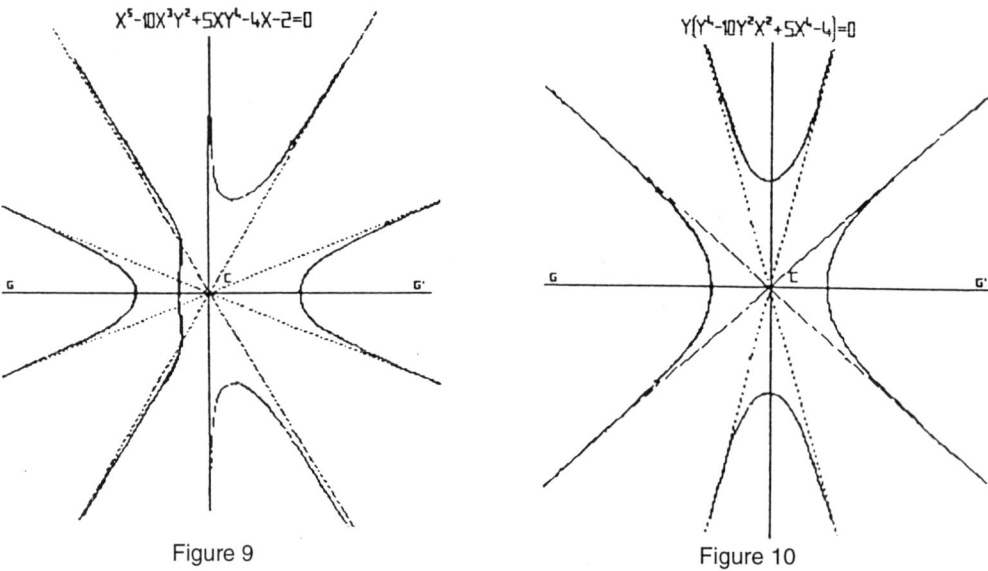

Figure 9   Figure 10

The asymptotes of C_U are given by $\phi = \dfrac{(2k + 1)\pi}{10}$ and those of C_T by $\phi = \dfrac{k\pi}{5}$; $1 \le k \le 5$

The main argument is now as follows: There is a circle (Γ) with centre at the origin on which there are 2m points of C_U and 2m points of C_T which can be located. Let S be the sum of the absolute values of the coefficients of X and for the radius R of the circle use whichever is the greater of 1 or S $\sqrt{2}$. We use the number $(2k + 1)$ to refer to the point of (Γ) corresponding to the angle $\dfrac{(2k + 1)}{4m}.\pi$, where $0 \le k \le 4m - 1$. These are, therefore, 4m equidistant points on (Γ). The 2m points of C_T lie in the intervals $[(8m - 1);(1)]$; $[(3);(5)]$; $[(7);(9)]$; ... $[(8m - 5);(8m - 3)]$ likewise the 2m points of C_U lie in the intervals $[((1);(3)]$; $[(5);(7)]$; $[(9);(11)]$; ... $[(8m - 3);(8m - 1)]$.

In each interval there lies one and only one point as is shown by the following reasoning, based again on the intermediate value theorem (fig. 11). At the point (1) we have

$$T = R^{m - 1}\left[R.\sqrt{\frac{1}{2}} + A\sin(m - 1)\phi + \left(\frac{B}{R}\right)\sin(m - 2)\phi + ...\right]$$

The sum $A\sin(m - 1)\phi + \left(\dfrac{B}{R}\right)\sin(m - 2)\phi + ...$ is in absolute value less than or equal

to S and so T is positive at point (1). A fortiori, T is positive between (1) and (3). Similar reasoning shows that T is negative between (5) and (7). Hence T must vanish at least at one point in the interval $[(3);(5)]$. There is no other value for which T vanishes in that interval since the derivative $\dfrac{dT}{d\phi}$ is negative between (3) and (5). A similar argument can be applied to C_U. We can therefore enumerate the points of C_T and C_U alternately on (Γ) from 0 to 4m – 1. Those of C_T correspond to even numbers and those of C_U to odd numbers (Figure 11):

C_T:	0	2	4	6
intervals	$[(8m - 1); (1)]$	$[((3); (5)]$	$[(7); (9)]$	$[(11); (13)]$

C_U:	1	3	5	7
intervals	[(1); (3)]	[(5); (7)]	[(9); (11)]	[(13); (15)]

The theory of algebraic curves shows, then, that each *"even"* point must be connected with one and only one *"even"* point by an arc of C_T. In the same way the *"odd"* points are linked by arcs of C_U. (These arcs cannot *"vanish"* at finite distance.) These arcs lie inside the circle (Γ); they enter the circle, coming from infinity, and leave it again to go to infinity.

It remains only to show that *there exists at least one point of intersection* of an arc of C_T with an arc of C_U. And this is done by a *reductio ad absurdum* argument.

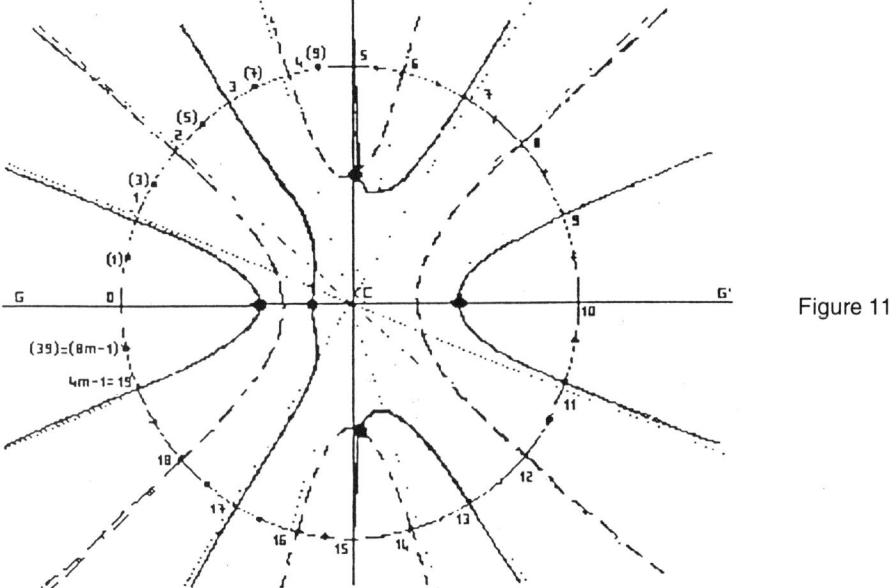

Figure 11

Suppose that there does not exist such a point. The real axis is a part of C_T so the point 0 is connected with the point 2m. The point 1 can only be connected with one of the points 3, 5, 7, ..., 2m – 1. Let that point be n (n < 2m). If the point 2 is connected with n' then n' < n, unless there happens to be an intersection point. If 3 is connected with n" then we have n" < n'. By repetition, it follows that at last we come to a point h connected with h + 2. The arc of C_U (or C_T).that starts from point h + 1 must necessarily cut the arc of C_T (or C_U) [h, h + 2]. This contradicts our hypothesis and so the curve C_U must necessarily cut the curve C_T (see fig. 12).

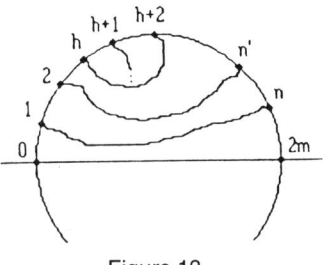

Figure 12

This last assertion, of the existence of at least one common point for C_T and C_U represents the core of the proof, and it remained uncontested for almost a century. Specialists in topology have only been able to prove it using the most subtle of arguments. Gauss himself was to remark in a note:

> It seems to be largely sufficiently proved that an algebraic curve can neither abruptly break off (as happens, for example, with the transcendental curve with equation $y = \dfrac{1}{\log x}$), nor get itself lost as a point after an infinite number of turns (like the logarithmic spiral), and as far as I know, no one has raised the least doubt about it.

However, if any one should raise the matter, I shall undertake on another occasion to furnish a proof which can not be doubted.

No one did ask for a proof of this result and Gauss never produced one. It was not until 1920 that A. Ostrowsky completed Gauss's proof (Gauss, Werke 10.2 Abh.3).

Gauss himself produced three further proofs, which shows the importance he attached to the theorem. These proofs were in addition to dozens of others which were produced from this time. All were obliged, sooner or later, to have recourse to arguments drawn from analysis. Nowadays, it is usual to use arguments connected with the theory of functions of a complex variable, such as the maximisation principle or Liouville's theorem, from which we know that a function that is holomorphic and restricted to **C** is constant. In all cases, one is convinced that no purely algebraic proof can exist, insofar as the fields **R** and **C** are constructs of analysis. The advantage of this first proof by Gauss is that it uses relatively elementary mathematics. We can also see how the young Gauss had mastered the geometric representation of complex numbers (even though he presented his proof in such a way that they did not appear explicitly). The functions T(r,φ) and U(r,φ) are none other than the imaginary and real parts respectively of the function X. In 1797 Gauss could not have known of the contributions made by his contemporaries to the geometric representation of complex numbers: for example that by Argand, published in 1806, *"Essai sur une manière de représenter les quantités imaginaires dans les constructions géométriques"*; and that by Wessel which, although published in Copenhagen in 1799, remained unknown until its French translation appeared in 1897 under the title *"Essai sur la représentation analytique de la direction."*

Present day thoughts
The set of complex numbers is revealed as the number field without equal for the greater part of mathematics, as much in algebra as in analysis or even in geometry

1. The field **C** of complex numbers, unlike Galois' imaginary numbers, has the property that every polynomial with complex coefficients has complex roots and can only have roots whose numbers are of this type, that is belonging to **C**. This property is indicated by saying that **C** is algebraically closed. From this follows the possibility of decomposing polynomials over **R** into linear or quadratic factors, which formed the basis of the first proofs. From this it can also be deduced that a polynomial over **C** of degree n has at least one root, and it can be shown that it has in fact exactly n roots if each multiple is counted.

   Linear algebra is closely dependent on the properties of **C**. If f is an endomorphism of a finite vector space over **C**, other than {0}, then f has at least one proper value.

2. The field of numbers has grown throughout the history of mathematics: integers, natural numbers, fractions, decimals, irrationals, reals and, finally, complex numbers. Is there any reason to stop here? In 1843 Hamilton constructed a system of hypercomplex numbers, the field **H** of quaternions. This set contains **C** but is greater than **C** in the same way that **C** is greater than **R**. The set of quaternions is a four

dimensional extension of **R** and in the first section we mentioned one of its curious properties.

Hardly two years later Cayley invented octonions (or octaves) whose set θ contained **H** as a subset and which is an extension of dimension 8 over **R**. These constructions were no simple idle past-times. They produced remarkable results. The octonions established the following identity in **R**:

$(P^2 + Q^2 + R^2 + S^2 + T^2 + U^2 + V^2 + W^2)(p^2 + q^2 + r^2 + s^2 + t^2 + u^2 + v^2 + w^2)$
$= (Pp - Qq - Rr - Ss - Tt - Uu - Vv - Ww)^2 + (Pq + Qp + Rs + Sr + Tu - Ut - Vw + Wv)^2$
$+ (Pr - Qs + Rp + Sq + Tv + Uw - Vt + Wu)^2 + (Ps + Qr - Rq + Sp + Tw - Uv + Vu - Wt)^2$
$+ (Pt - Qu - Rv - Sw + Tp + Uq + Vr + Ws)^2 + (Pu + Qt - Rw + Sv - Tq + Up - Vs + Wr)^2$
$+ (Pv + Qw + Rt - Su - Tr + Us + Vp - Wq)^2 + (Pw - Qv + Ru + St - Ts - Ur + Vq + Wp)^2$

This identity brings to mind two other identities which had been known for a long time, namely:

$(m^2 + n^2)(M^2 + N^2) = (mM - nN)^2 + (mN + nM)^2,$

and

$(m^2 + n^2 + p^2 + q^2)(M^2 + N^2 + P^2 + Q^2) =$
$(mM - nN - pP - qQ)^2 + (mN + nM + pQ - qP)^2 +$
$(mP + pM + qN - nQ)^2 + (mQ + qM + nP - pN)^2$

Can we go further? The answer is no, for Hurwitz showed in 1898 that the above identities can not be generalised for more than eight squares. And each new extension of **C** requires some essential property to be abandoned.

Hamilton's ambition was to construct for 3-dimensional space a system of hypercomplex numbers which could be used to deal with space in a similar manner to the way in which complex numbers can be used for the plane. For years he tried to find a satisfactory way of multiplying triplets of real numbers. He recalled this later in a letter he wrote to his son in 1865, shortly before he died:

> *Every morning, on coming down to breakfast, you would ask me, "Well, Papa, can you multiply triplets?" To which the answer was a sadly negative shake of the head: "No, I can only add and subtract them."* [1]

Nowadays it is easy for us to understand that there is not any multiplication in **R**, linear in the set **R**^{3}, which can extend **C** = **R**^{2}. In fact, if e = (1,0,0); j = (0,1,0); k = (0,0,1) is a canonical basis for **R**^{3}, let (1) ij = ρe + σi + τj, with ρ, σ, τ real. Using the hypotheses i^2 = –e and i(ij) = (i^{2}j) = –j, then multiplying (1) by i we get:

$-j = \rho i - \sigma e + \tau ij = \rho i - \sigma e + \tau(\rho e + \sigma i + \tau j) = (\tau\rho - \sigma)e + (\tau\sigma + \rho)i + \tau^2 j$

and, since e, i, j are linearly independent, this leads to $\tau^2 = -1$, contradicting the fact that τ is real.

---

1.   Hamilton, *Mathematical. Papers 3*, p. xv.

Hamilton can certainly be credited with two important successes:

– the addition of a *fourth dimension* through the use of a fourth element k independent of i and j.

– abandoning commutativity of multiplication with the rules

ij = – ji = k; jk = –kj = i ; etc.

As he himself explained:

> *The commutative character has been lost… However it is shown that another important property of the old multiplication has been preserved, or extended into the new, namely what might be called the associative character of the operation.*[1]

This is the first mention of associativity, a property that would itself be abandoned by Cayley in constructing his octonions.

The superb character of the field **C** of complex numbers is thus confirmed. It is, apart from isomorphisms, the unique commutative algebraic extension of the field of reals **R**. In particular, there does not exist any commutative field containing **C** except for **C** itself. That theorem was presented, without proof, by Weierstrass in 1863 in his course at Berlin and proved in 1867 by Hankel in his *"Theorie der complexen Zahlensysteme",* who proudly added:

> *And this resolves the question to which Gauss promised a reply in 1831 (Werke 2, p.178) but failed to give it: why relations between objects consisting of a multiplicity of more than two dimensions can not produce yet more types of magnitudes admissible in generalised Arithmetic.*

---

1.   *Ibid.*, p. 114.

Bibliography

Sources texts

BOLZANO, *Rein analyticher Beweis...* Ostwald's Klassiker n°153 Leipzig 1905. french tr. J. Sebestik. Revue d'Histoire des Sciences, **17** (1964), p. 136-164.

CARDAN, *Ars Magna sive de regulis algebraicis*, Nuremberg, 1545.

CAUCHY, *Cours d'Analyse algébrique*, Paris 1821.

CAYLEY, *The collected Mathematical Papers 1*, Cambridge University Press, 1889.

CLARKE, *2001 L'Odyssée de l'Espace*, d'après le scénario original de S.Kubrick et A.C. Clarke. Traduit de l'anglais par M. Demuth, Robert Laffont, Paris, 1968.

D'ALEMBERT, *Recherches sur le Calcul intégral*, Histoire de l'Académie Royale des Sciences et Belles Lettres, Année 1746 (Berlin 1748).

DEDRON, ITARD, *Mathématiques et Mathématiciens*, Magnard, Paris, 1960, English tr. Milton Keynes, Open University Press, 1978.

DESCARTES, *La Géométrie 1637*, rééd. Smith Dover, New York, 1964.

DHOMBRES, *Nombre mesure et continu*, Cedic F. Nathan, 1978.

EUCLIDE, *Les Œuvres d'Euclide* traduites littéralement par Peyrard. Repr. Blanchard, Paris, 1966.

FIBONACCI, *Scritti di Leonardo Pisano*, tome 2 – Ed. Boncompagni, Rome 1857-1862.

FUSS, *Correspondance mathématique et physique de quelques célèbres géomètres du 18e siècle*, 2 vols 1843, Johnson Reprint Corp. 1967.

GALOIS, *Ecrits et mémoires mathématiques d'Evariste Galois*, Ed. R. Bourgne, J.P. Azra., Gauthier Villars, Paris, 1962.

GAUSS, Thèse de doctorat Helmstedt, 1799 Werke III 3.

GIRARD, *Invention nouvelle en l'Algèbre*, Amsterdam 1629. Repr. Leiden, 1884.

HAMILTON, *Mathematical Papers 3*.

HANKEL, *Theorie der complexen Zahlensysteme*, Leopold Voss, Leipzig, 1867.

HEATH, *The Thirteen Books of Euclid's Elements*, Cambridge 1925. Repr. 3 vols., New York, Dover, 1956.

HIRSCH, SMALES, "On Algorithms for Solving f(x) = 0" in *Comm. Pure Appl. Math.*, **32**, 281-312.

HURWITZ, "Über die Komposition des quadratischen Formen von beliebig vielen Variablen", 1898, *Math. Werke* 2, 565-571.

KANT, *Critique de la Raison Pure* (French tr.), Bibliothèque de philosophie contemporaine, P.U.F., Paris, 1971.

KLÜGEL, *Mathematisches Wörterbuch*. Schwieckert, Leipzig, 1803-1836.

LEIBNIZ, *Acta Eruditorum*, 1702.

MUTAFIAN, *Equations algébriques et théorie de Galois*, Vuibert, Paris 1980.

NETTO, *Die vier Gauss' schen Beweise für die Zerlegung ganzer algebraischer Functionen...* (1799-1849), Ostwald's klassiker, Leipzig und Berlin, 1913.

STRUIK, *A Source Book in Mathematics 1200-1800*, Cambridge, Massachusetts, Harvard University Press, 1969.

WOEPKE, "Sur un essai de déterminer la nature de la racine d'une équation du troisième degré...", *Journal de Mathématiques pures et appliquées*, tome XIX, déc. 1854, p. 401.

Reference texts

Encyclopédie des Sciences mathématiques, I, 2, NETTO et LE VAVASSEUR, « Le Théorème fondamental », 189-205.

Zahlen, *Grundwissen des Mathematik*, 1, zweite, überrarbeitete und ergäntzte Auflage, Springer Verlag, 1988.

General works for further reading

ARGAND, *Essai sur une manière de représenter les quantités imaginaires dans les constructions géométriques*, Nouveau tirage de la 2e édition. Blanchard, Paris, 1971.

CASTELJAU, *Les Quaternions*, Hermès, Paris 1987.

EULER, *Recherches sur les racines imaginaires des équations*, Histoire de l'Académie Royale des Sciences et Belles Lettres, Année 1749 (Berlin 1751) – Opera Omnia Serie I, vol. 6.

– Complète introduction à l'Algèbre. Petersburg 1770 Ed. et compléments en français par Lagrange.

GILAIN, "Sur l'histoire du théorème fondamental de l'algèbre: théorie des équations et calcul intégral", *Archive for history of Exact Sciences*, **42**, (1991), 91-136.

KNESER, "Der Fundamentalsatz des Algebra und der Intuitionismus", *Math Zeitschr.*, **46**, p. 287-302, complété par son fils M. Kneser en 1981, *Math. Zeitschr.*, **177**, p. 285-287.

TAIT, *Traité élémentaire des quaternions,* tr. par Plarr, Gauthiers Villars, Paris, 1882.

WESSEL, *Essai sur la représentation analytique de la direction*, Ed. H. Valentino and T.N. Thiele trans. H.G. Zeuthen and others, Copenhague, 1897.

Abrégé d'histoire des mathématiques, sous la direction de J. Dieudonné, tome I chap. II, Hermann 1979.

TROPFKE, *Geschichte des Elementar-Mathematik Band 1*, 4. Auflage ed. K. Vogel, K. Reich, H. Gericke, D.E. Gruyter Berlin/New York, 1980.

How did you get on ?

Ex. 1   The coefficients a, b,c belong to E and a ≠ 0. There are two cases:

a = 1 : $ax^2 + bx + c = 0 <=> (x + 2b)^2 \equiv b^2 + 2c$ [mod 3]

whence $(x + 2b)^2$ can take the values 0, 1, 2

and so (x + 2b) = 0 or 1 or 2 or i or 2i, all in $\mathbf{F_9}$

and so there are two solutions: x = b + d or x = b – d = b + 2d

where d is the square root of $b^2 + 2c$

a = 2: $ax^2 + bx + c = 0 <=> (ix + ib)^2 \equiv 2b^2 + 2c$ [mod 3]

whence $(ix + ib)^2$ can take the values 0, 1, 2

and so   (ix + i2b) = 0 or 1 or 2 or i or 2i, all in $\mathbf{F_9}$

and so there are two solutions: x = 2b +2id or x = 2b – 2id = 2b + id

where d is the square root of $2b^2 + 2c$

Ex. 2

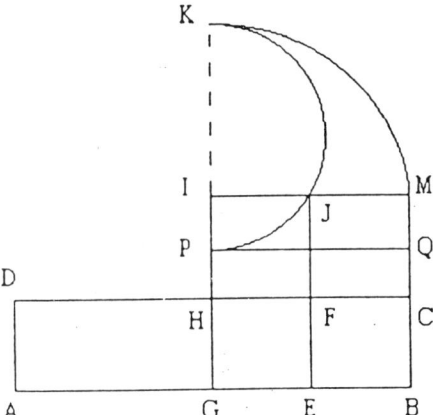

From the square GIMB of side 5 cm remove the rectangle GPQB of area 20 cm². The rectangle PIMQ remains which must be transformed into a square IJFH (in fact,

$IJ^2 = IP \times IK = IP \times IM$). see Euclid II. 14

AEFD and EFCB are the rectangle and square being sought.

Ex. 3   a) Let x = $\frac{p}{q}$ be a solution to $X^3 + 2 X^2 + 10 X = 20$ with p, q relative prime.

(Fibonacci would only be concerned with positive roots.)

Then $p^3 = q (20q^2 – 2p^2 – 10pq)$ and so q is a factor of p^3 and hence q = 1 and x is

an integer. But x = $2 - \frac{x^3 + 2x^2}{20}$ and hence x < 2. Since 1 is not a solution there can

not be a positive rational solution.

b) Let x = $\sqrt{n}$, with n a (non square) integer. Then since $\frac{2x^2}{10} = 2 - \left(x + \frac{x^3}{10}\right)$ we have

the rational $\frac{2n}{10}$ equal to the irrational $2 - \sqrt{n}\left(1 + \frac{n}{10}\right)$ which is impossible.

Ex. 4   Put X = $x - \frac{2}{3}$ and the equation becomes $x^3 + \left(\frac{26}{3}\right) x = \frac{704}{27}$ which can be solved by

Cardan's method to give, for X

$$\frac{\sqrt[3]{352 + 6\sqrt{3930}} + \sqrt[3]{352 - 6\sqrt{3930}} - 2}{3}$$

X is approximately equal to 1.368808107821 which in sexagessimals is
1 22' 7'' 42''' 33(iv) 4(v) 38(vi) 30(vii) 50(viii)

Ex. 5   Putting $X = x + 1$ eliminates the cubic term:

$X^4 - 4X^3 + 2X^2 + 4X + 4 = x^4 - 4x^2 + 7$

$x^4 - 4x^2 + 7 = (x^2 + \sqrt{7})^2 - (4 + 2\sqrt{7})x^2$

Let $a = \sqrt{4 + 2\sqrt{7}}$ to give

$x^4 - 4x^2 + 7 = (x^2 - ax + \sqrt{7})(x^2 + ax + \sqrt{7})$

substituting back $X = x + 1$ gives

$[X^2 - (2 + a)X + 1 + \sqrt{7} + a][X^2 - (2 - a)X + 1 + \sqrt{7} - a]$

Ex. 6   *Lemma 1*

For positive integer m, $P(x) = \sin\phi.x^m - \sin m\phi.r^{m-1}x + \sin(m-1)\phi.r^m$ is divisible by $x^2 - 2r\cos\phi x + r^2$

For $m = 1$, $P(x) = 0$

For $m = 2$, $P(x) = \sin\phi\,(x^2 - 2rx\cos\phi + r^2)$

For $m > 2$, $P(x) = (x^2 - 2rx\cos\phi + r^2)(\sin\phi.\,x^{m-2} + \sin2\phi.r x^{m-3} + \sin3\phi.r^2 x^{m-4} + \dots + \sin(m-1)\phi.r^{m-2})$

Lemma 2

Given that r and ϕ simultaneously satisfy

(1) $r^m\cos m\phi + Ar^{m-1}\cos(m-1)\phi + Br^{m-2}\cos(m-2)\phi + \dots + Kr^2\cos2\phi + Lr\cos\phi + M = 0$,

(2) $r^m\sin m\phi + Ar^{m-1}\sin(m-1)\phi + Br^{m-2}\sin(m-2)\phi + \dots + Kr^2\sin2\phi + Lr\sin\phi = 0$,

then, by use of lemma 1; all the following polynomials are divisble by $x^2 - 2rx\cos\phi + r^2$

$\sin\varphi.rx^m$	+	$\sin m\varphi.r^m x$	+	$\sin(m-1)\varphi.r^{m+1}$	
$A\sin\varphi.rx^{m-1}$	+	$A\sin(m-1)\varphi.r^{m-1}x$	+	$A\sin(m-2)\varphi.r^m$	
$B\sin\varphi.rx^{m-2}$	+	$B\sin(m-2)\varphi.r^{m-2}x$	+	$B\sin(m-3)\varphi.r^{m-1}$	
. . . .		. . . .		. . . .	
$K\sin\varphi.rx^2$	+	$K\sin2\varphi.r^2x$	+	$K\sin\varphi.r^3$	
$L\sin\varphi.rx$	+	$L\sin\varphi.rx$			
$M\sin\varphi.r$			+	$M\sin(-\varphi)r$	

Adding all these gives $\sin\phi.rX + 0 + 0$, using (1) and (2).

Ex. 7 C_T corresponds to $y = 0$ and $y^4 - 10x^2y^2 + 5x^4 - 4 = 0$

which can be solved to give $y = \pm\sqrt{5x^2 \pm 2\sqrt{5x^4 + 1}}$

C_U corresponds to $5xy^4 - 10x^3y^2 + x^5 - 4x - 2 = 0$

which can be solved to give $y = \pm\sqrt{x^2 \pm \sqrt{\frac{4}{5}x^4 + \frac{4}{5} + \frac{2}{5x}}}$

14

The Prime Numbers

François JABOEUF
IREM Montpellier

> *Mathematicians have searched in vain to discover some order in the sequence of prime numbers, and there are good grounds for believing that it is a mystery that will never be penetrated by the human mind. To be convinced of this, one has only to cast ones eyes over the tables of prime numbers, which some persons have taken the trouble to compile up to a hundred thousand, and it can be seen that no order or rule reigns there. This occurrence is all the more surprising in that Arithmetic has provided us sure rules by which we can continue the progression of these numbers as far as we wish, without however leaving us the least trace of any order...*

L. Euler, 1751 (commentatio 175, bibliothèque impériale 3, p. 10-31)

It is not possible to remain indifferent to numbers. They are rejected in irritation by some, yet handled with delight by others. How can we not be fascinated by their mysteries and their secrets, still unfathomed after twenty centuries of research and study, and by their seeming resistance to all attacks made upon them? To cite some of the well known problems that surround numbers:

– The irrationality of square roots: the diagonal and side of a square cannot be expressed as a ratio of integers: they are incommensurable.

– The quadrature of the circle (to construct a square having the same area as a given circle) and the transcendence of π (π is not the solution of any polynomial with integer coefficients).

– The general solution of algebraic equations (see Chapter 12), etc.

Certainly many of these problems have now been resolved, sometimes after many centuries of bitter struggle. The irrationality and then the transcendence of π were not established until the eighteenth century and nineteenth century respectively (Lambert, 1766; Lindemann, 1882). The four squares theorem, the possibility of writing every positive integer as a sum of four square integers, originally proposed by Diophantus in the fourth century, was not proved until the end of the eighteenth century by Lagrange. But a number of problems remain unresolved: the famous Fermat's Last Theorem

($x^n + y^n = z^n$ has no non-zero integer solutions for n > 2) remains an enigma[1]. The hopeless search, since the time of Euclid, for an odd perfect number (see Chapter 15); and the even perfect numbers are not completely known even with the list of Mersenne primes

($2^n - 1$). At the end of this chapter, we show that not only do we not know of any prime Fermat numbers ($2^{2^n} + 1$) for n greater than 5, but that the existence of such numbers remains unknown. As for identifying very large prime numbers, the most complicated theories, and the most powerful computers, are not always successful. And what are we to think of Gödel's incompleteness theorem, by which arithmetic possesses at least one undecidable proposition, that its truth or falsehood cannot be proved? Is it not the case that the whole of arithmetic is a vast mystery?

To close this introduction, here are five modest problems offered to the reader, all of which will be resolved by a reading of this chapter:

1.   The first six multiples of the number 142857 are written with the same digits and in the same order (but not starting with the same digit). Why is this?

2.   Every odd number not divisible by 5 has a multiple which can be written entirely using the digit 1: for example, $3 \times 37 = 111$.

3.   Let $\frac{u}{v}$, a fraction reduced to its lowest term, be the sum of the reciprocals of the integers from 1 to p − 1, then its numerator u is divisible by p^2, provided that p is a prime greater than 5.

4.   The spiral of Ulam: The numbers 41, 43, 47, 53, 61, 71, 83, 97, …, 251 found on the diagonal of the spiral shown in the diagram below, obtained by writing consecutive integers, starting at 41, in a spiral, are all prime numbers. Does this property continue along the diagonal, and if so how far, if the spiral is continued?

251	•	•										•	•	265	266
•	197	198	199	200	201	202	203	204	205	206	207	208	209	210	•
•	196	151	152	153	154	155	156	157	158	159	160	161	162	211	•
	195	150	113	114	115	116	117	118	119	120	121	122	163	212	•
	194	149	112	83	84	85	86	87	88	89	90	123	164	213	
	193	148	111	82	61	62	63	64	65	66	91	124	165	214	
	192	147	110	81	60	47	48	49	50	67	92	125	166	215	
	191	146	109	80	59	46	41	42	51	68	93	126	167	216	
	190	145	108	79	58	45	44	43	52	69	94	127	168	217	
	189	144	107	78	57	56	55	54	53	70	95	128	169	218	
	188	143	106	77	76	75	74	73	72	71	96	129	170	219	
	187	142	105	104	103	102	101	100	99	98	97	130	171	220	
•	186	141	140	139	138	137	136	135	134	133	132	131	172	221	
•	185	184	183	182	181	180	179	178	177	176	175	174	173	222	
•	•	•	•	•	•	231	230	229	228	227	226	225	224	223	

---

1.   Since this chapter was written, this enigma has been resolved by A. Wiles.

5. Decode the following claim by the philosopher G. Bachelard in his *Essai sur la connaissance approché* (1928), which is given here as:

13290165663   1354243070   1878655722   985339559   1865818302   430519627

1928020187   107345454   1776014312   474199866   1142306131   1041671881

689034962   1863263536

From Numbers … to Prime Numbers

Defining what a number is, and what a prime number is, has been a concern since antiquity. In the fourth century B.C. we find these definitions in Euclid Book VII[1]:

> *Definition 1. An **unit** is that by virtue of which each of the things that exist is called one.*
> *Definition 2. A **number** is a multitude composed of units.*
> *Definition 3. A number is **a part** of a number, the less of the greater, when it measures the greater.*

Today we say that "a number a divides (or is a factor of) a number b" if b is equal to the product of a and another integer (the number of times that a measures b). The first properties studied by Euclid were about comparing numbers, and finding common factors and the highest common factor (HCF). The two first propositions of Book VII describe the process of anthyphairesis (continued alternate subtraction), better known by the name of Euclid's algorithm, and which can be used to find the HCF of two numbers or to show that they are prime to one another. Definitions 11, 12 and 13 serve to explain the meaning of the terms prime and composite:

> *Definition 11. A **prime number** is that which is measured by an unit alone.*
> *Definition 12. Numbers **prime to one another** are those which are measured by an unit alone as a common measure.*
> *Definition 13. A **composite number** is that which is measured by some number.*

Today's definition is: a whole number different from 1 is prime if its only factors are 1 and itself, otherwise it is composite.

But why is it that, since they have been studied for more than two thousand years, we should still be interested in prime numbers? There are at least two reasons:

– Their Mystery: Compare them, for example, with the even numbers, which are also defined, just like the prime numbers, by a property about factors, here that they are all divisible by 2. It is easy to recognise them, and to provide a formula for them (2n), to determine the next even number, or to determine the nth even number. Also their distribution among the other numbers is perfectly regular and known (every other number is even). Now, as we shall see later, nothing at all like this can be said of the prime numbers.

– Their Importance: Besides their use in various branches of mathematics (in the theory of numbers of course, but also in geometry, in group theory, in analysis, and in cryptography as we shall see later), the prime numbers play an irreplaceable role in the construction of the integers, and this reinforces their mysterious aspect: they serve to construct other numbers, yet they themselves cannot be constructed: they could be said to serve Arithmetic in the way that the alphabet serves language. For instance, the

---

1.   Heath, *The Thirteen Books of Euclid's Elements*, vol. 2, p. 277.

number 1707 (the year of Euler's birth) can be obtained from the product of the two prime numbers 3 and 569. Other examples are: 1855 (death of Gauss) = $5 \times 7 \times 53$; 1596 (birth of Descartes) = $2 \times 2 \times 3 \times 7 \times 19$; 1665 (death of Fermat) = $3 \times 3 \times 5 \times 37$; 1601 (birth of Fermat) and 1777 (birth of Gauss) are both prime.

Every non-zero prime number different from 1 can be decomposed into a product of prime numbers, and this decomposition is unique.

This Fundamental Theorem of Arithmetic was first explicitly stated by Gauss in his *Recherches Arithmétiques* published in 1801. However, it seems to have been understood by Euclid, even if not stated in his *Elements*. These three propositions of Book VII relate to it.[1]

> *Proposition 30: If two numbers by multiplying one another make some number, and any prime number measure the product, it will also measure one of the original numbers.*

This proposition is better known today as a theorem of Gauss which is stated as: Every prime number that divides a product of numbers, will divide one of the factors of that product. The following proposition establishes that every composite number must have a prime factor.

> *Proposition 31: Any composite number is measured by some prime number.*
>
> *Let A be a composite number; I say that A is measured by some prime number.*
> *For, since A is composite, some number will measure it.*
> *Let a number measure it, and let it be B.*
> *Now, if B is prime, what was enjoined will have been done.*
> *But if it is composite, some number will measure it.*
> *Let a number measure it, and let it be C.*
> *Then, since C measures B, and B measures A, therefore C also measures A.*
> *And if C is prime, what was enjoined will have been done.*
> *But if it is composite, some number will measure it.*
> *Thus if the investigation be continued in this way, some prime number will be found which will measure the number before it, which will also measure A.*
> *For if it is not found, an infinite series of numbers will measure the number A, each of which is less than the other: which is impossible in numbers.*
> *Therefore some prime number will be found which will measure the one before it, which will also measure A.*
> *Therefore any composite number is measured by some prime number.*

A ————————   B ——————   C ————

An immediate consequence is:

> *Proposition 32: Any number is either prime or is measured by some prime number.*

Here we are not far from the Fundamental Theorem. All that is needed is to apply the argument of Proposition 31 to the strictly decreasing sequence of quotients obtained from the successive divisions; the last quotient found will necessarily be prime. The initial number will therefore be the product of all the prime factors (not necessarily distinct) that have been found.

---

1.   *Ibid.*, p. 331-333.

> **Exercise 1** Prove the uniqueness of prime factor decomposition by the use of Euclid's Proposition 30 of Book VII.

So the prime numbers are not only the "most simple" numbers for multiplication (they do not have any strict divisors), but moreover they constitute the foundations for the whole edifice of the integers. It is certainly worth drawing back the veil that hides some of their mysteries. To start with, how can we recognise a prime number?

The successive divisions algorithm

As an example, consider the number 19979. Is it prime? Is it composite? Simply applying the definition, without any preliminary investigation, will require 19977 divisions to find out, since every number from 2 to 19978 needs to checked, and if the number does happen to be prime then the process needs to be continued right to the end. But the least factor p, other than 1, of a number n, will be prime (if not, then it will not be the least), and so n is equal to pq where q is a factor of n different from 1 if n is not prime: so p is less than $\sqrt{n}$, and q is greater than $\sqrt{n}$. We therefore have an interesting result: every composite number n has a prime factor less than $\sqrt{n}$.

From which we get the successive divisions algorithm, by which we can determine whether a number (not too large) is, or is not, prime:

Divide a number n successively by all the prime numbers less than $\sqrt{n}$. It will be prime if, and only if, no division gives a zero remainder.

For the number 19979, only 34 divisions are needed since there are 34 prime numbers less than 141, the greatest integer less than $\sqrt{19979}$. Since none of these has a zero remainder, the number 19979 must be prime.

We can go further (see the law of scarcity), since there are "many fewer" prime numbers than integers, which is why this algorithm is so useful. It is easy to apply the algorithm if all the prime numbers less than $\sqrt{n}$ are known. If not, then since all primes (except 2 and 3) are of the form $6k \pm 1$, we could use these numbers instead of prime numbers, and the number of calculations needed will be rather more, in fact $\approx \sqrt{n}/3$.

> **Exercise 2** Prove that all primes, except 2 and 3, are of the form $6k \pm 1$, where k is an integer.

It should be noted that this method provides an algorithm for the decomposition into prime factors of any number whatsoever. Consider the number 84091: 2, 3, 5 are not factors but 84091/7 = 12013, which is not divisible by any of 11, 13, 17, 19, 23, 29, 31, 37 (see the list of prime numbers given below), but 12013/41 = 293, and since 41 is strictly greater than $\sqrt{293}$, then 293 cannot have any prime factor less than its square root and so is itself prime. Hence 84091 = 7 × 41 × 293 (using a total of 13 divisions).

> **Exercise 3** Use the successive divisions algorithm to find the prime decomposition of 52793.

The Sieve of Eratosthenes

Knowing the prime numbers between 2 and n allows us to find the prime number decomposition of all integers up to n^2, which is why tables of prime numbers are so useful. How, in practice, can such tables be constructed? For small numbers, we can proceed by successively eliminating, first the multiples of 2 (leaving only odd numbers), and then multiples of 3, multiples of 5, multiples of 7, and so on. Those numbers which remain will be prime since they have no exact divisors. Using the odd numbers, those removed are asterisked:

3   5   7   9*   11   13   15*   17   19   21*   23

This process of separating out the prime numbers is due to Eratosthenes of Syrene, of the third century B.C. and bears his name. The oldest text that we know of that describes the process is by the Neo-Pythagorean Nicomachus of Gerasa who lived about the close of the first century A.D. His *Introductio Arithmetica* contains a description of the Sieve of Eratosthenes:[1]

> *The production of these numbers is called a "sieve" by Eratosthenes because, in taking the odd numbers all together and without distinction, this method of production produces in them a separation as is done with a sieve, and we find on one side those that are prime and non-composite, and on the other side the second part which are composite, and these were mixed all together.*
> *The sieve is constructed like this: I write out all the odd numbers starting with three, in a list as long as possible, and, starting with the first, I examine all those that it measures. I find that it measures those that are obtained by passing by two intermediate numbers, going as far as one wishes; and they are not measured haphazardly or by chance, but the first number measures them (I should say the first number that is reached in passing over two intermediate numbers according to the quantity of every prime number in the list, that is its own quantity, namely three times); and the one that is reached on passing two intermediate numbers from that will measure according to the number which occupies the second place in the list, that is five times; and in the same way, the one that is found on passing two more numbers, will measure according to the quantity of the number which occupies the third place, that is seven times; and the one that is found going even further on passing two more numbers, according to the quantity of the one occupying the fourth place, that is nine times, and so on to infinity.*
> *Then, after that, taking another point of departure, I consider the second and I examine the numbers it will measure; and I find that those that it measures are reached each time on passing a group of four numbers; and as before, it measures the first according to the quantity of the first term of the list, that is three times; the second according to the second, that is five times, the third according to the third, that is seven times ... [...] ...*
> *And this continues according to the same analogy, without meeting any obstacle whatsoever, so that:*
> *– the numbers are in the same order as their places in the original list;*
> *– the quantity of the numbers that have to be passed is determined by the regular progression of the even numbers from 2 up to infinity; or, as twice the position occupied by the number that measures;*
> *– the number of times [that the number is found to be contained in the number it measures] is determined by the regular progression of the odd numbers, from three to infinity.*
> *If then you assign an index to the numbers [thus measured], you will find that those that play the role of measuring will never together measure the same number [of numbers]*

---

1.   Nicomaque de Gerase, *Introduction arithmétique*, I, 13.

(in fact, it happens that not even two numbers measure the same number [of numbers]), and that the numbers of the list are not all measured by some number, but certain of them escape entirely from being measured by any number at all; others are measured by a single number, others by two or even more. Those which remain not measured in any way, but escape measure, are the prime numbers and non-composite which are thus found separated from the rest as with a sieve.

What is necessary is to lay out the largest possible line of odd numbers. The first prime number is 3, and the process eliminates all multiples of 3, noting that they are separated by two terms, and the number of times that 3 measures each number is according to the successive odd numbers. The first term in what remains is 5, which is prime, and the process eliminates all its multiples (measured by the successive odd numbers). And so on ... All that are left are prime numbers. We see how the practical process is described by this text, while the Euclidean deductive approach is totally absent.

The sieve of Eratosthenes can be improved by use of the principle used in the successive divisions algorithm. 2, 3, 5, 7 being the only prime numbers less than 10, the elimination of all the multiples of these four numbers from the first 100 integers (or even up to $120 = 11^2 - 1$ integers) will leave behind only prime numbers, none having a factor less than its square root. Next, by taking out all multiples of the prime numbers found between 2 and 120, we find all prime numbers up to 14400, &c. Below we show a representation of the sieve for the first 120 integers. By arranging the numbers in 6 rows, the multiples of 2 and 3 can be easily recognised (the rows 2, 3, 4 and 6), and the multiples of 5 and 7 can be seen lying along diagonals shown by arrows. All the prime numbers are found on the first and fifth row. Is that a surprise? (see Exercise 2).

```
1   [7] [13] [19]  25  [31] [37] [43]  49   55  [61] [67] [73] [79]  85   91  [97] [103][109] 115
[2]→8 →14 →20 →26 →32 →38 →44 →50 →56 →62 →68 →74 →80 →86 →92 →98 →104→110→116
[3]→9 →15 →21 →27 →33 →39 →45 →51 →57 →63 →69 →75 →81 →87 →93 →99 →105→111→117
 4 →10→16 →22 →28 →34 →40 →46 →52 →58 →64 →70 →76 →82 →88 →94 →100→106→112→118
[5] [11] [17] [23] [29]  35  [41] [47] [53] [59]  65  [71]  77  [83] [89]  95  [101][107][113] 119
 6 →12→18 →24 →30 →36 →42 →48 →54 →60 →66 →72 →78 →84 →90 →96 →102→108→114→120
```

Fermat's algorithm

In 1643, in response to a challenge by Mersenne, Fermat described a method for factorising numbers which he used for the number 2 027 651 281 to find 46061×44021 after twelve steps. His idea was to factorise a number n in the form of the difference of two squares $x^2 - y^2 = (x - y)(y + x)$ by testing successive integers x starting from $\sqrt{n}$ to find if $x^2 - n$ is square. But can we be sure that the process will be successful and will the algorithm terminate? In fact, if n is composite and equal to pq, with p greater than or equal to q, the number of steps needed for the algorithm is about $\frac{1}{2}(\sqrt{p} - \sqrt{q})^2$ since $x = \sqrt{n} + \frac{1}{2}(\sqrt{p} - \sqrt{q})^2 = \frac{p + q}{2}$ and will be even less than this if n has a factor close to its square root. For example, the decomposition of 11021 only requires a single calculation which gives $11021 = 107 \times 103$: $\sqrt{11021} \simeq 104.9$ and $105^2 - 11021 = 4$, whereas the method of successive divisions requires 27 divisions by all the prime numbers less than 104. In the case where p and q are composite, the process is repeated to find the other factors. But what happens if the number n is prime? Fermat's algorithm will end in this case since if n is odd then $\left(\dfrac{n+1}{2}\right)^2 - \left(\dfrac{n-1}{2}\right)^2 = n$ and it has been known since 1891 (E. Lucas, *Théorie des Nombres*) that when n is prime this is the only way of writing n in the form of the difference of two squares.

Exercise 4	Show that every odd number n (> 1) is the difference of two squares $x^2 - y^2$ where $\sqrt{n} \leq x \leq (n + 1) / 2$ and that the pair of natural numbers x, y is unique if, and only if, n is prime.

Exercise 5	Find the prime factor decomposition of the number 52793 using Fermat's algorithm and compare the method with that used in Exercise 3.

These methods, requiring so many calculations, are unfortunately unsuitable for use with very large numbers. For example, one of the largest prime numbers yet known contains 227 832 digits, and the time needed for the most powerful computer yet built to test that it is prime by these methods, is greater than the age of the Earth. But if tests are not available, how can we know if there might be even greater prime numbers than this, say greater than $10^{1\,000\,000}$?

The sequence of primes is unlimited

Neither 2, 3 nor 5 divides exactly into the number $2 \times 3 \times 5 + 1$, and in the same way any number composed of the product of a number of primes numbers with 1 added, will not have one of these primes as a factor. But this number will, according to Euclid VII, Prop. 30, have a factor that is different from any of the factors of the product. There are therefore always more prime numbers than any finite list of prime numbers, no matter how large. Here is the argument presented by Euclid in Book IX, Prop. 20 The infinite does not appear here explicitly however: as usual at this time, the infinite is only potentially present.

Proposition 20: Prime numbers are more than any assigned number of prime numbers.

Let A, B, C be the assigned prime numbers;
I say that there are more prime numbers than A, B, C.
For let the number measured by A, B, C be taken, and let it be DE; let the unit DF be added to DE.
Then EF is either prime or not.

A ———— B ————— C —————— G ———————————
E ———————————————————————————— D ———— F

First let it be prime; then the prime numbers A, B, C, EF have been found which are more than A, B, C.
Next let EF not be prime; therefore it is measured by some prime number. Let it be measured by the prime number G.
I say that G is not the same with any of the numbers A, B, C.

For, if possible let it be so. Now, A, B, C measure DE. Therefore G also will measure DE. But it also measures EF. Therefore G, being a number, will also measure the remainder unit DF: which is absurd.

Therefore G is not the same with any of the numbers A, B, C. And by hypothesis it is prime.

Therefore the numbers A, B, C, G have been found which are more than the assigned multitude of A, B, C.[1]

This famous proof by Euclid is the one that is used today. But another proof, given by Euler in 1737 is not without interest. For the first time we see Arithmetic and Analysis, up to then two entirely separate domains of study, coming together and from then onwards the Theory of Numbers and Analysis are inseparable.

Euler knew the sum of the geometric series: $1 + \alpha z + \alpha^2 z^2 + \alpha^3 z^3 + $ etc. $= \dfrac{1}{1 - \alpha z}$; the

fraction $\dfrac{1}{(1 - \alpha z)(1 - \beta z)}$ can therefore be written as the product of the two sums $(1 + \alpha z$

$+ \alpha^2 z^2 + \alpha^3 z^3 + ...)$ and $(1 + \beta z + \beta^2 z^2 + \beta^3 z^3 + ...)$ and is equal to $1 + (\alpha + \beta)z + (\alpha^2 + \alpha\beta + \alpha\beta^2)z^2 + (\alpha^3 + \alpha^2\beta + \alpha\beta^2 + \beta^3) z^3 + ...$ etc. In this way Euler expanded the fraction

$\dfrac{1}{(1 - \alpha z)(1 - \beta z)(1 - \gamma z)(1 - \delta z) \text{ etc.}}$ in the form $1 + Az + Bz^2 + Cz^3 + Dz^4 + ...$ etc.,

where A is the sum of the coefficients α, β, γ, δ, etc., B is the sum of the products of the coefficients taken two at a time, C is the sum of the products of the coefficients taken three at a time, etc. Taking $z = 1$ and letting α, β, γ, δ, etc. take the values of the reciprocals of successive prime numbers 2, 3, 5, 7, etc., then thanks to the fundamental theorem of arithmetic, Euler obtains the infinite harmonic series $1 + \dfrac{1}{2} + \dfrac{1}{3} + \dfrac{1}{4} + ...$

which he claims is equal to the logarithm of $\dfrac{1}{1 - x}$ when $x = 1$ (!). The initial product fraction is therefore infinite, if we accept Euler's somewhat bold manipulation of infinite quantities taking no account of limits, and is the same as the list of prime numbers. Here, contrary to Euclid, is the infinite present in action.

Art. 272. Hence the series is always composed of an infinite number of terms, whatever the number of factors, finite or infinite. For example we have $\dfrac{1}{1 - 1/2} = 1 + \dfrac{1}{2} + \dfrac{1}{4} + \dfrac{1}{8}$

$+ \dfrac{1}{16} + \dfrac{1}{32} +$ etc. a series in which we only find numbers which are formed by multiplication by the number two; that is all the powers of two.

From this we have: $\dfrac{1}{(1 - 1/2)(1 - 1/3)} = 1 + \dfrac{1}{2} + \dfrac{1}{3} + \dfrac{1}{4} + \dfrac{1}{6} + \dfrac{1}{8} + \dfrac{1}{9} + \dfrac{1}{12} + \dfrac{1}{16} + \dfrac{1}{18}$

$+$ etc. Here we find only numbers that are formed by the combinations of 2 and 3, or which have no other divisors than 2 and 3.

Art. 273. Hence, if in place of α, β, γ, δ, etc. we write one divided by all the prime

numbers, and letting $P = \dfrac{1}{(1 - \frac{1}{2})(1 - \frac{1}{3})(1 - \frac{1}{5})(1 - \frac{1}{7})(1 - \frac{1}{11}) \text{ etc.}}$ we obtain

$P = 1 + \dfrac{1}{2} + \dfrac{1}{3} + \dfrac{1}{4} + \dfrac{1}{5} + \dfrac{1}{6} + \dfrac{1}{7} + \dfrac{1}{8} + \dfrac{1}{9}$ etc. a series containing all the numbers,

while it is the prime numbers under multiplication that produced the series. Now, since

---

1.   Heath, *The Thirteen Books of Euclid's Elements*, vol. 2, p. 412.

all the numbers are either prime numbers or numbers composed from them by multiplication, it is evident that we must find all the whole numbers here in the denominators... Example I. Since we have shown earlier that

$$l\left(\frac{1}{1-x}\right) = x + \frac{x^2}{2} + \frac{x^3}{3} + \frac{x^4}{4} + \frac{x^5}{5} + \frac{x^6}{6}\ etc.;$$

we have, in letting $x = 1$, $l\left(\dfrac{1}{1-1}\right) = l(\infty) = 1 + \dfrac{1}{2} + \dfrac{1}{3} + \dfrac{1}{4} + \dfrac{1}{5} + etc.;$

but the logarithm of an infinite number ∞ *is itself infinitely large;*

hence $M = 1 + \dfrac{1}{2} + \dfrac{1}{3} + \dfrac{1}{4} + \dfrac{1}{5} + \dfrac{1}{6} + \dfrac{1}{7} + etc. = \infty.$

$$[...]\ M = \infty = \frac{1}{(1 - \frac{1}{2})(1 - \frac{1}{3})(1 - \frac{1}{5})(1 - \frac{1}{7})(1 - \frac{1}{11})\ etc.} \cdot 1$$

Partition of the Prime Numbers

But whereabouts is this infinity of primes found within that other infinity of all the natural numbers? Are they distributed regularly? How frequently do they occur? Are there "more and more" of them or "fewer and fewer" of them as we run through larger and larger numbers? Are they widely dispersed, or are they concentrated into clusters, surrounded by greater and greater gaps? Notice that 2 and 3 are the only two consecutive primes. Two prime numbers differing by 2 are said to be pairs or twin primes and, except for 3 and 5, must be of the form $6k \pm 1$ (see exercise 3) like, for example, 5, 7, and 17, 19, or 1787, 1789, or 4001, 4003, &c. But are there an infinity of such prime number pairs? This was the 8th of the famous 23 problems posed by Hilbert at the 1900 Paris Second International Mathematical Congress, and a question that remains unanswered today. A number of the form $6k \pm 1$ will not necessarily be prime (25 or 35, for example) but it is possible, by a slight alteration in Euclid's reasoning of Book IX, Prop. 20 that there is an infinite number of primes of the form $6k - 1$:

Let n be any integer greater than 3 and P the product of all the prime numbers less than n. All the prime factors of the number $A = P - 1$ are greater than n since they cannot be both factors of A and P at the same time. But $P = A + 1$ is divisible by 6 so A cannot be of the form $6k + 1$ nor can it be the product of prime factors all of which are of the form $6k + 1$. A must therefore have at least one prime factor (Euclid VII. 31) of the form $6k - 1$ and greater than n. The set of prime numbers of this form therefore does not have an upper bound.

Exercise 6   1. Prove that there are an infinite number of primes of the form $4k - 1$.

2.* Prove also that there are an infinite number of primes of the forms $4k + 1$ and $8k + 5$, assuming that every prime factor which is the sum of two squares (of numbers which are prime to one another) is also the sum of two squares.

The Arithmetic Progression theorem: The results above are particular cases of an important and difficult theorem, first formulated in 1785 as a conjecture by Legendre in

1.   Euler, *Introduction à l'analyse infinitésimale*, vol. 1, p. 209-213.

the *Mémoires de l'Académie des Sciences* as: "Every Arithmetic Progression whose first term and common difference are prime to one another contains an infinite number of primes". In the *Théorie des Nombres* published in 1798, Legendre thought that he had proved the result, but it was not definitively proved until 1837 by Lejeune-Dirichlet who, at the instigation of Euler, set up a proof that depended upon the divergence of series.

There are then a great many primes that occur in arithmetic progressions, but what can be said about their distribution among other integers? Notice firstly that there are some intervals that contain quite a lot of prime numbers: 25% in [1; 100], 21% in [101; 200], 13.5% in [1001; 2000], and others that contain only a few: only 2% in [10 000 000; 10 000 100] which only contains the two prime numbers 10 000 019 and 10 000 079. And here is a particularly astonishing result: it is possible to find a list of consecutive integers, as large as is wished, which does not contain any prime. All that you need to do is to consider, for any arbitrarily large n, the n − 1 numbers from n! + 2 to n! + n, all of which will have, at least, one of the numbers from 2 to n as a factor.

Bertrand's conjecture: For any number n, there is a prime number between n and 2n This was first proposed by Bertrand in 1845 and was proved three years later by Tchebycheff, who finding convenient upper and lower bounds for the ratio $\pi(n) / (n/\log(n))$, established that $\pi(2n) - \pi(n)$ is strictly positive. ($\pi(n)$ stands for the number of primes less than n.) It is still not known whether or not Legendre's conjecture is true: that there exists a prime number between n^2 and $(n + 1)^2$.

The law of scarcity of prime numbers

In about 1785, Legendre gave an empirical approximation for $\pi(x)$ working from tables of primes in the form of $Ax/(B.\ln(x) + C)$.

In his *Essai sur la Théorie des Nombres*, he used tables of primes up to 1 000 000 to compare $\pi(x)$ with $\dfrac{x}{\ln(x) - 1.08366}$; the values corresponded well with relative errors of less than 1.5×10^{-3}. In the middle of the nineteenth century, Gauss counted the numbers of primes in intervals of 1000 integers and obtained an empirical result for the density of primes in the neighbourhood of a large number x, as $\dfrac{1}{\ln(x)}$. He gave a new approximation for the number of primes as $\pi(x) \approx \displaystyle\int_2^x \dfrac{du}{\ln(u)} = \text{Li}(x)$. (This is the logarithmic integral: ln is the natural logarithm.)

Exercise 7   Show that the two formulas for an approximation to $\pi(x)$ are equivalent as x tends to infinity.

In fact, these two approximations turned out to be in agreement with the theory developed a little later by Hadamard and De-La-Vallée-Poussin who proved the Prime Number Theorem in 1896: $\displaystyle\lim_{n \to \infty} \dfrac{\pi(n)}{n/\ln(n)} = 1$.

The proof is difficult, once again using analysis and makes use of the Riemann zeta function $\zeta(s) = \sum_{n \in \mathbb{N}*} \dfrac{1}{n^s}$. It was not until 1948 that an arithmetic proof was given, by Erdos and Selberg.

A consequence of the Prime Number Theorem is the law of scarcity of primes: $\lim_{n \to \infty} \dfrac{\pi(n)}{n} = 0$. The density of primes tends to 0 "at infinity". In other words, as we consider larger and larger numbers, the primes become more and more rare, even though there is always an infinity of them.

Formulas rich in prime numbers

The Arithmetic Progression theorem shows that all expressions $ax + b$, where a, b are prime to one another, will produce an infinity of primes. Is it possible to find polynomial formulas that only give prime numbers? The answer is no.

Exercise 8   Show that there is no polynomial P(x), not a constant, with integer coefficients, that will produce prime numbers for all values of integer x.

It is impossible to find a polynomial that only generates primes, but there are polynomials that produce a large number of primes. Examples are: $x^2 - x + 17$; $2x^2 + 29$; $x^2 - x + 41$; which produce, respectively, 16, 28 and 40 distinct prime numbers for successive integer values of x starting at $x = 1$. The polynomial $x^2 - 79x + 1601$ generates 80 primes (but each one twice) for values of x from 0 to 79. These polynomials are mostly due to Euler, who was a brilliant and indefatigable calculator.

> *This progression 41, 43, 47, 53, 61, 71, 83, 97, 113, 131, etc. whose general term is 41 – x + xx is even more remarkable in that the first forty terms are prime numbers.* (Letter from Euler to J. Bernouilli, c.1772).

And here an astonishing number appears: **163**. It is the numerical value of the discriminant of the Eulerian polynomial $41 - x + x^2$ $\{1 - 4 \times 41 = -163\}$, a formula rich in prime numbers (not only the first 40 values, but about a third of all the succeeding 10 000 000 values of x generate primes). And that is not all: the number **163**, a prime, possesses another remarkable property: the number $e^{\pi \sqrt{163}}$ differs from an integer by less than 10^{-12} (the value is 262 537 412 640 768 744 and the relative error is of the order of 5.10^{-30}). These two, apparently unconnected, properties followed a third property demonstrated by Gauss at the beginning of the nineteenth century, who was studying the representation of integers by biquadratic forms, that is $ax^2 + bxy + cy^2$, with a, b, c integers, and classifying them according to their discriminant $d = b^2 - 4ac$. Two forms are said to be equivalent if they are related by a bijective linear transformation $(x, y) \to (\alpha x + \beta y, \gamma x + \delta y)$ where α, β, γ, δ are integers and $|\alpha\delta - \beta\gamma| = 1$. Now, if two forms are equivalent, their discriminant d is the same. When d is negative, the inverse is not true except for the values: –3, –7, –11, –19, –43, –67, and –163. In fact it was not until 1952 (Heegner) and 1967 (Stark and Baker) that it was shown that this list is exhaustive. And -163 is precisely the discriminant of the form $x^2 - xy + 41y^2$.

Exercise 9

1) Using the identity
$(a^2 + b^2).(c^2 + d^2)=(ac + bd)^2 + (ad - bc)^2= ac - bd)^2 + (ad + bc)^2$,
prove that the set of numbers of the form $a^2 + 163.b^2$, where a, b are natural numbers, is closed under multiplication; and that if such a number is prime then (a, b) is unique.

2) Assume that it is true that every odd number that is represented uniquely by the form $\alpha^2 - \alpha\beta + 41\beta^2$ is a prime (α, β being prime to one another with $\beta > 0$). Show that $\alpha^2 - \alpha + 41$ is prime if the only integer value of x that makes the polynomial
$1 + 2(2\alpha - 1).x - 163.x^2$ equal to a square is $x = 0$.

3) Hence prove Euler's claim in his letter to Bernouilli ... and solve the problem 4) of the introduction regarding the spiral of Ulam. (Consider squares of side $2k + 1$ centred on 41.)

In fact many generative formulas have been more lately discovered and they do have a certain theoretic interest, even if they do not have any practical utility. For example, W. H. Mills has proved the astonishing property that there exists a real number k, strictly greater than 1, whose value is not known, which satisfies the following: for every integer n the number $[k^{3^n}]$ is prime. ($[x]$ means the integer part of x.)

In 1976 the American James P. Jones constructed explicitly a polynomial with 26 variables and of degree 25 which is a Diophantine representation of the set of prime numbers, that is that set is none other than the positive values taken by P while the variables describe N. Here is this monster[1]:

$(k + 2) \{1 - [wz + h + j - q]^2 - [(gk + 2g + k + 1)(h + j) + h - z]^2 - [2n + p + q + z - e]^2$

$- [16(k + 1)^3 (k + 2)(n + 1)^2 + 1 - f^2]^2 - [e^3 (e + 2)(a + 1)^2 + 1 - \sigma^2]^2$

$- [(a^2 - 1)y^2 + 1 - x^2]^2 - [16r^2y^4(a^2 - 1) + 1 - u^2]^2$

$- [((a + u^2(u^2 - a))^2 - 1)(n + 4dy)^2 + 1 - (x + cu)^2]^2$

$- [n + l + v - y]^2 - [(a^2 - 1)l^2 + 1 - m^2]^2 - [ai + k + 1 - l - i]^2$

$- [p + l(a - n - 1) + b(2an + 2a - n^2 - 2n - 2) - m]^2$

$- [q + y(a - p - 1) + s(2ap + 2a - p^2 - 2p - 2) - x]^2 - [z + pl (a - p) + (2ap - p^2 - 1) - pm]^2\}$

In fact this formula only gives positive numbers when the factor $(k + 2)$ is prime and the second factor $\{...\}$ is equal to 1 (i.e. for all the $[...]^2$ eaqual to zero).

In 1770, Edward Waring (*Meditationes Algebricae*, Cantab.) stated a theorem about prime numbers that was proved in 1771 by Lagrange and today is known as Wilson's Theorem. It states:

Every prime number p is a factor of $1 + (p - 1)!$

Lagrange expanded the product
$(x + 1)(x + 2) ... (x + (p - 1))$ in the form $x^{p - 1} + Ax^{p - 2} + Bx^{p - 3} + ... + Mx + N$ and showed that all coefficients A, B, C..., and M are divisible by p and that $N = (p - 1)!$ He finished by putting $x = 1$.

This result is all the more remarkable because its inverse is also true, something that is not true, for example, for Fermat's Theorem, which is only one necessary condition for the primality of a number, as we shall see later. In effect, if a number p greater than

---

1.   Jones, Sato, Wada, Wiens, "Diophantine of the set of prime numbers", *American Mathematical Monthly*, June-July, 1976, p. 449-464.

or equal to 2 is not prime, then it has a factor between 2 and p – 1, which will divide (p – 1)! and so cannot be a factor of (p – 1)! + 1, and so p cannot be a factor of (p – 1)! +1.

Exercise 10   Show that if p is not prime, and p > 4, then (p – 1)! is divisible by p. Explain the case where p = 4.

This provides a property of prime numbers which gives a simple test: does p divide 1 + (p – 1)! ? Unfortunately there is no good algorithm for calculating factorials and it is practically impossible to find (p – 1)! with any accuracy for large numbers because of the size of the product. All the same, Wilson's theorem does give rise to a function f of two variables x and y which produces only primes, all the primes occurring exactly once, except for 2 which appears frequently and an infinite number of times. The function is:

$$f(x, y) = \frac{1}{2}(y - 1)[|\, k^2 - 1\,| - (k^2 - 1)] + 2 \text{ where } k = x(y + 1) - (y! + 1)$$

Some values are f(1,3) = 2; f(5,2) = 2; f(103,6) = 7; f(329891,10) = 11, ... Clearly such a function cannot be calulated for very large numbers.

Exercise 11   Using Wilson's theorem show that k only vanishes when (y + 1) is prime. Hence deduce the properties of the function f stated above.

Large Prime numbers and Fermat's theorem

For any number N of n digits, the law of density of prime numbers tells us that there will be approximately $\sqrt{N} / \ln(\sqrt{N})$ prime numbers less than $\sqrt{N}$; now $(\sqrt{N} / \ln \sqrt{N})$ is at least of the order of $\frac{2.10^{(n-1)/2}}{(n-1).\ln(10)}$. To prove that N is prime by the method of successive divisions, even supposing that all the primes less than $\sqrt{N}$ are known (which is increasingly unlikely), would require at least about $\frac{2.10^{(n-1)/2}}{(n-1).\ln(10)}$ divisions. This is 3050 for n = 10; 5.6 x 10^{22} for n = 50; and 2.8 x 10^{47} for n = 100. A calculator capable of carrying out 10^{10} operations per second would require about 10^{30} years to identify a 100 digit prime number. Necessarily new tests are needed to begin to be able to study very large prime numbers, and the key to these starts with a theorem by Fermat.

> *Following this, it seems to me that it is important that I should tell you the foundation on which I base the proofs of everything that concerns geometric progressions, which is this: Every prime number infallibly measures one of the powers –1 of some progression, whatever it may be, and the exponent of the said progression is a sub-multiple of the prime number –1; and, after having found the first power that satisfies the question, all those whose exponents are multiples of the exponent of the prime will also satisfy the same question [...] And this proposition is generally true for all progressions and for all prime numbers; the proof of which I would send you, were it not that I fear it is too long ...*[1]

---

1.   Fermat, *Oeuvres*, t. II, Lettres à Frenicle du 18-10-1640, p. 206.

From this text we can state the theorem, later proved by Euler[1]: Every prime number p that does not divide a, divides $a^{p-1} - 1$.

Euler shows that the coefficients C_p^k in the expansion of the binomial $(a + b)^p$ are divisible by p prime for all k between 1 and $p - 1$ and so deduces (for b = 1) that p divides $(a + 1)^p - a^p - 1$; thus $(a + 1)^p - (a + 1)$ will be divisible by p if $a^p - a$ is divisible by p; since $1^p - 1$ is certainly divisble by p it will be the same for all $a^p - a = a(a^{p-1} - 1)$ by induction; if then p does not divide a, it will divide $a^{(p-1)} - 1$.

At the beginning of the nineteenth century Gauss set out the fundamental ideas of congruences that would considerably clarify calculations and deductions with integers. To state that a is congruent to b modulo (n), written $a \equiv b \bmod(n)$ means that n divides $(a - b)$, which is equivalent to saying that a and b have the same remainders when divided by n. This relation is compatible with addition and multiplication and so calculations with congruences are analogous to calculations with integers, and even with rationals if n is prime (all $a \neq 0 \bmod(n)$ has an inverse modulo (n), see later). For example, using Fermat's theorem, $2^{10} \equiv 1 \bmod (11)$, and $2^5 = 32 \equiv 1 \bmod (31)$. Squaring this last, we get $2^{10} \equiv 1 \bmod (31)$ and so $2^{10} - 1$ is divisible by 11 and 31 and so also by their product 341, so $2^{10} \equiv 1 \bmod (341)$. Raising this last to the power of 34 gives $2^{340} \equiv 1 \bmod (341)$. Thus 341 divides $2^{341-1} - 1$ even though it is not prime. So the converse of Fermat's theorem is false! The condition: p divides $a^{(p-1)} - 1$, a condition that is easy to test because there are· good algorithms for calculating powers modulo (p), is not sufficient to guarantee a prime. The numbers p, which for a number a between 2 and $p - 1$ divide $a^{(p-1)} - 1$ but are not prime, are said to be "pseudo-prime" for base a. So 341 is a pseudo-prime in base 2 and this is the smallest possible in this base (the next is 561). We shall see at the end of this chapter that those Fermat numbers $2^{2^n} + 1$ and Mersenne numbers $2^p - 1$ (p prime) which are not prime, are pseudo-prime in base 2. It can be shown that 286 is pseudo-prime in base 3, but is it the smallest pseudo-prime?

Exercise 12   Show that 2701 is a pseudo-prime in base 2. Find the least pseudo-prime in base 3

Is every composite number a pseudo-prime for at least one base? The answer is, yes for the odd numbers (we always have $(p - 1)^{p-1} \equiv 1 \bmod (p)$) but no for the even numbers 4, 6, 8, 10, 12, ... which are neither prime nor pseudo-prime for any base. The first pseudo-prime is 9 (in base 8) and the first even pseudo-prime is the perfect number 28 (in base 9 and base 25).

Are there many pseudo-primes for any given base? Yes, there are an infinite number of them. And only one is needed to produce an infinite number. For example, for a = 2, it is easy to show (see exercise 16) that if p is pseudo prime in base 2 then so is $2^p - 1$.

In fact, if a number a can be found between 2 and $n - 1$ such that $a^{n-1} \equiv 1 \bmod (n)$, Fermat's theorem proves that n is not prime (for example, 10537 is not prime because $2^{10536} \equiv 10022 \bmod (10537)$), but if $a^{n-1} \equiv 1 \bmod (n)$ nothing can be said *a priori*. To summarise, Fermat's theorem can be used to verify whether or not a given number is composite (with a good success rate) but it does not allow us to recognise prime numbers.

1.   Euler, *Opera Omnia 12, commentationes arithmeticæ, commentatio*, 134.

Exercise 13   1) Show that the sequence of remainders of 10^n modulo p; p prime other than 2 or 5, is periodic.

2) Deduce that the decimal form of $1/p$ will be periodic.

3) Examine the case $p = 7$ to solve problem 1 of the introduction.

Generalisation and refinement of Fermat's theorem

Returning to Fermat's text, there is a least value k of the strictly positive exponents m of a such that $a^m \equiv 1$ mod (p). The sequence of remainders a^n mod (p) is therefore periodic of period k, the k first remainders being all distinct and $a^m \equiv 1$ mod (p) only when m is a multiple of k.

The value k is called the gaussian of a for the modulus p and so divides (p – 1) when p is prime. We have seen that the inverse is false but when this gaussian has the value (p – 1), the remainders modulo (p) of the first (p – 1) powers of a will be all the integers from 1 to p – 1 and which are therefore prime with p ($a^{p-1} \equiv 1$ mod (p) produces the Bezout equalities between p and its remainders); hence p is prime. This result was established, by a slightly different reasoning, by E. Lucas in 1876 and is thus a partial inverse of Fermat's theorem: Let n be an odd number. If an integer a can be found between 1 and n – 1 such that $a^{n-1} \equiv 1$ mod (n) and $a^d \neq 1$ mod (n) for all proper divisors d of (n – 1) (i.e. d different from n – 1) then n is prime.

Lucas was able to establish that the number $F_4 = 2^{16} + 1 = 65537$ is a prime: the only divisors of $F_4 - 1$ are 1, 2, 2^2, ... 2^{15}, and 2^{16} and in taking a = 3, he was able to verify that none of the numbers 3^{2^m} has a remainder congruent to 1 mod (F_4) for m running from 1 to 16 and that $3^{2^{16}} \equiv 1$ mod (F_4). The calculation is considerably simplified by noticing that each remainder modulo F_4 is congruent to the square of the preceding remainder, i.e. $3^{2^{m+1}} \equiv (3^{2^m})^2$ mod (F_4).

In practice, this condition is only interesting if n – 1 is easily decomposable and if it has few divisors, and also if one value a can be found that is convenient.

In fact, since the *Recherches arithmétiques* of Gauss in 1801, it has been known that if n is prime, there is always such a number a, whose gaussian is n – 1 for modulus p, and also that there are as many such numbers a, as there are numbers prime with n – 1, between 1 and n – 1.

Exercise 14   For x a non zero whole number, the Euler phi function $\phi(x)$ is defined as the number of integers less than x which are prime to x.

1) Find $\phi(p)$ and $\phi(p^\alpha)$ for prime p.

2) Show that ϕ is multiplicative, that is $\phi(xy) = \phi(x)\phi(y)$ for all x, y prime to each other.

3) Show that if x is decomposed into prime factors $\prod\limits_{i=1}^{k} p_i^{\alpha_i}$,

then $\phi(x) = \prod\limits_{i=1}^{k} (p_i^{\alpha_i} - p_i^{(\alpha_i - 1)})$.

Exercise 15   1) Use Fermat's theorem to prove the following generalisation due to Euler: If a and n are prime to each other, then $a^{\phi(n)} \equiv 1$ mod (n), where $\phi(n)$ is the Euler phi function.

2) Let a be an odd number not divisible by 5. Show that $9a$ is prime with 10 and so solve problem 2) of the introduction.

The condition due to Lucas is not easy to put into practice but it was improved by D. H. Lehmer in 1927 who established that n is prime if there exists an integer a such that $a^{n-1} \equiv 1$ mod (n) and for all prime factors p of n − 1 we have $a^{(n-1)/p} \not\equiv 1$ mod (n). In fact this condition implies that the quotient of n − 1 by the gaussian of a does not have a prime factor: it is therefore equal to 1 and the gaussian of a is n − 1: n is prime.

Thus for F_4 with 2 as the only factor of $F_4 - 1$, a single test is required: $3^{2^8} \equiv -1$ mod (F_4) is sufficient since it leads to $3^{2^{16}} \equiv (-1)^2 \equiv 1$ mod (F_4). (See Pépin's test later.)

Here again, we do not know how to choose a value a that would be convenient, but Lehmer refined this result to the following form: if p is a prime factor of n − 1 and a an integer such that $a^{n-1} \equiv 1$ mod (n) and $a^{(n-1)/p} \equiv r$ mod (n) with r not 1, and with r − 1 prime with n, then all the prime factors of n are of the form $1 + kp^\alpha$, where p^α is the greatest power of p that divides n − 1. In fact. if q is a prime factor of n and b is the number $a^{(n-1)/p^\alpha}$ then since $b^{p^\alpha} \equiv a^{n-1} \equiv 1$ mod (q) and $b^{p^{(\alpha-1)}} \equiv r \not\equiv 1$ mod (q), then the gaussian of b modulo (q) is p^α and is a divisor of q − 1 by Fermat's theorem. Lehmer used this method to obtain the prime factor decomposition of

$$10^{24} + 1 = 17 \times 5882353 \times 9\,999\,999\,900\,000\,001.$$

To take a simpler example, choose n = 48593 then n − 1 = 48592 = 2^4 x 3037; with a = 3 we have $3^{(n-1)/3037} = 3^{16} \equiv 41\,916$ mod (48593) and $3^{48\,592} \equiv 1$ mod (48593) so all the prime factors of 48593 are of the form $1 + 3037k$ which are greater than $\sqrt{48593} \approx 220.4$ and so 48593 is prime.

A result due to Euler offers a further improvement: if p is an odd prime and does not divide a then $a^{(p-1)/2} \equiv \pm 1$ mod (p). In fact p divides

$$a^p - 1 = (a^{(p-1)/2} - 1)(a^{(p-1)/2} + 1)$$

and so must divide one of its factors. The inverse is yet again false (example: p = 341, a = 2) and a non prime number n which satisfies $a^{(n-1)/2} \equiv \pm 1$ mod (n) is said to be a strong pseudo-prime in base a (a lying between 1 and n − 1).

An unexpected consequence of this last refinement of Fermat's theorem is the probabilistic test by M. O. Rabin who showed in 1980[1] that if n is not prime, there are at least (n − 1)/4 bases a between 2 and (n − 1) for which n is a strong pseudo-prime. To recognise a (large) prime number n, make random choices of k numbers between 2 and n − 1; if they all satisfy the condition then one can state that n is prime with the possibility of an error of less than $4^{(-k)}$. This is not a proof, but for k of the order of 50 confidence in the result will be better than computer error.

There are several other tests of primality, in particular those that use the idea of quadratic residues, first used by Euler from 1755. A quadratic residue modulo n is an integer a congruent to a square modulo n. So in modulo 7 the quadratic residues are those numbers congruent to 1, 2 ($\equiv$ 9) or 4 and the non quadratic residues are those

---

1.   Rabin, *Probabilistic Algorithm for Testing Primality*, Journal of Number Theory 12, 1980, p. 128-138.

equivalent to 3, 5 or 6. We note that the 3rd powers of these numbers are equivalent to 1, for the first three, and to −1 for the second three, and this is a general result, called the Euler test: if p is an odd prime, then the number a not a multiple of p is a quadratic residue modulo p, if and only if $a^{(p-1)/2} \equiv 1$ mod (p). This condition follows from Fermat's theorem, for if $a \equiv b^2$ mod (p) then $a^{(p-1)/2} \equiv b^{(p-1)} \equiv 1$ mod (p). Furthermore, between 1 and p − 1 only x and (p − x) have the same square modulo p, so there are in the interval exactly (p − 1)/2 non zero quadratic residues. These are the (p − 1)/2 solutions to the equation $x^{(p-1)/2} \equiv 1$ mod (p), and the other numbers, of which there are also (p − 1)/2 are the non quadratic residues, which are solutions of the equation $x^{(p-1)/2} \equiv -1$ mod (p). When p is not too large, it is easy to determine the quadratic residues modulo p directly or by the use of the Euler test, but for very large numbers it is better to use the law of quadratic reciprocity stated by Euler in 1783 and by Legendre in 1785, but proved in 1801 by Gauss in his *Recherches arithmétiques*: p and q being two different odd primes, if $\dfrac{p-1}{2} \times \dfrac{q-1}{2}$ is even, then p is a quadratic residue modulo q if and only if q is a quadratic residue modulo p, otherwise p is a quadratic residue modulo q if and only if q is not a quadratic residue modulo p. This can be symbolised as: $\left[\dfrac{p}{q}\right] \times \left[\dfrac{q}{p}\right] = (-1)^{\frac{(p-1)}{2} \times \frac{(q-1)}{2}}$ where the Legendre symbol $[\dfrac{p}{q}]$ is simply the remainder of $p^{(q-1)/2}$ modulo q, and has the value 1 if p is a quadratic residue modulo q, and − 1 if not, from Euler's test.

In his *Recherches arithmétiques* Gauss uses the idea of quadratic residues to test for primality or to decompose a given number into its prime factors. If certain numbers a are known to be quadratic residues modulo n, this will allow us to eliminate all the prime numbers for which a cannot be a quadratic residue, and these will not be factors of n. Gauss treated the number 997 331 in this way, for which the numbers −6, 13, −14, 17, 37 and −53 are the quadratic residues and this allows him to eliminate all the prime numbers up to 127. He was then able to factorise: 997 331 = 127 × 7853.

Finally, a generalisation of Fermat's algorithm, that depends upon the decomposition of the number $\sqrt{n}$ into continuous fractions, is due to Legendre[1]. It consists in writing n, not in the form of $x^2 - y^2$ but as n divides into $x^2 - y^2$, that is $x^2 \equiv y^2$ mod (n). The factors of n can be found by calculating the HCF of n and $x \pm y$. To find x and y, Legendre's approach, taken up again by Lehmer and Powers in 1931, was to use the (periodic) expansion of $\sqrt{n}$ into continuous fractions: if a_i / b_i is the ith convergent, we have $a_i^2 - nb_i^2 = (-1)^{i+1}q_{i+1}$ where the ith quotient x_i will be

$\dfrac{q_{i-1}}{\sqrt{n} - p_{i-1}}$ (q_{i-1} and p_{i-1} being integers). By combining many of these conditions, we eventually arrive at the desired result $x^2 \equiv y^2$ mod (n). Lehmer and Powers were in this way able to obtain the factorisation 13 290 059 = 3119 × 4261, and by an improvement of this test Morisson and Powers in 1970 were able obtain a factorisation of the Fermat number F_7.

There are yet other tests for primality, in particular those of Adleman, Rumely, Pomerance, Lenstra and Cohen[2] which use the sums and tests associated with Gauss,

1.   Legendre, *Théorie des nombres*, p. 336-341.
2.   Adleman, Pomerance, Rumely, *On distinguishing prime numbers from composite numbers*, Annals of Mathematics, v. **117**, 1983, p. 173-206.

but they go beyond the limits of this chapter. Finally, there are specific tests for the very large Fermat and Mersenne numbers and we shall now turn to these.

Fermat numbers

The origin of Fermat numbers is found here, in his letter to Frénicle of August 1640:

1. Consider for example the double progression of the binary with its exponents written above:

1	2	3	4	5	6	...	13	16
2	4	8	16	32	64	...	8192	65536

I say that if you add a unit to all the numbers of the progression, and you make the augmented numbers of the progression 3, 5, 9, 17, etc., which do not have as exponents the numbers of the said double progression, [they] will be composite numbers.

2. Although it is possible to make a particular analysis which is too long to describe, for you to understand it is sufficient to understand, in the example that follows, what I have done: consider the number in the progression augmented by unity be 8193 which has the exponent a prime number 13. I say that if you divide 8193 by 3, the quotient cannot be divided by a number greater than unity or the double of the above stated exponent 13, or by a multiple of the said double of 13, etc., to infinity. While if the exponent is a composite number, which however can not be one of those of the double progression, I would be able to find all its divisors very easily.

3. But here is what I admire the most: it is that I am persuaded that all the numbers of the progression augmented by unity, whose exponents are the numbers of the double progression, are prime numbers like 3 5 17 257 65537 4294967297 and the following of 20 letters 18 446 744 073 709 551 617; etc. I do not have the exact proof, but I have excluded such a large quantity of divisors by infallible methods, and I have so many great luminaries who confirm my thoughts, that I hardly like to go back on my word.[1]

This extremely interesting text states several properties concerning numbers of the form $2^n + 1$:

1) they are composite if n is not a power of 2 [in fact, if $n = m2^s$ with m odd and > 1 then $(2^n + 1)$ is divisible by $2^{2^s} + 1$.

2) if n is prime the divisors of $(2^n + 1)/3$ are of the form $2kn + 1$.

3) the numbers $2^{2^n} + 1$, the Fermat numbers F_n, are prime, and Fermat was fairly certain of this but could not prove it.

A century later Euler proved that the sixth Fermat number (F_5) was composite and divisible by 641. Landry in 1880 showed that F_6, containing 20 digits, was also composite being divisible by 274 177. In fact it is now thought that the only prime Fermat numbers are 3, 5, 17, 257 and 65537; all the other Fermat numbers whose status is known turn out to be composite. F_{17} is the first Fermat number whose status is not yet known (it has 39457 digits); F_{1945} and F_{23471} are the largest known Fermat numbers, the second having more than 10^{7000} digits. If it turns out that no other Fermat numbers are prime, then the n-sided regular polygons that can be constructed by ruler and compass have been completely determined. Gauss in 1801 and then Wantzel in

---

1.   Fermat, *Œuvres*, X LIII, p. 205-206.

1831 succeeded in proving that such polygons correspond exactly with values of n of the form $2^k \times \prod_{i=1}^{m} F_i$ where the F_i are prime and distinct Fermat numbers.

Exercise 16   1) Show that if a divides b then $(x^a - 1)$ divides $(x^b - 1)$ for all integer $x > 1$.

            2) Deduce that F_n divides $(2^{(F_n - 1)} - 1)$, for all integer n.

The Fermat numbers are therefore either prime or pseudo-prime in base 2, and it may be this that led Fermat to his conjecture. In fact Fermat had only an "inner conviction" of his assertion, for he wrote in October 1640 to the same Frénicle:

> ... But I tell you clearly (for earlier I had warned you that, as I am not capable of claiming more than I know, so with the same frankness I tell what I do not know) that I have not yet been able to exclude all the divisors in that beautiful proposition that I sent you and which you confirmed, touching the numbers 3, 5, 17, 257, 65537, etc., because although I have been able to cut out most of the numbers and I have even got probable arguments for excluding the rest, I have not yet been able necessarily to prove the truth of that proposition, which however I do not doubt now anymore than I did before. [...]

Then on 25 December of the same year, he wrote to Mersenne:

> ... here are three questions that I put to him [Frénicle], about which my speculations do not entirely satisfy me:
> 1) The essential reason why 3, 5, 17, 257, etc. to infinity, are always prime numbers;

Fermat numbers rapidly become enormous: F_6 has 20 digits, F_7 has 39, F_8 has 78, F_9 has 155 and F_{10} has 309, ... How is it possible to find out if they are composite?

First of all, a result due to Euler[1] gives the potential form for divisors of F_n. (It is possible to approach this using the property 2) of Fermat's August 1640 letter): the divisors of F_n are of the form $2^{n+1} k + 1$. After examining the divisors of $a^2 + b^2$, then those of $a^4 + b^4$ and of $a^8 + b^8$, Euler deduced theorem 8 by induction: the only odd divisors of $(a^{2m} + b^{2m})$, with a and b integers prime with each other, are numbers of the form $2^{m+1} k + 1$. The case $a = 2$, $b=1$ gives the desired result.

Let now p be a prime (odd) divisor of F_n, if 2 is a quadratic residue modulo p we have $2^{(p-1)/2} \equiv 1 \mod (p)$ and consequently the gaussian of 2 modulo p does not divide 2^n but does divide 2^{n+1} since $2^{2n} \equiv -1 \mod F_n$, so it is 2^{n+1} and a divisor of $(p-1)/2$; p is therefore of the form $(2^{n+2} k + 1)$. Now Lagrange, and then Gauss, proved that 2 is a quadratic residue for all prime moduli of the form $8q \pm 1$, which is the case for p of the form $2^{n+1} k + 1$ whenever n is greater than or equal to 2. Hence the divisors of F_n are of the form $2^{n+2} k + 1$, as the product of all prime divisors of that form.

Exercise 17   Show that if p is prime and $p \equiv 1 \mod (8)$ then 2 is a quadratic residue modulo (p).

Hence, the divisors of $F_4 = 65537$ will be of the form $64k + 1$ but 193 is the only prime number of this form less than $\sqrt{F_4}$ and it does not divide 65537 which must therefore be prime. The divisors of $F_5 = 4\,294\,967\,297$ must be of the form $128k + 1$ and we quickly find the divisor $641 = 5 \times 128 + 1$; in the same way F_6 is found to be divisible by $274\,177 = 256 \times 1071 + 1$; and finally, F_{1945} is divisible by $5 \times 2^{1947} + 1$.

---

1.   Euler *Opera omnia*, vol. 12, p. 62-74.

The results established above considerably reduce the number of calculations needed to discover a divisor of F_n, especially when combined with other known properties, for example with a knowledge of the divisors of certain quadratic forms, which is what Legendre did in his *Essai sur la théorie des nombres*. We are able to find that certain Fermat numbers are composite, without however knowing what their factors are. This is the case for F_7, F_8, and F_{14}, ... where a specific test for Fermat numbers can be applied, the Pépin test (1877):

for non zero n, F_n is prime if and only if $3^{(F_n - 1)/2} \equiv (-1) \bmod (F_n)$

The proof rests on the previous results and on the law of quadratic reciprocity. The necessary condition: Suppose F_n is prime, here F_n is of the form $4k + 1$ for $n > 1$. So 3 is a quadratic residue modulo (F_n) if and only if it is one modulo 3.

But $F_n \equiv (-1)^{2^n} + 1 \equiv 2 \bmod (3)$ which is not a quadratic residue modulo 3; therefore 3 is not one modulo F_n and from Euler's formula, $3^{(F_n - 1)/2} \equiv (-1) \bmod (F_n)$; the condition is sufficient: if $3^{(F_n - 1)/2} \equiv (-1) \bmod (F_n)$ the gaussian of 3 modulo F_n is a divisor of $(F_n - 1) = 2^{2n}$ but not of $2^{2(n - 1)}$ hence $2^{2n} = (F_n - 1)$. Now we have seen above that this condition is sufficient for F_n to be prime. In fact we know that F_n is prime for n = 0, 1, 2, 3, 4 and that it is composite for values of n from 5 to 16, and then for n ∈ {18, 19, 21, 23, 25, 26, 27, 30, 32, 36, 38, 39, 42, 52, 56, 58, 63, 73, 77, 81, 117, 125, 144, 150, 207, 226, 228, 250, 267, 268, 284, 316, 452, 1945, 23471}. The status of the other F_{17}, F_{20}, F_{22}, F_{24}, ... etc., is unknown.

Exercise 18   Show that for n > 1, F_n in decimal form has final digits: 37 if n ≡ 0 mod (4); 97 if n ≡ 1 mod (4); 17 if n ≡ 2 mod (4) and 57 if n ≡ 3 mod (4).

Mersenne numbers

These are the numbers of the form $2^n - 1$ we met earlier, with Euclid and Nicomachus, when looking at the construction of perfect numbers (see Chapter 15). The nature of these numbers was not studied until the seventeenth century by Père Marin Mersenne in his book *Cogitata physica mathematica* (1644). According to him, the numbers $2^n - 1$ which we shall refer to as M_n are prime when n takes the values 2, 3, 5, 7, 13, 17, 19, 31, 67, 127, 257 and composite for all other values of n up to 257. A little before, in 1640, Fermat had written to Mersenne:

> ... *here are three propositions I have discovered, upon which I hope to erect a great structure: The numbers less by unity than those of the double progression, like:*

1	*2*	*3*	*4*	*5*	*6*	*7*	*8*	*9*	*10*	*11*	*12*	*13*
1	*3*	*7*	*15*	*31*	*63*	*127*	*255*	*511*	*1023*	*2047*	*4095*	*8191*

> *let them be called the radicals of perfect numbers, since whenever they are prime, they produce them. Put above these numbers, in natural progression: 1, 2, 3, 4, 5, etc., which are called their exponents. This done, I say:*
> *1) When the exponent of a radical number is composite, its radical is also composite. Just as 6, the exponent of 63, is composite, I say that 63 will be composite.*
> *2) When the exponent is a prime number, I say that its radical less unity is measured by double the exponent. Just as 7, the exponent of 127, is prime, I say that 126 is a multiple of 14.*

3) When the exponent is a prime number, I say that its radical cannot be measured by any other prime number except those that are greater by unity than a multiple of double the exponent or than double the exponent. [...]
Here are three beautiful propositions which I have found and proved without difficulty; I shall call them the foundations of the invention of perfect numbers. [...][1]

Exercise 19   Prove Fermat's claims:

1) M_n is not prime unless n is prime.

2) If n is prime ($\neq$ 2) then 2n divides $M_n - 1$, and the divisors of M_n are of the form $1 + 2nk$.

Exercise 20   Prove that if n is prime then M_n divides $2^{(M_n - 1)} - 1$: (M_n is in this case either prime or pseudo-prime in base 2.)

Note that Mersenne's claims are almost entirely correct since they are only in error for the five numbers M_{67} and M_{257} which are not prime, and M_{61}, M_{89} and M_{107} which are. Besides their importance in the construction of perfect numbers, Mersenne numbers are interesting in that they produce many large primes, which is not the case with Fermat numbers. This is how it has recently been proved that the numbers

$$M_{132049} = 2^{132049} - 1 \text{ and } M_{216091} = 2^{216091} - 1,$$

numbers having 39751 and 65050 digits respectively, are both prime. For such numbers, it is possible to take advantage of present-day powerful computers, but we can also use a test that was established in 1930, the Lucas-Lehmer test. E. Lucas demonstrated in 1878[2] the properties of the sequences $U_n = \dfrac{a^n - b^n}{a - b}$ and $V_n = a^n + b^n$ associated with the solutions a and b of the equation $x^2 = Px - Q$, where P and Q are prime to each other.

Exercise 21   Prove that $V_{2n} = (V_{2n-1})^2 - 2Q^{2n-1}$.

In the case P = 1, Q = 1, we have the Fibonacci sequences and Lucas obtained a condition for primality of $p = 2^{4q+3} - 1$ for the sequence (v_{2n}), then the case P = 4, Q = 1 leads to the primality of the numbers $p = 2^{4q+1} - 1$. Half a century later, starting from these ideas of Lucas, Lehmer was able to derive a test for primality of Mersenne numbers, a test that was suitable even for very large numbers:

Let (v_n) be a sequence defined by $v_0 = 4$ and $v_{n+1} \equiv (v_n)^2 - 2 \mod (M_k)$; then the Mersenne number M_k is prime if and only if $v_{k-2} \equiv 0 \mod (M_k)$.

So, for example, the number M_{11} is composite while M_{13} is prime: For M_{11} (= 2047) we obtain, modulo (M_{11}): $v_0 = 4$, $v_1 = 14$, $v_2 = 194$, $v_3 \equiv 788$, $v_4 \equiv 701$, $v_5 \equiv 119$, $v_6 \equiv 1877$, $v_7 \equiv 240$, $v_8 \equiv 282$, $v_9 \equiv 1736 \neq 0$. For M_{13} (= 8191) and modulo (M_{13}): $v_0 = 4$, $v_1 = 14$, $v_2 = 194$, $v_3 \equiv 4870$, $v_4 \equiv 3953$, $v_5 \equiv 5970$, $v_6 \equiv 1857$, $v_7 \equiv 36$, $v_8 \equiv 1294$, $v_9 \equiv 3470$, $v_{10} \equiv 128$, $v_{11} \equiv 0$. The number $M_{M_{13}} = M_{8191}$ is not prime, which puts paid to the conjecture that the Mersenne numbers M_n are prime when n is itself a Mersenne prime number. At the present time about thirty Mersenne primes are known: the 12 primes with Mersenne number $n \in \{2, 3, 5, 7, 13, 17, 19, 31, 61, 89, 107,$

---

1.   Fermat, *Œuvres*, XL, p. 198-199.
2.   Lucas, *Théorie des fonctions numériques simplement périodiques*, American Journal of Mathematics, 1878, p. 184-239; 289-321.

127} were recognised between the seventeenth century and the beginning of the twentieth century. It was not until after 1952 that the following values of n were discoverd for which M_n is prime: {521, 607, 1279, 2203, 2281, 3217, 4253, 4423, 9689, 9941, 11213, 19937, 21701, 23209, 44497, 86243, 110503, 132049}; in 1985 the number M_{216091} and in 1992 the number $M_{756\,839}$, which has 227 832 digits, were found to be prime.

Exercise 22 Prove that if $8q + 7$ is prime then $M_{4q+3} = (2^{4q+3} - 1)$ is divisible by $(8q + 7)$.

After this rapid and partial overview of the properties of prime numbers, we shall conlude by a recent and interesting application to cryptography, which makes use of congruences

The language of congruences has been set out and structured since Gauss. For a given natural number n, the set of integers having the same remainder under division by n, is called the class of congruence modulo (n) and is represented by any of its elements in the form $\overset{\bullet}{a}$ ($\overset{\bullet}{0}$ is therefore the class of all multiples of n; $\overset{\bullet}{a} = \overset{\bullet}{b}$ if and only if $a \equiv b$ mod (n)). Addition and multiplication carry over naturally to congruent classes:

$$\overset{\bullet}{a} + \overset{\bullet}{b} = \overline{a+b} \quad \text{and} \quad \overset{\bullet}{a} \times \overset{\bullet}{b} = \overline{a \times b}$$ the set of congruent classes is denoted by $\mathbb{Z}/_n\mathbb{Z}$; it consists of excatly n elements $\{\overset{\bullet}{0}, \overset{\bullet}{1}, \overset{\bullet}{2}, ..., \overset{\bullet}{n-1}\}$ and has the same structure as the set $\mathbb{Z}$ of numbers: it is said to be a commutative ring. The Bezout equality for two numbers a and b prime to each other (there exist integers u and v such that $au + bv = 1$) allows us to establish classes $\overset{\bullet}{a}$ which are invertible for multiplication (that is those for which there exists a class $\overset{\bullet}{b}$ satisfying $\overset{\bullet}{a} \times \overset{\bullet}{b} = \overset{\bullet}{1}$) containing numbers prime with n: there are exactly $\phi(n)$ of them, where $\phi(n)$ is the Euler phi function (see Exercise 14), and they form a multiplicative group. For an element a prime with n the set of powers of $\overset{\bullet}{a}$ constitutes a subgroup of this group the number of whose elements, called its order, is none other than the gaussian of a. It is esay to derive again Fermat's theorem and its generalisation by Euler, thanks to Lagrange's theorem that the order of a sub-group of a finite group divides the order of the group. Finally, where all the non zero elements of a ring are invertible, we say that the ring is a field (an example is the reals $\mathbb{R}$): thus, n is prime if and only if the ring of congruence classes modulo (n), $\mathbb{Z}/_n\mathbb{Z}$ is a field.

Knowing that in a field the polynomial $x^d + a_1 x^{d-1} + ... + a_d$ cannot have more than d roots and that if there d their sum is $(-a_1)$, solve the following exercise:

Exercise 23 1) Prove the Euler test for quadratic residues modulo (p) where p is an odd prime.
2) Hence deduce that for p prime, greater than or equal to 5, the sum odf the squares of the integers from 1 to $p - 1$ is divisible by p.
3) Solve problem 3) of the introduction.

Cryptography

Research into very large prime numbers and the decomposition of very large numbers into prime factors has seen a considerable expansion in recent years due to the development of powerful computers and also from an unexpected application, that of cryptography. In a world where security and rapid transmission of information is of paramount importance in certain areas (economics, politics, the military ...), "classical" methods for encrypting messages have proved ineffective. The basic principle of these methods, where a secret key is needed to decode a letter for letter substitution, is unable to resist attacks by powerful computers. Now a new method, using a "public key" was discovered in 1975 by Rivest, Shamir and Adleman. Its performance relies on the fact that it is easy to construct very large prime numbers but it is impossible (in any reasonable practical time) to discover the prime factors p and q of a number N = p x q where each p and q consist of, say, 50 digits. The key of the code, the number N, is public knowledge and available to all, but to decode a message the prime factors have to be known (and are kept secret). This is how it works in principle:

Encoding: To each receiver of a message is given a public key consisting of the numbers N (= pxq) and d (prime with $\phi(N) = (p-1)(q-1)$, the Euler phi function of N). By a "naive" (and non secret) method the sender encodes the message into one or several number M (M < N, M and N prime with each other), then into $C \equiv M^d$ mod (N). The numbers C are sent to their destination.

Decoding: This depends on Fermat's theorem, generalised by Euler (see Exercise 11): $M^{\phi(N)} \equiv 1$ mod (N). Now, the Bezout equality shows that when d and ϕ (N) are prime with each other, there are two integers e and k such that $e.d - k.\phi(N) = 1$ and so $C^e = M^{ed} = M^{1+k.\phi(N)} \equiv M$ mod (N). It is sufficient then to know C, which is public, and e, which is secret and known only to the recipient, in order to be able to recalculate M. Now, e can only be known, even if d is known, if $\phi(N)$ is known, and this needs p and q.

Signature: The coding being public, the recipient has to be able to identify the provenance of a message. Let N' and d' be the public keys and e' the hidden key known to the sender. He signs S and uses the code $T \equiv S^{e'}$ mod (N') (he is the only one who can do this). The recipient who knows the sender's public key (N', d') is therefore able to verify the signature by calculating $T^{d'} \equiv S^{e'd'} \equiv S$ mod (N').

Exercise 24   Resolve problem 5) of the introduction.

The code used is of the type above with N = 2 139 956 033; d = 48 611; each number $C \equiv M^d$ mod (N) where M represents a group of 5 characters. The characters are numbered from 0 to 50 according to the following alphabet order:

ABCDEFGHIJKLMNOPQRSTUVWXYZ ?, '.; : ! ( )1234567890=+-*/

The numbers M are calculated using base 51 (but units to the left); for example EULER = 115568656.

Bibliography

Sources texts

DEVLIN, *Mathematics: the new golden age*, Pelican, Penguin Books Ltd., Harmondsworth, 1988.

DIEUDONNÉ, *Abrégé d'histoire des mathématiques*, Hermann, Paris, 1978.

EULER, *Opera omnia, commentationes arithmeticae.*

EULER, *Introduction à l'analyse infinitésimale*, tr. J.B. Labey, Lib. Barrois, Paris, 1976.

FERMAT, *Œuvres*, P. Tannery and C. Henry (ed.) Gauthier-Villars, Paris, 1985.

GAUSS, *Recherches arithmétiques* , tr. Poullet-Delisle, Lib. Blanchard, Paris, 1807.

HEATH, *The Thirteen Books of Euclid's Elements*, second edition, 3 vols., Oxford, Cambridge University Press, 1926. Repr. Dover, New York, 1956.

HONSBERGER, *Joyaux mathématiques*, 2 vols.: tr. V. Glaymann and C. Nouguez, Editions Cedic, 1979.

LUCAS, *Théorie des nombres*, Paris, 1891.

NICOLAS, *Tests de primalité*, Expositiones Mathematicae, 1984.

NICOMACHUS, *Introduction arithmétique*, I, 13: Edition Hoche p. 29, 17-33, 18 tr. Michel Crubellier.

Reference texts

ELLISON & MENDES-FRANCE, *Les nombres premiers*, Hermann, Paris, 1975.

ITARD, *Arithmétique et théorie des nombres*, Que sais-je? n° 571, PUF.

ITARD, *Arithmétique et les nombres premiers*, Que sais-je? n° 1093, PUF.

SIERPINSKI, *Elementary theory of numbers*, Warsaw, 1964.

General works for further reading

ADLEMAN, Pomerance, Rumely, "On distinguishing prime numbers from composite numbers", *Annals of Mathematics*, vol. **117**, 1983.

HARDY, WRIGHT, *An introduction to the theory of numbers*, Oxford, Clarendon Press. Repr. 1977.

LEHMER, "Tests for primality by the converse of Fermat's theorem", *Bulletin of the American Mathematical Society*, **33**, 1927.

LEHMER, "Lucas's test for the primality of Mersenne's number"*, Journal of the London Mathematical Society,* 10, 1935.

LIONNAIS, *Les nombres remarquables*, Hermann, Paris, 1983.

LUCAS, "Théorie des fonctions numériques simplement périodiques", *American Journal of Mathematics*, 1878.

RIBENBOÏM, *The Book of Prime Number Records*, Springer, 1991.

RIVEST, SHAMIR, ADLEMAN, *A method for obtaining digital signatures and Public-Key cryptosystems*, com. A.C.M., Feb. 1978, vol. 21, No. 2.

How did you get on?

Ex. 1   Proposition 30 shows that the prime factors in a decomposition of a number a are the prime divisors of a, and that power of some factor p is the greatest integer k (k > 0) such that p^k divides a.

Ex. 2   Every integer can be written 6k + r, (k and r integers), where r is the remainder after division by 6. We have $0 \le r \le 5$ and r is divisible by 2 or 3 if $2 \le r \le 4$.

Ex. 3 $52793 = 13 \times 31 \times 131$

Ex. 4   Let $n = (x - y).(x + y)$ with $x > y \ge 0$ and $u = (x - y)$, $v = (x + y)$ so that $n = u.v$ and $1 \le u \le \sqrt{n} \le v$ where $x = (u + v) / 2$, $y = (u - v) / 2$ (x, y are integers since u and v are odd). Examining the function $u \to (u + n/u)/2$ shows that it is a decreasing bijection of $[1; \sqrt{n}]$ onto $[\sqrt{n}; (n + 1)/2]$; n is a prime number if and only if the only possible value of u is 1 (when $x = (n + 1)/2$).

Ex. 5 $\sqrt{52793} \approx 229.76$ and $52793 = 267^2 - 136^2 = 131 \times 403$ (needing 38 inspections of squares); then $\sqrt{403} \approx 20.07$ and $403 = 22^2 - 9^2 = 13 \times 31$ (needing 2 inspections of squares); $131 = 66^2 - 65^2$ (unique decomposition requiring 54 inspections of squares) and is prime. So $52793 = 13 \times 31 \times 131$. The method of Exercise 3 required 11 divisions by 2, 3, 5, 7, ... 31 (the last quotient 131 having no divisor less than 31, or $\sqrt{131}$, and so is prime).

Ex. 6   1) The reasoning is similar to the case for 6k − 1: here put A = 2P − 1, where P is the product of all primes smaller than n, with n > 5.
2) $A = P^2 + 1$ has a prime factor $p \times n$ and of the form $y^2 + z^2$ and so of the form 4k + 1 (argue from the parity of y and z); $B = P'^2 + 2^2$ (where P' = P/2) has all its prime factors > n and of the form 4k + 1 = 8x + 1 or 8x + 5 according to the parity of k, and they cannot all be of the form 8k + 1 since B can be written as 8Z + 5 (all prime numbers $p \ne 2$ are of the form $4y \pm 1$ and so $p^2 = 8z + 1$ and $P'^2 = 8Z + 1$). The assumed result is due to Fermat and Legendre. (The proof uses the method of infinite descent); it can be deduced (using Wilson's theorem) that the odd primes which are the sums of two squares must be those that are of the form 4n + 1.

Ex. 7 $\int_2^x \frac{du}{\ln(u)} = \frac{x}{\ln(x)} - \frac{2}{\ln(2)} + \int_2^x \frac{du}{\ln(u)^2} \underset{\infty}{\approx} \frac{x}{\ln(x)}$ (integration by parts).

Ex. 8   P(x), x an integer, will only produce primes if it is a constant polynomial. For any value of x that is a multiple of the constant term, P(x) will be divisible by, at least, the constant term. Hence no P(x) can produce primes for all values of x, except in the case where P(x) is constant.

Ex. 9   1) The result follows from direct multiplication (see also the product of the forms a + ib, c + id and also $a + ib\sqrt{163}$, $c + id\sqrt{163}$); Let $p = a^2 + 163.b^2 = c^2 + 163.d^2$, then $p(d^2 - b^2) = a^2 d^2 - b^2 c^2$ and so p, prime, is a factor of $ad \pm bc$: but $p^2 = (ac \pm 163.bd)^2 + 163.(ad \pm bc)^2$ is a factor of $(ad \pm bc)^2$ so $(ad \pm bc) = 0$ (if not we have $p^2 > (ad \pm bc)^2$). Hence $\frac{a^2}{c^2} = \frac{b^2}{d^2} = \frac{a^2 + 163.b^2}{c^2 + 163.d^2} = 1$ implying the uniqueness of the decomposition of p.
2) If $p = \alpha^2 - \alpha + 41 = \gamma^2 - \gamma\delta + 41\delta^2$ then 163 is a factor of $(2\alpha - 1)^2 - (2\gamma - \delta)^2$. Let $(2\gamma - \delta) = \varepsilon (2\alpha - 1) + 163x$ ($\varepsilon = \pm 1$), then in substituting back into the initial equality: $\delta^2 = 1 + 2(- \varepsilon x).(2\alpha - 1) - 163(\varepsilon x)^2$. Let x = 0 then $(2\alpha - 1) = \pm (2\gamma - \delta)$ and $\delta = 1$ ($\delta > 0$) hence $\alpha = \gamma$. The uniqueness of representing p (odd because $\alpha^2 - \alpha$ is even) shows that it is prime (see Legendre, *Théorie des nombres*, II-Chap. XIV).
3) It is easy to verify that the polynomial $1 + 2(2\alpha - 1)x - 163x^2$ is < 0 for all non zero integers x as α ranges from 0 to 40. The numbers $\alpha^2 - \alpha + 41$ will therefore be prime for all α between 0 and 40. For the spiral of Ulam, use induction to verify that the two numbers situated at the corners of the proposed square have the values

$41 - 2k + (2k)^2$ and $41 - (2k + 1) + (2k + 1)^2$: all the numbers will therefore be prime up to, but not including, $k = 20$ ($41 - 41 + 41^2$ is not prime).

Ex. 10 For $p = d.d'$ with $1 \le d < d' \le p - 1$, then $p = d.d'$ will divide $(p - 1)!$; if $p = d^2 > 4$, then $2 < d < 2d < p$, so $2d^2$ and so p will divide $(p - 1)!$; for $p = 4$, p is not a factor of $3! = 6$.

Ex. 11 If $k = 0$ then $p = (y + 1)$ divides $(y! + 1)$ and is therefore prime by Wilson's theorem. For such value y, k will be zero for the single value $x = (y! + 1)/(y + 1)$ and $f(x, y) = y + 1 = p$; in all other cases, $|k| \ge 1$ and $f(x, y) = 2$.

Ex. 12 1) $2701 = 37 \times 73$; $2700 = 36 \times 75$; $2^{36} \equiv 1 \bmod (37)$ and $2^9 \equiv 1 \bmod (73)$. Hence $2^{2700} \equiv 1 \bmod (73)$ and $\bmod (37)$: $73 \times 37 = 2701$ therefore divides $2^{2700} - 1$.
2) It is $91 = 7 \times 13$; $3^6 \equiv 1 \bmod (7)$ and $3^3 \equiv 1 \bmod (13)$; hence $3^6 \equiv 1 \bmod (7 \times 13)$ and $3^{90} \equiv 1 \bmod (91)$. The next two numbers are 121 and 286.

Ex. 13 1) 10 is prime with p, so divides 10^{p-1} (Fermat's theorem), the sequence of the remainders of $(10^n)_{n \in \mathbb{N}}$ modulo (p) is therefore $(p - 1)$ – periodic. $(10^{n + (p-1)} \equiv 10^n \bmod (p))$.
2) $1/p = 0. a_1 a_2 \ldots a_n \ldots$ where a_n is the quotient of the division of $10.r_{n-1}$ by p and r_{n-1} is the remainder of the preceding division (with $r_0 = 1$); therefore $r_n \equiv 10^n \bmod (p)$ and the sequence $(r_n)_{n \in \mathbb{N}}$ is periodic as well as the sequence $(a_n)_{n \in \mathbb{N}^*}$
3) For $p = 7$, we have $10 = 1 \times 7 + 3$, $30 = 4 \times 7 + 2$, $20 = 2 \times 7 + 6$, $60 = 8 \times 7 + 4$, $40 = 5 \times 7 + 5$, $50 = 7 \times 7 + 1$; these six divisions repeat themselves indefinitely and give the period decimal expansion of $1/7 = 0.142857 \ 142857 \ 142857 \ldots$ from which we obtain $2/7 = 2 \times (1/7) = 0.(2 \times 142857) (2 \times 142857) \ldots$and the decimal expansion of $2/7$ will have the same order, but starting at the third different position ($20 = 2 \times 7 + 6$). The same will be true for the other fractions $3/7$, $4/7$, $5/7$, $6/7$ but after $7/7$ the first division will not be one of the six preceding ones.

Ex. 14 1) $\phi(p) = p - 1$ and $\phi(p^\alpha) = p^\alpha - p^{\alpha - 1}$ (the numbers less than p^α and not prime with it are the $p^{\alpha - 1}$ multiples of p).
2) If x and y are prime to each other, then the mapping that associates all d in $[0, xy]$ with a pair (r_1, r_2) in $[0,x[\times [0,y[$, where r_1, r_2 are the remainders respectively of division of d by x and by y, is a bijection since it is injective in finite sets with the same number of elements xy. Since d is prime with xy if and only if r_1 is prime with x and r_2 is prime with y, it follows that $\phi(xy) = \phi(x) \phi(y)$.
3) This follows directly from 1) and 2).

Ex. 15 1) Euler's reasoning (p prime divides C_p^k for $1 \le k \le p - 1$) shows that if $x \equiv y \bmod (p^t)$ ($t > 0$) then $x^p \equiv y^p \bmod (p^{t+1})$ and (Fermat's theorem) for a, not a multiple of p, $a^{p^\alpha} \equiv a^{p^{(\alpha-1)}} \bmod (p^\alpha)$. If $n = \prod_{i=1}^{k} p_i^{\alpha_i}$ is decomposed into k prime factors, we get

$a^{\phi(n)} \equiv 1 \bmod (n)$ for a prime with n {since $\phi(n) = \prod_{i=1}^{k} (p_i^{\alpha_i} - p_i^{(\alpha_i - 1)})$ and each $p_i^{\alpha_i}$

will divide $a^{(p_i^{\alpha_i} - p_i^{(\alpha_i - 1)})} - 1$ and therefore $a^{\phi(n)} - 1$}.
2) 2 and 5 are not factors of a or 9 so $9a$ is prime with 10, so $10^{\phi(9a)} \equiv 1 \bmod (9a)$: therefore $(10^{\phi(9a)} - 1)/9 = 11 \ldots 11$ (with $9a$ digits 1).

Ex. 16 1) $x^a \equiv 1 \bmod (x^a - 1)$ so $x^b \equiv 1 \bmod (x^a - 1)$ because a divides b.
2) F_n divides $2^{2^n + 1} - 1 = F_n \times (2^{2^n} - 1)$ and $n + 1 \le 2^n$ implies 2^{n+1} divides $2^{2^n} = F_n - 1$ and so $2^{2^n + 1} - 1$ divides $2^{(F_n - 1)} - 1$.

Ex. 17 Let a be an element which has gaussian $(p - 1)$: $a^{p-1} \equiv 1 \bmod (p)$ and $a^{(p-1)/2} \equiv -1 \bmod (p)$; let $b = a^{p-2}$ then $ab \equiv 1 \bmod (p)$ and so, $(p - 1)/8$ being an integer, $(a^{(p-1)/8} + b^{(p-1)/8})^2 \equiv a^{(p-1)/4} + b^{(p-1)/4} + 2 \equiv 2$, since $(a^{(p-1)/4} + b^{(p-1)/4})^2 \equiv -1 - 1 + 2 \equiv 0$ modulo (p).

Ex. 18 Let $n = 4k$ $(k \geq 1)$; $16^2 \equiv 16$ mod (20) so by induction $16^k \equiv 16$ mod (20) and $F_n = 2^{16k} + 1 \equiv 2^{16} + 1 \equiv 37$ mod (25) since $2^{20} \equiv 1$ mod (25) and $F_n \equiv 1 \equiv 37$ mod (4), hence 100 divides $F_n - 37$. Similarly for $n = 4k + 1$: $F_n \equiv 2^{12} + 1 \equiv 97$ mod (25); for $n = 4k + 2$: $F_n \equiv 2^4 + 1 \equiv 17$ mod (25); for $n = 4k + 3$: $F_n \equiv 2^8 + 1 \equiv 57$ mod (25) (the two last results remain valid for $k = 0$).

Ex. 19 1) $2^q - 1$ divides M_n if q divides n (see Ex. 16).

2) n prime divides $2^{n-1} - 1$ (Fermat's theorem) and $2n$ divides $2(2^{n-1} - 1) = M_n - 1$. If p is a prime divisor of M_n, p is odd and $2^n \equiv 1$ mod (p) then the gaussian of 2 modulo p divides n prime, it is therefore n and it divides $(p - 1)$ (Fermat's theorem), whence $p = nx + 1$, but n and p being odd, x is even and $p = 2nk + 1$. Every divisor of M_n is of this form when produced by such p.

Ex. 20 Follows as a consequence of exercises 16 and 19.

Ex. 21 $(a^{2n} + b^{2n}) = (a^n + b^n)^2 - 2(ab)^n$, $(ab = Q)$, so $V_{2n} = (V_n)^2 - 2Q^n$ which gives the result substituting 2^{n-1} for n.

Ex. 22 $M_{4q+3} \cdot (2^{4q+3} + 1) = 2^{8q+6} - 1$ and $8q + 7$, prime, therefore divides (Fermat's theorem) $M_{4q+3} \cdot (2^{4q+3} + 1)$, it will divide M_{4q+3} if it does not divide $(2^{4q+3} + 1)$; now $2^{4q+3} \equiv 1$ mod $(8q + 7)$ since 2 is a quadratic residue of prime numbers of the form $8q \pm 1$.

Ex. 23 1) the polynomial $x^{p-1} - \overset{\bullet}{1} = (x^{(p-1)/2} - \overset{\bullet}{1})(x^{(p-1)/2} + \overset{\bullet}{1})$ has $p - 1$ roots: all non zero elements of the field Z/pZ, and $x^{(p-1)/2} - \overset{\bullet}{1}$ has also $(p-1)/2$ roots which are the quadratic residue classes modulo (p): therefore there cannot be any more so the $(p-1)/2$ other classes are roots of the polynomial $x^{(p-1)/2} + \overset{\bullet}{1}$.

2) $(p - 1)/2 \geq 3$ so the term of degree $(p-1)/2 - 1$ is zero as well as the sum of the roots, and p divides the sum of the quadratic residues modulo (p).

3)
$$\frac{u}{v} = \sum_{k=1}^{p-1} \frac{1}{k} = \sum_{k=1}^{p-1} \frac{1}{p-k}$$

so
$$2.\frac{u}{v} = \sum_{k=1}^{p-1} \frac{1}{k} + \sum_{k=1}^{p-1} \frac{1}{p-k} = \sum_{k=1}^{p-1} \left(\frac{1}{k} + \frac{1}{p-k}\right) = p.\sum_{k=1}^{p-1} \left(\frac{1}{k(p-k)}\right) = p.\frac{u'}{(p-1)!}$$

which gives:

$$(p-1)! \times \sum_{k=1}^{p-1} [k(p-k)]^{-1} = \overset{\bullet}{u'}$$

$$= -\sum_{k=1}^{p-1} [k(p-k)]^{-1} = \sum_{k=1}^{p-1} (\overset{\bullet}{k})^2 = \overset{\bullet}{0} \qquad \text{(from part 2)}$$

(In $Z/pZ - \{\overset{\bullet}{0}\}$ the squares and their inverses describe the same set, p therefore divides u' and p^2 divides $2u$ $(p - 1)!$ and so u (since p is prime with 2 $(p - 1)!$)

Ex. 24 To decode the message, take $N = 46819 \times 45707$; $\phi(N) = 2\,139\,863\,508$; $e = 1\,612\,237\,991$.

15

Looking for Perfect Numbers

Michel CRUBELLIER and Jacky SIP
IREM Lille

The nature of the Even Number manifests itself in the perfect number because of the evenly even numbers that are put together in it; and the nature of the Odd Number is also found there, since from the evenly evens come the odd numbers, and more particularly the primes and non-composite numbers. And we should not be astonished to find diverse properties attributed to the same number — like the number Six, for example, which is said to be perfect, and the first even-odd, or that it is a rectangular number, and that it was called Marriage by the Pythagoreans, because it is produced from the intermixing of the first meeting of male and female; and for the same reason this number is called Holy and represents Beauty, because of the richness of its proportions.

Iamblichus[1]

The whole numbers, which we are aware of from a very young age, are apparently simple and familiar mathematical objects making up a family produced by a simple law, that of repeated addition of one. However each one of them can be identified by remarkable properties which give the number a special character (for example: odd or even, prime number, square number, ...). The arithmetic of the Ancients, the forerunner of our number theory, attempted to establish the nature of these properties. Many problems in arithmetic are striking in the simplicity of their statement and the difficulty of their resolution. Some, such as Fermat's "Last Theorem" (there is no positive integer solution to $x^n + y^n = z^n$ for $n > 2$) are still left unresolved[2]. We shall see that there are also many unresolved propositions concerning perfect numbers.

---

1. Iamblichus, *In Nicomachi Arithmeticam Introductionem*, p. 34.
2. This was written in 1993, just a few weeks before Andrew Wiles proposed a proof of Taniyama's conjecture, wich implies the truth of Fermat's Last Theorem.

What is a perfect number?

The usual definition used today is that a natural number n is said to be perfect if the sum of its factors is 2n. This is how the definition appears in Euler's work, but the Ancients adopted a slightly different, if more striking, definition:

A perfect number is equal to the sum of its aliquot parts.

The aliquot parts of 6, for example, are 3 (half of 6), 2 (a third of 6) and 1 (a sixth of 6); 6 itself is perfect since it is equal to 3 + 2 + 1.

Exercise 1   Verify that 496 is a perfect number.

We should notice that "aliquot parts" is not exactly the same as divisors, or factors, as we understand it. The notion of divisor refers back to the operation that has to be undertaken; the notion of aliquot part refers to the result, the quotient if you like, but the Ancients considered that the aliquot part was present in the number itself, together with its own individuality, and that this contributed to the character of the number – like a piece of a jig-saw puzzle.

Interest in the concept of a perfect number illustrates well one of the most remarkable aspects of the mathematics of Classical Greece, that it was a set of doctrines, for the most part speculative, drawn more from considerations of harmony and beauty than for practical use or effectiveness in problem solving.

Perfect numbers occupied an important place in the arithmetic of the Ancients, as can be seen in the *Introductio Arithmetica* of Nicomachus, as well as in the three books of Euclid's *Elements* concerning arithmetic (Books VII, VIII and IX). (See Chapter 11 for more on Euclid's *Elements*.) These books end with a proposition (which we shall examine later) which gives an algorithm for finding perfect numbers. This was not accidental; the *Elements* do not correspond exactly with our concept of a deductive theory as the methodical exploration of possibilities contained in a set of axioms; what they do is to work towards a concluding construction which is remarkable; in the same way that the purely geometric books end with the construction of the five Platonic solids, so the number theory books end with a way of determining perfect numbers.

A "natural history" of numbers

The most detailed exposition on perfect numbers is to be found in the works of the neo-Pythagorean Nicomachus of Gerasa (i.e. 100). He describes three types of numbers: superabundant, deficient and perfect. A superabundant number is a natural number, the sum of whose aliquot parts is greater than itself; in other words it is a number rich in divisors. The Greek word *hyperteles,* which was used to refer to it, means literally what surpasses the measure of perfect. Certain mediaeval authors called it "more than perfect", and that is why we have decided to render the word used by Nicomachus as "superabundant", though in the rest of the chapter we shall conform to the more commonly used "abundant". On the other hand, a deficient number is one, the sum of whose aliquot parts is less than itself. The perfect number appears as a form that is an equilibrium between these two types, both of which are considered to be imperfect:

... Among simple even numbers, some are superabundant, others are deficient: these two classes are as two extremes opposed one to the other; as for those that occupy the middle point between the two, they are said to be perfect. And those which are said to be opposite to each other, the superabundant and the deficient, are divided in their condition, which is inequality, into the too much and the too little. In fact, it is not possible to conceive of any sort of inequality distinct from these two – neither vice nor malady, neither disproportion nor a lack of propriety nor anything else of this nature. In the case of the too much, is produced excess, superfluity, exaggerations and abuse; in the case of the too little, is produced wanting, defaults, privations and insufficiencies. And in the case of those that are found between the too much and the too little, that is in equality, is produced virtue, just measure, propriety, beauty and things of that sort – of which the most exemplary form is that type of number which is called perfect.

So the superabundant number is the one which, besides the parts which are suitable to it and fall to its lot, has others more numerous, as if an adult animal was formed from too many parts or members, "having ten tongues", as the poet says, and ten mouths, or nine lips, and provided with three lines of teeth; or with a hundred arms, or having too many fingers on one of its hands. In the same way with a number, if after taking a count of all of its parts, and they are brought all together, and compared with itself, it is found that its own parts are greater than itself, then this number is called superabundant; in fact it is greater than the measure of perfection with regard to its parts. Such numbers are 12, 24 and others; in fact 12 has a half, 6, a third, 4, a sixth, 2, and a twelfth 1, which brought all together make up 16, which is more than the original 12; and so its parts are greater then the whole. As for 24, it also has a half, a third, a quarter, a sixth, a twelfth, and a twenty-fourth, which turn out to be 12, 8, 6, 4, 3, 2, 1; brought all together they make up 36, which compared with the original 24, is found to be greater than it, although it is composed only of its parts; and so, here again, the parts are greater than the whole ... [1]

Exercise 2   Show that 60 is an abundant number.

The deficient number is the one whose character is the opposite of what has just been described; that it has parts, which when brought all together, turn out to be less than the number itself. This is as if an animal lacked members or natural parts, like in the verse: "a single round eye occupies the middle of their foreheads"; if someone is one-armed or one of his hands has less than five fingers, or if he does not have a tongue or something like that, he would be called deficient and in some way infirm, rather like a number whose parts are less than itself. Such a number is 8, or 14. In fact 8 has a half, a quarter and an eighth, which are 4, 2 and 1, which brought all together make 7, that is a number less than the original, so its parts are not enough to entirely make up the whole. Or again, 14 has a half, a seventh and a fourteenth, which are 7, 2 and 1; all together 10, less than the original number. So this one also is insufficient in its parts to entirely make up the whole which is formed from them. [2]

Exercise 3   Show that any power of 2, whose exponent is a non zero natural number, will be a deficient number.

In the text we have just read, numbers appear as given objects, which ancient arithmetic describes like a sort of natural history of numbers, as if the Ancients were describing and classifying biological or zoological specimens (and right up to the seventeenth century, descriptions of plants and animals not only referred to their observable physical characteristics, but also to their moral symbolism, their alchemy or

---

1.   Nicomachus, *Introductio Arithmetica*, chap. XIV.
2.   *Idem*, chap. XV.

their religious significance). These numbers then, possessed quasi-concrete properties, which could be discovered by calculation, but were also, more profoundly, characteristics of an essence. They were charged with meaning. The perfection of certain numbers, and the imperfection (by excess or default) of others, was seen in relation to good and evil, and also, by analogy – and this was most evident – with the idea of the finality of nature. Numbers which are abundant or deficient are described as monstrosities, while the perfect number reveals a kind of balance between matter and its constituents – between the aliquot parts and the number itself – or again, an internal completeness which comes from a good adaptation of the parts to the whole. In the Pythagorean tradition, this language is not simply symbolic or rhetorical, numbers are the substance of things, and the perfect number is "most exemplary" of the class of well adapted things, because it reveals the principles which are the cause of good adaptation.

Confident Assertions

Although Nicomachus was a relatively late author, his work allows us to view the intellectual climate in which an interest in perfect numbers developed. The most striking aspect of this passage, to the modern reader, is the way in which it is presented as fact. What we see here is not merely a feature of an archaic mathematical thought; while as a Pythagorean, Nicomachus stands in the line of an old tradition, he is also a well informed mathematician. He is well aware of Euclid's work, which was the basis of all mathematical teaching at the time, founded on the twin notions of construction and demonstration, that is that every result can be obtained by deduction, be it a theorem or the solution to a problem. But it is probable that these aspects did not interest him. For him, the essence of mathematical truth lay in revealing the objective properties of those ideal realities which constitute it. That is why he was not content just to make observations, but he made general assertions which appeared to be laws:

– *There exists a perfect number, and one only, in each interval of the type [10^i, 10^{i+1}[, i being a natural number;*

– *All perfect numbers are even, and they end in either 6 or 8.*

He continues:

These two numbers being opposed to each other as extremes, their middle is manifestly the number called perfect, which is to be found in equality; the parts coming together do not make more than itself, nor is it greater than them, but there is always equality; and in all cases the equality is discovered by intelligence in the interval between the too much and the too little, and it is like the just measure between the excessive and the insufficient, or like the consonant note between the sharp and the flat. And so when all the parts which are found to be contained in a number have been counted and brought together and, in comparison with that number, it does not surpass them in quantity, and it is not surpassed by them, that number is called absolutely perfect; such is the number which is equal to its parts, for example 6 or 28. For 6 has a half, a third and a sixth, which are 3, 2, 1; these parts brought together give 6; they are found to be equal to the original number, neither greater nor smaller. And 28 has for parts a half, a quarter, a seventh, a fourteenth and a twenty-eighth, which are 14, 7, 4, 2, 1 and which brought together as a single number produce 28, and so the parts neither surpass the whole, nor the whole the parts, but the comparison is an equality, which is the essence of perfection. It happens that, just as things which are beautiful and excellent are rare and easy to enumerate, while ugly and vile things abound, so superabundant and deficient numbers are plentiful and they can be found without a rule, their discovery being untidy, while perfect numbers are easy to count and are disposed according to a proper order. For one is found in the units, 6; again one in the tens, 28; a third among the hundreds 496; a fourth among the thousands, 8128; and it turns out that they all terminate in turn by the number six or the number eight, and they are always found among the evens. [1]

The algorithm of Nicomachus

The text goes on to state an algorithm, which can be expressed in modern terminology as: n being a natural number, numbers of the form $2^n.(2^{(n+1)} - 1)$ are perfect numbers, whenever the factor $(2^{n+1} - 1)$ is prime (this is called a Mersenne number (see Chapter 14)). It is not insignificant that the perfect number is identified as the product of a power of 2 with an odd prime. These powers of two, called in the text "evenly even numbers", are even numbers par excellence, since they contain no odd factor at all. In fact, the Pythagoreans taught that the order of the world is the product of the harmony between pairs of opposites, which they called, among other things, Even and Odd. An example of this sort of speculative writing can be seen the passage by Iamblichus (c. 325) quoted at the beginning of this chapter.

Nicomachus does not give his algorithm as the solution to a problem, constructed from data. Nor does he prove its validity; however, he is confident enough to declare that his process "does not leave out any perfect numbers and [...] does not include any that are not."

There exists an elegant and sure method of generating these numbers, which does not leave out any perfect numbers and which does not include any that are not; and which is done in the following way: first set out in order the evenly even numbers in a line, starting from unity, and proceeding as far as you wish: 1, 2, 4, 8, 16, 32, 64, 128, 256, 512, 1024, 2048, 4096; and then they must be "piled up" [totalled] each time there is a new term, and at each piling up, examine the result; if you find that it is prime and non-composite, you must multiply it by the quantity of the last term that you added to the line, and the product will always be perfect. If, otherwise, it is "second" and not prime, do not multiply it, but add on the next term and again examine the result; and if it is second and composite, leave it aside, without multiplying it, and add on the next term; if, on the other hand it is prime, and non-composite, you must multiply it by the last term taken for its composition, and the number that results will be perfect, and so on as far as infinity. In almost the same way, you will generate all the perfect numbers, one after the other, without leaving out any of them, and the process does not include any that are not. For example, I add 2 to 1, examine the result produced by the union of these two, which I find to be 3, conforming to what I showed earlier, that it is prime and non-composite; in effect, it has no heteronym part, but only the part that is paronym with it; [2]

Nicomachus uses the term *paronym* for the part which derives its name from the number itself (the third for three, the seventh for seven, etc.). In fact it is always the unit that is the paronym part, even though each time it appears under a different name: 1 is the third of 3, the seventh of 7, etc. He called all other parts heteronyms.

... that is why I multiply by the quantity of the last term taken for piling them up, that is 2, and I obtain 6; I affirm that this is the first perfect number in actuality, and that it has the parts which are seen in the numbers which make it up. For first of all, it has the unit of its paronym part, which is the sixth; 3 from the half (considered according to 2), and

1. Nicomachus, *Introductio Arithmetica*, chap. XVI.
2. *Ibid.*

reciprocally, the number two from the third. 28 is also formed by piling up the evenly even numbers up to 4, and is generated in the same way. In fact the recapitulation of the three makes 7, which is found to be prime and non-composite; for it admits of only the paronym part, the seventh. That is why I multiply it by the quantity of the last number taken for the piling up, and it produces 28, equal to its own parts and having its parts due to the numbers that were introduced previously: one half because of the number two, a quarter because of the number seven, a seventh because of the number four, a fourteenth because of being the matching of the half, a twenty-eighth because of its paronym part – that which is found in all numbers and which is the unit.[1]

Why is the fourteenth present in this complicated way ("the matching of the half")? It is because, for Nicomachus, the presence of each of the parts results from the way the number (28) is obtained: we add the evenly evens 1, 2, and 4, and we multiply their sum (7) by the last number of the list (4). This done, we have met all the proper divisors (or aliquot parts) of 28, with the exception of 14. Nicomachus has to give some reason for its existence, which is what he does by describing it as the "matching" (or "counterpart") of the factor 2. He continues:

Having found these two, in the units the 6, and in the tens the 28, you need to do the same thing for the following layers of numbers: having once again added the number that follows, 8, and all together they give 15; in examining this I find it is not a prime and non-composite number, and that in addition to its paronym part it has also a fifth and a third, which are heteronyms. That is why I do not multiply it by 8, but I add the next number, 16, and that gives 31. Since this is prime and non-composite, it must necessarily be multiplied – conforming with the general rule which defined this method – by the last term used for the piling up,16, and we obtain 496 in the hundreds, then in the same way 8128 in the thousands, and so on, as far as you have the strength to go on.[2]

Non-verified assertions

In addition to the confident assertions to which we have already referred, namely:

(1) there exists a perfect number, and one only, in each interval of the type $[10^i, 10^{i+1}[;$

(2) all perfect numbers are even;

(3) all perfect numbers end in either 6 or 8;

Nicomachus has just added two others:

(4) the algorithm will find all perfect numbers;

(5) there is an infinity of perfect numbers.

It is true that Nicomachus does explicitly claim the existence of the law of distribution stated in (1); he is content to let it be understood, but the context naturally implies it. Now, it appears that he could not have justified any of these five propositions. However, up to the beginning of modern times, the great majority of authors accepted them as being true without question, and they formed the basis of a chapter on perfect numbers in all arithmetics.

---

1.   Nicomachus, *Introductio Arithmetica,* chap. XVI.
2.   *Ibid.*

The first and third are simply false, as the reader can easily verify:

Exercise 4   Find the fifth and sixth perfect numbers according to Nicomachus's rule.

We see that there is no perfect number between 10 000 and 10 000 000, and that the succession of 6's and 8's for the final digit is irregular.

It took a long time before mathematicians became aware of this and stopped repeating these assertions. On the contrary, they were used as methods for predicting new "perfect numbers", and numbers were proposed like 130 816 (= 256 × 511) or 2 096 128 (= 1024 × 2047), which are in fact abundant numbers, since their odd factor is not prime: 511 = 73 × 7, and 2047 = 89 × 23.

Such errors must appear surprising on the part of experienced calculators, who were certainly capable of telling if 511 and 2047 are prime. Part of the problem must have been the desire for order and regularity, inherited from Pythagorean tradition; but there was also another problem, that Nicomachus had left unanswered. This concerns the necessary condition that restrcits the application of the algorithm: that the odd factor (a Mersenne number) should be prime. What are the conditions that guarantee that $2^{(n + 1)} - 1$ should be prime?

First, it is necessary that n + 1 should be odd; in fact if n + 1 is even then $2^{(n + 1)} - 1$ is immediately the difference of two squares, and so is the product of two integers. One is tempted to make this necessary condition a sufficient one, by asserting that:

(6)   for an exponent n + 1 odd, the Mersenne number is always prime.

This rule seems to have been implicitly accepted by writers such as Ibn Fallûs (13th century) or Fra Luca Pacioli (1496); it was explicitly stated by Charles de Boüelles (1509), who gave 511 as an example. It turns out that he rule is not valid, for example for 2^9, but its origin lay in the semblance of order that is decreed by assertions (1) and (3) of Nicomachus.

In fact, if for each odd value of n + 1, a "perfect" number exists, then they will occur in an almost regular pattern, since each will be equal to 16 times its predecessor, which gives a perfect number in almost every degree of powers of ten. Furthermore, the rule (6) gives, effectively, "perfect" numbers which terminate alternately in a 6 or an 8.

Exercise 5   Why should all even perfect numbers terminate in 6 or 8?

In fact, the distribution of prime numbers among Mersenne numbers, just as the distribution of primes among the integers, does not follow any known rule. Even today, they have to be found, one by one, testing their primality by use of computers (see Chapter 14). We only know 30 Mersenne numbers which are prime, and so we only know 30 even perfect numbers. And so the question of the infinity of perfect numbers (assertion 5) remains unresolved.

It is still not known if there are any odd perfect numbers (see later, p. 407-409).

We must therefore restrict the fourth assertion to even perfect numbers. But as far as we know, the Greeks made no attempt to prove it. The furthest they got in the theory of perfect numbers is to be found in Euclid IX, 36 (see later, p. 400-403), but what appears is the logically weaker inverse proposition. Euclid established that: "every number obtained from following the algorithm will be perfect", but does the algorithm produce *all* perfect numbers? It was not until two thousand years later that Euler proved that "every even perfect number is of the form described by the algorithm".

What justification is there for the algorithm?

We note that Nicomachus presents his algorithm by saying that it "generates" perfect numbers. The term is to be taken in a realist sense. It is not simply a process of discovery; the number owes its existence to the numbers used in its calculation, and which remain present within it – in the form of its aliquot parts. More precisely: the aliquot parts are present in the number "because of the numbers which precede it". This shows that each aliquot part is, either one of the terms of the series used to construct the number, or "the matching" of one of them. In the case of 496, for example, the numbers 62, 124 and 248 do not occur in the sequence that generates the number, but are, respectively, the matchings of 8, 4, and 2, which do occur there.

Furthermore, it appears to us the term "piling up" used by Nicomachus concerning the powers of 2, is precisely appropriate to the operation that takes place here, that is successive addition, according to an order imposed by nature of the objects themselves (since each power of two can only appear as the result of the duplication of its predecessor). The piling up consists, then, in a non commutative addition. Conversely, the addition of the aliquot parts is called "recapitulation" because it is done in the reverse sense of the steps taken to generate the number.

Hence Nicomachus is bound to have considered all the divisors of the perfect number in the course of its generation. The confidence thus achieved does not come from proof, but is the result of having followed a sequence which is the only one possible, and is exhaustive. It is, if you like, a confidence of an inductive type, but not a speculative induction, as is suggested by assertion (1) above; it is the result of a known and assured operational procedure.

Is 1 a perfect number?

Nicomachus ends his chapter with a discussion about the particular case of the unit:

> *Thus the unit is perfect in potentiality but not in actuality. In practice, if we take the unit, which is the very first of the list, for piling up, we shall examine it, conforming with the rule, to see what sort of number it is, and we shall find that it is prime and non-composite. In truth, this is not by participation like the others, for it is prime with respect to every number and the only one without parts. I shall multiply it then, by the last term taken for the piling up, that is to say itself, and I obtain the unit. For one times 1, is the unit. Thus the unit is perfect in potential: for it is equal to its own parts in potentiality, although the others are in actuality.* [1]

What is the problem here? For the Greeks, the mathematical status of the unit is uncertain: it is not a number (both Euclid and Nicomachus define the number as "a *plurality* composed of units"), although it is the ultimate constituent of the numbers and it possesses certain of the arithmetic properties of numbers. It can be added, subtracted,

---

1.   Nicomachus, *Introductio Arithmetica,* chap. XVI.

or multiplied, and appears as the quotient of a division (but it is neither a multiplier, nor a divisor).

The paradox relates to the fact that the algorithm gives 1 as the first solution, while 1, according to Greek usage, has no parts. Also, to admit 1 as a perfect number would be to contradict assertions (1) and (3), and it is noteworthy that Euler's definition does not apply to the number 1 (the sum of its divisors is not equal to 2).

Nicomachus found a conceptual solution to this difficulty in using the Aristotelian distinction between the potential and the actual – which finds its most famous development in the idea of the "potential infinite" (see Chapter 1). The theory of a "potential being", expresses the idea that the possible is in some way real, or that at least it is not a mere nothing : thus bronze is a statue "in potentiality", whereas the completed statue exists "in actuality", that is that it is plainly real. It should be said that for the Ancients, mathematical objects are immutable; thus the object in potentiality cannot, in this case, ever become actual.

If the unit is called "perfect in potentiality" only, it is that it is only potentially divisible. In actual fact, division by 1 is not a partition; furthermore, the Greeks did not recognise fractions with numerator 1 as an object that exists of itself. A quarter or a third is a quarter or a third of something (some number or some magnitude) and not of the unit.

A prehistory of perfect numbers

There is a question that the reader may have been considering from the outset. The relation between a number and the sum of its divisors is not very evident of itself, and seems devoid of practical applications. Why, then, was the attention of the Ancients drawn to such a special property?

The Greek mathematicians give no explanation about this. But some sort of answer may be found... in Egypt. The reflections of Fr. Hultsch on the calculation techniques employed by the ancient Egyptians allow us to formulate at least a hypothesis, and to sketch out a prehistory of perfect numbers. The Greeks made a distinction between arithmetic (our theory of numbers) and logistics or the art of calculating, used by technicians or for commerce. The texts that we possess contain only arithmetic, but commercial exchange existed between the two countries, and it is reasonable to assume that the Greeks used calculation procedures that were analogous to those of the Egyptians, a good many of whose writings are available to us. In what follows we shall speak of the methods of logistics employed by 'the Ancients' in general.

The numeration system used by the Egyptians is additive and has a base ten (see Chapter 1). A number written in such a system appears as the result of the addition of a unit to itself. The operations on the numbers are, therefore, variants of the act of counting. Addition and subtraction are done simply, since they are directly linked to the method of recording the numbers. Multiplication is the first operation to present any difficulty. The method used is repeated duplication. Its success depends on the decomposition of the multiplier into a sum of (zero or positive integer) powers of 2. This decomposition is possible for every integer (which amounts to saying that all integers can be written in base two). This is what Maurice Caveing has called "the fundamental theorem of Egyptian arithmetic":

Consider the unlimited sequence of increasing powers of 2: 2^0, 2^1, 2^2, ..., 2^i, ...
Every natural number either appears in the list of terms of that sequence, or is the sum of terms which belong to it. [1]

Although no such statement appears in any of the documents we have to hand, we may infer from these, that the Egyptian calculators possessed an operational certainty of this result. This, for example, is how the product 25 × 31 was found, 25 being the multiplier m and 31 the multiplicand M (the numbers here being written in our way):

>	**1**	**31**
	2	62
	4	124
>	**8**	**248**
>	**16**	**496**

m = 1 + 8 + 16

m.M = 31 + 248 + 496 = 775

In fact,   m.M = $(2^0 + 2^3 + 2^4).M = 2^0M + 2^3M + 2^4M$.

This technique dispenses with the need to learn multiplication tables; all that is required is to be able to double and to decompose in base two.

Division methods derive directly from the techniques for multiplication. With very few exceptions, the Egyptians only used unit fractions with numerator 1. To divide by any non zero integer is to multiply by its inverse (we shall use the notation $\overline{n}$ for the inverse of n). Since all multiplication depends on doubling, all that is needed is to know how to double a unit fraction of the type 1/n. In the Rhind Papyrus (c. 1650 B.C.) there appears a table of the duplication of fractions of this type, with odd denominators up to 101. The last of these calculations is 2 × $\overline{101}$ and is given as:

$2 \times \overline{101} = \overline{101} + \overline{202} + \overline{303} + \overline{606}$

This duplication rests on:

– the rule: the product of two unit fractions $\overline{m}$ and $\overline{n}$ is the unit fraction $\overline{m.n}$

– a property of the number 6: the sum of the inverse of its divisors is equal to 2.

This last property, used in the papyrus but not explicitly stated, is due to the fact that 6 is a perfect number. For every perfect number gives one, and only one, decomposition of the unit.

Exercise 6   Why is the sum of the inverses of the divisors of a perfect number equal to 2?

Did the Egyptians possess an idea of the perfect number? The documents that are in our possession do not allow us to give a definite reply to that question. What is certain, is that the practice of decomposing fractions into unit fractions must have led the calculators among the Ancients to search out those numbers that were rich in divisors, which would allow several different decompositions of the unit. (This was why,

---

1.   Caveing, *Essai sur le savoir mathématique dans la Mésopotamie et l'Egypte anciennes*, p. 253-254.

presumably, first the Babylonians, and then the Greeks, became interested in using base sixty, which subsists until our day for measuring time and angles, because of its many divisors). These numbers rich in divisors are precisely those "superabundant" numbers of Nicomachus. Numbers which are "deficient", on the other hand, do not provide for any decomposition of the unit. The perfect numbers appear here as the limit between the two types.

Justification for the algorithm

There remains the problem of knowing where the general form $2^n(2^{n+1} - 1)$ for even perfect numbers came from. We can only make a conjecture since, here again, there is no direct historical evidence.

To obtain a perfect number, we could look for ways of changing a deficient number (to obtain one that is richer in divisors), or of changing an abundant number (so as to restrict the number of divisors). It turns out that the powers of two, resulting from repeated duplication, are all deficient numbers, but extremely close to being perfect, since they only differ from the sum of their aliquot parts by one.

To be precise, the aliquot parts of 2^n are one and the powers of 2 of lower order than n; and this sum is one less then 2n:

$$2^0 + 2^1 + 2^2 + \ldots + 2^{n-1} = 2^n - 1$$

In order to obtain a number that is richer in divisors than some given number, all that is needed is to multiply it. A search must then be made of numbers produced by the product of 2n and another factor, which for simplicity should be prime and distinct from 2. Consider numbers of the form $2^n p$ where p is prime. We can then produce the sequences arising from duplication, which will generate the aliquot parts of $2^n p$, starting from one and from p. Then we can determine the sums of these sequences and the total sum of the aliquot parts of $2^n p$:

						sums
1	2	2^2	...	2^{n-1}	2^n	$2^{n+1} - 1$
p	2p	$2^2 p$	...	$2^{n-1}p$	–	$(2^n - 1)p$
						$S = (2^{n+1} - 1) + 2^n p - p$

A complete study of the different cases that are presented by p less than, equal to, or greater than $(2^{n+1} - 1)$ is to be found in the *Treatise on Amicable Numbers* by the Arab mathematician Thâbit Ibn Qurrâ (10th century). He shows[1] that we have:

– an abundant number if $S > 2^n p$, that is if $2^{n+1} - 1 > p$;

– a deficient number if $S < 2^n p$, that is if $2^{n+1} - 1 < p$;

– a perfect number if $S = 2^n p$, that is if $2^{n+1} - 1 = p$.

---

1.   Thâbit Ibn Qurrâ, *Treatise on Amicable Numbers*, proposition 5.

A perfect number is therefore generated by means of two successive duplicative series:

- the duplication of the unit until the sum of the series produces a prime number p,
- the continued duplication of p as far as the term $(2^{n-1})p$, where n satisfies:

$$2^{n+1} = p + 1.$$

The number will then be of the form $2^n(2^{n+1} - 1)$ with $(2^{n+1} - 1)$ prime.

Direct proof by Euclid

The justification for the algorithm of Nicomachus is to be found in Book IX proposition 36 of the *Elements* of Euclid. But the *Elements* preceded the *Introductio Arithmetica* by several centuries. Why did we choose to begin with the more recent text?

Nicomachus is in the direct line of a tradition that goes back to the Fifth century B.C. Although we do not have any direct historical evidence, it is very probable that speculations about perfect numbers started before Euclid, in the school of Pythagoras. It is reasonable to accept the work by Nicomachus as a witness to that much older tradition. Furthermore, the work by Euclid conforming more to our canons of mathematics, appears more complete. Its perspective is certainly different: it is written in the form of a rigorously deductive construction which concerns the theory of numbers. Here the algorithm is proved, that is to say that it is put to the test of objective criteria.

The text of proposition 36 follows a style that is constantly found in the *Elements*. The theorem is first stated in the most general form (the proposition), then reformulated in a way that appears to be a particular case (the exposition), in which the given values serve as references for the proof that follows. The theorem takes the form of a particular claim ("I say that FG is perfect"), which is then echoed by the final Q.E.D.

The term *double proportion* means here the ratio $\frac{2}{1}$. Also, it makes no difference if a number happens to be designated by a single letter (A or B, for example) or by two joined letters (for example FG or HK); Euclid has recourse to this latter notation when a number is going to be partitioned or if it is the result of some partition.

> *If as many numbers as we please beginning from an unit be set out continuously in double proportion, until the sum of all becomes prime, and if the sum multiplied into the last make some number, the product will be perfect.*
> *For let as many numbers as we please, A, B, C, D, beginning from an unit be set out in double proportion, until the sum of all becomes prime,*
> *let E be equal to the sum, and let E by multiplying D make FG;*
> *I say that FG is perfect.* [1]

---

1.   Heath, *The Thirteen Books of Euclid's Elements*, vol.2, p. 421.

For the convenience of analysis, we divide our discussion of the proof into three stages:

First stage: the exposition

The principle lying behind the proof is simple. Let us recall that the algorithm consists in finding a number p of the form $1 + 2 + 4 + \dots + 2^n$, and in multiplying it by 2^n.

It is sufficient therefore to construct two parallel sequences:

$\mathcal{A}$: 1, 2, 4, ..., 2^n

$\mathcal{B}$: p, 2p, 4p, ..., 2^np,

in which $p = 2^{n+1} - 1$.

The numbers that occur in the two sequences represent all the divisors of 2^np.

The two sequences are summed: their sum is equal to $2(2^np)$. Hence 2^np is a perfect number according to the Euler criterion.

Euclid's proof conforms globally to this principle, but its execution is complicated by the limited mathematical tools he has to hand, in particular:

a) for him, 1 (the unit) is not a number,

b) similarly, a number cannot be a part of itself,

c) it follows that Euclid does not recognise the two sequences $\mathcal{A}$ and $\mathcal{B}$ above, but the two staggered sequences:

$\mathcal{A}'$:   A, B, ..., D

$\mathcal{B}'$: E, HK, ..., M.

The sequence $\mathcal{A}'$ does not contain the term corresponding to the term E of sequence $\mathcal{B}'$ (it would be 1). Although he says that the numbers A, B, C, D are "continuously proportional beginning from an unit", the latter does not form part of the sequence, because it is not a number. The first term A is in fact the number 2. It is noticeable how Euclid is careful – mindful of the purity of the theory and the universality of its proposition – not to give a single numerical example. This is far from the descriptive and quasi-concrete presentation used by Nicomachus.

Also, the sequence $\mathcal{B}'$ contains no term corresponding to D (this would be the perfect number itself, which Euclid designates by FG).

d) he has no algebraic notation available to him, which would immediately show that the product of the last term of $\mathcal{A}$ by the first term of $\mathcal{B}$ (that is 2^n multiplied by p) is the last term of $\mathcal{B}$,

e) arguing from sequences of indeterminate length, he is unable to affirm immediately that this process has enabled him to find all the aliquot parts of the desired number.

In this first stage, he has to establish that the number FG (that is the perfect number 2^np) is indeed the double of M, the last term of the sequence $\mathcal{B}'$, and he does this by use of the theory of proportions.

> For, however many A, B, C, D are in multitude, let so many E, HK, L, M be taken in double proportion beginning from E;
> therefore, ex aequali, as A is to D, so is E to M.

Therefore the product of E, D is equal to the product of A, M.
And the product of E, D is FG; therefore the product of A, M is also FG.
Therefore A by multiplying M has made FG; therefore M measures FG according to the
units in A. And A is a dyad; therefore FG is double of M.
But M, L, HK, E are continuously double of each other; therefore E, HK, L, M, FG are
continuously proportional in double proportion. [1]

Second stage

Euclid establishes now that FG is the sum of the two sequences 𝒜' and ℬ', together
with the unit, which appears as the paronym part. For this he appeals to IX. 35: "*If as
many numbers as we please be in continued proportion, and there be subtracted from
the second and the last numbers equal to the first, then, as the excess of the second is to
the first, so will the excess of the last be to all those before it*" – a statement which can
be considered as a lemma, artificially detached from the main proof.

Now let there be subtracted from the second HK and the last FG the numbers HN, FO,
each equal to the first E; therefore, as the excess of the second is to the first, so is the excess
of the last to all those before it.
Therefore, as NK is to E, so is OG to M, L, KH, E.. And NK is equal to E; therefore OG is also
equal to M, L, HK, E.
But FO is also equal to E, and E is equal to A, B, C, D and the unit.
Therefore the whole FG is equal to E, HK, L, M and A, B, C, D and the unit;
and it is measured by them. [2]

Third stage

Finally it remains to establish that no aliquot part exists other than the numbers
which have been constructed. For this, he argues by contradiction. It is interesting to
note that this is a strategy that he constantly uses whenever he is faced with the infinite
or the unbounded. Examples are to be found in his proof of the irrationality of $\sqrt{2}$ (X.
117) and in the proof of the infinity of prime numbers (IX. 20).

In order to follow the detail of the complex proof, which makes frequent appeal to
the properties of proportions, it would help to bear in mind the underlying idea of the
proof. If there should exist, in addition to the numbers of the sequences A' and B',
another divisor P, say, then there must be a corresponding divisor Q, so that P.Q = FG
(this is what Nicomachus calls the "matching" of the number P). One of these two
divisors must necessarily belong to the sequence 𝒜'. Euclid proves this by
proportionality:

I say also that FG will not be measured by any other number except A, B, C, D, E, HK, L,
M and the unit.
For, if possible, let some number P measure FG, and let P not be the same with any of the
numbers A, B, C, D, E, HK, L, M.
And, as many times as P measures FG, so many units let there be in Q; therefore Q by
multiplying P has made FG.
But, further, E has also by multiplying D made FG; therefore, as E is to Q, so is P to D.

1.   *Ibid.* p. 421-422.
2.   *Ibid.* p. 422.

> *And, since A, B, C, D are continuously proportional beginning from an unit, therefore D will not be measured by any other number except A, B, C.*
>
> *And, by hypothesis, P is not the same with any of the numbers A, B, C; therefore P will not measure D.*
>
> *But, as P is to D, so is E to Q; therefore neither does E measure Q.*
>
> *And E is prime; and any prime number is prime to any number which it does not measure.*
>
> *Therefore E, Q are prime to one another.*
>
> *But primes are also least, and the least numbers measure those which have the same ratio the same number of times, the antecedent the antecedent, and the consequent the consequent.; and, as E is to Q, so is P to D; therefore E measures P the same number of times that Q measures D.*
>
> *But D is not measured by any other number except A, B, C; therefore Q is the same with one of the numbers A, B, C.*
>
> *Let it be the same with B.*
>
> *And however many B, C, D are in multitude, let so many E, HK, L be taken beginning from E.*
>
> *Now E, HK, L are in the same ratio with B, C, D; therefore, ex aequali, as B is to D, so is E to L.*
>
> *Therefore the product of B, L is equal to the product of D, E. But the product of D, E is equal to the product of Q, P; therefore the product of Q, P is also equal to the product of B, L.*
>
> *Therefore, as Q is to B, so is L to P.*
>
> *And Q is the same with B; therefore L is also the same with P: which is impossible, for by hypothesis P is not the same with any of the numbers set out.*
>
> *Therefore no number will measure FG except A, B, C, D, E, HK, L, M and the unit.*
>
> *And FG was proved equal to A, B, C, D, E, HK, L, M and the unit; and a perfect number is that which is equal to its own parts; therefore FG is perfect.*
>
> *Q.E.D.[1]*

The proof set out above is a synthesis, but Euclid gives no indication of his analysis, as Ibn al-Haytham (11th century) remarks, who himself attempted one.

The inverse proof: Euler

It remains to be known whether it is true, as Nicomachus holds, that the algorithm "does not leave out any perfect numbers". This is what Euler (1707 – 1783) proves in two manuscripts which were not published until after his death: the *Tractatus de numerorum doctrina* and the *De numeris amicabilibus*. We give here his exposition in the *Tractatus*, which is more detailed and clearer, although the proof is incomplete. Let us see, first of all, how he states the problem (the symbol $\int$ means in what follows the sum of the divisors of a natural number):

> *Here, it is customary to propose the following problem: find a number which has to the sum of its divisors a given proportion; that is, such as* $N : \int N = n : m$, *or* $\dfrac{\int N}{N} = \dfrac{m}{n}$, *in which however it is above all necessary that* $m > n$; *in particular, if we have* $m = n$, *then we have* $N = 1$.

---

1. *Ibid.*, p. 422-424.

Once the ratio m : n *is expressed in the least terms possible, the number* N *will be equal, either to* n, *or to one of its multiples. Suppose* N = an, *we shall have* $\int$N = $\int$an = am. *But, except for the case where* a = 1, *we have* $\int$an > a $\int$n, *and from that* m > $\int$n. *Consequently, if we should have* m < $\int$n *there will be no solution; if on the other hand* m = $\int$n, *there is a unique solution, namely* N = n.

Unless, then, m = $\int$n *or* m > $\int$n, *the problem does not have a solution. In the first case, the number looked for,* N *will be equal to* n *itself, and no other solution is possible. As for the second case, where* m > $\int$n, *the number* N *will be equal to one at last of the multiples of* n, *say* N = an, *if indeed there is a solution. For there exist at least ratios* m : n *for which the problem is impossible, even if we should have* m > $\int$n...[1]

Euler relates the problem of perfect numbers to the general problem of the relation between a natural number and the sum of its divisors. This effectively demystifies the idea of the perfect number, which now becomes simply a special case in a more general problem of number theory:

We could add to this several other problems, in which other relations are proposed between numbers and the sums of their divisors; I shall not mention these, since with the principles made clear there is no difficulty in finding a method for solving them.[2]

This sober view of the problem reflects the history of a change in the way of thinking about arithmetic since the rather mystical Pythagorean approach exemplified in the passage by Iamblichus which we quoted at the beginning.

Let us see how Euler continues with his proof. This depends partly on the technique of decomposition into prime factors, and partly on a result he had established earlier:

If N = $p^\lambda q^\mu r^\nu$ ... s^ξ, then $\int$N = $\int p^\lambda . \int q^\mu . \int r^\nu$ ... $\int s^\xi$

This is a particular case of what we today call the multiplicativity of the function "sum of divisors", that is that $\int$(mn) = $\int$m.$\int$n, on condition, nevertheless, that m and n are prime with each other.

Exercise 7 Verify the muliplicativity of the function "sum of divisors" with the example: $\int$120 = $\int$8.$\int$15.

It is true to say that this property had been noticed by other mathematicians before Euler, such as Thâbit Ibn Qurrâ al-Fârisî (13th century), and Descartes; but they were still thinking in terms of the aliquot parts. For example, the same property stated in this way is:

If two numbers prime to each other are known, together with their aliquot parts, then the aliquot parts of their product is also known: for example, let one of the numbers be a *and its aliquot parts be* b, *let* c *be a second number, and its aliquot parts be* d, *then the aliquot parts of* ac *will be* ad + bc + bd.[3]

---

1.   Euler, *Tractatus de Numerorum Doctrina*, § 103-105.
2.   Euler, *Tractatus de Numerorum Doctrina*, § 110.
3.   Descartes, *De partibus aliquotis numerorum*, p. 301.

Written in this form it is rather less remarkable, and we can understand why these mathematicians had not thought of using it in the way Euler did – which allowed him, entirely through the power of algebra, to come to the form $2^n(2^{n+1} - 1)$, which is the key to the algorithm.

> A perfect number is a number, the sum of whose divisors is twice the number itself. Thus, if $\int N = 2N$, N will be a perfect number. If it is even, then it will be of the form 2^nA, A being an odd number, either prime or composite. Given that $N = 2^nA$, we have
>
> $$\int N = (2^{n+1} - 1)\int A, \text{ which gives } \frac{\int A}{A} = \frac{2^{n+1}}{2^{n+1} - 1}.$$ Since the numerator of this
> fraction is greater than the denominator by just one, it cannot be greater than the sum of the divisors of the denominator; it must therefore be equal to, or less than it. In the second case, there cannot be a solution; the first case can only exist, if $2^{n+1} - 1$ is a prime number. So each time $2^{n+1} - 1$ is a prime number, A must be taken, equal to it, and we shall have a perfect number $= 2^n(2^{n+1} - 1)$.
>
> Thus all even perfect numbers are contained in the formula $2^n(2^{n+1} - 1)$, on condition that $2^{n+1} - 1$ is a prime number, which cannot happen unless $n + 1$ is prime; although it is not true that every prime number, equal to $n + 1$, makes $2^{n+1} - 1$ prime.[1]

In the last remark, we come back to the question of prime Mersenne numbers, not resolved even now. In this sense, the question of even perfect numbers is not resolved since, as we have seen, their discovery is tied to being able to recognise when Mersenne numbers are prime (see Chapter 14).

Exercise 8   In what sense is the proof given above incomplete?

Do any odd perfect numbers exist?

Euler continues:

> ... As for knowing whether, in addition to even perfect numbers, there exist any odd perfect numbers, nobody up till now has demonstrated it.[2]

In actual fact, it appears that throughout the centuries, the assertion by Nicomachus that all perfect numbers are even, has been accepted without discussion. However, in 1638, in a letter to Mersenne, Descartes declared:

> ... I think I am able to prove that there are no even numbers which are perfect apart from those of Euclid; & that also there are no odd perfect numbers, unless they are composed of a single prime number, multiplied by a square whose root is composed of several other prime numbers. But I see nothing that would prevent one from finding [numbers] of this sort: since, for example, if 22 021 were prime, in multiplying it by 9 018 009, which is a square whose root is composed of the prime numbers 3, 7, 11 & 13, one would have 198 585 576 189, which would be a perfect number. But, whatever method one might use, it would require a great deal of time to look for these numbers, & perhaps the shortest would have more than 15 or 20 digits.[3]

---

1.   Euler, *Tractatus de Numerorum Doctrina*, § 106-108.
2.   Idem, § 108.
3.   Descartes, *Letter* 149, p. 429-430.

What remains today of these optimistic opinions? It is still not known today if there are any odd perfect numbers. One of the mathematicians who gave a great deal of time in researching the subject was J. J. Sylvester, who wrote at the end of the last century:

> *the existence of [an odd perfect number] – its escape, so to say, from the complex web of conditions which hem it in on all sides – would be little short of a miracle*[1]

It is not that this quest has been entirely in vain. In default of finding any odd perfect numbers, a number of, more and more numerous, conditions surrounding its existence have been established. In this area, the first significant results are due to Euler. This is how he states the problem, with reference once again to the multiplicativity of the "sum of the divisors" function:

> *If an odd perfect number exists, then all its factors must be odd. If it should be* = ABCD etc., *we must have* $\int A. \int B. \int C. \int D...$ = 2ABCD ..., *as a number which is evenly odd.*[2]

A number which is *evenly odd* (a term which comes from Plato and still in use in the eighteenth century) is the double of an odd number. Since the sum of the divisors of a perfect number is double that number, the sum of the divisors of an odd perfect number is necessarily evenly odd. Also, the letters A, B, C, D, stand for either prime numbers or the powers of such numbers. Today we would write $a^\alpha b^\beta c^\gamma \dots d^\delta$.

> *So, amongst the sums of the divisors* $\int A. \int B. \int C. \int D, \dots$, *one, and only one, must be evenly odd, and all the others odd; thus all the factors* A, B, C, D, $\dots$, *except for one will be even powers of prime numbers.*[3]

"One, and only one, must be evenly odd", since the sum of the divisors must have one and only one factor 2.

Why "even powers of prime numbers" (that is the exponents are even)? All the divisors of a power b^β of a prime number b, are the numbers b^0, b^1, b^2, ..., b^β. The sum of the divisors is made up of $(\beta + 1)$ odd terms. If this is to be an odd number, then β must be even.

Let us pick up again the thread of the argument:

> *... all the factors* A, B, C, D, *... except one will be even powers of primes, and that one will be either a prime number of the form* 4n + 1, *or a power of that number whose exponent is* $4\lambda + 1$.[4]

This particular factor is the one that makes the sum of the divisors evenly odd; in what follows we shall refer to it as a^α. From what has gone before, its exponent α is necessarily odd; and the number a is also odd.

An odd number must be of the form 4n + 1 or 4n – 1. Euler uses this alternative in order to place some restriction, both on the number a, and on the exponent α. However, he does not explain his proof. Here is one, in which it is necessary first to establish the result concerning the exponent.

Suppose $\alpha = 4\lambda - 1$, the sum of the divisors of a^α will be composed of 4λ terms, of which 2λ will have even exponents and 2λ will have odd exponents.

1.   In Wagon, "Perfect Numbers", p. 66.
2.   Euler, *Tractatus de Numerorum Doctrina*, § 109.
3.   *Ibid.*
4.   *Ibid.*

The sum of the 2λ terms with even exponents is of the form $(4m + 2\lambda)$. Recall that a is odd and that, for example, $(2n + 1)^2 = 4(n^2 + n) + 1$.

For the sum of the 2λ terms with odd exponents, it is necessary to consider two possible cases: since a is of the form $(2n + 1)$, we can have:

– either n is even $(n = 2k)$,

– or n is odd $(n = 2k + 1)$

In the first case, the sum of the terms having odd exponents will be of the form $(4m' + 2\lambda)$.

In the second case, it will be of the form $(4m'' - 2\lambda)$.

Therefore the sum of all the divisors of a^α will, in both cases, be a multiple of 4, which is contrary to the initial condition: that this sum has to be evenly odd. Hence α is of the form $4\lambda + 1$.

It remains to show that a is of the form $(4n + 1)$. The sum of the divisors of $a^{4\lambda + 1}$ is composed of $(4\lambda + 2)$ terms, of which $(2\lambda + 1)$ have even exponents, and $(2\lambda + 1)$ have odd exponents. (The reasoning here is of a similar type to the earlier argument.)

If $a = 4n - 1$, the expansion of the terms with even exponents will terminate in "+1" and that of the terms with odd exponents will terminate in "–1". All the other terms in the sum of the expansion will be multiples of 4. The sum total will therefore be divisible by 4 (since there will be as many "+1" terms as there are "–1" terms). Here again, the result is incompatible with the fact that the sum of the divisors of an odd perfect number must be evenly odd, that is double an odd number.

Euler concludes:

> *Therefore, such a perfect number will have the form* $(4n + 1)^{4\lambda + 1} PP$, *in which P is an odd number and* $4n + 1$ *a prime number.*[1]

In using PP here (that is P^2, Euler used both forms concurrently), he intends the product of all the factors $B, C, D, ..., $ etc. which we have designated by $b^\beta, c^\gamma, ..., d^\delta$. This result validates the conjecture made by Descartes in the letter we quoted above, but with two qualifications concerning the nature of the particular factor: the prime number must be of the form $(4n + 1)$; and the particular factor has also to be a power of a.

To be continued ...

Euler's formula expresses a positive condition (if an odd perfect number exists, it must necessarily be of this form). Against that, the main strategy of researchers, since the middle of the last century, has been to establish restrictive conditions. The most fruitful line of research up to now has been to establish remarkable results concerning the minimum number of distinct prime factors that an odd perfect number would have to possess. It is known now that there would have to be at least 9.

Exercise 9   Establish that an odd perfect number must have at least 3 distinct prime factors.

---

1.   Euler, *Tractatus de Numerorum Doctrina*, § 109.

This type of proof requires the successive examination and elimination of a very great number of cases, that can only be explored with the aid of powerful computers. This has allowed mathematicians to push back the lower bounds for an odd perfect number. This is achieved by calculations combining all the diverse known restrictions. The table below (after Stan Wagon[1]) provides an idea of the successive stages in this research, and also of the rapidity of progress since the development of computers:

Lower bounds on the number of distinct prime factors of an odd perfect number:	
3	Nocco (1863)
4	Servais (1888), Sylvester (1888)
5	Sylvester (1888)
6	Gradstein (1925)
7	Robbins (1972), Pomerance (1972)
8	Hagis (1975)

Lower bounds on an odd perfect number:	
2 000 000	Turcaninov (1908)
1.4×10^{14}	Kanold (1944)
10^{18}	Muskat (1948)
10^{20}	Kanold (1957)
10^{36}	Tuckerman (1967)
10^{50}	Hagis (1973)
10^{160}	Brent and Cohen (1989)

And so we see that the set of numbers within which odd perfect numbers cannot occur grows more and more, and more and more rapidly, and we may well wonder if any can exist at all. The smallest odd perfect number, if there is one, will need more than 160 digits to write it in decimal form. But we need to add, that this magnitude, impressive though it is, is relatively small compared with the greatest even perfect number that is known: $2^{216\,090}\,(2^{216\,091} - 1)$, a number that would require more than one hundred and twenty thousand digits to write out in full (needing more than thirty pages of this book). And in the interval $[0, 10^{160}[$, there are not even more than a dozen even perfect numbers.

On the other hand, it cannot be seen how it might be possible to give a direct proof of the impossibility of the existence of an odd perfect number. We may then ask ourselves, as it has been asked about Fermat's last theorem, whether the question is undecidable, that is that it could be impossible within the axioms of arithmetic employed today to give either an affirmative or a negative answer to the question of the existence of odd perfect numbers.

---

1.   Wagon, "Perfect Numbers", p. 67.

Twenty-three centuries to discover thirty numbers? And what efforts and ingenuity employed on these numbers which after all have no use! But let us leave it to Euler to have the last word:

> *Among all the problems which we are used to dealing with in Mathematics, none for certain, are judged by the majority of modern mathematicians, to be more sterile or more detached from all possible use, than those which concern speculation about the nature of numbers and research into their divisors. In this judgement, the mathematicians of today differ greatly from the Ancients, who were accustomed to accord a great value to these speculations. And for all that the Ancients knew that this research into the nature of numbers might have only a little use in that part of Mathematics that is called applied, or for investigation of matters that particularly concern Physics; nevertheless they dedicated a great deal of energy and effort into examining the properties of numbers. For as well as it seeming to them that investigation of the truth was in itself laudable and worthy of human consciousness, they judged also, rightly, that by these researches the art of investigation could be extended, and that the faculties of the mind would become better able to deal with important questions. And in this opinion they were not deceived, for we have manifest proof of this in the considerable developments that have enriched Analysis since that epoch; in fact it appears entirely to be the case that that science would never have achieved such a degree of perfection had the Ancients not put so much zeal into developing questions of this type, which the greater part of modern mathematicians despise so much on account of their sterility. And for that reason, we can be even less in doubt that in further pursuing the study of these matters in the future, remarkable developments will be brought to Analysis.* [1]

---

1.   Euler, *De numeris amicabililus*, § 1.

Bibliography

Sources texts

DESCARTES, *Letter 149 to Mersenne*, 15-11-1638; Letter 153 to Frenicle, 9-1-1639, in *Correspondence*, vol. II, Adam-Tannery, Paris, 1898.

DESCARTES, *De partibus aliquotis mumerorum*, in Œuvres, vol. X, Adam-Tannery, Paris, 1908, p. 300-302.

EULER, *Tractatus de Numerorum doctrina*, in *Opera Omnia*, vol. I-V, Leipzig, 1944, p. 179-283.

EULER, *De numeris amicabilibus*, in *Opera Omnia*, vol. I-V, Leipzig, 1944, p. 353-365.

HEATH, *The Thirteen Books of Euclid's Elements*, 2nd edition, 3 vols., Cambridge University Press, 1926. Repr. Dover, New York, 1956.

IAMBLICHUS, *In Nicomachi Arithmeticam Introductionem*, ed. Pistelli, Leipzig 1894.

NICOMACHUS, *Introductio Arithmetica*, ed. Hoche, Leipzig, 1866; French translation (J. Bertier), Nicomaque, *Introduction Arithmétique*, Paris, 1978. English tr. D'Ooge, *Introduction to Arithmetic*, New York, 1926.

THABIT IBN QURRA, *Traité des nombres amiables* (Risâlat fî-l-'a'dâd al-mutahâbbat), manuscripts: Paris, B.N. ar. 952.2; Istanbul, Aya Sofia 4820, p. 110-122; partial French tr. and analysis, Woepcke, "Notice sur une théorie ajoutée par Thâbit ben Korrah à l'arithmétique spéculative des Grecs", *Journal Asiatique*, **20** (1852), p. 420-429.

[The texts quoted above, with the exception of Euclid, were translated by M. Crubellier. The Greek text of *Introductio Arithmetico* which we followed differed sometimes from that of Hoche.]

Reference texts

DICKSON, *History of the theory of numbers*, vol. 1: Divisibility and Primality, New York, 1952.

STEWART, "Diviser pour régner", *Pour la science*, 157, 1990, 150-156.

WAGON, "Perfect Numbers", *The Mathematical Intelligencer*, 7, 2, 1985, 66-68.

General works for further reading

CAVEING, *Essai sur le savoir mathématique dans la Mésopotamie et l'Egypte anciennes*, Lille, P.U.L., 1994.

HULTSCH, "Die Elemente der Ägyptischen Theilungsrechnung", *Abhandlungen der philologisch-historischen Classe der königlich-sächsischen Gesellschaft der Wissenschaften*, Leipzig, **17**-1, 1895, 3-192.

RASHED, "Ibn al-Haytham et les nombres parfaits", *Historia Mathematica*, **16**, 1989, 343-352.

How did you get on?

Ex. 1   The divisors of 496 are:

1	496
2	248
4	124
8	62
16	31

Their sum is
$$1 + 2 + 4 + 8 + 16 + 31 + 62 + 124 + 248 + 496 = 992 = 496 \times 2.$$

Ex. 2   The divisors of 60 are :

1, 2, 3, 4, 5, 6, 10, 12, 15, 20, 30, 60.
The sum is:
$$168 = 2 \times 84 \text{ and } 84 > 60.$$

Ex. 3   For non zero natural number n, the divisors of 2^n are : $1, 2, 2^2, 2^3, \ldots , 2^n$. These are the first $(n + 1)$ terms of a geometric sequence with common ratio 2 and so their sum is $\dfrac{1 - 2^{n+1}}{1 - 2} = 2^{n+1} - 1 < 2(2^n)$. And a number n is deficient if the sum of its divisors is less than 2n.

Ex. 4   When $(n + 1)$ is even, the numbers of the form $2^{(n+1)} - 1$ are never prime (they are the difference of two squares). Further:
$$2^9 - 1 = 511 = 73 \times 7$$
$$2^{11} - 1 = 2047 = 89 \times 23$$
But $2^{13} - 1 = 8191$, which is prime. The fifth even perfect number is therefore given by:
$$2^{12}.(2^{13} - 1) = 4096 \times 8191 = 33\ 550\ 336$$
We next find:
$$2^{15} - 1 = 32\ 767 = 7 \times 31 \times 151$$
Then: $2^{17} - 1 = 131\ 071$, which is prime. The sixth even perfect number is therefore:
$$2^{16}.(2^{17} - 1) = 65\ 536 \times 131\ 071 = 8\ 589\ 869\ 056.$$

Ex. 5   The successive powers of 2 with integer exponents > 0, terminate in the digits 2, 4, 8, 6, 2, 4, 8, 6, ...
Let x_n be the final digit of 2^n, y_n the final digit of $(2^{(n+1)} - 1)$ and z_n the final digit of $2^n.(2^{(n+1)} - 1)$, we have:

values of n	x_n	y_n	z_n
1, 5, 9, 13, ...	2	3	6
2, 6, 10, 14, ...	4	7	8
3, 7, 11, 15, ...	8	5	0
4, 8, 12, 16	6	1	6

The cases $x_n = 2$ and $x_n = 8$ correspond to even values for $(n + 1)$ and so cannot generate perfect numbers. But if we accept the claim by Charles de Boüelles that all odd values of $(n + 1)$ will produce perfect numbers, then we can see that they will end alternately in 6 or 8, which was predicted by Nicomachus. But even in this case, the first perfect number 6 arises as a coincidence, since it is derived from the case n = 1: (2×3); and not as do the other 6's from n = 4, 8, ... : (6×1).

Ex. 6   Let n be a perfect number, whose divisors are: 1, a, b, ..., i, j, ..., l, m, n, where $1 \times n = a \times m = b \times l = \ldots = i \times j = n$. It should be noted that a perfect number will necessarily have an even number of divisors since it cannot be square (for even perfect numbers, this comes from the formula that produces them; for odd perfect numbers, see p. 405-407).

The sum of the inverses of the divisors of n is therefore:

$$\frac{1}{1} + \frac{1}{a} + \frac{1}{b} + \dots + \frac{1}{i} + \frac{1}{j} + \dots + \frac{1}{l} + \frac{1}{m} + \frac{1}{n}$$

$$= \frac{n}{n} + \frac{m}{n} + \frac{1}{n} + \dots + \frac{i}{n} + \frac{i}{n} + \dots + \frac{b}{n} + \frac{a}{n} + \frac{1}{n}$$

$$= \frac{n + m + l + \dots + j + i + \dots + b + a + 1}{n}$$

$$= \frac{2n}{n}$$

$$= 2.$$

Ex. 7   The divisors of 8 ($=2^3$) are : 1, 2, 2^2, and 2^3 ; $\int(8) = 1 + 2 + 4 + 8 = 15$.

The divisors of 15 (= 3 × 5) are 1, 3, 5, and 3 × 5; $\int(15) = 1 + 3 + 5 + 15 = 24$.

If we expand the product $(1 + 2 + 2^2 + 2^3).(1 + 3 + 5 + 15)$ we obtain the sum of all the divisors of 120, since $120 = 2^3 \times 3 \times 5$.

(Example given by Stewart, p.151.)

Ex. 8   Euler establishes that $\dfrac{\int A}{A} = \dfrac{2^n + 1}{2^n + 1 - 1}$. The right hand side is an irreducible fraction. Euler seems to assume that the same is true for the left hand side, from which he concludes that $\int A = 2^n + 1$ and $A = 2^n + 1 - 1$. He omits to consider the case where $\int A = k (2^n + 1)$ and $A = k (2^n + 1 - 1)$.

In his *De numeris amicabilibus*, this latter possibility is dealt with by the following argument:

If $A = k (2^n + 1 - 1)$, then $\int A \geq 2^n + 1 + k + A$ (since A will have divisors of at least itself, k, $(2^n + 1 - 1)$ and 1; even more if k and $(2^n + 1 - 1)$ are not self prime). Therefore we have:

$$\frac{\int A}{A} \geq \frac{2^n + 1 + k + A}{A} \geq \frac{2^n + 1(1 + k)}{(2^n + 1 - 1)k} = \frac{2^n + 1}{2^n + 1 - 1} \times \frac{1 + k}{k} > \frac{2^n + 1}{2^n + 1 - 1}$$

which contradicts the earlier result that was established.

Ex. 9   Here is the proof given by Sylvester (from Stan Wagon):

Suppose $n = a^\alpha b^\beta$ is perfect, where a and b are distinct odd primes, and α and β are natural numbers. Then a contradiction is reached as follows:

$$2 = \frac{\int n}{n} = \frac{\int(a^\alpha)}{a^\alpha} \frac{\int(b^\beta)}{b^\beta} = \left(1 + \frac{1}{a} + \frac{1}{a^2} + \dots + \frac{1}{a^\alpha}\right)\left(1 + \frac{1}{b} + \frac{1}{b^2} + \dots + \frac{1}{b^\beta}\right)$$

$$< \left(1 + \frac{1}{3} + \frac{1}{9} + \frac{1}{27} + \dots\right)\left(1 + \frac{1}{5} + \frac{1}{25} + \frac{1}{125} + \dots\right) = \frac{3}{2} \times \frac{5}{4} < 2$$

The twenty first perfect numbers (plus unity) from the Book of Perfect Numbers by Charles de Boüelles (1509). This table, based on the supposition that $(2^p - 1)$ is prime whenever p is odd, is false. Only the numbers found in lines 2, 3, 4, 5, 8, 10, 11, 17 are perfect; all the others are abundant. The way the table is laid out suggests that the author wanted to demonstrate the diagonal, which reveals that (nearly) one perfect number occurs in every power of ten, as Nicomachus had wished to be the case. It can be noticed that in lines 6, 11, 16 and 21, there is a jump of two cases, and the initial digit 1 is put in brackets. Might it be that Ch. de Boüelles thought he had to mask, as much as he could, this trying discordance between doctrine and truth?

[Source: Caroli Bouilli Samarobrini, *Liber de Perfectis Numeris* (1509), p. 157]

Bibliography

- Commission Inter IREM Histoire et Epistémologie des mathématiques:

  La Rigueur et le Calcul, Documents historlques et épistémologiques. Cedic-Nathan, Paris, 1982.

  Actes du Colloque Inter l.R.E.M Montpellier 31 Mai - 1er Juin 1985. Histoire et Epistémologie des Mathématiques. Rôle des problèmes dans l'histoire et l'activité mathématique. IREM Montpellier, Montpellier, 1986.

  Mathématiques au fil des âges. Textes choisis et commentés. Gauthier Villars, Paris, 1987.

  Pour une perspective historique dans l'enseignement des mathématiques. Bulletin Inter IREM, IREM-Lyon, Lyon, 1988. English tr. Guillerme C. and Weeks C., The IREM Papers, *History in the Mathematics Classroom*, Fauvel J. ed., The Mathematical Association, London, 1990.

  La démonstration mathématique dans l'histoire. Actes du 7ᵉ Colloque Inter-IREM Epistémologie et Histoire des Mathématiques. Besançon 12 et 13 Mai 1989. IREM Besançon, Besançon, 1990.

  La figure et l'espace, Actes du 8e colloque inter-IREM Epistémologie et Histoire des mathématiques, IREM de Lyon, Lyon, 1993.

  Histoire d'infini. Actes du 9ᵉ colloque Inter-IREM Epistémologie et histoire des mathématiques. Landernau 22-23 mai 1992, IREM Brest, Brest, 1994.

  Histoire et épistémologie dans l'éducation mathématique. History and Epistemology in mathematics education. Actes de la Première Université d'Eté européenne. First European Summer University Prooreeding Montpellier 19-23 juillet 1993, IREM Montpellier, Montpellier, 1994.

  Quatrième Université d'été d'Histoire des Mathématiques, Lille, juillet 1990, IREM Lille, Lille, 1994.

  Contribution à une approche historique de l'enseignement des mathématiques. Actes de la 6ᵉ université d'été interdisciplinaire sur l'histoire des mathématiques. Besançon, 8-13 juillet 1995, IREM Besançon, Besançon, 1996.

- A.D.E.R.H.E.M *(Association pour le Développement des Etudes et des Recherches en Histoires des Mathématiques,* IREM de Poitiers, 40 Avenue du Recteur Pineau 86 022 Poitiers Cedex):

  Destin de l'art et dessin de la Science. Actes du colloque ADERHEM. Caen 1986. IREM-Caen, 1991.

- A.P.M.E.P. (Association des Professeurs de Mathématiques de l'Enseignement public, 26, rue Duméril, 75013-Paris):

  n°41 *Fragments d'histoire des mathématiques.* 1981.

  n°48 *Présence d'Evariste Galois.* 1982.

  n°65 *Fragments d'histoire des mathématiques 11.* 1987.

  n°83 *Fragments d'histoire des mathématiques. Emergence du concept de groupe.* 1991.

  n°86 *Quadrature du cercle. Fractions continues et autres contes.* 1992.

- BKOUCHE R., CHARLOT B., ROUCHE N., *Faire des mathématiques: le plaisir du sens.* Armand Colin, Paris, 1991.

- BOLL M., *Histoire des mathématiques.* Que Sais-je ? n°4. Presses Universitaires de France, Paris, 1979.

- BOURBAKI N., *Eléments d'histoire des mathématiques.* Hermann, Paris, 1980.

- BOUTROUX P., *L'idéal scientifique des mathématiciens dans l'Antiquité et dans les temps Modernes.* Alcan, Paris, 1920. Reimp. Gabay, Paris, 1992.

- BOUVERESSE J., ITARD J., SALLE E., *Histoire des mathématiques.* Larousse, Paris, 1977.

- BOYER C.B., *A history of mathematics.* Wiley and Sons, New York, 1968.

- BREHIER E., *Histoire de la philosophie.* (2 tomes). Presses Universitaires de France, Paris, 3rd ed., 1985.

- BRETON P., *Une histoire de l'informatique.* Points Sciences. Editions du Seuil, Paris, 1990.

- BRUNSCHVICG L., *Les étapes de la philosophie mathématique.* Alcan, Paris, 1912. Reprint Blanchard, Paris, 1972.

- BUNT L., JONES P., BEDIENT J., *The historical roots in elementary mathematics.* Dover, New-York, 1988.

- CAJORI F., *A history of mathematical notations* (2 vol). The Open Court Publishing Company, La Salle, 1928-29. Reprint Dover, New York, 1993.

- CANTOR, M., *Vorlesungen uber Geschichte der Mathematik.* (4 tomes). Teubner, Leipzig, 1892-1902.

- CAVAILLES J., *Philosophie mathématique.* Hermann, Paris, 1962.

- CAVEING M., *Essai sur le savoir mathématique dans la Mésopotamie et l'Egypte anciennes*, Presses Universitaires de Lille, Lille, 1994.
- CELNIKIER L.M., *Histoire de l'astronomie*. Lavoisier, Paris, 1986.
- CHABERT J.-L., BARBIN E., *et al.*, *Histoires d'algorithmes. Du caillou à la puce*, Belin, Paris, 1994, english tr. to appear.
- CLERO J.P., LE REST E., *La naissance du calcul infinitésimal au 17ᵉ siècle*. Cahiers d'histoire et de philosophie des sciences n°16. Centre de documentation en sciences humaines, Paris, 1980.
- COLLETTE J.P., *Histoire des mathématiques* (2 tomes). Vuibert, Paris, 1973-1979.
- *Concise dictionary of scientific biography*. Charles Scribner's sons, New York, 1981.
- DAHAN-DALMEDICO A., CHABERT J.L., CHEMLA K., *Chaos et déterminisme*. Points Sciences, S80. Editions du Seuil, Paris, 1992.
- DAHAN-DALMEDICO A., PElFFER J., U*ne histoire des mathématiques. Routes et dédales*. Points Sciences, S49. Editions du Seuil, Paris, 1986.
- DAUBEN J.W., *The history of mathematics from antiquity to the present; a selective bibliography*. Garland, 1985.
- DAUMAS M., *Histoire de la Science*. Gallimard, Paris, 1957.
- DAVIS P., HERSH R., *The mathematical experience*. Birkhauser, Boston, 1982. Trad. Chambadal L., *L'Univers mathématique*. Gauthier-Villars, Paris, 1985.
- DEDRON L., ITARD J., *Mathématiques et mathématiciens*. Magnard, Paris, 1960.
- DHOMBRES J., *Nombre, mesure et continu. Epistémologie et Histoire*. Cedic-Nathan, Paris, 1978.
- *Dictionary of scientific biography* (16 volumes). Charles Scribner's sons, New York, 1970-1980.
- DIEUDONNE J., et coll., *Abrégé d'histoire des mathématiques. 1700-1900.* (2 tomes). Hermann, Paris, 1978.
- DUHEM P., *Le système du monde: historique des doctrines cosmologiques de Platon à Copernic*. Hermann, Paris, 1913-1917.
- *Encyclopedic dictionary of mathematics* (2 tomes). The Mitt Press, Cambridge, 1977.
- *Encyclopédie des mathématiques pures et appliquées*, (Dir. MOLKE J.). GauthierVillars, Paris, Teubner Leipzig 1904-1914. Reimp. Gabay.
- EUCLIDE D'ALEXANDRIE, *Les Eléments*, vol. 1, livres I à IV, vol. 2, livres V à IX, trad. VITRAC B., Presses Universitaires de France, Paris, 1992-1994.
- FAUVEL J., GRAY J., *The history of mathematics: a reader*. Mac Milan, Open University, 1987.
- FLEGG G., *Numbers Trough the Ages*, The Open University, 1989.
- FLOCON A., TATON R., *La Perspective*. Que Sais-je ? n°1050. Presses Universitaires de France, Paris, 1963.
- FOWLER D.H., *The Mathematics of Plato's Academy. A New Reconstruction*, Clarendon Press, Oxford, 1987.
- GHEVERGHESE G.J., *The crest of the peacock. Non-European Roots of Mathematics*. Tauris, London, 1991.
- HOCQUENHEIM M.L. et coll., *Histoire des mathérnatiques pour les collèges*. Cedic-Nathan, Paris, 1980.
- HOFSTADTER D., *GODEL, ESCHER and BACH: an eternal golden braid*. Basic books, New York, 1979. Trad. fr. Henry L., French R., *GODEL, ESCHER, BACH: les brins d'une guirlande éternelle.* Inter Editions, Paris, 1985.
- HOUZEL C., et coll., *Philosophie et calcul de l'infini*. Maspero, Paris, 1976.
- IFRAH G., *Histoire universelle des chiffres*. Seghers, Paris, 1981.
- JACOMY B., *Une histoire des techniques*. Points Sciences, S67. Editions du Seuil, Paris, 1990.
- JAROSSON B., *Invitation à la philosophie des sciences*. Points Sciences, S74. Editions du Seuil, Paris, 1992.
- KLEE V., WAGON S., *Old and New Unsolved Problems in Plane Geometry and Number Theory*. Mathematical Association of America, 1991.
- KLINE M, *Mathematical thought from Ancient to Modern Times*. Oxford University Press, New York, 1972.
- KLINE M., *Mathématiques: la fin de la certitude*. Christian Bourgeois, Paris, 1989.
- KOYRE A., *Etudes d'histoire de la pensée scientifique*. Presses Universitaires de France, Paris, 1966, ou Gallimard, Paris, 1973.
- LAKATOS I., *Proofs and refutations*. Cambridge University Press, 1976. Trad. fr. Balacheff N., Laborde J.M., *Preuves et réfutations. Essais sur la logique de la découverte scientifique*. Hermann, Paris, 1984.

- LE LIONNAIS, F., *Les grands courants de la pensée scientifique*. Blanchard, Paris, 1948 Reprint Cahiers Rivages du Sud, Paris, 1985.
- LEVY T., *Figures de l'infini. Les mathématiques au miroir des cultures*. Editions du Seuil, Paris, 1987.
- LEIBNIZ G.W., *La naissance du calcul différentiel*. Introduction, traduction et notes par Parmentier M., Vrin, Paris, 1989.
- LLOYD G.E.R., *Methods and Problems in Greek Science*. Cambridge, New York, 1991.
- MARCHALL P.E., *Histoire de la géométrie*. Que sais-je ? n°109. Presses Universitaires de France, Paris, 1968.
- MARTZLOFF J.C., *Histoire des mathématiques chinoises,* Masson, Paris, 1987. English tr. Wilson S., *A History of Chinese Mathematics*, Springer, Berlin, 1995.
- *Matin des Mathématiciens (Le)*. Entretiens sur l'histoire des mathématiques présentés par Noel E. France Culture, Belin, Paris, 1985.
- *Mathématiques en Méditerranée : des tablettes babyloniennes au théorème de Fermat*. Edisud, Marseille, 1988.
- MAY K.O., *Bibliography and research manual of the history of mathematics*. Toronto University Press, 1973.
- MENNINGER K., *Numbers Words and Numbers Symbols. A Cultural History of Numbers*, Dover, New York, 1992.
- MONTUCLA J.E., *Histoire des mathématiques*. Agasse, Paris, 1799-1802. Réed., Blanchard, Paris, 1968.
- NATIONAL COUNCIL OF TEACHERS OF MATHEMATICS (THE), *Historical Topics for the Mathematics Classroom*. Washington, 1969.
- NEUGEBAUER O., *The exact sciences in Antiquity*. Dover, New York, 1969. Trad. Souffrin P., *Les sciences exactes dans l'antiquité*. Actes Sud, Arles, 1990.
- PANZA M., PONT J.C., *Espace et horizon de réalité. Philosophie mathématique de Ferdinand Gonseth*. Masson, Paris, 1992.
- PIAGET J., *Logique et connaissance scientifique*. La Pléiade, Paris, 1967.
- POPPER K.R., *The logic of scientific discovery*. Hutchinson, London, 1973. Trad. fr. Thyssen-Rutten N., Devaux P., *La logique de la decouverte scientifique*. Payot, Paris, 1984.
- REY A., *La science dans l'antiquité*. Albin Michel, Paris, 1930-1948
- RIVENC F., DE ROUILHAN P., *Logique et fondements des mathématiques. Anthologie (1850-1914)*. Payot, Paris, 1992.
- ROBIN L., *La pensée grecque et les origines de l'esprit scientifique*. Albin Michel, Paris, 1963.
- RUSSELL B., Introduction à *la philosophie mathématique*. Payot, Paris, 1991.
- RUSSO F., *Eléments de bibliographie de l'histoire des sciences et des techniques*. Hermann, Paris, 1969.
- SARTON G., *Introduction to the history of science*. Baltimore, 1927.
- SERRES M., *Eléments d'histoire des sciences*. Bordas, Paris, 1989.
- SMITH D.E., *History of mathematics* (2 tomes). Dover, New York, 1958.
- SMITH D.E., *A Source Book in Mathematics*, Dover, New York, 1959.
- SINACEUR H., *Corps et modèles*. Vrin, Paris, 1991.
- STRUIK D.J., *A Source Book in Mathematics, 1200-1800,* Princeton University Press, Princeton, 1986.
- TATON R., *Histoire du calcul*. Que sais-je, n°198. Presses Universitaires de France, Paris, 1969.
- TATON R., *Histoire générale des sciences* (4 tomes). Presses Universitaires de France, Paris, 1957-1964.
- TROPFKE J., *Geschichte der Elementarmathematik*. De Gruyter, Berlin, 1902-1939. 4eme éd. Band 1, *Arithmetik und algebra* (VOGEL K., REICH K., GERICKE H.) De Gruyter, Berlin, 1980.
- VERDET J.P., *Une histoire de l'astronomie*. Points Sciences, S62. Editions du Seuil, Paris, 1990.
- VUILLEMIN J., *La philosophie de l'algèbre* (tome 1). Presses Universitaires de France, Paris, 1962.
- YOUSCHKEVITCH H., *Les mathématiques arabes (VIIIe-XVe siècles)*. Trad. fr. Cazenave M., Jaouiche K., Vrin, Paris, 1976.

Index of Names

Index of Topics

Contents

Imprimé en France
par MAME Imprimeurs à Tours
Dépôt légal Octobre 1997 (N° 97090195)